S0-FDF-435

THE DAWN OF MODERN GEOGRAPHY

PART II

THE DAWN OF
MODERN GEOGRAPHY

VOL. II

A HISTORY OF
EXPLORATION AND GEOGRAPHICAL SCIENCE
FROM THE CLOSE OF THE NINTH TO
THE MIDDLE OF THE THIRTEENTH
CENTURY (c. A.D. 900–1260)

By C. RAYMOND BEAZLEY, M.A., F.R.G.S.

FELLOW OF MERTON COLLEGE, OXFORD

WITH REPRODUCTIONS OF THE PRINCIPAL MAPS OF THE TIME

NEW YORK
PETER SMITH
1949

This volume was originally published by Mr. John Murray in 1901.

*Reprinted 1949
By Permission Of
The Clarendon Press
Oxford, England*

PREFACE

This volume carries on, into the Central Middle Ages, that History of Mediaeval Geography—in exploration, travel, and science—of which the first instalment was published in 1897. The former period, which must be regarded as the Dark Age of Geography, was taken as reaching from the Conversion of the Roman Empire to the close of the ninth century; the present, to which the name of 'Dawn' is more strictly applicable, covers the period from the close of the ninth to the middle of the thirteenth century; the two succeeding centuries, from the middle of the thirteenth century to the death of Henry the Navigator, in 1460, may be taken as completing the story of geographical action and research in the Middle Ages.

In the present volume we have to deal with the central Mediaeval Period, and especially with (1) the explorations and enterprises of the Northmen down to about A.D. 1066, the eve of the Crusades; (2) the history of pilgrim travel to the Holy Land, and of other intercourse between Syria and the Christian West, down to the middle of the thirteenth century; (3) the Hebrew geographers and travellers, especially the twelfth-century Benjamin of Tudela; (4) the diplomatic and missionary travellers of the thirteenth century—especially Carpini and Rubruquis—who took advantage of the new routes and new conditions established by the Mongol Conquests to penetrate further afield than any one had yet ventured from Christian Europe; (5) the commercial intercourse, especially of the Mediterranean basin, and the extension of Christian trade-adventure to more distant regions; (6) certain typical geographical writings and plans of the Central Mediaeval time, and especially those of the Byzantine Emperor, Constantine VII. (Porphyrogennetos), of the German chronicler, Adam of Bremen, and of the map-makers, Beatus of Valcavado and his copyists, Henry of Mainz, Lambert of St. Omer, and various anonymous designers. A fuller treatment of the details of these map-schemes is given in an *Appendix on Maps*, pp. 591-642. Under these six main heads an account, it is hoped, will be found of every important geographical enterprise or speculation in Christendom, and especially in Latin Christendom, during the Central Middle Ages (c. A.D. 880-1260).

Two alterations from the original plan have been made necessary by want of space. It has not been possible to insert even an incomplete survey of Arab and Chinese exploration and 'earth-study' in this time; and it has only been possible to treat very partially the geosophical theorising of the Scholastics. Thus Roger Bacon and Albertus Magnus, Plato of Tivoli and Gerard of Cremona, have been of necessity almost entirely omitted from this volume; but this is not of any great moment, as all that is valuable for our purpose in the writings of the Crusading period finds a sufficient place in chapter vii.; and the geographical mythology of the Middle Ages in general has been already considered at some length (*Dawn of Modern Geography*, i. ch. vi.). In the centuries that now concern us (from the tenth to the thirteenth) no fact is more significant (though none has been more frequently missed) than the steady but gradual waning of fanaticism, the ever-growing change from the conditions of an earlier period, when religious conceptions appear as of overwhelming importance. For, in spite of the fact that religious wars are officially the great interest of all classes, *e.g.* in the twelfth century, the ambitions and ideals of the merchant are really beginning to supplant these of the priest and the devotee; it is under the spur of commercial as much as of political designs that Christian Europe sets about retrieving the position that had been lost during the Dark Ages; and the discovery of a new world for trade is the permanent result of the Crusades and the sign and abiding symbol of a true Mediaeval Renaissance (eleventh to thirteenth century).

The maps of this time (c. A.D. 900-1260) are numerous and interesting, some of them showing a great advance on any older Christian designs in the West. Our illustrations (all taken from the original manuscript sources) include three examples of the Spanish 'Beatus' group, the 'Cotton' map of the tenth century (a possibly Irish work of singular merit), the Madaba Mosaic (a striking example of early Byzantine cartography), and several plans of Matthew Paris, including a map of England, which ranks among the best designs executed in any country before the fourteenth century and the beginning of the Portolani or compass coast-charts.

I have to express my cordial thanks to various friends, especially Professor York Powell, Professor W. R. Morfill, the Rev. W. H. Hutton, Fellow of St. John's College, Oxford, Mr. Herman Cohen, and above all, Mr. W. Ashburner, late Fellow of Merton College, for their kind assistance in reading over proof sheets and making suggestions.

CONTENTS

CHAPTER I

INTRODUCTORY

The Central Mediaeval Period the subject of this volume—This period marked by two movements : (α) of the Northmen ; (β) of the Crusaders — Sub-divisions of the period — Leading figures in travel and geographical literature, 900-1260—Advances in geographical science—Maps of the time—Moslem and Chinese geography of this period 1–16

CHAPTER II

THE NORSEMEN IN THE HISTORY OF EXPLORATION, TO THE BEGINNING OF THE CRUSADES (CIRC. A.D. 800–1070)

Chief lines of Norse expansion—Value of Norse discoveries and explorations—Ubiquity of the Norsemen—Formative, provocative, and invigorative effects of their invasions, etc.— The oldest Norse Geography, in the *Ynglinga Saga*, etc.— Harald Fairhair, his journeys oversea—Discovery of Iceland and Greenland—Voyages of Norsemen to Western Europe and the White Sea—Ohthere and King Ælfred— Eric Blood-axe on the Dvina — Other journeys to the Lapland and White Sea coasts —Vikings in the Baltic— Connections between the East and West Baltic Lands— Novgorod, the Russian Principalities, and the Scandinavian peoples—Novgorod as a centre of trade ; its commerce with the far North and North-East — Novgorod and the Hanseatics—Olaf Tryggveson at Novgorod—Olaf in Wendland and Western Europe ; his marriage and residence in Ireland ; his return to Norway—Intercourse of Iceland and Norway in later tenth century—Last years of Olaf Tryggveson—

The discovery of Helluland, Markland, Vinland, and other 'American' regions by the Norsemen—Rival narratives of the Vinland Voyages—The enterprises of Biarne Heriulfson and Leif Ericson—Leif Ericson's settlement in Vinland—Thorvald Ericson; his voyage and death in Vinland—Fruitless venture of Thorstein Ericson—Voyage and settlement of Thorfinn Karlsefne—His contest with the Skraelings; he abandons his purpose of Colonisation—Freydis, Red Eric's daughter; her Vinland voyage — Additional details supplied by *Red Eric's Saga*, especially as to Thorfinn Karlsefne, the 'Wonder Strand,' 'Hop,' the 'Unipeds,' etc.—Later voyages to Vinland—'White Man's Land' and 'Great Ireland'—Vinland Controversies—Possible identifications for Vinland and the Skraelings—Olaf Haraldson; his early voyages in Western Europe; succession to the crown of Norway—Thorir Hund in the White Sea—Olaf's intercourse with the Orkneys, Iceland, etc.—Olaf in the Far North; his war with Cnut and flight to Novgorod—Olaf's fall; minor travels, of Sighvat, etc.—Magnus the Good and Harald Hardrada; early wanderings of the latter—Harald in Russia, at Constantinople and Jerusalem, and in the Mediterranean—His later enterprises and Polar projects 17-111

CHAPTER III

PILGRIM TRAVEL (CIRC. A.D. 900-1260)

Christian pilgrimage from the close of the ninth century to the time of Pope Gregory VII.—The outlook at the close of the First Millennium, especially in Latin Europe, Ireland, the Byzantine Empire, the Eastern and Western Caliphates—Intercourse of the Latin world with Palestine—Growing Byzantine influences in the Levant, A.D. 880–1030—Latin journeys of the tenth and eleventh centuries to Syria—Destruction of the Holy Sepulchre, A.D. 1010—French, German, and other pilgrimages of 1035, 1038, 1053, 1054, 1058, 1064—New Era with Hildebrand; the outlook about 1070 — Causes and results of the Crusades — Pilgrim travellers and writers at the time of the First Crusades—The *Innominati*; the first of their anonymous narratives described—Saewulf of Worcester; his outward journey—

CONTENTS

PAGE

Saewulf's travels in Greece, the Archipelago, and Asia Minor—Saewulf in Palestine; storm at Jaffa—Saewulf in Jerusalem; marvels related by him—Saewulf's description of Hebron, the Jordan, Galilee, and Lebanon—Saewulf's return journey—Abbot Daniel of Kiev; peculiar value of his journey and narrative—Daniel's outward journey, by Constantinople, Rhodes, and Cyprus, to Jaffa—Daniel in Jerusalem—Daniel's description of Mar Saba Convent, the Jordan, various sites of Galilee, etc.—Daniel's account of the Holy Fire; his return to Russia — Scandinavian pilgrims and pilgrim routes—King Sigurd of Norway—His sea-journey to Palestine along the coast of Spain—Sigurd's exploits in Spain and the Balearics—Sigurd in Palestine; the capture of Sidon—Sigurd at Constantinople; his return-journey overland—Adelard of Bath; his literary tour in the Levant—Later pilgrim journeys and narratives; their great number and comparative unimportance—Chief sub-divisions — Fetellus; his description of Jerusalem, Mount Sinai, etc.—Marvels and myths—John of Würzburg—John's merits; vividness of his narrative—His protest on behalf of the Germans in the crusades — Theoderich; artistic and observant character of his record—His probable connection with John of Würzburg—Johannes Phokas; his references to Manuel Comnenus—The later *Innominati*—Fragmentary pilgrim notices down to 1187—Pilgrim narratives, 1187-1260—The *City of Jerusalem*—Ernoul— James de Vitry—Antony of Novgorod—Wilbrand of Oldenburg—Thietmar — Sabbas of Servia — Fragmentary pilgrim notices, 1187-1260 112-217

CHAPTER IV

BENJAMIN OF TUDELA AND OTHER JEWISH TRAVELLERS, TO THE MIDDLE OF THE THIRTEENTH CENTURY

Importance of Jewish travel in Middle Ages—Hebrew wanderers and geographers before Rabbi Benjamin—Jewish missionaries in Yemen and among the Khazars — Tenth century Correspondence between Chasdai at Cordova and King Joseph in Khazaria—Date of Benjamin's travels—His historical allusions—His profession as a merchant—Sub-divisions of his journey and narrative—Benjamin's outward

xiv CONTENTS

PAGE

route, through Eastern Spain, the Narbonnese, Provence, Italy, Greece, etc. — His description of Constantinople —His voyage to Syria—Merits of his notes on Syria ; examples — His route from Damascus to Bagdad ; his account of the latter—Position of the Abbasside Caliph and the Jews of Bagdad, as portrayed by Benjamin—More legendary character of Benjamin's Records, after Bagdad— His stories of the Jews of Arabia, Persia, and Central Asia, of Babylon, of the Tombs of the Prophets, of David El Roy, and of the Assassins and Turks—Trade of the Persian Gulf, and Indian Ocean—Malabar, Ceylon, and China—Aden, Nubia, the Nile, and Egypt—Benjamin in Cairo and Alexandria—Catalogue of Merchants trading to Alexandria— Benjamin's return by Sicily—His remarks on Germany and Slavonia—Rabbi Moses Petachia of Ratisbon and Prague— His journey to Syria through Eastern Europe and the North of the Black Sea — His notes on 'Kedar' and 'Meshech' in Modern South Russia—Petachia in Georgia, Armenia, and Mesopotamia—His journey from Bagdad to Damascus—His account of Hebron and other Palestine sites —Later Jewish travellers and geographers, 1200-1260   218-274

CHAPTER V

DIPLOMATIC AND MISSIONARY TRAVEL OF THE THIRTEENTH CENTURY : CARPINI, RUBRUQUIS, ETC.

New era created by the Mongul conquests—Christian attitude towards the Mongols—Embassies despatched to the Tartar Courts in 1245, 1247, 1249, 1252, etc.—the Mission, journey, and narrative of John de Plano Carpini—His account of the Tartar country, manners, history, etc.—Virtues and vices of the Nomades—Widespread destruction effected by the Mongols—Their skill in war ; their cruelty and avarice —Universal ambitions of Ghenghiz's successors ; Carpini's warning to Christendom—Carpini's outward route through Bohemia, Poland and Russia ; arrival at Kiev—Journey from the Dniepr to the Volga ; arrival at Batu's camp— Carpini's description of the Steppes and their great rivers —Journey from the Volga to Mongolia—Carpini's classification of the regions surrounding Batu's camp and the Lower Don—Arrival in the valley of the Syr Daria—The ruined towns of Moslem Central Asia—Route by Lake Ala-Kul over the dividing range west of Mongolia—Arrival at the Great Khan's 'Horde'—Ceremonial of the Court ; election of Kuyuk Khan—Carpini's audiences of the Mongol Emperor—Meetings with other Christians and Europeans, envoys, prisoners, hostages, etc.—Carpini receives Kuyuk's reply to the papal letters—His return by Batu's camp to Kiev—Italian traders on the Dniepr—Mission of Andrew of Longumeau—His ill-success—Peculiar route followed by Andrew—Mission of William of Rubruquis—His companions—Outward route by Constantinople and the Crimea —Rubruquis' Black Sea Geography—His route from Sudak to the Isthmus of Perekop—His account of the Tartar manners and customs, the dwellings, food, fashions, laws, and ceremonies of the Nomades—Rubruquis' journey to the Don and Volga ; his account of the Finns of the Forest Region to the North, and of the hill tribes of the Caucasus to the South—Rubruquis at the camp of Sartach near the Volga—Journey to the Camp of Batu—Passage of the Volga —Arrival and reception at Batu's Court—Journey from the Volga, over the Ural river, to the region of Lakes Balkhash and Ala-Kul — Rubruquis' description of the Uigurs, Tanguts, Tibetans, Cathayans, etc. ; his opinion of Nestorian Christians—Journey from Lake Ala-Kul, over the dividing range, into Mongolia—Arrival at the Great Khan's Court, and reception there—Europeans at Mangu's Court— Visit to Karakorum ; Rubruquis' account of the tribes of the Far North—Controversies with Nestorians, Moslems, Buddhists, Manichaeans, etc.—Mangu dismisses Rubruquis with letters to St. Louis—Numerous envoys from the distant parts at Mangu's Court—Rubruquis' return to Batu's camp— News of Hayton, King of Little Armenia—Journey from Batu's camp, through the Caucasus, to Armenia—Through the Sultanate of Iconium to the coast of Cilicia—Comparative merits of Rubruquis and Carpini ; recent controversy—King Hayton of Little Armenia ; his journey to the Mongol Courts—Previous journey of the Constable Sempad ; his stay at Samarcand—Hayton's route through Kars, Derbent, and the Caucasus, to Batu on the Volga— Thence over the Ural to the Great Khan Mangu—Hayton's return through Eastern and Western 'Turkestan,' North Persia, South Armenia, etc.—Hayton's stories of Buddhist customs, etc. 275-391

CHAPTER VI
COMMERCIAL TRAVEL (CIRC. A.D. 900-1260)

Mercantile activity of Mediaeval Latin Europe centred in Italy —Commercial position of Constantinople—The results of the Crusades in the commercial opening of the world— Amalfi; its early development — Wide extent of its eleventh-century trade — Decline after the first Crusade, especially from the Norman Conquest of 1131—Amalfi and maritime law—Venice; origins of its trade and influence— Early development of its constitution and prosperity — Close connection of Venice with Constantinople — Clandestine commerce with the Saracens — Intercourse of Venice with Germany and the Levant — Importance of maritime routes to Syria during the Crusades— Concessions granted to Western traders by Crusading Princes; various types of these concessions—Instances of maritime aid furnished by Venice, etc., to crusading armies —Venice and the Comneni—Close alliance with Alexius Comnenus against the Normans—Venetian quarrel with Manuel Comnenus—Struggle of Venice and her commercial rivals for ascendancy in Byzantine trade—The 'Crusade' of 1202-4—Period of Venetian ascendancy under the Latin Empire; distant developments of their commerce—Venetian intercourse with Moslem States in thirteenth century—Decline of Venetian influence in the Crusading States—Genoa; its origin, early history, and especial sphere of energy in the Western Mediterranean—Active share of Genoa in the Crusades—Intercourse with Constantinople; treaties with Manuel Comnenus — Intercourse with Egypt and North Africa, with Spain, the Narbonnese, Provence, etc.— Triumph of Genoa in Byzantine commerce at end of this period, through restoration of Greek Empire—Pisa; origin and early history — Great prosperity in eleventh and twelfth centuries— Extreme activity in Crusades—The concessions of the Pisans in the Crusading States—Their rivalry with Venice, even at Constantinople — Intercourse with Egypt, North Africa, Spain, and the Provençal lands —Decline of Pisa in the thirteenth century—Marseilles; its share in promoting pilgrimages and Crusades to Syria— Its prosperity in eleventh and twelfth centuries—Rivalry with Genoa — Intercourse with coasts of Western Mediterranean—Montpellier, Narbonne, and Barcelona; their

CONTENTS xvii

PAGE

early prosperity—Narbonne as an intermediary between England and Egypt—Influence of the Catalan seamen on commercial and maritime law—Invention of the Portolani perhaps due to Catalans—Biscayans, Castilians, Galicians, and Portuguese; their share in Mediaeval trade — Commercial advances eastward from the Crusading States of the Levant—Centres of trade in the kingdom of Jerusalem, the Counties of Tripoli, Edessa, etc. — Damascus and Bagdad as emporia—Pisan, Venetian, and other Christian merchants in the Moslem Levant—Trebizond, Nicaea, and Little Armenia as gates of trade-intercourse between East and West—Trade route crossing Asia Minor, Mesopotamia, Syria, and Arabia—Christian merchants on the Euxine coasts, in the Steppe Country, and in Russia — The chief emporia of the Black Sea; Kaffa, Matracha, Sudak, Tana, Kherson or Sevastopol, Trebizond, etc.—Little Armenia as a centre of trade—New trade routes opened by the Mongols—Importance of Egyptian commerce in central mediaeval period— Continued prosperity of Alexandria—Intercourse between Christendom and the Further East through Egypt 392-464

CHAPTER VII

GEOGRAPHICAL THEORY AND DESCRIPTION (CIRC. A.D. 900-1260)

Two chief examples of geographical writing taken: (1) Byzantine, of tenth century; (2) German of eleventh—With these we must also notice (3) maps of the Xth, XIth, XIIth, and XIIIth centuries—(1) Constantine Porphyrogennetos; his position as a geographer—The work *On the Administration of the Empire*; its high value and interest—The writings *On the Themes, Ceremonies*, etc., of Byzantine Empire—Summary of the *Administration;* its main subdivisions—The Byzantine revival under the Macedonian Dynasty—Byzantine influence on the surrounding nations, especially the Slavonic—Range of Byzantine geography in tenth century — The Neighbours, friendly and hostile, of the Empire—The Bulgarians and Constantinople—The Hungarians and Constantinople—The Russians and Constantinople — The Petchinegs and Constantinople—The Khazars and Constantinople—The Arabs and Constantinople—Vassals of the Empire, especially in

xviii CONTENTS

PAGE

Italy, Dalmatia, the Crimea, the Caucasus, and Armenia—Geographical information, as to many of these, from Byzantine sources — Subjects of the Empire—The City of Constantinople itself the Empire—Its universal attractions — Employment of all races in its service—(2) Adam of Bremen; his importance in mediaeval geography—Adam at the Court of Archbishop Adalbert of Bremen; his opportunities there — Adalbert and Svein, King of Denmark, Adam's chief patrons and informants—Adam's other sources—Adam's general geographical theories—His description of the countries of Northern Europe — His 'Scythia'—His account of the Scandinavian lands, of the Baltic coasts, and of the chief ports of these seas, especially Jumna and Birca—Adam's treatment of 'islands' in the Northern Ocean—His notices of Iceland, Greenland, and Vinland — The lands around Bremen and 'Saxonia'—Christian and Pagan Slaves—Russia and its metropolis of Kiev, as treated by Adam—Finnish tribes of the Far North —Harald Hardrada's venture in search of the Northern limit of the world—(3) Maps of the Central Mediaeval Period — The Beatus group, 'St. Sever,' 'Osma,' etc.— Leading characteristics; the Apostolic Pictures, etc.— Illustrative details; connection with cartography of later imperial time — The so-called 'Jerome' Maps — The 'Cottoniana' or 'Anglo-Saxon' Map; its merits, possible origin, and special purpose—Its possible connection with the Itinerary of Archbishop Sigeric of Canterbury—The Map of Henry of Mainz; connection with the 'Imago Mundi' of Honorius of Autun, and with the 'Psalter,' 'Ebstorf,' and 'Hereford' Maps—The 'Psalter' Map—Illustrative details of all these related designs—Maps of Lambert of St. Omer; their connection with the 'Macrobius,' 'Sallust,' and T-O-sketches—Illustrative details of all these—The leading T-O maps; their connection with classical schools — Map of Guido 'of Ravenna' — The Madaba Map and the 'Situs Hierusalem' as examples of detailed 'Chorography'—Illustrative details—Connection of the 'Situs' with other works — Lost maps of this period 465-590

APPENDIX ON MAPS . . 591-642

LIST OF ILLUSTRATIONS

	TO FACE PAGE
EXAMPLES OF 'BEATUS' CARTOGRAPHY—	
(α). The Map of 'St. Sever,' of c. A.D. 1030	550
(β). The 'Turin' Map of c. A.D. 1150	552
(γ). The British Museum or so-called 'Spanish Arabic' Map of A.D. 1109	554
THE 'COTTON' OR SO-CALLED 'ANGLO-SAXON' MAP OF C. A.D. 990	560
THE WORLD-MAP OF HENRY OF MAINZ, OF C. A.D. 1110	564
THE LARGER 'JEROME' MAP (SO-CALLED) OF THE TWELFTH CENTURY	566
THE 'PSALTER' MAP OF THE THIRTEENTH CENTURY	568
THE MACROBIAN 'COTTON' ZONE-MAP OF THE TENTH CENTURY	574
THE CLIMATE MAP OF PETRUS ALPHONSUS, C. A.D. 1110	576
THE MADABA MOSAIC MAP OF THE SIXTH CENTURY	580
THE 'SITUS HIERUSALEM,' OF C. A.D. 1100	582
EXAMPLES OF MATTHEW PARIS' CARTOGRAPHY (MIDDLE OF THE THIRTEENTH CENTURY)—	
(α). The 'England,' of the most finished 'Cotton' form	584
(β). The 'World Map,' from the London copy	586
(γ). The 'Itinerary' from London to Apulia (South-Italian section)	588
(δ). The 'Palestine,' from the best London copy	590
THE WORLD-MAP OF GUIDO, C. A.D. 1119	632

THE DAWN OF MODERN GEOGRAPHY

CHAPTER I

INTRODUCTION.

IN a former volume we attempted to trace the story of geographical movements and the records of travellers and men of science from the conversion of the Roman Empire to the close of the ninth century, thus arriving at one of the darkest periods of the Middle Ages. In the present, or second, part of this survey of mediaeval geography our task will be a different one. For here we begin upon a ground which seems well-nigh entirely barren; we cross that desert which separates the earlier and later Middle Age, the 'Romanesque' and the 'Gothic'; and we follow the streamlet of western enterprise as it leads us into a country far richer and more abundant than any we have traversed since the age of Constantine (A.D. 900-1260).

Before the close of the eleventh century Christian Europe has rallied from its decadence and apparent torpor, and is on every side passing beyond its former bounds, reconquering ancient possessions, and seeking new fields for its energy. The Crusades are the central expression of this revival, which, though defeated in some of its immediate objects, was entirely successful in kindling a spirit of patriotism, of practical religious fervour, and of boundless enterprise, whereby our Western World finally attained to the discovery, conquest, colonisation, or trade-dominion of the best portions of the earth.

The pilgrims, traders, and missionaries of the earlier Middle Age have already shown us how far Christian Europe, up to the close of the ninth century, was in touch with the outside world. In the central mediaeval era which now occupies us we find, on one side, a certain contraction, on another an immense extension, of geographical enterprise. In Central Asia and in the far East and South of that Continent the Nestorian Missions slowly decline as Islam advances, and as the Syrian starting-point of these same missions becomes more and more a centre of Mohammedan life. But to the north and west of Europe the Scandinavian mariners (first as heathen, then as Christian, explorers), discover, ravage, or settle new lands beyond the furthest limits of ancient knowledge. And as these northern adventurers scatter themselves over the whole shore-line of the older Christendom, as well as the coasts of Greenland, Vinland, and North-Eastern Europe, they are slowly brought into the federation of the Western Church, and thus the pilgrim spirit is translated, as it were, into new and more powerful forms. Christendom had seen a wonderful expansion of the heathen North; now that it had won the Northmen to itself it was ready to imitate their example. The deliberate purpose of ecclesiastical statesmen gave direction to a universal feeling of capacity and ambition chafing under restraint. Even the struggle which, under Gregory VII., broke out between the Empire and the Papacy did not prevent the combined efforts of Christian peoples and warriors to win a new vantage ground in the Levant, and to reopen for European enterprise a way to the more distant East.

The discoveries and conquests of the Northmen fall mainly within that first barren period to which we have referred; during much of this time they go on quite apart from the life of Christendom, and at first appear in bitter hostility to it. But the gradual association, incorporation, or alliance of the Scandinavians with the nations they came to plunder or to destroy, is perhaps the most decisive fact in the story of the Christian Middle Ages, and affords a basis and starting-point for every subsequent development. The Crusades, the commercial and territorial expansion that follows the Crusading movement, and the extension of European spirit and influence towards ultimate predominance in the outside world, are all related to the formative, provocative, and invigorative influence of the northern invasions. The travels and adventures of Ohthere and Wulfstan, Eric Blood-axe and Eric Haraldson (c. A.D. 890, 918, 930), on the coasts of Lapland, the White Sea, and the Baltic; of Olaf Tryggveson and Olaf the Saint in Russia and Western Europe (c. 965-1015); the voyages of Thorir Hund to the Dvina in the track of Ohthere (c. 1025); of Nadodd, Gardar, Ingolf, Leif, Gunnbiorn, Red Eric, Leif Ericson, and Thorfinn Karlsefne to Iceland, Greenland, and Vinland (c. 865-1010); and of Harald Hardrada in Eastern Europe, in the Mediterranean, and in the Polar Seas (c. 1035-1055), stand out as the leading incidents in the history of this early Scandinavian exploration of the outer world, and give the outlines of a picture unmatched in the story of mediaeval geography.

Apart from the work of the Norsemen, which must be considered as introductory to the rest, the geographical movements of the Latin world within this time (900-1260) fall into three main divisions. First, we have the pre-Crusading journeys and writings of the tenth and eleventh centuries; secondly, those which fall within the first century of Crusading activity; thirdly, those which belong to the time that follows Saladin's overthrow of the Latin Kingdom of Jerusalem. In the first of these epochs we have material of little importance in itself, and chiefly interesting as a preparation for what was to come; in the second a notable extension of the horizon is obvious; but it is not until the third that we see men from Catholic Europe travelling beyond the ancient limits, crossing the heart of Asia, and bringing to the Court of China first-hand evidence of the lands 'near the sunsetting.'

Among the chief figures of these three periods we have in the first-named the famous Gerbert (Pope Sylvester II.), whom a later tradition makes an early preacher of Syrian Crusade, but whose alleged visit to the Holy Land is more than doubtful; Duke Robert of Normandy, who certainly visited Jerusalem in 1035-1036; and his great contemporaries, Fulk the Black, Count of Anjou, Ealdred, Archbishop of York, Siegfried of Mainz, Hermann of Bamberg, William of Angoulême, and Svein son of Godwine But there is no pilgrim journey of this time which stands out in any marked degree, or has left records of such importance as to bring into relief the personality of any particular traveller or writer. This pilgrim literature, of which we have seen so many examples in the earlier Christian centuries, begins again with the second division of our subject, the century of the first three Crusades (1096-1191). And in this literature the first names of note are Saewulf of Worcester, Adelard or Athelard of Bath, and Daniel of Kiev, three of that host of peaceful wanderers who followed the conquerors of the First Crusade. Saewulf and Daniel have left us their note-books, and both of these belong to the new time, a time full of life and personal interest, in sharp contrast to the generations immediately preceding, when history is almost inarticulate, and when even the bare outlines of the world's leading figures are scarcely discernible. The hordes of tenth and eleventh century pilgrims give no sign of anything but the simplest penitential purpose, though every fresh conversion of the northern nations [1] (especially since the end of the First Millennium) had been bringing a new stream of devotees to Italy and the Levant. But now, when Mediaeval Christendom had begun to take shape, when a revival of intellectual life had shown itself, and when religious passion had become more steady and less unworldly, the discoverer, the scientific observer, and the merchant, are blended with the pilgrim pure and simple in an ever larger number of those records that have come down to us.

Thus Saewulf, the layman and the merchant; Adelard, the investigator of nature; Daniel, the Russian monk and patriot; all show us something of the new face of things, and suggest changes still greater. And it is here, even more than in knowledge, or extent of travel, that these later pilgrims show so marked an advance on the chief of their predecessors. This is also the main interest in the journey of King Sigurd of Norway, at this very time (1107-1111), for, though his voyage added nothing definite to geographical knowledge, he represents, in his person as a Crusading Norseman, the meeting of two forces whose union was of infinite promise for the future. For here is dramatically combined the militant Christianity of the Western Church, and the daring spirit of that race which had restored the energies of European nations, and opened so much of land and sea to European enterprise.

[1] In which Hungary must be included.

Our second period closes with the records of Benjamin of Tudela (1159-1173), and other less important travellers, whose horizon has broadened from the nearer East of the Levant to the Middle Orient, to the lands beyond Syria, the Euphrates, and the Caucasus. Thus the European is beginning to enter a world outside the influence of his own races and centres of power, even outside the purview of the contracted Christendom of the Dark Ages; a world where the visitor becomes gradually conscious of a new light radiating from the far distant, self-contained, and long hidden[1] civilisations of India and China.

The third period that here concerns us (c. 1190-1260) brings the West into direct connection with those distant regions, incredibly rich and populous and attractive as they seemed to the Latin world of that time. Meantime, that Latin world had committed a criminal blunder, and had witnessed with apathy, almost with satisfaction, the fall of two great disasters (one self-inflicted, the other unforeseen and incalculable, but still more terrific) upon the common interests of European nations. On one side, the Crusaders of 1202-1204 had turned away from their promised work to seize Constantinople, and found a Latin empire on the ruins of the shrunken Byzantine State; on another side, the Roman Church had seen (from the year 1213) the rise of the Tartar Empire of Ghenghiz Khan and his successors, without raising a finger to save the Christians of the Russian States, of Poland, or of Hungary. All else was forgotten except the partial truths that the Greeks had broken communion with Rome, and that the Tartars had weakened Islam; wild hopes of Mongol conversion were entertained in the West; and Latin missions crossed Europe and Asia to the court of the Great Khan. Thus John de Plano

[1] *I.e.* from Mediaeval Europe.

Carpini (1246-1247), William de Rubruquis (1253-1255), Nicolo and Matteo Polo,[1] uncles of Marco (1255-1269), with many others, made their way to the headquarters of the Nomade princes, to Sarai on the Volga, to Karakorum near Lake Baikal, and to the summer capital of Kublai Khan, beside the wall of China.

Thus, by about 1260, we have reached a state of things which is, in all respects, an advance on the Dark Age Period, as shown in the seventh, eighth, or tenth centuries. The outlook of the European has greatly extended; the all-absorbing ambitions of the first Saracens have been thwarted; it is not likely now that Christendom, however endangered from time to time, will finally, and as a whole, succumb to Islam. In particular, its geographical outlook, its knowledge of the world, both practical and scientific, has been widened and deepened, while that of its great rival is already giving signs of an autumn, which, however splendid, is an autumn still. We must not be deceived by exceptions. No greater Moslem traveller ever lived than the fourteenth-century Doctor of Tangier; yet it is undeniable that, in the fourteenth century, Moslem geography and Moslem science are declining. Turks and Tartars by successive invasions and devastations, Spanish Christians by their incessant attacks, were slowly ruining the higher culture of the Mohammedan world. We may, perhaps, see the beginnings of this ruin even from the time of the great encyclopaedist Yakut († 1229); the slowly dying flame blazes up once more in a brilliant farewell with the work of Ibn Batuta and Abulfeda; but after this there is little but cold and darkness.

On the other hand, the close of the Crusading Era leaves us at the end of the great scholastic age, at the abortive

[1] Whose journey lies just outside the period of this volume

conclusion of all efforts for Latin domination in the Levant; but it also brings us to the completion of a great and fruitful chapter of social and political growth, and to the beginning of an even more decisive advance. For, as the warlike spirit of Christendom seems to grow somewhat weary, and as its efforts at founding new kingdoms (in Antioch, Cyprus, Jerusalem, or Constantinople) appear more and more fruitless, a still greater extension of European life and knowledge begins. The missionaries and merchants who followed the Crusading armies to the Euphrates, and crept on to Ceylon and the China Sea, brought Further and Central Asia—*Thesauri Arabum et divitis Indiae* — within the ken of the Frankish and Latin nations. Again, though baulked in the Eastern Mediterranean, and disappointed in their hopes of permanent advance along the Asiatic land-routes, the European races turned, with an energy and daring impossible at an earlier time, to discover a maritime route to India and China. As early as 1270 these attempts began with the first voyages to the Canaries, and after two centuries of slow endeavour the way round Africa was laid open. We may notice that the Vinland voyages of the Northmen were discoveries rather of special than general interest or usefulness; unconnected as they were with the main lines of trade, or with the general life of Christendom, they were never realised or understood by the Western World at large. It was otherwise with the explorations of Carpini, of Rubruquis, of the Polos, and of the early Italian voyagers who first sought the 'way by sea to the ports of India to trade there.'

And here we may repeat that the renaissance of Western Europe in the Crusading age was fully as much an intellectual as a material and political movement. Science was at last touched and changed by the new life as completely as the arts of war, of travel, of trade, or of missionary enterprise. And among other branches of science none made greater progress than Geography. Roger Bacon and Albert of Bollstadt made distinct advances on all previous mediaeval workers in this as in other fields. Plato of Tivoli, Gerard of Cremona, Adelard of Bath, represented the new spirit in Christian learning, which gladly confessed its obligations to Moslem teachers, and did good service in the translation and adaptation of Arabic texts for the use of Latin Christians. Descriptive geography shows a remarkable improvement, and may indeed be considered to have recovered no small part of the spirit of Strabo. The purely theoretical labours of Bacon, Albert, Adelard, and the rest, must be left to a subsequent part of this enquiry; but they may fairly be alluded to here, as technically falling within the present period. On the other hand, some detailed reference is given in this volume to the Byzantine geography, so admirably represented by the 'Porphyrogennetos,' Constantine VII. (who, in the middle of the tenth century, described the neighbours, vassals, and subjects of the Graeco-Roman state), as well as to the principal Western rival, in pre-Crusading times, of the Imperial student of Constantinople. For, little more than a century after the appearance of the royal author's treatise on the *Administration* of the Eastern Empire (c. A.D. 953), Adam of Bremen gave a better survey of 'Northern Parts' and of recent discoveries in the same, than had yet been attempted, better indeed than had yet been possible. The explorations of the Vikings had passed almost unnoticed by the main stream of European thought, but the Annalist of the Church of Hamburg (c. A.D. 1070) was a bright exception to a rule of ignorance and neglect. For even if mediaeval Catholicism showed little practical interest in the north-western movement of the tenth and eleventh centuries, at least it saved itself, by Adam's work, from the reproach of totally forgetting that movement, or of omitting to enter on its register the existence of a fact so memorable.

Once more, the diplomatic and missionary travellers of the thirteenth century were not only explorers of a high order, but did excellent service also to anthropology by their careful observations of manners and customs, to philology by their judicious enquiries into language, and to ethnology by their truly scientific classification of the chief races of Eastern Europe and Northern Asia. It must not be forgotten that they were the first Christian writers, with the exception of some Byzantines, to give a truly literary character to geographical writings; and in the person of the encyclopaedist, Vincent of Beauvais, their work is closely connected with the Christian philosophy and universal science of the schoolmen's ideal. Roger Bacon also is deeply indebted to Rubruquis, just as Rubruquis and all other practical travellers are indebted to Bacon for his fruitful study of nature, a study which included some of the earliest investigations upon the polar properties of the magnet. In the West, the Oxford scholar is only preceded, in this last-named discovery, by Alexander Neckam, the monk of St. Albans, by Guyot de Provins, the Languedoc satirist and man of letters, and, more doubtfully, by some of the daring voyagers of the Icelandic Sea.[1]

By the close of the period now in view (A.D. 900-1260) the magnet, fitted with the compass-card, has been fully adapted, at least in the Mediterranean, to the use of seamen; an elementary but approximate method of ascertaining positions has been secured by means of the Astrolabe; and

---

[1] Cf. *Landnamabok*, as referred to in Nordenskjöld, *Periplus*, 50; but this passage does not belong to the original text of *Landnamabok*, but is an insertion of about 1290-1300.

no less important advances seem to have been made in the delineation of the coast-lines of the world-surface. For, though no specimen has survived of the new Portolani, or Handy Outline-Maps, anterior to 1300, there can be no doubt that the drawing of such maps commenced some years before the *Carte Pisane* was executed; and in these maps the use of the compass, the beginnings of modern survey-method, and the proper fixing of positions by astronomical determination, are for the first time observable in cartography. Within the limits of the Crusading time, indeed, no survivals have yet been discovered of this new type of map; on the contrary, our documents furnish us only with designs of the normal mediaeval types. But even here we have to notice a marked advance, both in quantity and quality, upon the map-work of the former or Dark Age period (A.D. 600-1000). The 'Beatus' examples, of which only one, and that the rudest of all, could in any sense be reckoned as falling within the first Christian Millennium, continue through the eleventh, twelfth, and thirteenth centuries to be reproduced in monkish schools; and though none of these copies are good in form, they occasionally contain a fair amount of matter. Especially in the 'St. Sever,' 'Osma,' and 'Gerona' manuscripts, this school of map-making shows to some advantage by the side of the usual Arabic monstrosities of this time; and no student of mediaeval geography can afford to neglect the Spanish variety of pictured chart, for no one is more characteristic of the earlier Middle Ages. The eighth-century cosmography of the Anonymous Ravennese Geographer finds a certain, though slight, cartographical explanation in the twelfth-century maps of the so-called Guido or his copyist. Another important family of Western maps is to be found in the allied designs of Henry of Mainz, of the 'Psalter,' of Hereford, and of Ebstorf, all probably descendants of an eleventh-century original, and related also to certain twelfth-century editions of two ancient sketches illustrating the works of St. Jerome. Better in form than any of these, and in all probability a work of Irish scholarship, is the tenth century 'Cottoniana,' or 'Anglo-Saxon' map, one of the geographical surprises of a period whose obscurity and barbarism are not without glimpses of light from quarters outside of, and neglected by, the Teutonic and Latin worlds. With this map we may probably connect the itinerary of Archbishop Sigeric of Canterbury from England to Rome (A.D. 990-994); for the 'Cottoniana' was perhaps the work of an Irish scholar known to Sigeric, or resident in his household. Again, rough sketches with an educational purpose, exhibiting the various zones or climates and the distribution of the great land areas, or illustrating the geographical references of some favourite classical authors, are exceedingly common; and examples of these may be found in Lambert of St. Omer, in the so-called T-O and 'climate-maps,' and in the Macrobius and Sallust-Designs. The work of Lambert, connected as it is with all these families of elementary geographical world-plan, is also a mappe-monde of considerable scope and content, clearly antique in its detail, and recalling in the twelfth century the pre-Christian theories of Greek speculators on the other land masses supposed to exist beyond the ocean to the south, west, or east of the Habitable World known to man. Lastly, the English chronicler Matthew Paris has left, at the very close of this era, a local map of England of unusual merit, along with an ordinary and disappointing mappe-monde, and several smaller essays in cartography. The same attempt at specialising and so improving draughtsmanship, by a rigid limitation of view to an area with which the designer was well acquainted, is observable in the twelfth-century Plan of Jerusalem or *Situs Hierusalem*. It is humiliating, if not surprising, to find that the Byzantine geographers of Justinian's age had executed an even better representation of the Holy City and its environs more than six hundred years before, as shown in the recently discovered mosaic of Madaba. Less ancient than this, and of far less value, but probably anterior to the first Crusade, and remarkable as, for the first time, showing us Jerusalem in the exact centre of a round world-sketch, is another Byzantine map of **T-O** pattern, the most elaborate example of its kind, which has survived in a Latin transcript of 1110.

The Moslem Geography of this period (c. 950-1260) must be held over to a subsequent volume for want of space, but here we may very briefly notice how it continues during these centuries, as during the eighth and ninth, to pour out an enormous mass of work; the greatest of Mohammedan encyclopaedists, Yakut the Greek, belongs to the later twelfth and earlier thirteenth century; a century earlier, Edrisi of Ceuta and Palermo accomplishes the most famous of Arabic descriptions of the *Oikoumené* or Habitable World. In the tenth century Mukaddasi and Albiruni (or Abu Rihan) show themselves to be worthy successors of Masudi and Ibn Haukal; the travels of Ibn Mohalhal recall the distant wanderings and hair-breadth escapes of Sinbad the Sailor; Nasir-i-Khusrau of Merv, on the eve of the first Crusade, wins a reputation as a master in the literature of geography, if not in the annals of adventure; Al Heravi of Herat, an earlier Ibn Batuta, traverses almost all the lands of Islam, 'scribbling his name on every wall'; and these are only the leading figures of a great company. On that more purely scientific side of geography, where the Moslem students of earlier time had accomplished so much, Albiruni is perhaps the greatest name of any Mussulman age or nation; but as our period advances, and the Spanish successes of the tenth century are followed by the disasters of the eleventh, the re-action of the twelfth by the disruption of the thirteenth, it becomes ever more apparent that Islam, as it recedes politically, is also intellectually on the decline. For the age of Philip Augustus and Louis IX. (118-1260) witnesses the overthrow of all the great Mohammedan States which had arisen on the disruption of the undivided Caliphate; witnesses the extinction of the chief line of the Prophet's successors, who for 500 years had wielded a papal, and often an imperial, power at Bagdad; and finally witnesses, as the complement of these calamities, the ruin of the chief schools of Moslem culture.

China,[1] during most of this period, has much less to do with the Western world than before the revolution of A.D. 880. Internal troubles, partly caused by the older foreign intercourse, serve as the excuse for a long reign of comparative exclusiveness; but even during this time relations are not wholly broken off between Islam and the Middle Kingdom. The earlier Christianity of Nestorian type, so prominent in the eighth century and the time of the Singanfu inscription, fades almost entirely out of sight; on the other hand, the faith of the Koran gains many proselytes. The incessant warlike activities of Central Asia and 'Chinese Turkestan' become at times a barrier to any overland travel: the unity of China itself is broken up;

---

[1] The detailed account of Chinese explorations and geographical study during this time, from the tenth to the thirteenth centuries, has also to be kept over from this volume for want of space; the following summary, however, gives the chief points of celestial activity.

dynasty succeeds dynasty in a series of bewildering revolutions; both the North and South, both the Ho-ang-ho and the Yang-tse-kiang, are desolated by anarchy; and the weakness of the divided nation exposes it to subjugation in detail by the arms of Ghenghiz Khan and his successors.

After the Mongol conquest an extraordinary freedom of intercourse is restored; even European and Christian travellers now penetrate to the Far East, and, what is still more noticeable, repeated embassies make their way from China to the Western world. Just as in the remote past Chang-kiang (B.C. 135) and Pan-chao (A.D. 94-102) had travelled or conquered as far as the shores of the Caspian and the borders of Persia and India, so now in the earlier thirteenth century Chang-chun and others journey at the command of the new Tartar sovereigns of Asia to the Mongol camps in Afghanistan, Persia, and Georgia. Chinese spies are sent by the Mongol Government into the countries of Western Europe [1] in the same way that Carpini, Rubruquis, Ascelin, and their companions are despatched by the Pope or the temporal sovereigns of Christendom to the courts and camps of the Tartar Empire. The opening of China and Central Asia (in a new and fuller sense) to the knowledge of the Christian West is the work of the unifying, levelling, and anti-fanatical Mongols, who break down the old walls of partition at the same time that they erase not a few of the old landmarks of former civilisation. With the restoration of a native Chinese dynasty, and with the expulsion of Kublai Khan's descendant from Pekin, the old exclusion, the old fanaticism, the old barriers, are restored; but for

---

[1] Cf. the later embassy of the Nestorian monk, Rabban Çauma, an Uigur, born at Pekin, who visits (c. A.D. 1282) Rome, Paris, and Bordeaux, which he calls the capital of England.

Europeans the former ignorance, fear, and imprisonment of spirit has for ever passed away; and among the practical objects of the new time none is more ardently pursued than the re-discovery and exploitation of Cathay and India. The overland missions end in failure, but their place is taken by the explorers, traders, and preachers who find new ways across the ocean; to follow in the steps of Marco Polo becomes the dream of every adventurous spirit; and the revelation which the thirteenth century brings to European civilisation has permanent and vitalising results for every Christian nation.

CHAPTER II

THE NORSEMEN IN THE HISTORY OF EXPLORATION TO THE BEGINNING OF THE CRUSADES

THE discoveries, conquests, and colonies of the Norsemen from the White Sea to America, throw the first glimpses of light upon many parts of that 'barbarian' world of which ancient science, trade, and politics had known little or nothing.[1] And, indeed, it was natural that here, if

[1] The numerous *Foreign Coins in Norse Countries* of course do not prove that the Roman world had a first-hand knowledge of North-Eastern Europe, but they point to a considerable trading intercourse in the later classical time and the earlier Middle Age, such as is further illustrated by the notes of Pythèas upon the amber commerce, by the order of Germanicus to his fleet to sail to the amber island (Glaesaria), and by the Knight sent to the amber coast by Julianus, Nero's manager of gladiatorial shows. Cf Pliny, *Natural History*, Book xxxvii., ch. ii. The 'Knight' here mentioned visited the amber markets of the Baltic coasts, traversed the regions of Germany abounding in amber, and brought back such a quantity that the nets used for protecting the 'podium' (*i.e.* the spectators) at the amphitheatre against the wild beasts were studded with this article.

Among the coins may be noticed *e.g.* — (1) Roman; of the following Emperors; Augustus, Claudius, Nero, Galba, Otho, Vitellius, Vespasian, Titus, Domitian, Nerva, Trajan, Hadrian, Antoninus Pius, Marcus Aurelius, Commodus, Pertinax, Septimius Severus, Caracalla and Geta, Macrinus, Elagabalus, Alexander Severus, Valens, Valentinian I., Gratian, Gordian, Gallienus, Aurelian, Tacitus, Decius, Probus, Carus, Numerian, Carinus, Diocletian, Maximian, Constantius Chlorus, Licinius, Constantine the Great, Constantine II., Constantius, Constans, Honorius, Valentinian III., Majorian, Libius Severus, Anthemius, Romulus Augustulus, etc.

(2) Byzantine or of the Eastern Empire after Theodosius the Great; Theodosius II., Marcian, Leo I., Zeno,

anywhere, in 'Polar' lands, rather than in Mediterranean countries, European enterprise should first show signs of new life after the paralysis of the Dark Ages. From the rise of Islam the Saracens controlled the great trade routes of the south and east. It was only on the west and north that the coast was clear—of all but natural dangers.

In the Moslem world men 'of the market-place' were now busy in following up old lines of trade, or, as in Eastern Africa, were discovering and exploiting new and more promising routes; men of science were commenting on the ancient texts of Greek and Latin learning, or adapting them to enlarged knowledge. But in Christendom, in the long atrophy of mental and physical activity which covers most of the seventh, eighth, ninth, and tenth centuries of our era, the practical energy of heathen enemies (for the Northmen were not seriously touched by Christian influence till long after they had become a terror to Europe) was the first sign of permanent revival.

Anastasius, Justin I., Justinian, Tiberius II., Maurice, Constantine Copronymus, Michael III. Several others belonging to co-emperors, pretenders, the lesser princelings of the last years of the Western Empire and consorts of various emperors have also been brought to light; besides a vast number of barbaric imitations and of defaced and unreadable coins. The earliest coins found in Gottland are those of Augustus. Then follow those of Nero, and coins of all the different emperors to Alexander Severus; the greatest number are those of Trajan; Hadrian; Antoninus Pius; Faustina, wife of Antoninus Pius; Marcus Aurelius; Faustina, wife of Marcus Aurelius; and Commodus. In Hagestaborg the most numerous were the three Antonines; the earliest was one of Nero, the latest one of Septimius Severus.

The Byzantine coins from Constantine the Great are mostly ornamental. A few coins of the Republican Period have also been found in Norse countries; *e.g.* of the families of Lucretia, Naevia, Sulpicia, Funa, Publicia, Postumia, Procelia, Tituria, Veturia, Sicinia.

Many Arabic and other mediaeval coins have also been found in Scandinavian lands (see pp. 462-4n. of this volume), as well as an immense number of other objects from the Southern world.

With the awakening of physical energy and daring came that of mental and spiritual vigour; the whole life of the Middle Ages began afresh with the conversion of Scandinavia, of Russia, and of Hungary; but, in the abundant and brilliant force of the mediaeval renaissance, we must recognise the work of those Norsemen who had breathed something of their own spirit into every Christian nation.

The Norse discoveries were not only the first, they were beyond comparison the chief, victories of European travel and enterprise, in the truly Unknown World, between the time of Constantine and the Crusades. The central fact of European expansion in the Dark Ages is the advance of the Scandinavian adventurers and settlers to the west, the south, and the east in the last two hundred years of the first Christian Millennium.

Of the three main lines of northern advance, the first, the western or north-western, running by Caithness, the Orkneys, the Shetlands, and the Faröes, reached Iceland, Greenland, Baffin's Bay, and the shores of Newfoundland, Labrador, and Nova Scotia;[1] but, from the settlements on the coasts and islands of Northern Britain, a fresh wave of pirates and colonists swept down south-west into the narrow seas of St. George's Channel, and beat upon the north and east of Ireland and upon the western coasts of England and of 'Bretland.'

The second invasion ran along the North German shore, and, reaching the Straits of Dover, fell upon both sides of the Channel, only varying in strength as the resistance was stronger or weaker in England or in France. Pushing on along this sea-path, the more daring adventurers united (off

---

[1] Perhaps even Massachusetts, but as Dr. Storm has shown (*Vinland Voyages*, 344-67), it is not necessary to bring Vinland so far South.

Cornwall and the Scilly Isles) with their Ostmen brethren from Ireland, and sailed on to the plunder of the Bay of Biscay and of Northern Spain. It was not long before some of them had found out the wealth and weakness of the Moslem Kingdom of Cordova, and were trying to force their way up the Douro and the Tagus. From the great Northman settlement on the Seine new and ever more daring ventures brought little groups of 'Normans' to South Italy and Sicily, who formed a kingdom of the two Sicilies, carried their ravages to the Saracen ports of North Africa, and gave to a Norman prince the momentary lordship over much of Tunis; just as from the same Normandy a greater stream of conquest flowed across the Narrow Sea to overwhelm the Old English State.

On the third or eastern side of the Norse expansion there are likewise two main divisions; one taking its road across the Baltic, and proceeding north-east to Finland and due east to Novgorod and Russia, the other coasting along Norway to Finmark and the North Cape of Europe, to the Murman coast of Lapland and the country of the Biarmamen[1] on the White Sea and the Dvina.

The field of this activity was indeed immense, but it was covered with a rapidity no less astonishing. It was in the later years of Charles the Great (790 to 814) that the Viking Keels first alarmed the Frankish coasts. In 797 they made their first descent on England;[2] before the middle of the ninth century they were threatening the whole shore-line of Christendom, from Galicia to the Elbe; in 840 they founded their first kingdom in Ireland, and even earlier than this they had become the terror of the Hebrides,

---

[1] 'Perm.'

[2] Cf. the Anglo-Saxon Chronicle, under the year 787, 'In his days' [*i.e.* the West-Saxon King Berhtric's, A.D. 785-802] . . . 'first came three ships of Northmen. . . . The reeve rode to the place and would have driven them to the king's town and they slew him.'

the Shetlands, and the Orkneys;[1] from 860 commenced their attempts upon Iceland; in 862 Rurik and his men founded the Central State of mediaeval Russia. Pressing on into still wider fields, they sighted Greenland soon after 877;[2] in the last quarter of the ninth century they brought their ships to the neighbourhood of the present Archangel, and in the first years of the tenth they won Normandy from Charles the Simple by the treaty of Claire-sur-Epte (922).[3] This advance is almost as rapid as that of the early Saracens. Within a hundred years from the first disturbance of the old local independence by the levelling power of the new national kingdoms, within three generations from Halfdan the Black, the Vikings, first as ocean rovers and buccaneers, then as rebellious emigrants, and finally as founders of new states, had reached and passed beyond the limits of the Old World to north and west, from 'Jacob's Land'[4] to Finmark, and from Ireland to the Baltic.

Much of this of course was only the passing of a more active race over ground, which had once been well known, even if much of it was now being forgotten. But in the far north-east and north-west, the Norsemen clearly enlarged the horizon of knowledge—in the plains of Eastern Europe, in the shorelands of North America, and along the coasts and islands of the Arctic Seas.

On the side of America indeed the Norse discoveries were not only an advance upon all that the older geography had known; they remained, throughout the Middle Ages, an

---

[1] Cf. Dicuil, *De Mensura Orbis terrae* (written in 825), vii. 15 (= vii. 2, 3, Letronne).

[2] The traditional date of Gunnbiorn's discovery, which the best modern critics would bring down a generation later, to the early years of the tenth century.

[3] The 'Norman' settlement began c. 912; the treaty was c. 922.

[4] Galicia in North-West Spain, so named from the Shrine of St. James (Santiago) at Compostella. On the Scandinavians in Spain, see pp. 111, 176-9, of this volume; Dozy, *Recherches*, ii., especially pp. 295, 329-340.

unique achievement. The Venetian, Welsh, Arabic, and other claims to have rivalled the Scandinavians in this field are utterly unsubstantial and untrustworthy. It is a fairly certain fact that about the year of Christ 1000, the north-west line of Viking advance reached at least as far as the north-east shores of New England. Against all other mediaeval voyages to the Western Continent one verdict only can stand, a verdict of Not Proven.

The other lines of Norse advance, though marked by equal daring and far greater military prowess, have less originality, and occupy a much smaller place in the history of exploration. There was roving and fighting in plenty, the giving and taking of hard knocks with every nation, from the White Sea to the Straits of Gibraltar, and from Limerick to Constantinople; and, as the Vikings discovered new lands (new at least to themselves), they re-named most of the capes and coasts, the rivers, islands, and countries of Europe, North Africa, and Western Asia. Thus, Galicia became 'Jacob's-Land'; Islam, outside 'Spanland,' passed into 'Serk-' or 'Saracen-Land'; the Negro or even the Moorish countries were the 'Lands of the Bluemen.' The Hebudes or Hebrides were the 'South Isles'; Northern and Western France 'Valland'; North Germany 'Saxland'; the East Slav or Russian country 'Gardar,' 'Gardariki,' or the 'Garth-realm.' The city of Constantine was the Great Town or 'Miklagarth'; Novgorod was 'Holmgarth'; but this was a State which owed its greatness to a Viking migration, and was long connected by ties of peculiar intimacy with the Scandinavian lands. Moreover, it belongs to the field where Norse enterprise broke fresh ground; to the south and south-west, on the contrary, no rebel Vikings, or royalist hunters of such rebels, sailed the seas beyond the limits of the classical geography; and, as pilgrims, traders, travellers, and conquerors in the Mediterranean, their work was of course not one of exploration. They bore a foremost part in breaking down the Moslem rule in Southern Europe; they visited the Holy Sites—

> 'When sacred Hierosolyma they'd relievèd
> And fed their eyes on Jordan's holy flood,
> Which the dear body of Lord God had lavèd,'—

they fought as 'Varangian' bodyguards in the armies of the great Byzantines, Nikephoros Phokas, John Tzimiskes, Basil II., or Maniakes; but in all this they discovered for themselves rather than for Europe.

One of the most typical references to Geography in the Sagas is in the famous chapter at the beginning of the mythology of the *Heimskringla:*—'On the situation of countries.' 'It is said that the earth-circle which the human race inhabits is much cut asunder with bights and bays, and that great seas run into the land from the outer ocean. Of a certainty, it is known that a sea goes in at Norva Sound,[1] right up to the land of Jerusalem; and from that sea, again, a long bay, which is called the Black Sea, goes off to the north-east, and it divides the two World-Ridings, that is to say, Asia on the east from Europe on the west. To the north of the Black Sea lies Sweden the Great, or the Cold; and this is reckoned by some as not less in size than the Great Saracen-Land, or even the Great Land

---

[1] Straits of Gibraltar, so named traditionally from one Norva, supposed to be the first Norseman who passed through this channel, in the tenth century. But they had certainly gone through the Straits in the ninth and even in the eighth century, as Dozy has shown, *Recherches*, vol. ii. p. 295. On the other hand, *Magnus Barefoot's Saga*, ch. xxii., gives this honour to Skopti Ogmundson, at the beginning of the twelfth century. Skopti sailed with five long ships in harvest to Flanders; was there the winter; early in spring went on to Valland; in summer passed through Norva Sound; in harvest-time arrived at Rome. There Skopti died. 'Men said he was the first Norseman to sail through Norva Sound,' etc.

of the Bluemen.[1] And the northern parts of this Sweden lie unpeopled by reason of the frost and the cold, just as the southern parts of Blue-Land [2] are waste because of the sun's burning. Mighty lordships are there in this Sweden, and peoples of manifold kind and speech; there are giants and there are dwarfs—aye, and Bluemen,[3] and folk of many kinds and marvellous, and wild beasts, and dragons wondrous great. Out of the north, from those mountains which lie outside of all peopled lands, there comes a river which is rightly called Tanais,[4] and this falls into the ocean at the Black Sea, dividing Asia from Europe.'

It seems impossible to identify with present-day sites the famous places of the Norse Mythology; even Asgard,[5] Asgarth, or Gods-town, can only be conjectured, and is, perhaps, no place-name at all, but a term like the 'Age of Gold,' expressive of a supposed condition of the race at an early period. Here and there, however, even the Mythical Sagas plainly refer to natural facts. Thus the great mountain barrier from north-east to south-west, which parted Great Sweden from other kingdoms, may well be an allusion to the Ural Range.[6] Again, we are told that Odin wandered westward to the Garth-realm or 'Gardariki,' and south to 'Saxland,'[7] where he made himself a great kingdom. These are the terms in regular use, by the Norsemen of the historical time, for the mediaeval Russian and German lands. Once

[1] Or Negroes.
[2] Negro-Land.
[3] Here perhaps Samoyedes.
[4] The Don. Of old it was called Tanabranch or Vanabranch, adds the Saga; the land between the Vana-mouths was then called Vanland or Vanhome; cf. ch. i. of the *Ynglinga Saga*.
[5] Those who attempt identification make Asgard either (1) Assor, or (2) Chasgar in the Caucasus, the Aspurgum of Strabo. *Ynglinga Saga*, ch. ii., fixes Asgard east of Tana-branch in Asia, in the land of Asland or Ashome, of which it was the chief burg.
[6] Ch. v. of *Ynglinga Saga*.
[7] Ch. v. and ch. xxxii. of *Ynglinga Saga*. The term 'Garth-realm' is, though quite exceptionally, used also for the Byzantine empire.

more, it is clear that a principal object of the mythological narratives is to tell us about Odin's settlement in Sweden, and the daring exploits of the oldest Norsemen upon the ocean.[1] Even in those days there were many 'Sea-Kings who ruled over men, but had no lands'; when 'he alone might be called a Sea-King who never slept beneath sooty roof timbers, and never drank at the hearth nook.'[2] Even then, too, it was known that Sweden had 'such great wild woods' that it was many days' journey across them. From this Sweden it was that one of the earliest and most important Viking movements proceeded just a thousand years ago. In 860-862 the Russian Slavs of the North,—or a Scandinavian colony dominating these Slavs, and living about old Novgorod on Lake Ilmen, about Lake Ladoga, and about Isborsk,—asked for help against their hostile brethren of the South, living about Kiev; and in answer to this appeal Rurik and his war-band arrived from the lands of Upsala, and founded those mediaeval Russian states, which in the tenth and eleventh centuries were first the stealthy plunderers, then the open enemies, and finally the allies in faith and in arms, of the Byzantine Empire. So the oldest Russian annals[3] gave the story; and in its main lines it may be accepted as fairly representing the truth of an early pre-Christian intercourse, between Scandinavia and the Eastern Slavs, which had results of the first importance, and practically created a new European national centre.

From the ninth to the twelfth century the intercourse

---

[1] Ch. vii. of *Ynglinga Saga* mentions Odin's ship *Skidbladnir*, 'wherein he would fare over mighty seas,' and which he could 'fold together like a napkin.'

[2] Ch. xxxiv. of *Ynglinga Saga*.

[3] Cf. Nestor's *Chronicle*, chs. xiii. xv.; under the years, A.M. 6368-70 (=A.D. 860-862). In A.D. 850, according to Nestor, the Russian land 'began to take its name.'

of Swedes, Danes, and Northmen with the Russian principalities of 'Gardariki' was constant and close; and perhaps most of all in the time of the Vinland voyages, when Vladimir and Yaroslav reigned in Novgorod, and when the two Olafs—the son of Tryggve and the Saint—found refuge at that Court from the troubles of their own distracted Norway. Marriage alliances between the Princes of Gardar and those of Sweden are recorded[1] at least half a century before the expedition of Rurik; and in the later development of the Russian people this Scandinavian element perhaps counted for nearly as much as in some other countries where it has been more fashionable to trace, and sometimes to exaggerate, the 'influence of the Northmen.'

Considered as an organised and united state, the Norse Kingdom began with Halfdan the Black and his son Harald Fairhair, at the same time that Danish unity was first achieved under Gorm the Old (860-935). These great chiefs attacked, shook, and finally brought to an end the old local independence; and one result of this was a strengthening of the Scandinavian movement of expansion and aggression upon the outer world. For now, in the discontent that arose 'when Harald seized on the lands[2] of Norway, the Outlands were found and peopled, and especially the Faröes, and Shetland, and Iceland,' and the rebel Vikings who fled before the new monarch took to piracy in the Western Sea; in winter they made their home in the Orkneys and the Hebrides,[3] and in summer they pillaged as far as Norway. But after some years Harald grew tired of chasing

---

[1] Cf. Icelandic *Landnamabok*, p. 195, in the genealogy of Thorbiorn. 'Gorm was a duke in Sweden . . . his son married a daughter of Burislav, King of Gardar, in the East' (c. A.D. 820). Cf. also Morris and Magnusson, *Saga Library*, vol. i., Pref. p. xviii.

[2] *I.e.* the tribal kingdoms.

[3] Sudreyar, the 'South isles' or the Hebrides, ch. xx. of *Harald Fairhair's Saga*.

the pirates away from his coast, and sailed out into the North Sea in pursuit of them.

He drove them out of Shetlands and Orkneys and Hebrides, and even out of Man; he ravaged the coasts of Scotland; and in the islands that he subdued he left a deputy to govern in his name as Earl. Thus was founded, about 920, the Norse Principality of the Orkneys, whose troubled life was prolonged almost to the close of the fifteenth century, when it was finally absorbed by the Scottish Kingdom. From these Islands the Norsemen raided Caithness and Sutherland, winning easy triumphs over the Scots, and riding off to their ships with the heads of Scottish chieftains tied to their horses.[1] But the Orkney settlements in the end gave more trouble than profit to the central kingdom, and in exploration (as in civilisation) they seem hardly ever to have become a real starting-point of further progress.[2]

Far more useful in these respects were the early Viking colonies in the Faröes. From these islands, as early as about 860, one of the Norse leaders, 'Nadodd,' pushing into the ocean to the far north-west, sighted Iceland. He gave it the name of 'Snowland,' and reported that he had found a desert land easy of settlement; but his was not the first visit. As early as 795 Irish hermits had passed at least one winter on its shores, and the traces of their cells and chapels—bells and ruins and crosses—were quickly noticed by the Norsemen when they began to settle there.[3] Nadodd was followed in 864 by Gardar

[1] See example in *Harald Fairhair's Saga*, ch xxii.
[2] Though rovers who haunted the Orkneys constantly went on to Iceland.
[3] The 'Westman's Eyar' or Irishman's Islands, off the Iceland coast, are called after these 'papar' or priests. The early Norse settlements in the Shetlands are curiously commemorated, among other things, by the name 'Sholto' or 'the Shetlander'—only to be found in Scotland and Iceland.

the Swede, who called the new land after himself; by Raven Floke, who in 867 was guided by the flight of birds[1] to the same country; and by the permanent settlers of 874,[2] led by Ingolf and Leif and other sheep-farmers of the Faröes.

Even this was not enough. Three years later, in 877, at the very time of the furthest Danish advance in England, when Guthorm had driven the English king into the Isle of Athelney, the Norsemen reached another stage in their north-western advance; for now Gunnbiorn[3] sighted a reef or skerry and a great land in the ocean beyond Iceland, which he called 'White Shirt' from its snowfields, and which Red Eric, a century later, re-named 'Greenland,' in the hope of a 'good name' attracting settlers. By this discovery the Old World had almost, as it were, stumbled upon the New.

Geographically, this portion of the Arctic Continent falls to the share of North America; and once its fiords had become in their turn centres of colonisation and of further progress, the reaching of Labrador and Newfoundland was natural enough. The real voyage lay between Cape Farewell and the European Mainland; it was indeed a stormy and dangerous passage from the Bays of South Greenland to the western regions of 'Helluland' or

---

[1] Ravens. These birds, says an early addition (c. A.D. 1300) to the Icelandic *Landnamabok*, originally a work finished before 1148, were of great use as guides to the sailors of the northern countries, who had not yet any lodestone. Cf. Nordenskjöld, *Periplus*, p. 50 (E. trans).

[2] Some think all these dates are about thirty years too early.

[3] Some think, however, that Gunnbiorn's voyage was of later date, about 910-920, and that the skerry which he saw was only part of the high reef or ridge in the sea between Iceland and Greenland, now under water, but in the ninth and tenth centuries possibly above the sea in certain parts. One record says that Eric went to Greenland to discover the land which Gunnbiorn, son of Ulf the Raven (Kraka), saw, when he was driven west to Iceland and discovered Gunnbiorn's reef.

'Markland,' but the journey was not long, and, as far as can be judged from scanty records, it was neither so cold nor so ice-bound as at present.

But exploration had outrun settlement; the pace of the Norse advance on this side slackened for a hundred years; and it was not till 984 or 986 that Eric the Red, exiled for manslaughter by the Icelanders, first led his countrymen to plant themselves in 'Hvitserk.'

In this venture Eric set out from Snowfield-Jokul in Iceland, and sailed west till he reached land—on the east coast of Greenland—which did not appear very suitable for a colony. From this landfall he coasted along to the south, rounded a Cape (which he named 'Hvarf'[1] or 'Wrath,' and which moderns call 'Farewell'), and wintered on an island henceforth known by his name. Three years in all he spent on this coast, prospecting for his intended settlement; then returning to Iceland, he brought back so good a report of the land, that twenty-five ships full of emigrants followed him on his final journey to the Green Country. Only half of these persevered to the end; and with them the leader colonised Eric's Fiord, the starting-point of a still more distant and memorable series of adventures in the next generation.

In the same age that witnessed the conquests of Halfdan the Black and Harald Fairhair, the fiercest attacks fell upon England, France, and the North German lands from the coasts of Norway and Denmark. From the first descent upon Wessex, in 797, their invasions of Britain were incessant and terrible, but for a long time spasmodic and impermanent. But in 855 the raiders for the first time wintered in Northumbria, and in the next twenty years they pressed continuously, and at last

[1] Properly 'Turn.'

overwhelmingly, upon a large part of England. By 870 they had mastered nearly the whole of Northumbria and Mercia; and the following decade was marked by a desperate fight for the possession of Wessex, which resulted in the premature triumph of Guthorm, the final victory of Ælfred, and the peace of Wedmore, which confined the invaders' rule to North and North-Eastern England (877–878).

In 876 the Northmen had already made their first settlement in France, after many years of warfare;[1] and it was from this side that Hasting drew the main part of his strength in the repeated onslaughts which he led upon Wessex at the end of Ælfred's reign (891–896).

The rule of Harald Fairhair, by its ruthless suppression of Viking piracy near home, also brought about the Norse conquest of Normandy. Hrodulf, Rholf, or Rolf 'the Ganger,' the son of Rognvald the 'Mere Earl,' was an incorrigible raider, and so huge that no horse could bear him; 'wherefore he went always on foot, and from this got his name.' By his constant plundering on the Baltic coasts,[2] he so enraged the king, that he made Rolf an 'outlaw from all Norway.' His father was Harald Fairhair's dearest friend, and his mother was a poetess;[3] but neither the one nor the other could save their son from exile. He was driven west over

---

[1] Marked by the sack of Limoges, Bordeaux, Tours, Saintes, Nantes, Toulouse, Orleans, Bayeux, Evreux, Rouen, and other places, the great simultaneous attack of 845 upon all the three Frankish kingdoms, and the various sieges of Paris, the first of which was in 845 (followed by those of 855, 861, and 885).

[2] 'Eastlands.'

[3] The *Saga of Harald Fairhair*, ch. xxiv., gives the verses that she sang to the king to dissuade him:—'Thou hast cast off Nefia's namesake; As a wolf from the land thou driv'st him; 'Tis ill to be wild in quarrel With a wolf of Odin's warboard. If he fare wild in the forest He shall waste thy flock right sorely.' Rolf's mother, Hild, was the daughter of Rolf Nefia.

sea to the Hebrides;[1] thence he made his way to the North Coast of France and conquered Normandy. From Normandy, in later time, started the free lances, who built up a Norse Kingdom in South Italy, as well as the host who, under a certain Bastard William, Rolf's greatest descendant, conquered the Island Kingdom to the north.

The ambition of Harald Fairhair was not content with one expedition to Britain; for soon after he went a second time to the Orkneys to crush the ever-turbulent pirates of the Western Islands; but a far longer and more daring voyage was that of his son Eric in about 930. When the boy was twelve years old his father gave him five long ships,[2] and he went on an eight-years' war-cruise in the Baltic, and along the Danish and German, the French and Flemish, the British and Irish coasts. Finally he sailed away north to Finmark, and right up to Biarmaland. This is one of the earliest voyages that the Sagas record to the region of the White Sea—the land of Perm, or the heathen Biarma-men—a country which in after days was often visited and plundered by Norse raiders and merchants, and became famous also for its furs, its idol-temple of Iomala, and the magic of its people. But early as it is, this voyage of Eric Haraldson was anything but a venture into the Unknown. Before the close of the previous century (c. A.D. 890) Ohthere of Halogaland (who afterwards described his journey to King Ælfred of Wessex) sailed from his home to the mouth of the Dvina round the North Cape of Europe and along the Lapland or Finmark coast and the shores of the White Sea. He lived, as he thought, to the north of all the Northmen, on the mainland by the Western Sea; to the North of him was a waste, peopled only by a few scattered Finns, hunters and fishers;

[1] 'South Isles.'   [2] Ch. xxxiv. of *Harald Fairhair's Saga.*

with the inquisitive instinct of his race he determined to find out for himself how far this waste extended, and whether any men lived beyond it. He was probably not ignorant of the doings of Bjorgolf of Halogaland or Brynjolf his son,[1] who about this time collected the tribute of the Finns for Harald Fairhair; but even if such a rival had preceded Ohthere to the North Cape or beyond, there seems no clear evidence of any earlier Norseman picking out the whole of the Biarmaland route. His journey was first towards the pole for six days, till he had passed beyond the furthest haunts of the whalers and come to where the land turned eastward, running north no longer. Four days in the new direction brought him to a point where the shore began to run southward into the 'inland sea,' and for five days more he followed the coast as it bent inwards to the mouth of a great river, where the land was well tilled and all peopled by the Biarma-men. All the way he had the waste upon his right and the open sea upon his left; he did not dare to enter the Biarmaland, but he opened up a certain intercourse with the natives, and learnt from them many things about their own country and the regions near; from his own observation he gave an excellent description of the long and narrow district in which the Scandinavians of the Far North were hemmed in between the wild mountains of the Finns and the ocean. Ohthere was not only an explorer but a trader; his White Sea voyage was partly in search of walrus-ivory, and hides, and some of the 'noble bones' or tusks of this 'smallest of whales' he probably displayed in later years among his trophies at the English Court.

Nor was Ohthere the only precursor of Eric Haraldson on the Biarmaland route. In 918-99 Eric Blood-axe made

[1] Cf. *Egil Skallagrimson's Saga*, chs. vii.-ix.

a different sort of journey to the Dvina—an armed foray which ended in a great battle with, and triumph over, the Biarma-men, who mustered in great force to resist him.[1] Meantime the fur trade of the Finmark coast had been growing, in the latter years of Harald Fairhair, into a high-prized monopoly of the crown; and an impetus to Icelandic migration from Halogaland and other Arctic provinces of Norway was given by the King's quarrel with the house and friends of Bjorgolf, who had so long acted like little satraps on the Lapland coast. Thorolf, the heir of Bjorgolf's grandson Bard, not only travelled far and wide in Finmark as the royal steward,—collecting tribute, holding fairs, and killing or expelling rivals,—but made alliances and armed forays on his own account.[2] Thus, in A.D. 897, in company with Faravid, King of the Kvens, he attacked 'Kirialaland,' in which we may possibly see the modern Korelia, to the west of Biarmaland proper;[3] but soon after this his schemes of northern trade and dominion were abruptly checked by the suspicion and hostility of his overlord.

Shortly before the introduction of Christianity into Iceland (from A.D. 996) we find Hallvard the Easterling, a trader from the Bay of Christiania, urging the hero Gunnar of Lithend to sail abroad with him, and boasting how he had been to all lands between Norway and Russia, as well as to Biarmaland.[4] To this same mysterious region many later journeys were made from the Norse Fiords throughout the tenth and eleventh centuries; and here, where in after

[1] *Fgil's Saga*, ch. xxxvii.
[2] *Egil's Saga*, chs. x., xiii., xiv., xvii.
[3] Korelia, according to Engelhardt, *Russian Province of the North*, ch. ii., had been already attacked by the Scandinavians of King Eric Emundson, before A.D. 833. This depends on the identification of the Kvens whom Eric invaded with Korelia. Anyhow, Kvenland and Kiriala-land together may be taken to stretch over the region between the Gulf of Finland and the White Sea.
[4] *Burnt Njal's Saga*, ch. xxviii.

time the traders of Great Novgorod gathered peltries for the Hanse markets, and where still later the Tsars of Muscovy gained their first outlet to the sea, the Northmen (in the age of our Æthelstan and Eadmund and Æthelred) seem to have done a profitable business. Thus King Harald Greycloak, or Greyskin (c. A.D. 970), sailed in the long days of summer to the banks of the Dvina, and 'gat exceeding wealth.'[1] A much more famous journey was that of Thorir Hund, about 1020; this is one of the most dramatic incidents of the Sagas; and from its context we may gather that these voyages round the North Cape of Europe are typical of a much larger number, as to which we have only bare allusions or no record at all.

The same, of course, is true of Norse exploits in other fields like the Baltic, where Halfdan the White and the Black followed Rolf Ganger's steps to Esthonia, and gained a famous name among the chief pioneers in one of the favourite scenes of Viking adventure.[2]

Harald Fairhair's successor, Hakon the Good, was in England at the Court of Æthelstan, when the first great King of Norway died (934). There was a struggle for the crown between Hakon and his brother Eric, and the usual

---

[1] Ch. xiv. of *Saga of Harald Greycloak;* the same event has a song to commemorate it:—'The conqueror of the boastful Reddened the firebrand east there, All northward of the township, Where saw I Biarm folk running; Good word the men's appeaser Found on the side of Dvina.'

[2] In this region the voyage of Wulfstan from Haddeby in Denmark is specially interesting, as apparently enterprised by an Englishman (c. A.D. 890) at a very low time of English trade and marine. Ælfred describes this journey, in his *Orosius,* as complet- ing a sort of periplus from the White Sea round to Esthonia. Ohthere's first voyage to the Dvina is followed by a second to Haddeby; then comes Wulfstan, who, in seven days, reaches Truso, on or near the river Elbing, and describes the Wendland coast, the river Vistula, the Frisches Haff and Esthonia,—where stood many towns, and in every town a king; where the rich drank kumiss and the poor mead; where honey and fish abounded, and where strange burial customs were to be witnessed.

refuge was sought by the pretender when his schemes had failed. Eric sailed west over sea to the Orkneys, and then south to England and Scotland; and there he harried 'all about.' Æthelstan gave him Northumberland, 'the fifth part of England,' to hold in fief; and Eric had his seat at York, where Ragnar Lodbrok's sons had formerly dwelt, in a country mostly peopled by Northmen.

But when King Eadmund came to the throne of Wessex, Eric was driven out, and had to take to piracy on the British and Irish coasts.

Meantime, the earlier Viking descents on the Baltic shore-lands brought after them, as usual, the Norse king's fleet; for now King Hakon sailed along Skaney,[1] 'Selund,' and Gautland, slaying Vikings, both Danes or Wends.[2] Here, again, a casual mention of the Saga opens a long chapter of history. These Wends, or Slavs, settled on the south Baltic coast, for centuries waged an incessant warfare with the Scandinavian kingdoms on the north-west, sometimes in alliance with one of their enemies against another. Many Viking rovers, such as the half-legendary hero Palnatoki, planted themselves along the Wendish or Pomeranian shore, and one of their nests at Jomsborg became especially famous. The struggle of ages between Scandinavians, Germans, Finns, and Slavs, for the control of the southern and eastern Baltic, opens in this remote period, and, passing through many changes, does not reach its present settlement till the time of Napoleon (1790-1815);—when the Swedes, representing Scandinavian influence, were finally thrust out; when the Finns, who for six hundred years had been their

---

[1] Scania (hence 'Scandinavia'), during most of the middle ages a Danish possession, in the extreme south of the modern Swedish Kingdom.

[2] Germanic name for Slavs; *Saga of Hakon the Good*, ch. viii.

subjects, were transferred by the agreement of Tilsit and the armed force of Alexander I. to a new allegiance; and when Prussia representing Germanic, and Russia representing Slavonic elements, divided the whole region as it is to-day.

Apart from this Baltic voyage, Hakon the Good's reign does not give much evidence of Norse expansion or intercourse with the outer world; and the same is true of the time of Harald Greycloak. We hear only of forays in Ireland and Britain; of the voyage to Biarmaland above mentioned; and of a poem by a celebrated skald[1] on the men of Iceland. But with the age of Olaf Tryggveson the range of Norse interests and adventures is suddenly widened, and we hear of exploits in all countries from Novgorod to Iceland, and from Spain to the Dvina. From its beginning the life of young Olaf was that of a wanderer. His first recollections were of exile, when his mother had fled through Sweden to the Baltic, where Vikings captured and enslaved them both, and where Olaf was sold to a man named Klerk 'for a right good he-goat.' But this time of misery did not last long; for an uncle of Olaf's was then living in Russia, at the Court of St. Vladimir,[2] and, coming into Esthonia to collect the taxes for the Lord of Novgorod, he was able to free his nephew in dramatic fashion, and to take him back to Holmgarth, keeping his kinship a secret. One day, however, Olaf was standing in the gate of Novgorod, already one of the greatest marts of the north, looking at the traders as they came and went, and seeing an old enemy, he brained him with an axe in the free and simple fashion of the Norsemen. The

---

[1] Eyvind Skald-spiller. See the *Sagas, of Hakon*, ch. ix.; *of Harald Greycloak*, chs. xiv. and xviii.

[2] 'Valdimar,' otherwise Vladimir the Great (and Saint), Grand Prince of Kiev, 980–1015. Earlier than this the Saga shows him reigning as 'king' at Novgorod.

barbarous Russian customs of the time put illiberal restrictions on the blood-feud, and Olaf was charged with murder. But his uncle took him to the 'queen,' and the 'king' made peace and fixed the blood-money. Then the latter was told about Olaf's race and history, and at last he took him into his service. 'Olaf was nine winters old' when he came to Novgorod,[1] and he stayed nine years more with 'King' Vladimir (A.D. 972-981).

It is perhaps worth notice in this narrative, not only how close is the connection[2] between the Scandinavian and Baltic Lands on one side, and the Russian Principalities on the other, but also how Novgorod—'Lord Novgorod the Great,' on the Volkhov—is now, at the close of the tenth century, the principal centre of power and trade and

---

[1] 'Garth-realm.'

[2] On this connection, besides the references in the *Heimskringla*, we may compare *Laxdale Saga*, ch. xii., where Gilli the Russian, apparently a Scandinavian trading on the East Baltic coasts, sells the Icelandic Viking and traveller Hoskuld, a girl slave, Melkorka, a daughter of Myrkiartan the Irish King. Gilli is noticed as wearing a Russian hat. Perhaps the Greek hat which Thorkel, brother of Gisli, was wearing when slain was likewise Russian: *Gisli the Outlaw's Saga*, ch. xv. Also cf. *Saga of Burnt Njal*, ch. xxx., where Gunnar of Lithend harries in Esthonia, near Revel, and in the isle of Ösel, off the Esthonian coast; ch. xxxi. of same, where among Gunnar's presents to the King of Denmark a Russian hat is recorded; ch. lxxx. of same, where Kolskegg the Icelander, baptized in Denmark, fares east to Russia (Garth-realm) and stays there one winter on his way to Constantinople; ch. cxviii. of same, where the half-fabulous exploits of Thorkell Foulmouth on the east Baltic coasts are recounted, *e.g.* the slaying of a wild man of the woods (a half-human monster) as well as of a fire-dragon in Adalsyssla, etc.

Again in *Egil's Saga*, ch. xlvi., Egil Skallagrimson harries in Courland: and in *Eyrbyggia Saga*, ch. xxix., Biorn, afterwards the Champion of the Broadvikings, banished from Iceland for manslaughter, takes service with Palnatoki, captain of the Jomsborg Vikings, and makes himself a great name in Baltic warfare. Once more in the *Heath-Slayings Saga*, ch. xli., Bardi the Icelander goes to Russia, and becomes one of the Russian Prince's Vaerings or Scandinavian guard, and wards the Prince's realm, but is killed at last in a naval battle, apparently while still in the Russian service.

national intercourse among those Russian principalities in their relation to the Baltic. We shall see that it keeps this position all through the period of these Sagas; long after their records cease Old Novgorod is still one of the greatest cities, and especially one of the greatest markets, in Northern Europe. Its traders spread over all that we now call the North of Russia, to the coast of Lapland, the North Sea, the Dvina, and the Petchora; some of them appear to have reached and crossed the Ural Range, and carried their wares into the lower valley of the Ob, a country often known as Yugria. This far-reaching commerce must not be too hastily construed as a political dominion; such an idea would probably be false, if applied to the lands east and north of Lake Onega; but, in the absence of any other organised power throughout these vast districts, this mercantile ascendency prepared the way for the political sovereignty of Moscow. At the end of the ninth century, or the beginning of the tenth, the men of Novgorod had already penetrated into the basin of the Northern Dvina;[1] from about the year 1000 they begin their visits to the more distant Petchora; and in the course of the next few years they make repeated attempts to cross the dividing ridge into the land of 'Yugria' and the valley of the Ob. In 1032 they were repulsed by the natives in trying to cross a certain pass in the Urals, called the Iron Gates; between 1096 and 1187 the trade dominion of the

---

[1] See Charles Rabot, *A travers la Russie boreale*, 1894, pp. 160-170; Tchulkoff, *Geschichte des russichen Kommerzes*; and S. Sommier, *Sirieni, Ostiacchi, e Samoiedi dell' Ob*, 1887; also Engelhardt, *Russian Province of the North*, E. trans. 1899; chs. ii., v.; pp. 42, 54, 55, 107-8; and *Early Voyages to Russia*, etc., Hakluyt Society, vol. i., p. 23. Kholmogori, the chief Slav settlement in the neighbourhood of the White Sea, before the foundation of Archangel in 1584, is mentioned by the Grand Duke Ivan Ivanovitch (Ivan II. of Moscow) in a letter to the *Posadnik* (Governor) of the Dvina in 1355 or 1356.

Novgorod republic claimed to have reached a firm establishment both in the Petchora country and in Yugria; but its tenure was really of a more uncertain kind. It was utterly extinguished between 1187 and 1264; the attempts of the Hanse merchants to recover their position in 1193 was futile; but in the later thirteenth century we find them again enjoying their older access to these regions, their older so-called tribute from the same. This mercantile ascendency, intermittent as it probably was, appears beyond dispute to have opened the way by which Ivan III. of Moscow, on his conquest of Novgorod, advanced to the Arctic Ocean (1471-1478).

Novgorod merchants also penetrated to the Murman, or north, coast of Lapland, at least as early as the opening of the thirteenth century; even in the eleventh and twelfth it has been claimed that their settlements were to be found on the banks of the Dvina, Onega, and Mezen, and by the shores of the Northern Ocean. Hordes of freebooters from the great city plundered and subjected the Finnish tribes of Biarmaland, and the modern Pomors of Russian Lapland are descendants of these adventurers, or *ushkuiniki*. In the eleventh century the men of Novgorod are allied with the Korelians, to the South-west of the White Sea, against the Finnish Emi; Kola is mentioned in 1264 by the Russian Chronicles; and in 1323 Prince Yuri Danilovitch of Novgorod concluded a treaty with Magnus, King of the Swedes, by which Sweden engaged to respect the trade-ascendency of Novgorod in Lapland, east of the Varanger Fiord. Even in the eleventh century[1] Yaroslav the Wise and St. Olaf of Norway concluded an arrangement about these same countries, leaving to the Novgorod sphere of trade-influence all the lands east of Lyngen Fiord.

[1] According to Eric Berner (Works, Stockholm, 1740).

On another side Novgorod faced Western Europe. Along with Bruges and London it was from the thirteenth century the greatest Mart of the Hansa, or Hanseatic League, in extra-German lands. When other Russian or semi-Russian peoples and states were submerged by the Tartars, Novgorod alone escaped. As time went on, it became more and more German and Hanseatic; until at last Russians were laid under heavy disabilities and treated as a subject race within its territories; and Novgorod became, instead of a centre of Russian life and trade, instead of an outlet to the Baltic, the firmest barrier between that Slavonic trade and the outer world. Thus from the *Skra* or code of the Hanseatic settlement, Court of the Germans, or Court of St. Peter, at Great Novgorod, we see that German trading settlements were fixed there as early as the eleventh century; but the German ascendency dates from the privileges granted in 1269. The Hanseatic merchants enjoyed numerous advantages; for example, if a native Russian were bankrupt German creditors must be paid first; further, they had the right of banishing such a bankrupt from the city with his family. As tax, the Hanse merchants at first paid a piece of cloth and a pair of gloves for their monopoly of the whole trade of Livonia, afterwards nothing. No foreigner, except a Hanseatic, might learn Russian, and so endanger their monopoly; the whole province[1] was held responsible for the crime of such an interloper. Finally, no Russian, under a penalty of 100 marks, was to be suffered to live in Livonia; Russians were strictly forbidden to trade on the sea, and punished with fine, imprisonment, and confiscation of goods, for attempting to do so. This policy proved the ruin of Novgorod; it became the ambition of the rising power of Moscow to break through this barrier, to crush

[1] Of Livonia.

the power of that Republic which, once Russian, had become the resolute enemy of Russia; and the internal anarchy of the great city destroyed the last hope of effective resistance. In 1478 Tsar Ivan III. took it; a century later Ivan IV. destroyed it utterly, carried away most of its people, and removed its treasures, above all, its great bell that once had called the citizens to public meeting or to riot; at this day it probably contains scarce a tenth of the numbers that crowded its streets in the fifteenth century.[1] The ruthlessness of the destruction proves the depth of the hatred it had roused; it had become in Russian eyes a traitor city, which now strangled the commerce of its brethren, forbade them to expand, to reach the sea, to navigate, or to learn of other nations; *delenda est Carthago*.

But in the time of Olaf Tryggveson and of St. Vladimir, Novgorod was still Slav, still Russian[2]; its transformation into an anti-Slav out-post of Germany was yet to come; nor was it accomplished till the end of the period covered by this volume.

In about 995 young Olaf left Novgorod and sought his fortune on the sea, a fortune that might bring him at last to his own kingdom again. Vladimir had shown him all honour; he was made a Captain in the standing army of defence; with his own private wealth he formed a powerful war-band; but after a time the king grew jealous, and Olaf took to the life of a Viking rover.

First he put out into the Baltic, the 'East Salt Sea,'

---

[1] Without adopting the exaggerated figures of some chroniclers (*e.g.* 400,000), yet it seems probable that Novgorod, in the time of its greatest prosperity as a Hanseatic centre, must have contained nearly a quarter of a million of people; in 1897 it had 20,000. Moscow, Riga, Revel, Libau, all profited by the depression of Old Novgorod; but above all the foundation and prosperity of St. Petersburg fatally diverted the course of trade.

[2] Or rather Russo-Scandinavian.

and, sailing south, came to Wendland, where he married the king's daughter and became 'ruler of that realm with her.'[1] When the Emperor Otto II. attacked Denmark to make it 'take christening and the right troth,' Burislav, King of the Wends, followed him with Olaf Tryggveson, who had just come back from a Viking voyage to Gothland and Sweden.[2] Olaf stayed in Wendland three years; then his wife died, and he started again on his wanderings.[3]

From the Baltic he sailed to Friesland and Flanders, to England and Scotland and Ireland. He raided in the Hebrides and in Man, and thence made his way to the Scilly Isles, Brittany, and the north-west of France.[4] Four years passed in these cruises, during which Olaf was christened,[5] and married again. His second wife was the sister of the King of Dublin, Olaf Kuaran, of partly Norse, partly Irish family; and this marriage brought Tryggveson again to Ireland. At this point happened the great change in his life. From the career of a Viking, a pirate chief, the possible founder of a new Norse kingdom in the west, he was suddenly recalled to his old home, to that scene of fierce dynastic and domestic strife which he had left while an infant of a few days old.

After the fall of Harald Greycloak Norway had become a sub-kingdom of Denmark, and was ruled by Earl Hakon the Wise, one of the most capable and subtle statesmen that

[1] Probably an exaggeration of the Saga. King Burislav, Olaf's father-in-law, evidently reigned down to the time of Olaf Tryggveson's departure from Wendland. See chs. xxi., xxii., xxix. of *Olaf Tryggveson's Saga*.

[2] Ch. xxv. and xxvi. of *Olaf Tryggveson's Saga*

[3] Ch. xxx. to xxxv. of *Olaf Tryggveson's Saga*.

[4] 'Valland' in which Normandy was often included.

[5] Before this he had called himself Oli the Garthman (Novgorodian), but in Scilly a hermit told him he should be a glorious king, 'and many other great works of Almighty God did he reveal'; on which Olaf 'would fain be christened, and so was he with all his men': ch. xxxii. of *Olaf Tryggveson's Saga*.

men had seen in the north. This Hakon had long sought far and wide for young Tryggveson, and at last he heard news of him in Ireland. So he sent out his friend Thorir Klakka, who had 'gained a good knowledge of countries' by piracy and by trade, and ordered him, under pretext of a merchant voyage to Dublin, to find Olaf, and entrap him.

So Thorir came west from Norway to Ireland—a common voyage enough in those days—and soon he fell in with Tryggveson, who asked him many questions about Norway and Earl Hakon. And Thorir said the earl was indeed mighty, but all the people would rather have a king of the blood of Harald Fairhair. Olaf caught at the treacherous suggestion with the eagerness of an exile, and set sail at once with five ships, taking Thorir in his company. Passing through the Hebrides and Orkneys, he soon arrived in Norway, and hurried northward, night and day, so as to come unawares upon Earl Hakon, who was well informed of every movement of his foe. The earl would now have been sure of his prey but for one thing. Since he sent Thorir Klakka to the west, the great men of Norway had risen up in revolt; and now his own craft had supplied the malcontents with a formidable leader. The whole country rallied round the young prince; Hakon was forced to fly; and the usurper was soon after murdered by a serf, who brought his head to Olaf, and lost his own head for his pains.[1]

[1] 'And now so great,' says Snorri, ch. lvi. of *Olaf Tryggveson's Saga*, 'was the enmity against Hakon that none durst name him other than "The Evil Earl" for long after. Yet for many things was he worthy to be lord; for the great stock he came of; for the wisdom and insight, yea, and the bounty wherewith he dealt; for his high heart in battle and his good hap withal. As says Thorleif Redfellson:—Of no Earl ever heard we 'Neath the Moon's highway, Hakon, More famed than thou.' But to Snorri his ruin was the decree of fate, the doom of his obstinate paganism, and his unbridled license.

In the time of Hakon's rule we have a curious reference to Pagan Iceland, its Norse settlers and their trade. 'It had been made a law in Iceland that for every nose in the land a scurvy rhyme should be made on the Dane King' (Harald Gormson). For an Iceland ship had been cast away in Denmark, and the Danes had plundered it as lawful wreckage. To this a wild story is added of Harald's magic against the Icelanders. He sent out a wizard who could change his shape at will, and would have done infinite mischief, but that he was baffled by the spirits of the land— so, at least, the Iceland Sagas proudly recorded of their beloved country, where the fells and the hills were all full of guardian monsters, and where mighty dragons and eagles, bulls and giants, watched over the shore that no enemy might enter.[1]

This was one repulse for Denmark; it was a more serious matter when the Jomsborg Vikings from Wendland[2] and Sweden, under King Svein, the son of Harald Gormson, failed to drive Hakon out of Norway[3] as they had sworn to do, and thus, breaking with their old ally, left him unaided before Olaf Tryggveson, whose victory was for Denmark the loss of a dependency.

From this time we hear less of any dealings with the outer world, till, in referring to the conversion of Iceland, the Saga gives us the famous narratives of the Vinland, Markland, and other north-western voyages. We may notice

---

[1] Chs. xxxvi. and xxxvii. of *Olaf Tryggveson's Saga*.

[2] Wendland may be roughly taken as equal to the modern Pomerania.

[3] The Jomsborg Vikings had already helped Svein against the father Harald. Afterwards they had kidnapped him, and forced him to be their go-between in a peace with the Wends. This peace was connected with marriage alliances between the families of King Burislav and King Svein. Cnut the Great was a son of this Svein and his Wendish wife, and thus partly Slav. See chs. xxxviii.-xlvii. of *Olaf Tryggveson's Saga*.

the eventful Baltic voyage of the merchant Lodin[1] which brought about the freeing of King Olaf's mother from slavery; the journey of Dangbrandt[2] or Thangbrand, the 'Saxon' priest, to Iceland, to preach the Christian Faith with the help of some timely violence; the meeting of the King with the hero Kiartan and other men from Iceland; and the visit of Leif the Greenlander, son of Red Eric, to Norway;[3] but the chief interest of the poet-chronicler is fastened upon the conversion of Norway. It is no business of ours here to follow Olaf Tryggveson along that track of blood and fire,[4] by which he brought his subjects to enter the church; along the incidents of his contests with the warlocks, or with ghosts, 'quickened in men's bodies by cunning of the Finns';[5] or along the intricacies of the political struggle which at last crushed the King of Norway before the League of Danes, Swedes, Wends, and Norse rebels. But the Saga's references to Icelandic affairs, scanty as they are, begin to assume more importance, and are especially interesting as a preface to the record of that triumphant advance of Norse discovery, which for a time linked Iceland with America.

[1] Ch. lviii. of *Olaf Tryggveson's Saga*. The reality of this voyage is disputed by some.

[2] Ch. lxxx. of *Olaf Tryggveson's Saga*; chs. xcvii., xcviii. of *Burnt Njal's Saga*; and ch. xli. of the *Laxdale Saga*, which calls him a 'Court Priest.'

[3] It was on his return from this visit that Leif found Vinland, according to *Red Eric's Saga*.

[4] See ch. lxxxiii., lxxxvii., xcii. of *Olaf Tryggveson's Saga:* 'the grimmest of all men was he in his wrath, and marvellous pains laid he on his foes. Some he burnt in the fire, some he let wild hounds tear, some he gave to serpents, some he stoned, some he cast from high rocks.' Yet Olaf was also the 'gladdest and gamesomest of men, kind and lowly, exceeding eager, bountiful and glorious of attire, and before all men for heart in battle,' ch. xcii. of *Olaf Tryggveson's Saga*. On the King's intercourse with the Icelanders in Norway, and especially with Kiartan, and the famous swimming contest of these two strong men in the Nid, cf. the *Laxdale Saga*, chs. 40, 41, 42, 43, 46.

[5] Ch. lxxxiii. and lxxxv. of *Olaf Tryggveson's Saga*.

The German Thangbrand[1] was hardly the right man to deal with independent folk like the Icelanders. He may have been a good clerk, and a doughty man, but he was headstrong and quarrelsome, and the effect of his preaching was impaired by his frequent duels. Thus two skalds of the Island paid with their lives for some scurrilous verses; and after three years the unlucky missionary came back to King Olaf, saying that his task was hopeless. But already the Icelanders who were in Norway had mostly 'taken christening' at the King's wish,[2] and with the baptism of Leif Ericson[3] the conversion of Greenland was also begun.

Thus Olaf's reckless proselytism triumphed at last in the north-west,[4] but in the home kingdoms it seemed to rouse up foes on every side. Among the rest, Queen Sigrid of Sweden, a valuable ally to Norway, though infamous for her murder of King 'Vissavald' of Novgorod, and others who had trusted to her hospitality, was mortally affronted by Olaf over this very matter;[5] and her marriage with Svein Forkbeard of Denmark added to the growing embarrassments of the new King of Norway. Olaf's marriage with Thyri, the fugitive wife of the Wend King Burislav, and the bitter enemy of her brother King Svein, was a crowning stroke of imprudence. Moreover, he was not yet rid of all danger from the race and friends of Earl Hakon, and the Saga has some details of interest about the wanderings of Eric

---

[1] Ch. lxxx., xci. of same.

[2] Ch. lxxxviii., lxxxix., xc., xci. of same.

[3] Ch. xciii. of same. Apparently A.D. 999. In ch. xciv. 'this summer,' is defined as 'when Olaf had ruled over Norway four winters.'

[4] Ch. ciii. and civ. of same. Leif's old pagan father was fond of saying that all his son's good fortune in his Vinland voyage was hardly a set off to the bad luck he had brought on the land with his jugglers and their jugglery; ch. civ.

[5] Ch. xlviii. of same. The King's refusal to marry this 'faded heathen bitch,' and his blow in the face with a glove, after the formal betrothal had taken place, was accepted, as it was clearly meant, for an open act of political defiance.

Hakonson, the leader of this party, who now joined the alliance against Tryggveson. Eric had made friends with the King of the Swedes, another Olaf, always ready to aid those who threatened his western rival; but at first, though 'much folk resorted from Norway to Earl Eric,' the time had not come for a decisive struggle. So he went on a Viking cruise to Gothland,[1] waylaying ships of traders or of other Vikings; and he harried wide along the borders of the sea. He sailed also south to Wendland, and to the East Baltic coast, and raided the country of King Valdimir, where he stormed the town of Aldeigia, on Lake Ladoga.

At last the storm broke. In the year of Christ 1000 Olaf Tryggveson,[2] returning from a Wendland voyage, was attacked off Svold, by his enemies, in overwhelming numbers, and perished in the struggle. The words of his careless scorn were long remembered. 'We fear not the Danes; in those blenchers there is no heart; and, for the Swedes, better were they at home licking their blood bowls, than coming here under our weapons.' In his great ship, the *Long Serpent*, he grappled with Eric Hakonson,[3] and the fight was long; but the forces of the League were overpowering. Even the great archer, Einar Thambarskelvir, could not save the day. A Finn bowman of Earl Eric's split Einar's bow asunder with an arrow, and at that loud sound, the Saga would have us think, men stopped their fighting for a moment, as in reverence for an omen of the approaching end. 'Then spake King Olaf, "What brake there so loud? Norway, King, from thy hands," said Einar.'

The rebel Norsemen now threw heavy beams on to the bulwarks of the *Long Serpent*, so as to heel it over, and

---

[1] The Island.
[2] Chs. cvi.-cxx. of same.
[3] After beating off the Swedes and Danes.

give their weapons play; then at last they cleared the deck with repeated boarding rushes; and Olaf leapt into the water, throwing up his shield over him, and disappeared.

The Saga of Olaf Tryggveson inserts in the body of its narrative,[1] shortly before the fatal venture of the hero in the Baltic, a record of the Norse voyages in the far north-west. These voyages have naturally become the most famous, as they are the most brilliant and suggestive of Scandinavian explorations; and as their chief events refer to the last years of this reign, it will be best, perhaps, to take them in this place. Various criticisms have been made upon portions of the narrative, and it seems probable that in the long accepted text of these chapters 'on Vinland Faring' we have at least some interpolations, and even more displacements of time-order and of the proper sequence of events. In all examination or criticism of the Vinland tradition, we have to distinguish between the two forms in which this tradition has come down.[2] The first, earlier, and shorter form, is in the so-called *Saga of Eric the Red* (and of Thorfinn Karlsefne); the second, later, and more elaborate text, is that of the *Flatey Book*, which has been adopted in certain manuscripts of the *Heimskringla*, and is the narrative most commonly known.

The *Saga of Eric the Red* survives in two manuscripts, the older of about A.D. 1334, the later of about 1400. The former is commonly known as the *Book of Hauk* (*i.e.* of a man who claimed descent from Thorfinn Karlsefne). In its original form, the narrative was probably compiled about A.D. 1300, and the surviving texts are transcripts

---

[1] After ch. civ. of *Olaf Tryggveson's Saga*, on the 'Christening of Greenland.'
[2] Cf. Storm, *Studies on the Vinland Voyages*, especially pp. 319–325; Reeves, *Wineland the Good*, especially pp. 19–52, 181–5, 79–83.

THE VINLAND VOYAGES: RIVAL TEXTS

of one which the scribes do not seem to have always understood. Those who have praised the Vinland narrative in this form as always preferable to that of the *Flatey Book*, conveniently forget certain absurdities and contradictions which are sometimes as troublesome as in the despised version represented by the *Heimskringla*. But whatever its difficulties, the *Red Eric Saga* has at least, like the old Lord Lafeu, the 'privilege of antiquity upon it'; and in many parts, if not in all, it represents a tradition of considerable weight. On the whole it must be considered as of superior authority where its record clashes with *Flatey*; but the case is different where the latter gives us matter not represented at all in *Red Eric*. We cannot treat the omissions of the earlier manuscript as decisive against details recorded in the later; arguments from silence are always dangerous, and doubly so in mediaeval enquiries; and the agreements between the rival texts are sufficiently numerous to warrant the belief that the once-received text of *Flatey* in the main presents a trustworthy record. First of all, then, let us follow the story of the fuller though later chronicle.

In the beginning we hear of the voyage of Biarne, the son of Heriulf. From his earliest youth he desired to go abroad—that is from Drepstok in Iceland, where his home was. At last, one year, when Biarne possessed a merchantship of his own, Heriulf went to Greenland with Red Eric[1]

---

[1] On this voyage Heriulf was accompanied by a 'Christian man, from the South isles' (Hebrides), who composed the *Sea-rollers Song*, evidently in the stress of terrible weather; one stave is quoted in *Flatey*: 'Faultless searcher of monks' hearts, He who Heaven's halls doth rule, Hold the hawk's seat ever o'er me' (*i.e.* guard me with thy hand). Just before this *Flatey* describes Leif's journey to Norway, and his conversion by Olaf Tryggveson, 'sixteen winters after Eric the Red went to colonise Greenland,' which was 'fifteen winters before Christianity was made law in Iceland.'

and settled at Heriulfsness, while Eric planted himself at Brattahlid.[1] Both settlements were in the extreme south-west of Greenland, near Cape Farewell; while hard by, at Gardar, where afterwards the Bishops of Greenland had their seat, lived Eric's daughter Freydis with her husband Thorvard.

Now, after a time, Biarne determined to go and spend the winter with his father, though neither he nor his men had ever been in the Greenland Sea before. He set out accordingly, and sailed for three days. Then a north wind with fog set in, and this lasted a great while. At the end they saw the sun once more, and could distinguish the quarters of the heaven; and again they sailed on for a day and a night,[2] and came in sight of land. But Biarne thought this could not be Greenland. They sailed close up to it and saw that the country was level, without mountains, and covered with wood, and that there were small hillocks inland. Then they left this land on the larboard side and sailed two days and nights, till again they sighted firm ground. But Biarne thought this was no more Greenland than what they had seen before. For this was flat and covered with trees, and in Greenland there were great ice-mountains. The ship's people would have landed here, but Biarne forbade it. They put to sea again with a south-west breeze and sailed for three days and nights. Then they saw a high mountainous land with ice-mountains, which was clearly an island, so Biarne would not land here either. Once more they stood out to sea with the same breeze, and after three days they came to Heriulfsness in Greenland.[3]

[1] A.D. 984 or 986, according to tradition.

[2] Lit. a *doegr*, which in the Sagas always means a period of twenty-four hours, as Dr. Gustav Storm demonstrates.

[3] This is reckoned to have been in the year of Christ 986, in the vulgar chronology, which is very uncertain.

The next section introduces Leif Ericson. After some time[1] Biarne came over from Greenland to Norway on a visit to Earl Eric, and told men about his journey. 'But folk thought he had not been eager to get knowledge,' as he could not give any account of the new lands he had visited. Soon he went (in the next summer) back to Greenland, having set all to talk about the discovery of unknown countries. One Norseman especially—and he a Greenlander—was anxious to do more than talk. Leif of Brattahlid, a son of Red Eric, a large and powerful man and of most imposing presence, went over to Biarne and bought his ship. The loyal son urged his father to come in command, 'for with him they were most likely to have good luck'; but, when they were close to the ship Eric's horse stumbled, and the old heathen drew back. ''Tis not fated,' said he, 'that I should discover more lands than this one of Greenland.' So Leif went without him. Yet he had in the ship among his thirty-five companions, at least one trusty counsellor, his foster-father from the South Land,[2] named Tyrker. They came first to the country which Biarne had seen last of all, and here they went on shore. There was no grass to be seen, but huge snowy mountains in the upland, and from the sea to those mountains the land was all like a table of flat rock, and it seemed to them a region of no profit. But they went on shore all the same, and Leif said: 'It has not been with us as with Biarne, that we have gone not upon the land; I will give it a name, and the name shall be Helluland'[3] (Slab-land). Then they put to sea and found another land, and there also they went on shore. Now this was flat and over-

---

[1] The interval is left absolutely uncertain in *Flatey*, and there is no authority for the year 994, which some have conjectured for this visit.
[2] Germany.
[3] From 'hellu,' a flat stone.

grown with wood, low-lying towards the sea, and the beach had broad stretches of white sand. And Leif said: 'We will give this a name according to its kind, and call it Markland' (Woodland). Again they put to sea with an on-shore wind from the north-east, and after two days and nights they came to land once more. But first they went on to an island which lay on the north side of this land, and looked about them, for the weather was good. There was dew upon the grass, and they thought they had never tasted anything so sweet as this. Afterwards they sailed into a sound or strait which lay between this island and a ness that jutted out from the land on the north. Here was shallow water at ebb-tide, and the Norsemen ran their ship on land, but were not able at once to bring it right into port. They were too eager to wait for it to float again, so they rushed to the shore and came to a place where a river flowed out of a lake. And as soon as the vessel floated, they towed her up the stream into the lake, and there they cast anchor, carrying their beds out of the ship, and building themselves booths there. Then they put all things in order for wintering, and built a large house; for the land seemed to them very good. They did not want for fish food; there were salmon both in the river and in the lake, larger than they had ever seen. There was no frost in winter, and the grass was not much withered, so that no winter fodder, they thought, would be needed for cattle. Day and night here were more equal[1] than in Iceland or Greenland.

---

[1] 'For on the shortest day the sun was in the sky between Eyktarstad and Dagmalstad,' lit. 'had Eyktarstad and Dagmalstad on the short day.' According to Vigfusson (Dictionary, under the word *Eykt*), who had fresh calculations made by two eminent astronomers, this implies a day of seven hours; according to Hildebrand, of six hours; and according to the editors of the *Antiquitates Americanae*, a day of nine hours is

When they had finished their house-building, Leif said to his men: 'I will divide the crew in two parts, and so explore this country. Half shall stay here and do the work, while half shall search the land; but these last shall come back at night and not wander from each other.' And this they did for some time. Leif himself, like his men, took turns in staying at home and exploring the country. One day there was a great discovery. Tyrker did not return at night,[1] and Leif started with twelve men to find him; but they had gone only a little way when Tyrker met them. He was very merry, and in the gladness of his heart he could only speak at first in German, smiling and rolling his eyes, but no one could understand him. After a while, growing calmer, he said in Norse: 'Something new have I found—vines and grapes; believe me, it is true, for I was born where there is no lack of vines and grapes.' And Leif said: 'Now we have to divide our work, day and day about. We must gather grapes on the one, and on the other we must cut vines and fell wood to load our ship.' So a cargo of wood was hewn, and the stern boat was filled with the grapes that they had cut [and of the trees of that Vineland they took samples, and of the self-sown wheat of its fields, and of its maple wood[2]], and in the

meant. The last reckoning is endorsed by Laing and Anderson, *Heimskringla, Preliminary Dissertation*, and by the much higher authority of Dr. Gustav Storm, *Studies on Vinland Voyages*, pp. 307-12, and of Reeves, *Wineland the Good*, pp. 181-5.

[1] Dr. Storm doubts the whole of the Tyrker story, and considers it an interpolation, like the narrative of Biarne; but the former he regards as much more damaging to the credibility of the *Flatey* narrative. On the other hand, Tyrker is described in a vivid way that does not wholly suit this view. He had a prominent forehead, restless eyes and small features, was diminutive in stature, and though nothing much to look at, had great skill at handicraft.

[2] 'Mosur.' These words are treated as an interpolation by Reeves, *Wineland the Good*, p. 67.

spring they sailed away, and Leif called that country Vinland [the Good, from the produce of its grapes]. With a fair wind they came to Greenland,[1] and its fells below the glaciers; and thereafter was Leif called the Lucky, and gat much wealth and fame, and much talk was there of that Vinland journey of his.

The third voyage was that of Leif's brother Thorvald. When all other men praised the deeds that had been lately done, he declared that it was not enough, and that the new country should be more narrowly explored; and for that end he asked his brother to lend him his ship. Leif granted it forthwith, and Thorvald put to sea with thirty men. Nothing is told of their journey until they came to 'Leif's Booths' in Vinland, where they remained all the winter. In the spring Thorvald ordered some of his men to go westward in the ship's long-boat and explore the coast. They did so, and found the land beautiful and well-wooded, with but a small distance between the sea and the woodland, great numbers of islands and shallows, and a beach of white sand. They saw no sign of dwelling for man or beast, save a wooden corn-barn upon an island far toward the west, and in the autumn they returned to Leif's Booths, where Thorvald passed another winter. The next spring (or summer) he went out himself, first eastward and then northward, to examine the land, apparently with part only of his crew. After a time the Norsemen came to a ness or cape,[2] where a storm drove them upon land and broke their keel, and they stayed a

---

[1] On the way home he rescued fifteen Norsemen including one Thorir, and Gudrid his wife. This Gudrid afterwards married Thorstein Ericson and the explorer Thorfinn Karlsefne; hence the notice of this incident, which helps to link on the achievements of Karlsefne to the family story of Red Eric's house.

[2] 'Kjalarness,' often taken to mean Cape Cod, is perhaps better understood as a point in Nova Scotia.

long while to repair the damage. And when they left, they raised the keel upon the ness and named the place Keel-Ness [and so it was called by others after them]. Then they sailed along the land eastward and into the mouth of the adjoining firth, till at last they sighted a beautiful headland, covered with wood, where Thorvald moored the vessel and laid out gangways to the beach, and went on shore, saying: 'I would gladly set up my farm here.'

But now they came upon the first traces of other men. Far off, upon the white sandy beach, three distant specks were sighted; and these proved to be three skin boats of the natives,[1] with three men hiding under each. All these people, save one, were taken and killed; but he escaped with his boat to a place within the fiord, where there seemed to be several dwellings 'like little lumps on the ground.' A heavy drowsiness now fell upon the Norsemen, and they neglected to keep any watch, till a sudden cry aroused them, 'Awake, Thorvald, awake, thou and all thy comrades, if thou wilt save thy life.' And at the same moment came a countless host from up the fiord in skin-boats, and laid themselves alongside. The Vikings put up their shield-wall along the gunwale, and kept off the arrows of the *Skraelings* till the foe was weary of the assault, and made away as fast as they could, leaving Thorvald with a mortal wound under the arm. He had but time to bid his men carry him to the point where he had wished to settle—for it was fated that he would stay there awhile, but with a cross at head and feet—when his speech failed and he died, the first victim of these hazardous western voyages. With his last breath Thorvald warned his men to retrace their steps and rejoin their companions; but they stayed in that country

[1] Elsewhere called Skraelings.

another winter, gathering wood and grapes, and loading their ship; next spring the united company returned to Eric's Fiord in Greenland, bringing heavy tidings to Leif.

Another fruitless venture followed. Thorstein Ericson[1] had married Gudrid, the widow of Thorir the Easterling, whom Leif his brother had rescued from a rock in the ocean. The misfortune of another brother made him take a keener interest in the new discoveries; and he now bethought him that he would go to Vinland for Thorvald's body. He equipped Leif's vessel, and with his wife and twenty-five men he put to sea. But the whole summer he drove about in the ocean without finding land, or knowing where he was; and in the first week[2] of winter he came back without result to Lysu Fiord in the 'Western Bay' of Greenland.

Then follows in the Saga a long and highly fabulous story of Thorstein's death and the prophecies of his corpse. The practical issue of all this was to be found in the third marriage of Gudrid to an 'Iceland man,' as foretold[3] by the ghost—in other words, to the famous Thorfinn Karlsefne, whose journey is next described.

He came in a ship[4] from Norway to Greenland in the year of Thorstein's death; and stayed with Leif at Brattahlid, the home of Red Eric's clan. In the same winter he married Gudrid, 'being a man of great wealth.' At this time, as before, much was said about a Vinland voyage; and both Gudrid and others urged Thorfinn to

[1] At that time, adds *Flatey* [i.e. when Thorvald was buried at 'Crossness'], Christianity had become law in Greenland, but Red Eric had already died.
[2] The Icelanders reckoned winter from the first Saturday after the 14th of October.
[3] The ghost also foretells that Gudrid will in later days 'go abroad and to the south,' i.e. (probably) make a pilgrimage to Rome.
[4] No hint is given in *Flatey* of more than one ship in this expedition.

undertake the same. But he determined, if he went at all, to do more than reconnoitre and return. If possible, he was resolved to settle and colonise, if not to conquer. With these larger ambitions, an ampler force and greater resources were necessary. So he took with him [in his own ship] sixty men, five women, and cattle of all kinds; with his followers he made agreement that he and they should have equal shares in any gain that might be won; Leif once more lent his Booths in Vinland; and Thorfinn sailed out to the West. Probably he had with him men who had already been on the voyage, for he escaped Thorstein's ill-luck, and came direct to the Vine Country with all his men and goods. The new settlers were cheered to find that a storm had lately driven a whale on shore at Leif's Booths, and thus provided them with abundant food. They were soon busy gathering the grapes of the favoured land, and they felled wood diligently for shipping.[1]

The first winter passed quietly enough, but in the summer the Norsemen became aware of the presence of the Skraelings. Suddenly a great troop of men issued from the woods with skin bundles, making as if they would trade in grey furs, sables, and all kinds of peltries, but coming dangerously near to the cattle. Frightened by the bellowing of a bull, they vanished again into the forest; but ere long they returned, once more offering to barter. Now they walked right up to the houses of Karlsefne's people, and even tried to come in, but this was not permitted.

[1] Placing it upon a cliff to dry, where the garrison of ten men was afterwards sent. *Flatey* adds various details which seem to point to a true record—the cattle being turned out on the land, the males soon become restless and vicious; game and fish are found in the country. One of the Skraelings, who seemed their chief, kills one of his men with an axe belonging to the Northmen,

Neither folk understood the speech of the other; but the Skraelings showed by signs that they would like to have weapons in exchange for their skins. Thorfinn forbade his men to sell their arms, but he told the women to take out milk and dairy stuff, and when the savages saw these things they were eager to buy them. So the Skraelings 'carried away their winnings in their stomachs,' and parted with their furs to small advantage. But the Norsemen, full of suspicion of their visitors, made a strong fence round the habitation, and put all ready for defence, when the time of need should come.

At the beginning of the next winter reappeared the Skraelings in much greater number than before, with the same wares. And again Karlsefne told the women to carry out the same food as was best liked before, and nothing else.

But this meeting was not destined to pass off so peaceably, and the unseen powers showed plainly that a threatening future hung over the colonists. A child had just been born to Thorfinn and Gudrid—Snorre, the earliest American of European parentage whom history records. While the mother sat by the cradle within the door there came a shadow to the entrance, and a woman went in with a black kirtle and a snood or fillet around her head; she had bright chestnut hair and was very pale, and her eyes were larger than are ever seen in the head of man. And the goodwife put out her hand to her that she might sit down beside her, 'and lo, there was a great noise and the woman was no longer there.' And at the same moment was one of the Skraelings slain because he would take the weapons of the Norsemen; and all the savages fled. Then Karlsefne ordered that some should go to the woods and make a clearance for the cattle against the time that the Skraelings might come out of the forest, and that ten other men should go to a ness hard by.[1] And just where he had foretold, at a place between the forest and a lake, the Norsemen fought with the Skraelings, and slew many of them, and the rest fled into the forest.

With this outbreak all things were changed. The hope of a peaceful settlement was gone; and Thorfinn seems to have despaired of settling in the new country on other than peaceful terms. The numbers and the subtlety of the barbarian foe; their deadly arrows; the warning fate of Thorvald; the vast distance of the colonists from their nearest base in Greenland; the scanty numbers of his people:—all these must have weighed with Karlsefne and led him to abandon the enterprise, so carefully prepared, so vigorously begun. He remained in Vinland the rest of that winter, but in the spring he returned with a heavy load of wood and grapes and peltries, but also with the weight of a great failure. For an enterprise had miscarried which, as we now can see, might have altered the course of history, which even at the time aroused great hopes, and which, if successful, would certainly have added a new chapter to the Middle-Age development of mankind.

According to *Red Eric Saga* (although this is not recorded in the *Flatey Book*) one of Thorfinn's lieutenants was Thorvard of Gardar, the husband of Freydis, a bastard daughter of Red Eric and a woman of infinite villainy. He and his wife thus form a link between the voyage just recorded and that, of doubtful and highly mystical character, which follows next and last in *Flatey*. In the same summer that Karlsefne returned from Vinland, a vessel arrived in Greenland from Norway under two brethren, of Iceland race, Helge and Finnboge, and they remained all that winter. 'And now men began to talk again about a Vinland

[1] Probably a strategic point commanding the settlement.

voyage, as both a gainful thing and an honourable;' and Freydis, coming from Gardar, proposed to these newcomers that they should go with her to Vinland in their ship and share equally in all the profit they might make. Each side, Freydis and the brothers, had the right to take thirty men—besides the women—but the cunning daughter of Eric did not abide by the compact; for her warriors were thirty-five in number. Leif again lent his Booths, and they sailed; Helge and Finnboge having the larger ship, but fewer men; Freydis and Thorvard the smaller vessel, with a stronger crew. When they came to Vinland, Freydis declared Leif's houses were for the sole use of herself, Leif's sister, and her people; the others had no right there. Therefore Helge and his brother built them a shed further from the sea, on the borders of a lake. At first the settlers had some sports in common, the brethren doing all in their power to promote concord; but as the leader of the other party was steadily in search of a quarrel, the games soon ceased, and none even went from one house to the other.

The final catastrophe is among the most mysterious passages of the Saga, if the whole narrative of this voyage be not rejected as a legendary and obviously unhistorical addition. Thorvard, we are told, enraged to madness by the false charges of his wife against Helge and Finnboge, surprised and slaughtered them with all their followers, and among them five women, whom Freydis murdered with her own hand. The explanation is easy, that the massacre was an act of jealous vengeance; that Freydis was an adulteress who urged Thorvard to this as a screen for her own guilt, and that her accusations were not wholly false; but the Saga's extreme reticence and obscurity puts a certain difficulty into this reading of the story, and this difficulty is increased by the wholesale character of the butchery, and the fate of the women. On the whole, passion does not seem to be the complete explanation of the tragedy, whose slow approach the Saga indicates with all the mystery, but little of the subtlety or the coherence, of a Greek dramatist.

Early in the next summer the survivors returned to Eric's Fiord, pretending that their companions had stayed behind in Vinland. Freydis had bribed her people to conceal the crime, but most of it soon came out; Leif discovering part by the argument of torture as applied to three of his sister's followers. The story ends with an inconclusiveness which excellently suits the narrative. Leif, after all his vigorous efforts to ascertain the truth, does no more. 'I do not care to use severity,' he declares, more in the style of a modern humanitarian than of an old Norseman, 'and therefore I will not treat Freydis as she deserves, but this I will foretell of her, that her posterity can never thrive.'

There is but little more recorded in the *Flateyarbok* of these Vinland voyages and their leaders, but a few words are added on Thorfinn Karlsefne. Freydis had only just returned to Eric's Fiord when he sailed from Greenland back to Norway, 'and it was common talk that never had a richer ship sailed on this voyage than the vessel that he steered.' Being both rich and prosperous, it is not astonishing that Thorfinn was now 'held in great esteem,' even among the landed men of the home kingdom. Next year he came again to Iceland, but soon after this he died; his wife Gudrid became a nun; Snorre, his Vinland-born son, seems never to have returned to his first home; and so darkness falls upon the chief record of early European discovery in the Atlantic Ocean.

The *Saga of Eric the Red*, in so far as it differs from

the *Flatey Book*, or contains a fuller narrative, may be regarded (as we have seen) in the light of a primary text; and it therefore remains for us to see what it adds to, or alters in, the narrative of Vinland voyages already summarised. Beginning with a reference to Leif Ericson as the discoverer of the western country, not direct from Greenland but on a return voyage from Norway to Eric's Fiord, and wholly ignoring the previous journey of Biarne recorded in *Flatey*, the *Eric Saga* adds a few points to the history of Thorstein's fruitless adventure. No separate voyage of Thorvald Ericson is admitted; his fate is here associated with the great enterprise of Thorfinn Karlsefne; and Thorstein's voyage is made an altogether earlier occurrence. Leif's account of the western lands had given rise to much talk of exploration, and both Red Eric his father, and Thorstein his brother, made ready to follow up his discoveries. Some little time before there had arrived at Brattahlid[1] a noble Icelander, Thorbiorn Vifilson, and his daughter Gudrid or Thurid. He was an old friend of Eric's, and a man of good name and family, but his fortune had decayed, and he now came to Greenland in a ship that plays a prominent part in the Saga. With it Eric and Thorstein now made their fruitless quest of Vinland, and with the same vessel (among others) Thorfinn Karlsefne renewed the enterprise some years later. Thorbiorn probably arrived at Eric's Fiord during Leif's absence in Norway and 'America'; and both he and all his household had already become Christians, before the new faith had been made law in Iceland.

Eric and Thorstein, then, borrowed Thorbiorn's ship and set out with twenty men; but they could not steer the course they wished, and were tossed about upon the ocean, once coming in sight of Iceland, and at another time seeing

[1] First at Heriulfsness.

birds from the Irish coast. At the beginning of winter they came back, exhausted and disappointed, to their home, where Thorstein consoled himself with Thorbiorn's daughter. Both texts agree in describing his death and the ghostly wonders [1] that accompanied it, as well as the speedy arrival of Thorfinn Karlsefne; but from this point the *Eric Saga* passes into a *Thorfinn Saga*, and is clearly based upon better and fuller tradition than the *Flatey* narrative. Thorfinn is described as a successful and far-travelled merchant, but no mere chapman, being of good family and a great grandson of the famous Icelander Thord the Yeller, a son of the no less respectable Thord Horse-Head, and a descendant on his great-grandmother's side of the Irish King Kiarval. In his own ship he brought forty men from Norway to Greenland, and he was accompanied by three friends, afterwards prominent in the Vinland explorations, Snorre Thorbrandson, Biarne Grimolfson, and Thorhall Gamlison; the two latter together commanded another vessel. He was cordially received at Brattahlid, married the widow [2] of Thorstein Ericson, and adopted (with his new relationship) the Vinland ambitions of Red Eric's house. With the aid of Snorre he fitted out his own ship for the passage; Biarne and Thorhall made ready theirs; and a third, Thorbiorn's vessel, in which Thorstein had ventured, was equipped by Thorvard, the husband of Freydis, Thorvald Ericson, and Thorhall the huntsman, Eric's old steward and trusted adviser. This Thorhall was a poor Christian, adds the Saga, but he had a wide knowledge of unsettled lands; in appearance he was dark and of giant build; he was elderly, overbearing, and taciturn, underhand in his dealings, offensive in his language, always ready to stir up evil—in a word, Eric's bad genius and the curse of Thorfinn's enterprise.

[1] Though differing in various details.
[2] Gudrid, Thorbiorn's daughter.

The Saga suggests that he was no better than a wizard; he boasted openly that his patron, Thor the Trusty, was of more avail than Christ; he opposed Karlsefne in his plans, and composed ribald verses on the poor success of the enterprise and the lack of vines in 'Vinland.'

The sixty men and five women of Thorfinn's company, as given in *Flatey*, may have been the crew of his own ship; but they evidently formed only a part of the expedition. The whole fleet carried one hundred and sixty persons. They sailed first from Eric's Fiord, near Cape Farewell in Greenland, northwards to the 'Western Bay,' just within Davis Straits, and so to 'Biarne' [or Disco Island.[1]] From this point they came south two days and nights to Helluland, where they found many Arctic foxes and great slabs of stone, some a dozen ells across. Two days further to the south they came to Markland,[2] which they described as well-wooded and stocked with animals; thence, sailing on to the south for a long space, they had land to starboard, and at last came to a cape where they found a ship's keel, and so named Keel-Ness. Here there were great strands and long sand-beaches, and they called them the *Wonder Strands*, because they were so long to sail by. After this the shore was much broken with inlets; and here they put on land two of their swiftest people to explore the country. Both these runners were Gaels (or Irish); they were named Haki and Hakia; and King Olaf Tryggveson had in former days given them [3] to Leif Ericson. In three 'half days' the Gaels

[1] The 'Bear isle' would be in N. Lat. 60°, if this identification can be maintained.

[2] An island, adds the Saga here, lay off this Forest land to the South-East, and there they found a Bear, and they called this Bear island (Biarne), the same name as the first place they reached after leaving the West Bight or *Vestribygd*.

[3] They were no doubt among the captives he took on his Viking cruises. The woman's name is also written Hekia and Haekia; these Gaels were fleeter than deer, we are

returned from the south with grapes and ears of wheat[1] which grew wild, as they said. So the Norsemen sailed on till they came to where the shore was broken with inlets; and they put into a fiord running into the land. And in the mouth of this was an island swarming with eider ducks, and round the island and high into the fiord there ran strong currents. So from these currents they named them 'Stream Island' and 'Stream Firth.'[2] Here Karlsefne settled for the winter, enjoying the 'fine hilly country thereabouts.' In the spring he divided his people; northwards he sent Thorhall with eight men to seek for Vinland,[3] beyond the Wonder Strands; while he himself, with one hundred and fifty followers, went south. Thorhall's party passed the Wonder Strands and Keel-Ness, intending to cruise westward around the Cape, but they were driven by westerly gales away from the coast and out to sea. Finally, they were thrown ashore in Ireland, where they were seized and enslaved, Thorhall himself perishing. Meantime Thorfinn, with the main body, cruised south for a long time until they came to a river that flowed down from the upland into a lake and so into the sea. At the mouth of this stream were great bars, and it could only be entered at the height of the flood-tide. They called the place 'Hop' (a small landlocked bay), and in the lowlands hereabouts they found self-sown wheat fields, and vines upon the higher ground. Every brook was full of fish, and the settlers caught halibut in pits upon the shores. In the woods around were many

---

[1] In another reading, 'A bunch of grapes and an ear of new-sown wheat.'

also told; they were on board Karlsefne's own ship; they each wore a garment [Irish?] called *Kiafal*, open at the sides, sleeveless, with a hood at the top, and fastened between the legs with buttons and loops.

[2] Straum-ey and Straum-fiord.

[3] Here in the narrower sense, apparently: viz. the country at Leif's Booths. No distinction of this kind appears in *Flatey*.

wild animals of various kinds. They stayed here a fortnight with their live stock in great contentment; in their fancied security they kept no watch. But early one morning there came to them a fleet of skin-canoes, full of ill-looking folk, of sallow colour, with large eyes, broad cheeks, and ugly heads of hair.[1] These people surveyed the strangers for a time,[2] and then paddled away to the southward, and were lost to sight behind a cape that was near.

Thorfinn Karlsefne decided to winter at 'Hop,' and put up dwelling-houses for his men a little above the bay, some of their booths being near the lake, and others further away. The weather was mild, without so much as one fall of snow, and the Norsemen's cattle were able to live in the open field. [On the shortest day the sun was above the horizon during the watch before and after that of mid-day.]

After a time, when spring opened, the skin-canoes came again from the south in such numbers that it was as if coals had been scattered broadcast out before the bay, their crews steering up close to the habitations of the colony. At first they bartered grey furs for bits of red cloth and milk soup, but soon they endeavoured to obtain weapons for their peltries, and above all, spears and swords. Thorfinn and Snorre, however, took care to prevent this exchange. As the barter went on, an ever larger stock of furs was sold for the same amount of red cloth, which the Skraelings would often bind round their heads. Suddenly a bull that belonged to the Northmen ran out of the woods bellowing, and the Skraelings

---

[1] This has, of course, been termed a 'pretty accurate description of Esquimaux, and, unless an interpolation, almost conclusive on the Skraelings' identity.'

[2] The Saga adds that on their appearance they brandished staves with a noise like flails in the same direction as the sun moves. The Norsemen took this for a sign of peace and held up a white shield in answer. On the Skraelings' second visit this was repeated; but on the third, the natives, now becoming hostile, waved their staves in a direction contrary to the sun, and were answered with red shields.

took fright and rowed away to the south along the coast. For three weeks no more was seen of them; but then they approached again from the south in great numbers, as if a stream were pouring down, brandishing staves in a menacing manner, and uttering loud cries. Thorfinn, taking all these signs as hostile, hoisted a red shield, and prepared to fight: while the Skraelings sprang from their boats and discharged missiles from slings upon Karlsefne's men. After a time a weird thing happened. The natives raised on a pole a thing like a ball, blackish in colour, and about the size of a sheep's belly, and hurled this upon the ground above the Northmen. It fell with a frightful noise and struck a panic into the colonists, who fled up along the river-banks to some jutting crags, where they fought again. The terror and confusion of the Northmen was complete; in vain Red Eric's daughter attempted to rally them; one and all took refuge in the forest. But the Skraelings in their turn were frightened by the action of Freydis, who appeared to them like a supernatural being. Thorbrand Snorreson lay dead on the ground, his skull broken by a flat stone (the Saga here conveys the impression that the Skraelings mysteriously surrounded and destroyed their foes), but Freydis snatched up his sword, pregnant as she was, bared her bosom, and struck herself on the breast with it. At this portent the natives rushed back to the shore, jumped into their boats, and rowed away.

After the battle, Karlsefne's people returned to their homes, and began to 'weigh carefully what throng of men that could have been, which had seemed to descend upon them from the land.' They came to the strange conclusion that this party of their assailants was only a delusion of their own[1] and that the real attack came from one

[1] Apparently effected by witchcraft. A story is added about the savages finding an axe and trying it first upon wood, then upon stone; as it failed to cut the latter they threw it away.

side only, from the sea and skin-canoes; yet the settlers were so dispirited by the enmity of the natives that they resolved to abandon Hop. So they sailed away along the coast, where, to the northward, they found and killed five Skraelings, looking like banished men, and clad in skin doublets, who were lying asleep on the beach, with their boat beside them, containing some animal marrow mixed with blood. Proceeding onwards, Thorfinn and his men came to a cape that looked like a cake of dung from the animals that lay there at night. Arriving again at Stream Firth, they found abundance of all necessaries, and here Biarne and Freydis remained behind (as some say) with one hundred of the settlers; while Thorfinn Karlsefne, and Snorre Thorbrandson went to the south with forty others; this last force stayed at Hop about two months, or a little less, and returned again to the Stream Island and Bay in the same summer. After this, Thorfinn left most of his people at Stream Fiord, going on himself with one ship to look for Thorhall and the eight lost men. He sailed round Keel-Ness, keeping northward at first, and afterwards west, but could find no one. All this time he had the land to larboard of him, covered with thick forests and fringed with hills; and these hills, he thought, were all one range with the high ground at Hop, Stream Fiord being about equidistant between the two.

Karlsefne and his men coasted along this forest shoreland till they came to a river which flowed from east to west; into the mouth of this they sailed, and lay to by the southern bank of the same. Here a marvellous adventure occurred: a *Uniped*, or one-legged man, came up close to the ships and killed Thorvald Ericson, shooting an arrow into his vitals, as he sat at the helm. Thorvald drew out the arrow before he died, and exclaimed: 'There is fat around my paunch; we have hit upon a faithful country, and yet we are not like to get much profit by it.' The first clause of this dying speech is exactly the same as part of the last song of Thormod the Coal-brow Skald after the battle of Stiklestad; and the whole adventure is probably a compilation from the fashionable *mirabilia* of the Middle Ages, which placed the One-Legged Folk in various outlandish regions, usually in the south of Africa. Some have thought the story is introduced to give embellishment to the death of Thorvald Ericson, who was too important a person to be dismissed in an ordinary manner, and who, as we may fairly assume from the general coincidence of this narrative with *Flatey*, really perished in the new world.

The explorers now sailed away from the land of the Unipeds; they spent yet a third winter in the new world, at Stream Fiord; but here quarrels broke out between the colonists over the women of their company, and in the spring they returned to Greenland. Sailing with southerly winds they came first to Markland; and here they fell in with five Skraelings. Two of these, who were boys, they took with them, and taught the Norse tongue; and from them they learned much about the Skraeling folk—as, for instance, that they had a father named Uvaegi, a mother called Vaetilldi, and two kings, Avaldamon and Valdidada, that they had no houses, that they lived in holes and caves, and that they knew of a land 'on the other side' over against their country, inhabited by a race clothed in white, who had a way of yelling loudly, and often carried poles before them with rags attached. The Northmen believed this country must have been White Man's Land or Ireland the Great. Karlsefne had, in the early part of his venture, become separated from Thorhall; now, on his return, he lost the company of one of his other captains, Biarne Grimolfson, who was driven by storms into the Irish Ocean, into waters so infested with worms that his ship was like to sink, and it went hard with the crew to save even a part of their number in the long-boat.[1]

So much for the narrative of the Western voyages enterprised by Red Eric's House and friends, as told in *Flatey* and the *Eric Saga*; these give us by far the most important and ample treatment of 'America' in Scandinavian literature; and with their conclusion it only remains to add a few scattered data to the most remarkable chapter in mediaeval exploration. There are, however, some later references to a continued but extremely slender intercourse between Vinland and the Greenlandic and Icelandic settlements of the Norsemen. Thus we hear of Bishop Eric going to the Grape Country from Eric's Fiord in 1121; and, more doubtfully, of clergy from the Greenland diocese of Gardar, in the Austribygd, sailing to lands in the west, far north of Vinland, in 1266; of the two Helgasons discovering a country west of Iceland in 1285; and of a voyage from Greenland to Markland in 1347, undertaken by a crew of only seventeen men.

From 1112 to 1409 there was a regular succession of Greenland bishops, and the names of seventeen have been preserved, beginning with Arnold in 1124.[2] Earlier than this, in 1075, Archbishop Adalbert of Bremen had been requested, by the Norse colonists of this remote Christian

---

[1] Thorfinn's son Snorre, says *Red Eric Saga*, was born during the first autumn of Thorfinn Karlsefne's expedition, and was three winters old when they took their departure.

[2] See S. Laing and Rasmus Anderson, *Preliminary Dissertation to the Heimskringla*, especially p. 186; and Rasmus Anderson, *America not discovered by Columbus*, a good study of the Vinland Voyages, with an absurd title, 1883.

outpost, to send them clergy to baptize, bury, and perform the sacraments of the Church.[1] At the end of the Crusading Period these Greenland colonists, in their two chief settlements of the Eastern and Western Bays, were still maintaining themselves against the climate and the savages, against the ice and the Esquimaux. In the Austribygd or East Bight, which was by far the more important, was placed the Cathedral and the Bishop's home, eleven other churches, two monasteries, and one hundred and ninety farms or groups of dwellings, composing two little boroughs or towns; in the Vestribygd were three churches and ninety farms. At other places, probably a good way to the north of these plantations, definite traces of the old European life and religion have been found,[2] as at Kakortok and elsewhere in the latitude of 60°—60° 55'; and the existence of these well-built stone churches certainly confirms the Saga's references to possible missionary efforts of the Greenland priesthood in Vinland. But in the island of Kingiktorsoak, to the north-west of Disco, and in north latitude 72° 55', west longitude 56° 5', there is a more specific evidence of the ultimate American direction of the more daring Scandinavian keels. Here, on this barren islet at the entrance of Baffin's Bay, an inscription in runes, dis-

---

[1] So Adam of Bremen, iii. 11 (ch. 127 and Appendix).

[2] Among the chief remains of ancient settlement in Greenland are (1) At Kakortok or Karkortok, on Igalikko Fiord, N. Lat. 60° 50', W. Long. Greenw. 44° 37', near the settlement of Juliana Hope; here the ruined church measures 51 feet by 25, with its stone walls, 4 feet thick, rising to heights of 16 and 18 feet. Two round arched windows, four windows not arched, and two doorways, can still be traced. (2) At the extremity of Igalikko Fiord, N. Lat. 60° 55' the foundations of a building, probably a church, have been discovered (in 1830); these foundations measured 96 feet by 48; here was also a Runic inscription in memory of one Vigdis, a woman. (3) Foundations of buildings 120 feet by 100, apparently the base of rows of dwellings; foundations of single houses; and sepulchral slabs, have also been brought to light in and about Lat. N. 60°. Cf. Laing and Anderson, *Prelim. Dissert.*, pp. 188-190.

covered in 1824, tells how 'Erling Sigvatson and Biarne Thordarson and Eindrid Oddson, in the Saturday before Ascension Week, raised these marks and cleared ground, in the year of Christ 1135.' Now, unless we can assume an entire change of climate, the sea in the neighbourhood of this island could not be navigated at any time in the early summer within which Ascension Day must fall; and the inscription in this case would show that the men who raised the marks and cleared the ground had also wintered there.

Besides these references to, or evidences of, later movements towards the American strand, we have a little group of stories, variously understood as bearing on Vinland, on the far south of North America, or on the Canary Islands. Thus from a certain Rafn, a trader to Limerick, his great-great-grandson, Are Frode,[1] derived a tradition which described how, about the year of Christ 983, one Are Marson, of Reykianess in Iceland, was driven by storms out into the Western Sea, and came at last to a country which he called White Man's Land, and others termed Great Ireland. Thither he was followed in 999 by Biarne Asbrandson, and in 1029 by Gudleif Gudlaugson, surnamed 'the wealthy,' a settler at Stream Firth in Iceland, and a great sea-farer, who owned a large trading ship. This Ireland the Great — which men said was originally peopled by Christians from Ireland—'lay westward, in the Main, near

[1] Are, the Learned, 'the first man in Iceland who wrote down in the Norse tongue both old and new narratives of events,' c. A.D. 1067-1147 (8). He wrote '240 years after the first settlement of Iceland by the Norsemen,' c. A.D. 1117; but early writer as he was, another preceded him, viz. Isleif, the first bishop of Iceland (appointed by Adalbert of Bremen, and died c. 1047), who compiled Histories of the Norse kings, from Harald Fairhair to Magnus the Good. Are's contemporary Saemund (A.D. 1056-1133) was the author of similar works. On the story of Are Marson, see *Landnamabok*, ch. xxii. Rafn was distantly related both to Are Marson and Leif Ericson, *Landnama*, Part II., ch. xxi.

Vinland the Good, and it was called six days' voyage westward from Ireland thereto.' Are Marson was not able to leave that land, as he desired, but was forced to stay, 'and there, being among Christians, he was baptized, and held in honour.' Rafn, of Limerick, was the first who told of this; but a certain Thorkel Gelleson also had somewhat to relate about the same. For certain Icelanders told him that they had heard from Thorfinn,[1] Earl of the Orkneys, how the aforesaid Are Marson had been recognised by some of their countrymen in White Man's Land, and how he had sent back a token by which his friends knew that it was he indeed.

Of the discovery of Gudleif Gudlaugson the *Eyrbyggia Saga* gives a more or less detailed account. Towards the end of the reign of St. Olaf (c. A.D. 1030), Gudleif went to Dublin, on a trading voyage from Iceland to Ireland; on his return he sailed round Ireland by the west, and was overtaken by a heavy storm from the east and north-east, which drove him far out to sea in a south-westerly direction. Neither he nor any of his men knew where they were, and they drove about in the Ocean for the greater part of the summer before they came to land. Here they were seized by the natives, who came in crowds to the vessel, and spoke a language which none of the Norsemen understood, but which some thought a little like Irish. The visitors observed that their captors were disputing whether to make slaves of them or to put them to death. But meantime a grand old man, of majestic presence, with white hair, came riding along, and all the natives received him with the greatest respect. He accosted the Icelanders in the Norse language, and asked if Snorre, one of the chief persons in Iceland, was still alive, and his sister Thurid.

[1] Probably the great Earl Thorfinn II. (†1064) see p. 95.

He would not tell his name, and advised the strangers not to come there again, as the inhabitants were fierce and inhospitable, and the land had no good harbours. He gave them a gold ring to give to Thurid, whom he loved much better than Snorre, her brother, and a sword for Thurid's son. Gudleif brought these back to Dublin, and the next summer home to Iceland, and it was thought that this friend in distress was a certain Biorn,[1] a skald and warrior much respected, who had left Iceland in a ship about 998, and had been lost sight of since. He had been in love with Thurid, and was therefore persecuted by her family, and especially by her brother and her husband; and this, concludes the story, was the 'whole truth known concerning Biorn.'[2]

It was in 1697 that Peringskjöld published the first edition of the *Heimskringla*, with Swedish and Latin translations of the original Icelandic. The manuscripts he used are not now extant, but they were merely transcripts of other manuscripts which we still possess.[3] In 1777 Schöning commenced a new issue, under the auspices of the Danish government, from the collation of three manuscripts and from Peringskjöld's edition.[4] The eight chapters which contain the main Vinland narrative are often supposed to be an interpolation in one of the manuscripts used by the first editor. They do not occur in the three extant manu-

---

[1] Biorn Breidvikinga Kappa, the 'champion of the Broadvikings,' who first won himself a great name in Baltic Viking warfare. Here he joined the men of Jomsborg; cf. *Eyrbyggia Saga*, chapters xlvii. and lxiv.

[2] Thurid's son (by Biorn, as commonly supposed) was the famous Kiartan who tried strength with Olaf Tryggveson in a swimming contest in the Nid.

[3] See Torfaeus, *Historia Vinlandiae Antiquae*.

[4] In six volumes folio. The manuscripts were those in the Arne Magnaeus collection. This edition was finished by Thorlacius and Werlauff in 1826.

scripts of the *Heimskringla*, but they must certainly be earlier than the end of the fourteenth century, for they may be found verbatim in a Codex[1] written between 1387 and 1395.

The narrative of the Vinland voyages has unfortunately next to no confirmation from monuments, but draws it authority almost entirely from documents, of which some were committed to writing in the latter, some in the earlier years of the fourteenth century. The inscribed rocks, the buildings, and the skeletons, which have been supposed to witness to the truth of the Saga on American soil, have been, with one exception, abandoned as evidence by all but enthusiasts. Thus the Dighton Writing Rock on the Taunton River, near Berkley in Massachusetts (in N. Lat. 41° 44'), once claimed as a Runic inscription recording the voyage of Thorfinn Karlsefne, is now generally supposed to be of Indian origin, though it has certainly been tampered with in very modern times.[2] The Old Stone Mill at Newport in Rhode Island is a still more doubtful relic; and it seems probable that, even if part of the structure is Indian and

---

[1] The *Codex Flateyensis*, which numbers them as chs. cv.-cxiii. of *Olaf Tryggveson's Saga*, the same position which they occupy in Peringskjöld's edition, and one which they could not have held in Snorre's original text. The *Codex Flateyensis* is not an original work of one author, but a collection of Sagas transcribed from older manuscripts. The arrangement is chronological, and the various narratives are artificially grouped together under the reign in which the several events happened, even though not otherwise connected. This most important manuscript is so named from the Island of Flatey in Breidafiord in Iceland, where it was long kept. In 1650 Bishop Brynjolf Sveinson of Skalholt bought it from the owner, Jonas Torfason, for King Frederick III. of Denmark, who gave in exchange perpetual exemption from land-tax to Jonas for his small estate. The annals of the *Codex Flateyensis* end at A.D. 1395, and they were written out by the Priest Magnus Thorhalson.

[2] Cf. Laing and Anderson, *Preliminary Dissertation*, pp. 214-223. These editors believe the marks on the Dighton Rock are not letters at all, but scratches or primitive drawings made at various times.

only in part due to Governor Benedict Arnold,[1] none of it can with any certainty be attributed to the Norsemen. The Skeleton in Armour, discovered in the early part of the present century at Fall River in Massachusetts, carries with it, in spite of Longfellow's poem, no decisive evidence of Norse origin. This, in fact, can only be asserted of one such monument or material object west of Greenland—the rock inscription of Kingiktorsoak near the entrance of Baffin's Bay.[2]

In examining the narratives we must distinguish between the general drift, which is credible enough, and the particular details, which are sometimes consistent and historical, but sometimes the reverse. In other words, we cannot doubt that the Norsemen, as recorded in this narrative, really discovered lands to the west of Iceland and Greenland which must be identified with parts of the American Continent; but we may well doubt whether the record of this great discovery is free from legendary and unhistorical admixture. One point of some interest and importance, though not expressly noticed[3] in these Sagas as a reason of the explorations, may be brought to corroborate their evidence on the general facts they record. The scarcity of wood in Iceland and Greenland not improbably exercised an influence on the search for new lands. In the famous Book of the Iceland settlers, known as the *Landnamabok*, it is told as a wonderful thing that one of the primitive

---

[1] Later half of the seventeenth century. Cf. Laing and Anderson, *Preliminary Dissertation*, pp. 226-230.

[2] All these relics are discussed by Torfaeus and Arne Magnaeus, who both incline to fix Vinland near the mouth of the St. Lawrence, or in Newfoundland. Storm appears to argue with greater force for Nova Scotia. See *Studies on the Vinland Voyages* in *Mémoires de la Société royale des Antiquaires du Nord*, nouvelle série, 1888, pp. 344-346, 350-357, etc.

[3] It is, however, in a sense alluded to in the story about Karlsefne's Vinland Wood, quoted below, and in the frequent mention of shiploads of wood brought from Norway and the south and west by various mariners, traders and explorers.

colonists, named Avang, found 'such great plenty of wood where he settled, that he built himself a long-ship therewith'; and in another Saga[1] it is mentioned, with the same astonishment, that the famous Hialte Skeggieson made for himself at his own home a vessel so large that he sailed therein to Norway. As a rule, the men of Iceland and Greenland had to get their sea-going vessels from the forests of Scandinavia; for even in Iceland trees were rare, and the drift-wood on their shores was an uncertain material both in its supply and durability. Now, when once the wooded countries of Vinland, Markland, and Helluland had been discovered, it was natural enough to go in quest of these western lands from which came so much of the precious jetsam to the northern coasts of the Viking settlements. Every explorer in the *Flatey* narrative loads his ship with wood; it is the first thing men seek on landing; to the last they are busied in adding to their stock of the same. Thorfinn Karlsefne, on his return to Norway, even disposes of some of this wood at a high price; he sold, we read, a piece of mosur or maple wood of Vinland for a door bar or a besom, and a Bremen merchant paid half a mark of gold for the same.

Some of the other details, however, as given in *Flatey*, offer easy material for criticism and objection. For instance, Leif and his successors arrive in Vinland in the spring, and we can hardly suppose that a later month than June can here be understood. It would indeed be a remarkable land which was then producing grapes, ripe or almost ripe, and ears of wheat. At the same time that the Norsemen were gathering these precocious fruits, the eider ducks had just laid their eggs on the Stream Isle,[2] which would throw us back to an early date in spring.

[1] The *Kristni Saga*.   [2] Straum-ey.

The ducks and the grapes raise other difficulties. In modern times the eider is hardly ever seen lower than 60° north latitude; yet here we have it in proximity to a country that in many respects may be said to answer to New England, or at least to the south of Nova Scotia.

Again, the whole story of Tyrker the German, tipsy with eating grapes, reads like the fiction of an ale-drinker of the north, who knew no more of wine than that it was the juice of the grape.

To some the language of the Saga about the Skraelings presents equal difficulties, and is no more credible than its suggestions that grapes would intoxicate without vinous fermentation, that wheat would grow without being sown, and that a corn—or kiln—barn could be found in a land without dwelling-houses. The question of these mysterious people would seem, however, to belong to quite a different category, and to be incapable of such hasty and contemptuous treatment. We must return to this point later on, only expressing here our belief, first, that the Skraelings may rather be identified with some vanished races of North American Indians (such as the Beothuks or the Micmacs) than with the Esquimaux; and secondly, that some of the tribes of the latter race were living in the tenth and eleventh centuries much further to the south than they have been found in modern times. It is unfortunately obvious, as various critics have pointed out, that the Saga's account of the time occupied in sailing from land to land in the Vinland voyages leads to no satisfactory result, for we do not know how much was usually accomplished in the 'day' of twenty-four hours. The action of strong currents from the south-west might often make them lose in one part of the 'day' almost as much as they had gained in another. Thus it is by no means easy either to estimate the exact value of the Saga's language on this point[1] or to calculate the position of the new discovered lands from such shadowy data.

Again, the descriptions of these new lands are in many cases equally vague and uncertain. The stony soil, scattered with fragments of slate and affording little vegetation—this applies equally to all the coast-line from Newfoundland northwards; and the name of Helluland or Slab-land would suit almost any part of Labrador just as well as the Island of *Terra Nova*. Once more, the notice of Markland, level and covered with thick forests, might be written about countless points of the North American shore; and even some of the details which seem conclusive do not really take us much further. An island with a sound between it and the mainland, a low shore with white sandy cliffs and shallow water, a fiord or inlet of the sea, a river running out of a lake, a bay between two headlands, these are facts which look promising at first sight, but appear more and more evasive when one seriously tries to fix the position of the locality. For they are features common to nearly all sea-coasts, and discoverable everywhere.

On the other hand, more definite conceptions may be drawn from the general picture of the Norse settlements in this region, of their mode of life, and of the natural products which they discovered. Putting aside for the moment some inconsistent details of the winter climate, obviously exaggerated in expression, it is clear that we have to deal with countries which cannot be placed much further south than the latitude of 41° North, or further north than the latitude of 49° 55′, according[2] to the estimate of the shortest

---

[1] 'They sailed three days,' 'two days,' etc.

[2] According to the best interpretation of the perplexing words, 'The sun was in the sky between Eyktarstad and Dagmalastad on the shortest day,' lit., 'The sun had there Eyktarstad and Dagmalastad on the short day.'

day given us in the narrative of Leif's voyage and plantation. On this day, it is said, the sun was above the horizon in the watches immediately before and after that of midday. Now in Iceland, on the shortest day, the whole course of the sun was included in the one mid-day watch, which began at 10.30 in the morning and ended at 1.30 in the afternoon. The preceding watch would begin at 7.30 in the morning; the succeeding watch would end at 4.30 in the afternoon. According to the most natural reading of the Saga's words, we must surely understand, not merely that the sun rose *sometime* in the earliest and set *sometime* in the latest watch, but that it was above the horizon *throughout* these three watches. This would give us a day of nine hours (7.30 A.M. to 4.30 P.M.) and a latitude of 41° 24′ 10″, lying between Seaconnet and Judith Points, which form the entrance to Mount Hope Bay in Massachusetts. This Bay, and its river the Taunton, correspond well with the Hop of the Saga and its neighbourhood.

It may or may not be conceivable that the name of Hop, given by the Norsemen, was preserved in native tradition,[1] and so passed to the Puritan settlers of the seventeenth century; if so, this would be one of the most curious survivals in history; but neither this verbal identification, nor the various local correspondencies of the Mount Hope Bay district with the Norse descriptions, seem so convincing, in reference to the general position of Vinland, as another detail. The Saga tells us with the utmost distinctness that there was no[2] frost in winter; that winter fodder was not[3] needed for the cattle; that day and night

[1] In the form of *Haup*.
[2] Admitting exaggeration here, we may read 'little,' *i.e.* nothing that could be named in comparison with Iceland or Greenland.

[3] 'Would not be needed as they thought' may surely be understood as a positive statement, for we are never told afterwards that they were wrong in their supposition.

here were more equal than in Iceland or (the extreme south of) Greenland; and that wild vines grew in profusion, and were such a feature of the land, that the Norsemen named it from this very circumstance. It can hardly be maintained that if these points are thrust aside, we can any longer attach definite value to the narrative, which would then become, at the best, an account of a true discovery recorded in an entirely mythical form. And if these details are accepted it seems useless to talk any longer of looking for Vinland to the north of the St. Lawrence, where a winter fully as severe as that of Iceland prevails. In other words, we are absolutely compelled, either to reject the whole body of the narrative, allowing truth only to a general fact which the particulars have utterly distorted; or else to seek our Vinland at least as far south as Southern Nova Scotia, where the northern limit of the wild vine is to be found. The exact place matters little; claimants as good as Mount Hope Bay may be found to the north, and have this advantage, that here, as in Nova Scotia, it is more possible to bring together in one locality the wild vines of Vinland and the skin-canoes of the Skraelings, the self-sown wheat of the fields, and the sables, grey skins, and white furs of the forest animals. What we are concerned with is the impossibility of a far northern locality, so often and so confidently assumed, for the site of Leif's Booths and Thorfinn's Colony.

A word more on the question of the Skraelings. It has been suggested that the terror of these folk at the bellowing of Karlsefne's bull belongs to the emotions of an island people who had never seen such an animal before; whereas any tribe living on the Continent would have known the bison or musk ox.

Those who support the identification of these natives

with the Esquimaux are entitled to argue rather strongly from the fact that, although now confined to the far north, it is tolerably certain that they were once spread over a wider and more temperate area. From this, as from many other points of view, it is truly unfortunate that the North American Indians are extinct as a race in these regions, and have left no written records. Among collectors of their traditions it is often said that these same Indians were once acquainted with the Esquimaux as near neighbours, and regarded them with exactly the same feelings of dread as the Norsemen, Danes and Swedes felt for the wizard and dwarfish Finns and Lapps of the North of Europe.

From the details given in the Saga about the skin-boats of the savages and other matters, the same school derives still more confidence. We have here, they say, a simple choice. In other words, we must either take the Skraelings of the narrative to be Esquimaux, or we must look upon the whole description here given of these strange people as untrustworthy and beyond verification. As to one reference we may be positive. Unquestionably the ugly, sallow, large-eyed, broad-cheeked, shock-headed natives, who came in skin-canoes to reconnoitre, and then advanced to trade, and finally to fight, with Thorfinn's men, were Esquimaux and nothing else.

Again, as to the insular ignorance of this people, who had never seen cattle, and who seemed so dwarfish to the Northmen, what better terms could be chosen to draw a life-like picture of the Esquimaux? Some additional support is given to this theory by the tradition which Are Frode professed to give from the recollections of Thorkell Gelleson of Helgafell (c. A.D. 1050-1150) who knew Greenland well, and had spoken with one of Red Eric's original comrades. The Norse colonists, according to Thorkell, found in Greenland traces of human dwellings, fragments of boats, and stone implements, from which 'one might guess that people of the kind formerly inhabiting Vinland, and known among the Greenlanders as Skraelings, must have passed there.'[1]

But the matter is not so easy of settlement. The words preserved from the speech of the Skraeling boys taken in Markland are entirely inconsistent with the tongue of the Esquimaux, but agree somewhat better with the relics preserved of the Micmac and Beothuk languages spoken by savages whom the sixteenth-century explorers discovered in Newfoundland, Nova Scotia, and New Brunswick. The Saga's picture of the physical characteristics of the Skraelings, of their skin-canoes and cave (or boat) dwellings, would suit the Beothuks or Micmacs quite as closely as the Esquimaux, to judge from the descriptions of the former that have been collected. While the fierce and treacherous nature of the enemies of Thorvald Ericson and Thorfinn Karlsefne, and their eagerness to trade in skins and furs, seem almost decisive against the exclusive Esquimaux theory. May it not be possible to hold that the Icelandic records here contain references to both the races whose respective claims to be the true and only Skraelings have been so vigorously set in opposition.[2]

[1] Cf. Are Frode's still extant *Libellus Islandorum*; and Storm, *Studies on Vinland Voyages*, pp. 315, 316.

[2] Undoubtedly the Skraelings of Greenland were Esquimaux. But the Norsemen probably applied the name without much discrimination to very dissimilar natives of the new-found western lands 'over against' Greenland. The word Skraelings was used in the Sagas for decrepit people, of low physique and culture. On the skin-boats, also called ships, and said to be rowed with several oars, the Saga obviously mixes up details suitable in some cases to Esquimaux, in others to American Indians and birch-bark canoes. The 'big-eyed' appearance of the Skraelings agrees well with that of the Micmacs, as described by Lescarbot; while the narrowing forehead of the Indians gives the broad-cheeked appearance. The 'malicious look' does not at all suit the merry, bland, Esquimaux. Cf. Storm, *Studies on Vinland Voyages*, pp. 360-7.

After the fall of Olaf Tryggveson the *Heimskringla* leaves Norway and Norwegian matters, as it has already left them more than once, to follow the early Viking career of a future King. This was Olaf the younger, the later St. Olaf, a son of Harald of Greenland, and the central figure in the conversion of Scandinavia. Around him obviously gathers the main interest of the narrative of the long series of Norse rulers whom Snorre Sturleson describes; only two others appear in the Royal Sagas as comparable to him;— Olaf the son of Tryggve and Harald Hardrada. These two share indeed in the first class of Snorre's heroes, but not in the first place with St. Olaf. His early adventures, his reign —officially described as fatal to heathendom—his tragic end, the malice, witchcraft, and subtlety of his enemies, the miracles of his relics, these form the largest canvas on which the poet and the annalist has exercised his skill throughout the whole of the Sagas.

The pre-eminent position thus assigned to Olaf Haraldson is plainly due to Christian influence. He is first because no other place can be given to the Royal Saint, the saviour of his people, their guide within the fold of the Catholic federation. Like St. Stephen of Hungary, like St. Vladimir of Russia, like Æthelberht of Kent, Olaf the Holy begins the history of his nation as a Christian, and therefore as a civilised, state. What had gone before Olaf Tryggveson was glorious but unhallowed; the son of Tryggve had begun the work but had not been able to finish; and his cruelty was so enormous that the Church shrank from canonising such a man of blood. But the milder virtues of his successor, firm as a king, gentle as a Christian, won the homage of those to whom the Catholic religion meant light and heathendom darkness.

Judged by other standards, neither the reign nor the

person of St. Olaf were quite so important. Politically, his figure is insignificant by the side of his great rival, Cnut of Denmark, who ousts him from his kingdom. He can scarcely be thought to equal the daring genius of Harald Hardrada. Still less can an impartial judgment place him alongside the first of Norsemen, the greatest of their statesmen and empire builders, William the Conqueror or Robert Guiscard.[1]

Geographically, that is in relation to the side of history with which we are here concerned, St. Olaf's reign is likewise of secondary importance. It is true we have a fair number of details, from this period, of undoubted value in the chronicle of northern travel; but of these perhaps the chief item is the Biarma-land voyage of St. Olaf's wizard enemy, Thorir Hund. Here again the life of Harald Hardrada offers us a more representative figure, and a wider outlook on the great world beyond Norway.

When he was but twelve years old, young Olaf Haraldson, like the son of Tryggve, started on a Viking voyage. First he sailed to Sweden and Gothland; then to Finland,[2] where at this early age the future saint fought victoriously against the wizardry of these dangerous heathens; then to Friesland and England,[3] where the kingdom of Wessex was now in its death-struggle with the Dane. 'At that time Svein Forkbeard and his men had gone wide over England, and King Æthelred had fled to Valland,' or Normandy. Olaf took up the cause of the English and fought bravely for Æthelred. First he made for London, which the Danes

---

[1] Or even the more shadowy Rurik, if we look at the future importance of that chieftain's life and work.

[2] Chs. iv. to ix. of *St. Olaf's Saga*.

[3] Chs. x., xi. of same. When Olaf was coasting along Friesland he lay off '*high* Kinnlimside,' says the Saga, in heavy weather. Some have identified this with the Texel, in spite of the epithet 'high.'

held; and here Olaf and Æthelred went up the Thames and made an onset on the bridge across the river, between the city and the cheaping town of Southwark. Now this bridge was so broad that waggons might pass each other thereon; and here were castles and bulwarks looking down the stream; and under the bridge were piles stuck into the bottom of the river. And when Æthelred could not take the bridge, Olaf, with a great raft cunningly devised, broke it down, and so won the city. He had flake-hurdles made of willow twigs and green wood, and laid all over his ships sheds of wicker-work, and underneath the sheds he planted staves so thick and high that it was a good palisade to fight from. So he forced his way right up under London Bridge, through showers of missiles, lashed cables round the piles that upheld the structure, and then rowed down stream, dragging the piles till they were loosened from under the bridge, and all on top fell in.[1]

Four years[2] he stayed in England after this and won battles for Æthelred; but when the old King of Wessex died he went south over sea, apparently to Normandy[3] and

[1] Ch. xii. of *St. Olaf's Saga*.
[2] Seemingly four; though the Saga says 'three' cf. ch. xiii. beginning, xv. beginning, and xiv. end, of *St. Olaf's Saga*.
[3] Chs. xv., xvi. of *St. Olaf's Saga*. In 'Valland,' he visited 'Ringfirth,' 'Grislapool,' 'Williamsby,' 'Fettlefirth,' 'Seliupool,' 'Gunvaldsborg,' a 'great and ancient' town, and 'Charles' Water.' In nearly all these places, which mostly appear hopeless of identification, he fough and won battles against other Vikings,—a fresh proof, if any were needed, of the universal activity of his race at this time. Ringfirth and its castle on the heights, 'where Vikings sat,' were perhaps in the neighbourhood of Coutances (?) or at Mont St.'Michel (?): cf. Dozy, *Recherches sur l'histoire... de l'Espagne*, edition of 1881, ii. pp. 308-14, according to which 'Charles' Water' is the Bay of Cadiz. The fragments of an older redaction of *St. Olaf's Saga* declare that at the 'Karlsar' or 'Charles' Water,' Olaf found idolaters, whose deities, a Siren and a Boar, he killed; and here he got a wind favourable for passing the Strait, but was turned back by the vision. Charles, or Karl, is here viewed by Dozy as merely a Norseman's version of Hercules ('the great man'), and this eminent scholar contends with

Brittany, and the coasts of the Bay of Biscay and western Spain, and after many victories determined to sail on to Norva Sound or Gibraltar Straits, and so to Jerusalem. But now there came to him in a dream a man, noble-looking and well-favoured, yet awful, who bade him give up his purpose, and go back to the lands of his birthright, 'for thou,' said he, shalt be a King of Norway for time everlasting.'

We may suppose that Olaf had now reached beyond the Garonne and the Bay of Biscay as far as Cadiz and the Pillars of Hercules; but after this vision he turned back to the valley of the Loire, and ravaged Poitou and Touraine, raiding as far as the market town of Parthenay, a little west of Poitiers.[1] Thence he slowly retraced his steps to Normandy. 'And now thirteen years had passed from the fall of Olaf Tryggveson. And in Normandy there were then two earls, William and Robert, Lords of Rouen and sons of Richard, who was the grandson of William Long-sword,[2] who was himself son of Rolf Ganger, who won Normandy. And Emma, wife of Æthelred, King of the English, was sister of these earls; who were akin to the lords of Norway, and at all times the greatest friends of the Northmen. So Olaf tarried the winter in Seine-water, and had there a land of peace.'

Meantime Cnut the Dane had won[3] England from Æthelred's sons; and these last came to Rouen, while Olaf

much skill and persuasiveness for the reality of the tradition that as late as A.D. 1145 a colossal statue of Hercules stood in the Bay of Cadiz on a lofty pedestal; at last it was destroyed by the Moslems, to whom (*e.g.* Masudi) we owe the most exact description of the marvel which gave name to the 'Strait of the Idols of Copper.'

[1] See chs. xvii. and xviii. of *St. Olaf's Saga*. The appearance in the vision was afterwards supposed to be the spirit of Olaf Tryggveson.

[2] The Saga calls him Long-*spear*, ch. xix. of *St. Olaf's Saga*.

[3] Earl Eric of Normandy was his ally in the conquest, and died 'of blood letting in England,' just as he was setting out on pilgrimage to Rome. Ch. xxiii. of *St. Olaf's Saga*.

was there, fresh from his western voyage. 'And they were all together that winter in Normandy, and agreed that Olaf should have Northumberland if they could get England from the Danes.' In the spring the allies made a descent upon England, but Cnut's men were too many for them; so Olaf, after winning a battle in Northumberland,[1] sailed off to Norway, where all the people welcomed him as their king.[2]

The rest of this long Saga—though containing a goodly number of allusions to foreign lands, and especially to Iceland and Greenland, the Orkneys and Shetlands, the north of Scotland, Ireland, and Russia—has little direct bearing on exploration, travel, or war, in distant parts. Thorir Hund's Biarma-land or White Sea journey is the chief glimpse afforded us of the outer world within this period;[3] and the story of this is as follows. One winter (probably of the year 1023 or 1024) Olaf sent a trusty servant of his to the north. This servant was Karli of Halogaland, the most northerly province of Norway proper, to which Ælfred's Ohthere also belonged. Karli had more to do than collect the King's moneys and leave his messages in the uplands of Norway; he was to sail right away north to the White Sea and the Dvina. It was before all a commercial venture; the King and Karli joined in fitting out a ship, and each was to have one half of the profits. So Karli steered north to Halogaland early in the spring with Gunstein his brother, and coasted up to Finmark. Going

---

[1] The raid was at 'Youngford,' the battle at 'Wald'; ch. xxvi. of *St. Olaf's Saga*.

[2] Chs. xxvi., xxvii., xxviii., xxxvi. of *St. Olaf's Saga*. We may notice in ch. xxxii. the *Cordovan* boots and hose which King Sigurd Sow wore at his meeting with Olaf— a trace of the early intercourse of the Northmen with Southern Spain. Cf. Dozy, *Recherches sur l'histoire de la littérature de l'Espagne pendant le moyen age*, third edition, 1881, vol. II. pp. 250-371.

[3] 1015-1030. See ch. cxliii. of *St. Olaf's Saga*.

thus, they passed close by the home of Thorir Hund of Birch-Isle,[1] who had a blood-feud to settle with Karli. Thorir accordingly sent messages of the friendliest kind to the passing ship, saying that he also was minded to fare that summer to Biarma-land in their company and share evenly in all winnings. So Karli and his brother bade him join them with a force of the same strength as they had (five-and-twenty men), but he brought a huge buss, manned with his own house-carles, nearly eighty in number. Then Gunstein became suspicious and would have turned back, but Karli persisted, only remonstrating with Thorir about his vessel and its crew. But Thorir declared in such a venturous voyage one could not have too many good men.

All the summer they sailed on in company, sometimes one ship being ahead and sometimes the other, along Finmark and round the North Cape and past the Murman Coast of Lapland and over the White Sea to Biarma-land, where they found a good market; at this 'cheaping stead' Thorir got some grey wares and beaver-skins and sables, and Karli, with his well-filled purse, bought many peltries. When the bartering was over they turned their prows for home and sailed down the river Dvina. For all the traffic was at a place some way from the mouth of the stream, probably near the later Kholmogori, so famous in the time of Queen Elizabeth and Tsar Ivan IV., of Chancellor and Jenkinson and Burrough. While business lasted there had been, as usual, a 'peace with the folk of the land'; but now this was proclaimed to be at an end, and, for any more profit, plunder was needed. So when they came out to the sea, the crews met together, and Thorir asked if any one was willing to go on land and gain wealth for himself;

[1] 'Biark-ey.'

for wealth there was to be had, buried in the woods, or under mounds, 'according to the wont of that land.' Thorir was well acquainted with the grave hoards and other buried treasures of the Biarma-men; probably he had often made the White Sea voyage before. And all said they would venture if wealth was to be looked for. Thorir bade them prepare to start at night; they left a few men behind to guard the ships, and the rest went on land and set out.

'At first there were flat fields and then a thick woodland.' Thorir went ahead, and ordered his followers to move on silently, barking the trees to show the way back. At last they came out into a great clearing in the wood; and in the clearing was a treasure-mound with a high palisade and a door securely fastened. But Thorir had timed it well; one watch had just gone home, and the next had not yet arrived. The Norse leader hastened up to the fence, and, hooking his axe on the top, hauled himself over, hand over hand, and opened the door from within. He warned his men only to dig up the gold and silver, 'for here also stands Iomala the God of the Biarma-men, and let none dare to rob him.' But when they had loaded themselves with booty, Thorir turned back to Iomala, and took a silver bowl which stood in his lap, full of silver pennies, and poured them into his cloak, while upon his arm he slipped a bow which was over the bowl.

By that time all the men had got out through the faggot fence, and became aware that Thorir had stayed behind. So Karli turned back to look for him, and they met inside the gate, and Karli saw that Thorir had got the silver bowl. Then Karli himself ran into Iomala, and saw that a thick collar was about his neck; with his axe he smote asunder the string at the back of the neck and took the collar; but the stroke was so mighty that Iomala's head fell off, and a marvellous crash was heard.

'And straightway when that befell, the warders came forth into the clearing and blew their horns; and now trumpets were heard on every side. So the Northmen rushed into the wood, but behind they heard constant whooping and crying, for the Biarma-men had come.' They were lost but for Thorir. He had guided them to the place, and now he guarded the rear, guarded it above all with his wondrous witchery. Before him walked two men carrying a great bag, 'and what was therein seemed most like to ashes.' And Thorir dipped his hand in it, and sowed it about their path, and at times he cast it forth over his company, making them invisible.

'Thus they fared out of the wood and into the open fields, while the Biarma-men rushed after them, behind and on both sides, crying and yelling; yet saw they never the Norsemen and their weapons came not near. At the ships Karli and his people went first aboard, and cast off into the main'; then Thorir too sailed across the White Sea.[1]

The nights were still bright (it was the time of the white nights of midsummer), and the Northmen sailed the whole 'day' and every day (of four-and-twenty hours) till they came one evening to certain islands. There they struck sail and cast anchor, waiting for the ebb, as there was a strong current before them. And Thorir called upon Karli to deliver up the collar of the idol Iomala, 'for then ye had to thank me,' said he, 'for your lives; and thou, Karli, didst run us into the greatest peril.' But Karli replied that the collar was King Olaf's. 'Thereafter Karli tried to get away from Thorir, sailing as fast as he might; but he could not; and so they came both to Geirsver. And

[1] 'Witch-wick.'

this is the first place where one who comes from the North may lie to at a pier.' Thorir lay up the haven and Karli further out. For the last time Thorir asked that they should share the booty; and on a final refusal, he killed his rival and seized the precious collar; he would have sent Gunstein and all the rest after Karli, but they escaped to King Olaf, abandoning their ship and its rich cargo. 'And thus Thorir Hund became one of the chief foes of the king.'

The full effect of this was seen later, when Cnut the Great's ambition threw him across Olaf's path, and when every element of disaffection in Norway gathered round the claims of the Danish king. Meantime Thorir remained in his far northern Finmark home, carefully watching events; ever on his guard against king's messengers from the South; and again, more than once (if we may trust some later allusions), going on the White Sea voyage, buying reindeer skins, and gaining a deeper knowledge of the witchcraft of the mysterious North.

St. Olaf's relations with Iceland are not of special importance, but such as they are, they may be briefly noticed. For instance, one Sighvat the Skald, whose father met the king while he was on his western Viking voyage, became a favourite poet of the Court. Both he and his father were typical Icelanders in their keenness for long trading journeys,[1] and the king's favourite talk was with old rovers (such as he himself had once been) about the distant countries they had visited, such as Iceland and Shetland and Faröe and Orkney;[2] but in spite of his fondness for these topics he did not fail to lay heavy

---

[1] Ch. xli. of *St. Olaf's Saga*.
[2] Ch. lvi. of the same. See also chs. lxix. end, lxxi. Cf. moreover in chs. xlv., xlvii., the Saga's remark on Olaf's contingent of '100 men in *Welsh Helms*,' on his ship the *Karl's Head*.

dues upon Iceland ships, and to threaten Icelandic independence.

At times St. Olaf used these distant islands as convenient places of exile for his foes; thus he sent his rebellious vassal Roerek [1] to Leif Ericson in charge of Thorarin the Icelander, 'of all men the sagest,' and a great seafarer, who had been 'long in the Outlands, and was then dighting a merchant ship of his own' for Iceland. Greenland and Eric's Fiord, however, he had never visited, and indeed he seemed loath to take so great a voyage; reluctantly yielding in words to the king's wish, he found in stormy weather a good excuse for keeping to his beaten track, and never delivered his royal prisoner to Leif.

From all this we can better understand the Saga when it tells us how much the king had to do with the western colonies of the Northmen, and how steadily he aimed at binding them more closely to Norway itself. His political schemes were foiled in Iceland, but he completed the conversion of that country; he subjugated the Orkneys afresh; and in all the Outlands he put down, at least for a time, the eating of horse-flesh and the exposure of infants, practices viewed with about equal horror, as decisive marks of paganism.

The history of the Orkneys during this time offers the same savage and repulsive aspect as before. Nowhere did the Viking nature degenerate into more aimless brutality than in their settlements on the coasts and islands of Northern Britain. The very names are significant. Eric Blood-axe, Thorfinn Skull-cleaver, Ottar the Swart, Eyvind

---

[1] Otherwise 'Roderic' or 'Rurik,' an 'under-king,' afterwards famous as the only sovereign buried in Iceland, whom Olaf had blinded; in revenge the victim watched his time; sat by Olaf in church; felt that the saint had no armour on; and aimed a dagger-stab at his heart, narrowly missing his prey. Chs. lxxxv., lxxxvi. of *St. Olaf's Saga*.

Urochs-horn, Einar Wry-mouth, these are among the most prominent personages in the endless squabbles of kites and crows which fill the chronicle of the Norse Earldoms in Caithness, the Hebrides, the Orkneys, and the Shetlands. Before Harald Fairhair the Orkneys were but a Viking lair;[1] in his days they were 'builded,' that is, reduced to order and made prosperous, according to the cheerful optimism of the *Heimskringla*. In reality, they remained almost as lawless and piratical as before; their dependence on Norway was vague and slight, and was openly disavowed whenever it was found convenient to do so; their efforts to extend Norse rule in Scotland ended at last in complete failure; and, in contrast to Iceland, as we have already noticed, they do not seem to have formed a starting-point (to any appreciable extent) either of new exploration or of new literature. Here, about 995, Olaf Tryggveson, passing round these islands on his way home from his early wanderings,[2] seized the reigning Earl Sigurd; and, in his usual hearty manner, offered him christening or death. Sigurd's son, 'the Whelp,' he took with him as a hostage to Norway; but old Sigurd, when the storm had passed, paid no more attention to his overlord, married the daughter of Malcolm, King of Scots, and busied himself with raiding in Ireland. Just then hopes of conquest were high among the western Norsemen—hopes that from their Ostman kingdoms on the east coast they might conquer the whole of Erin; but in 1014, Brian Boru[3]

---

[1] See chs. xcix.-cix., of *St. Olaf's Saga*. On Norse travel to and from Orkney, Shetland, etc., cf. also *Saga of Burnt Njal*, chs. ix., lxxxiv., lxxxviii.-ix., cli.-iii., cliv.-clvi ; *Eyrbyggia Saga*, chs. i., v., xxix. ; *Egil's Saga*, chs. iv., xxxiii., lxii.; *Grettir's Saga*, chs. i.-iii., iv.-v. ; *Laxdale Saga*, chs. iv., xi., li.

[2] Ch. lii. of *Olaf Tryggveson's Saga*.

[3] The Munster-man, High King of Ireland. His victory was won largely by statesman-like alliance with the Ostmen of the South and West.

overthrew the invaders at the great battle of Clontarf, and among them fell Sigurd. His children inherited, not only Orkney, but Caithness and Sutherland, as fiefs of the King of Scots; one son, Einar Wry-mouth, was a famous raider 'about Ireland and Scotland and Bretland,' and suffered a great defeat at the hands of the Irish under their King Konofogor.[1]

Like a cuckoo in the nest, another and more famous son, Thorfinn Skull-cleaver, as descendant of the great Scot king Malcolm, devoted his restless energies and high abilities to the ousting of his brothers from their share in the inheritance, until at last one of the injured appealed to St. Olaf. The king seized the chance to assert in a more formal way his supreme lordship in Orkney and Shetland; but his disastrous struggle with Cnut of Denmark again changed the face of things; Thorfinn became more powerful than of old, and rose to be the 'noblest of all Earls of the Islands,' winning Shetland, and Orkney, and the Hebrides, and a great dominion in Scotland and Ireland. 'From the Giant Isles to Dublin,' sang a skald, 'each man was counted Thorfinn's.'[2]

The Faröe Islanders, at this time, also gave their allegiance to the King of Norway; and there it was settled that such law and land-right should be held as he should frame, and such tax paid as he should settle. But in practice the gathering of Norway dues and the enforcing of Norway laws proved to be a harder thing.[3]

---

[1] Ch. lxxxvii. of *St. Olaf's Saga*. It was in the same autumn that Eyvind Urochs-horn, a Norseman high in favour with Olaf, was murdered by Einar on his way back to Norway from Ireland. See ch. ciii. of *St. Olaf's Saga*.

[2] Ch. cix. of *St. Olaf's Saga*. As to Olaf's first successes, a skald sang: 'The Shetlanders grown leal now, To thee for thanes are counted, Ne'er yet was seen an Yngling In East-land who beneath him Brake down the Isles of West Land,' ch. cviii. of *St. Olaf's Saga*. Harald Fairhair's exploits are conveniently forgotten.

[3] Ch. cxxxvi. of *St. Olaf's Saga*.

St. Olaf's reign was largely passed in a struggle against three enemies, Cnut of Denmark, Olaf of Sweden, and the malcontents of his own country, led by Thorir Hund of Birch Isle and Harek of Thiotta. In this connection we hear again of Holmgarth or Novgorod, and of King Yaroslav. Olaf of Sweden, resolved at all costs to prevent the marriage of his daughter with Olaf of Norway, gave her to the Russian Prince; and as her jointure she had 'Ladoga[1] and the earldom thereto appertaining' (A.D. 1019). It is now that Sweden, perhaps the oldest of Scandinavian Kingdoms, begins to play a larger part in Scandinavian history; and the *Heimskringla* stops to give it a short description,[2] telling us of its 'Gautland' Bishop and his 1100 Churches, and of Upsala, with the King's Seat and the Archbishop's Chair, and the Hoard or Treasury, and the place of the National Assembly. 'For in all matters the Swedes yield to Upsala law, and the King's revenue is called " Upsala wealth." St. Olaf's fate was, after all, to be allied with the House of Sweden, while its King continued his bitterest foe. One Princess was snatched from him, but he was at last married to another, a daughter of the Swede King by a Wendish or West-Slav mistress.[3] Another link with Baltic and Russian lands in this same time is the story of Gudleik—Gudleik of Novgorod, as he was called, though a Norseman, from his frequent journeys and large business in the North-Slav countries. King Olaf, ever ready,

---

[1] Already mentioned as 'Aldeigiaborg.' See chs. xcii., xcv. and lxxxviii.-xci. of *St. Olaf's Saga.* Sighvat, ch. xcii., hints perhaps that Novgorod's money was decisive:—'I learned of many a matter Of the good gold ward in Garth-realm. . . .'

[2] Ch. lxxvi. of *St. Olaf's Saga.*

[3] Ch. lxxxix., xciv. of *St. Olaf's Saga.* Swedish 'public opinion,' was hotly in favour of the alliance with Norway. For our object their promise is of interest, that, if Olaf the Swede gives his daughter to Olaf 'the Thick' (St. Olaf), they will help him to win Estland, Finland, Courland, *Kirialaland*, and wide about the East Lands (Baltic Coasts). Ch. lxxxi.

as we have seen him before, to do some trading on his own account, employed Gudleik to buy him things that were hard to get in Norway, choice clothes for robes of state, and furs of great price, and a rich table service.

Let us now turn from east to north. After he had been king five years, Olaf took a journey to Halogaland, with the double purpose of fixing his own power more firmly, and of promoting the Christian Faith in those outlying regions.[1] Here he stayed most of the summer, and the chief landowners submitted outwardly to his power and became his men. All the people of the region 'took christening' also at the king's order; but after he had gone it was seen that the new creed was but skin-deep in Halogaland.

Olaf was now recalled to the South by a pressing danger, and the storm that had long threatened from the West broke at last upon his head. King Cnut 'the Rich,' 'the Ancient,' or 'the Great,' Lord of Denmark by inheritance, and of England by conquest, set himself to realize the full scope of his ambition—a Scandinavian and Anglo-Saxon Empire of the North, which should finally embrace all lands possessed or settled by Danes, Swedes, and Norsemen from Ireland to Finland, from the Eyder to the North Cape of Europe, and from the Channel to Greenland. The Lordship of Norway was necessary for his plans, and he had little scruple in 'craving the whole for his own.' He sent a splendid embassy out of England to Norway, and 'right gloriously was their journey arrayed'; they argued long with Olaf about their Mission—how Cnut's forefathers had the realm in old time, how the Scot kings in Fife had become his men, and so forth—but they could get no submission. Then the war broke out; just in time to

---

[1] Chs. cx., cxi. of *St. Olaf's Saga*.

save the men of the Faröes, of Iceland, and of Halogaland from an open struggle with Olaf.[1] For that very spring a ship had sailed from Norway to the Faröes with orders which would have been difficult to evade, for the instant despatch of a mission to Court. Similar, but even more stringent, demands had been made upon Iceland, demands of absolute submission, the acceptance of the laws of Norway, and the payment of thane-tax and nose-tax, 'for every nose a penny'; but the Icelanders with one accord had refused all such talliages. Once more, Finn Arnison was despatched to Halogaland, where he levied taxes and fined Thorir Hund for the slaying of Karli and the seizing of goods that were in his charge for the king. But Thorir was again too skilful for his foes; he cheated Finn Arnison with a mere fraction of the fine, and fled to Cnut with nearly all his Biarmaland wealth.[2]

Curious figures flit about the stage as the crisis approaches; Stein the Icelander, who fared abroad in a ship of his own to the western coast of 'Gizki,'[3] was outlawed for a murder and fled to Cnut; Thorod, of the same land, who went to gather taxes for Olaf in the border country between Sweden and Norway, and afterwards returned to his Island-home after a terrible adventure with a witch-wife; Sighvat the Skald, who wandered from Norway to Rome, and from Rouen to England, on a merchant voyage; Karl-o'-mere,[4] 'one of the greatest of lifters,' who turned from his Viking life to become a firm friend and servant of St. Olaf, and was killed

---

[1] Chs. cxlv., cxlvi., cxlviii., cxlix. of *St. Olaf's Saga*.

[2] Thorir was forced to hand over the gold collar of Iomala, but he carried off great treasures of grey skins and beaver. The story of his crafty fencing with his enemies in this difficulty is one of the most humorous and dramatic incidents in the Sagas; ch. cxlix. of *St. Olaf's Saga*.

[3] An isle off the Norwegian coast, in N. Lat. 62° 30'. Ch. cxlviii. of *St. Olaf's Saga*.

[4] Chs. cli., clii., cliii. of *St. Olaf's Saga*.

in the Faröes, gathering in the Royal dues. All these incidents show us the wide range of Norway's power at this time, and show us, too, how feeble and uncertain was the tenure of that power in the saintly but impolitic hands from which the rule of the North was now slowly passing.

Cnut's first attack was indecisive; but in the second campaign a universal defection, added to the immense armaments of the invader, threw Norway into the hands of the new 'Emperor of the Arctic,' and Olaf fled, where so many Norse chieftains had fled before, to Novgorod.[1] Here King Yaroslav welcomed him heartily, and offered him the government of Old Bulgaria on the Volga, over which the Russian Grand Prince now[2] claimed dominion. Here, in the country of the Modern Kazan, the people were not yet Christianised; and the exile thought well of this offer; but his men were all loath to follow him so far. It was better, they said, for him to try his fortune again in his own kingdom before attempting anything more distant. It would not have been an easy task to convert the Moslems of 'Bolghar,' which had formed a part of Islam from the close of the ninth century, and which was never thoroughly conquered by the Russian princes of the Middle Ages, though on occasion it acknowledged the vague suzerainty of some of the most powerful East Slav rulers.[3] For his own part St. Olaf would have preferred to go on pilgrimage

[1] Chs. clxxx., cxci. of the same, A.D. 1020.

[2] Ch. cxcviii. of same; the 'dominion called Volgaria, which is one part of Garth-realm, where the folk was heathen.'

[3] On this Old Bulgaria of the Middle Volga and Lower Kama, cf. pp. 105n., 292n., 303, 335, 341-2, 462-4n. of this volume; Ibn Dasta in Rossler's *Romanische Studien*, pp. 359-362, etc.; Saveliev in Erman's *Archiv.*, vi. pp. 91-8; Frähn's Ibn Foslan, pp. 226, etc., 147, 168, 258, 266; Frähn in the *Mémoires de l'Academie de St. Pétersbourg*, série vi. tom. i. pp. 183-199; Masudi, ii. 15, ed. Barbier de Meynard and Pavet de Courteille.

right away to 'Jerusalem world,' and leave the struggles of this life to others; but the heavenly powers themselves turned him from this abdication. A vision of Olaf Tryggveson,[1] coupled with the persuasion of Biorn the marshal, determined the king to try once more the chances of a restoration; and in the face of all dissuasion from the prudent Yaroslav, he persisted in this resolution. He crossed the ice, made his way to the open Baltic, and by the time of the spring thaws appeared in Sweden.

Meantime Thorir Hund had been prospering since his return with Cnut's host. He had made the Finn-journey two winters and been long on the fells; measureless wealth had he won by chafferings with the Finns. Most notable of all, he had made for himself twelve coats of reindeer skins of Lapland, with such wizardry, that no weapon could bite on them, 'aye, less than on ring-mail.' And both he and Harek of Thiotta made ready to fight against Olaf if he should come from the East. On the other hand, some of the nobles stood for the old king, and chief among the last was his young half-brother, afterwards famous as Harald Hardrada, but as yet only a boy, fifteen years of age.

With Swedish guides St. Olaf moved slowly through the Ironstone Land and Iamtland to Norway. And as he moved he had a vision, the vision of his race; he saw over Thrandheim and all Norway, and 'ever wider over all the world, both land and sea; those homesteads where he had been, and just as clearly those which he had never seen, aye, and those of which he had never heard tell, both peopled homesteads and unpeopled wastes, as wide as is the world.'

The two armies met at Stiklestad; and there Thorir

---

[1] Cf. *St. Olaf's Saga*, chs. cxciv.-cxcviii., ccii.-cciii. The same vision had already turned him back from his early Viking life. See above, p. 87.

Hund, with his reindeer skins and his witchcraft, proved too strong for the king. So he fell before the 'Hound whom steel would not bite'; yet the same man was among the first, after the death of Christ's anointed, to proclaim his saintship and to attest his miracles. It was during the time of his exile at Novgorod that Olaf really laid aside the ruler and became the saint; and the national feeling of Norway, outraged by the rule of the Danish Usurper, made the victim of Stiklestad a royal martyr, and the reverence for his memory became the emblem of patriotism.

When St. Olaf left Novgorod and the kingdom of Yaroslav to make another struggle for the crown, his young son Magnus remained in the care of the Russian prince; and after the battle of Stiklestad his step-brother Harald, the future Hardrada, sought a like refuge at Holmgarth.[1] It was not long before one of the royal martyr's greatest foes also came to Ladoga and Novgorod, and swore allegiance to Magnus Olafson. Kalf Arnison[2] had been one of the three who cut down the old king at Stiklestad; now, like the great archer Einar,[3] he was in revolt against King Cnut and had to fly to Sweden and Gardariki.[4] Sighvat the Skald, St. Olaf's friend and favourite singer, was another prominent Norseman on whom Magnus could count. He came home soon after the death of his old patron, and, staying in Sweden during Magnus' exile, asked news of Olaf's son from all the traders that he met on their way to or from Novgorod. The distant wanderings of his race were typified in Sighvat; whose home was in Iceland; who had traded in England and Valland;[5] who had been to Rome on pilgrimage; and who heard of Olaf's death while he was on his way from the

[1] Novgorod.
[2] Ch. ccxl. of *St. Olaf's Saga*.
[3] Thambarskelfir.
[4] Russia.
[5] North and West France.

South across the Alps.[1] Now he had travelled down into Sweden with thoughts of a journey to Russia, for, as he sang, news from the East was full of the praises of the young king in his exile, and even the birds of passage spoke of him.[2]

Before many months came a far more decisive change of fortune; for Cnut the Great, so long 'held in worship' by all men of the North, died in England. But even earlier than this the Loyalists were beginning to move once more. From Novgorod to Ladoga, from Ladoga to the Baltic, and across the narrow sea to Sweden, by the same route that his father had gone on his last journey, Magnus Olafson made his way back to Norway with very different prospects. He was welcomed by almost every class and every district as universally as his father had been rejected. The disappearance of the Saint's rebel enemies soon followed. Thorir Hund, the repentant chief of them, went on pilgrimage to Jerusalem,—never, as some said, to return again. Harek of Thiotta was killed; Kalf Arnison had to fly the country; while others became Vikings in the Western Seas, and long plundered the coasts and isles of Scotland, Ireland, and the Hebrides.[3]

Magnus' reign was a brilliant one for Norway. Under the misrule of Harald and Harthacnut, the Great Cnut's work was all undone; Norway was now able to gain the better of Denmark;[4] and St. Olaf's son even made a claim on that realm of England, from which Cnut had ruled his

---

[1] 'On Alps by a burg one morning, I stood and thought of the King.' Ch. ix. of *Magnus Good's Saga*.

[2] 'E'en though there fly betwixt us The smallest fowls air-cleaving, I hear of the small king's son As he fares and my hope is holpen.' Ch. ix. of *Magnus Saga*.

[3] See chs. xii., xiii.-xv. of *Magnus Saga*, A.D. 1035-1047.

[4] So far, at least, as to force Harthacnut to an agreement that whoever outlived the other should inherit his kingdom. Under this treaty Magnus ruled Denmark, 1042-1047, chs. vii., xix. of *Magnus Saga*.

MAGNUS GOOD; HARALD HARDRADA

Empire, but which had now passed back to the West Saxon House in the person of Edward the Confessor. At the peaceful answer of the English king, who would offer no resistance, but simply appealed to the common justice of princes, Magnus 'the Good,' with a sort of shame, waived his demands. Now that England had cast out the race of Cnut there seemed no shadow of quarrel; and meantime there was work enough in the new dominions on the Baltic and in wars with Wends,[1] Courlanders, and Saxons. What Magnus abandoned was left for a more daring spirit and a more grasping ambition twenty years later. That spirit and that ambition were incarnate in Harald Hardrada. His life is perhaps the most perfect example of the Norse type, as pirate-wanderer, free lance, and conquering king. William the Conqueror, Cnut the Ancient, Robert Guiscard, Roger of Sicily, are all in a sense greater men of the same race—greater as organisers, law-givers, administrators, and masters of statecraft. But among all the Scandinavians there is no rover, no adventurer, like the man who, after fighting in well-nigh every land of Christians, or of the neighbours and enemies of Christendom, yet hoped for time to sail off to the new-found countries 'under the pole,' and so perfect a life of unmatched variety by unmatched discovery. Fate, which granted him many things, denied him the chance of being a Barentz or a Hudson of the eleventh century. He had fought with wild beasts in the Hippodrome of Constanti-

---

[1] See chs. xxiv., xxv., xxvii. of same. In these campaigns Magnus captured Jomsborg, the famous stronghold of Baltic Vikings, which the Dane kings had so dreaded, and won a great victory over his Slav and Lettish adversaries at Heathby near Skotborg-Water on Lyrshaw Heath. Here 'the Wends fell as thick and fast as if they lay in wave drifts,' ch. xxix. Duke Otto of Brunswick 'in Saxland,' son-in-law of St. Olaf, was with him in this battle. Certain (two) Iceland men were also present, ch. xxix. of same.

nople; he had bathed in the Jordan and cleared the Syrian roads of robbers; he had stormed eighty castles in Africa; he had succoured the Icelanders in famine; and men knew and feared him alike in Russia and England. He lived as a prince, a conqueror, or a guest, in Norway and in Novgorod, in Sicily and in Yorkshire; in his own songs he boasts that he had sailed round Europe; yet he never led his followers into that unknown world where his countrymen had shown the way. A true forerunner, in spirit, of those explorers who have most greatly dared, he fell without one discovery. Men of his own nation and time had been before him everywhere; but he united in himself the deeds and the adventures, the conquests and the enterprises, of many. Never more clearly than in him may be recognised that northern spirit which was the very ground and impulse of the movement of Europe towards the outer world, in which lay, hidden yet certain, the future triumph of the Christian nations over every rival.

To Snorre Sturleson, St. Olaf the Christianiser—the weak, obstinate, religious king, who thought it sin even to cut chips from a block of wood on the Sunday—is the central figure in his country's history. And in a sense, for good or for evil, Snorre was right. The conversion of the Scandinavian people was indeed a turning point in their history. They ceased to be the terror, they became the leaders, of the Christian world. Yet the Norse genius for exploration, though not for war, seems to decay with the conversion of the race, and the subtlety and craft that is so marked in the pagan Scandinavians is often replaced by an exceptional simplicity and want of foresight among the later Christian types of Norse nationality. The fierce contempt of life, the unflinching pursuit of material gain, alike by strategy or by force, never turned from its object by discouragement or resistance, disdaining no means to reach the goal, never more dangerous than when repulsed, never more watchful than when reposing, generous to a fault with friends, but implacable with foes, waiting any time for vengeance, yet never sacrificing, for a sentimental vengeance, solid gain— these qualities, which mark so unmistakably Thorir Hund or Harald Hardrada, Robert Guiscard or William of Normandy, appear slowly to die out in the later generations of Christian Northmen. In truth, men like Gustavus III. or Charles XII. recall the prominent type of heathen Norsemen far less than do some of the empire-builders of another country where the Scandinavian blood had also left its mark. Ivan III., the 'Collector of Russian land,' and Harald Fairhair, the Unifier of Norway; Ivan IV. and Cnut of Denmark; Yermak, the Conqueror of Siberia, and Rolf the Conqueror of 'Valland,' offer us parallels of no little suggestiveness. In both race-histories there is a similar spirit at work, but working on very different material. Out of a people all composed of adventurers, it was perhaps harder to form a perfectly unified state; leadership was more difficult where so many could dispute the lead, and where so few were willing to be led. But on the submissive and ductile nature of the Slav there could be more easily impressed a sense of subordination and of discipline, which was not without its value as a servant of high political ends.

Harald Hardrada was wounded in the battle where Holy Olaf[1] fell, but he was hidden by friends, healed of his wounds, and sent east through Sweden to 'Garthrealm.'

[1] See ch. i. of *Harald Hardrada's Saga* and the verses:—'The Burner of the Bulgars There well availed his brother,' etc. As to his life in Russia, the Saga says nothing more of his wars against the (Volga) Bulgarians (ch. ii.), except to quote the lines of a skald (Bolverk):—'But the next year east in Garthrealm Ne'er heard we of peace-waster Waxing more famed than thou . . . Until at ship's bows Harald Saw Miklagarth's bright metals.' 'Miklagarth,' the 'Great City,' is, of course, Constantinople.

There he remained some years with King Yaroslav, and
'fared wide about the East Ways,' till at last he went on
to seek his fortune in the service of the Byzantines, the
richest paymasters of Mediaeval Christendom, to whose
banner so many of Harald's countrymen had already flocked.
Zoë the Wealthy then reigned at Constantinople in the
name of Constantine Monomach, and employment was
at once given to the new-comer. First he fought for the
Empire in the 'Greek Sea,' and put down the Corsairs
that infested the Archipelago; then he performed brilliant
deeds in more distant regions. After a time all the
Varangians of the Foreign Guard were put under Harald's
command, and he became the trusted leader of the motley
host, a Norse general not only over Norsemen, but over
Slavs, English, Germans, and men of Latin speech. But
he and his mercenaries claimed to be independent of the
Greek troops and their officer, declaring they were 'masters
of their own matters and bound in service only to the
king.'[1]

[1] Cf. chs. ii.-iv. of *Harald Hardrada's Saga*. As to earlier journeys of Northmen to Constantinople, cf. e.g. (1) *Grettir the Strong's Saga*, chs. lxxxviii.-xcv., where Thorbiorn Angle, after slaying Grettir in Iceland, goes abroad to Norway and Miklagarth, in the time of Michael Kataλaktos, when 'many Northmen' were going to Byzantium and taking 'war-wage' there. Like Thorbiorn Angle, Thorstein Dromund, Grettir's avenger, enters the Vaering Guard, wherein he discovers Thorbiorn. A long story follows of his intrigues with 'the lady Spes,' wife of one Sigurd, but apparently herself a Greek. 'In those days Harald Sigurdson (Hardrada) was at Miklagarth.' Thorstein and Spes go to Norway, and finally on pilgrimage to Rome. (2) *Slaying Stir's Saga*, under A.D. 1008-9, records how Guest Thorhallson, after a manslaughter, flies to Norway and Miklagarth, where he prospers and remains. Thorstein Stirson pursues but never finds Guest. (3) *Laxdale Saga*, chs. lxxii.-iii., lxxvii., tells how Bolli Bollison goes abroad into southern lands, 'for a man he thought was deemed benighted who learnt nothing further a-field than what was to be seen in Iceland.' Bolli journeys to Norway and Denmark, and so to Miklagarth, where he stays many winters. He enters the Vaering Guard, and is the first Northman to 'take war-wage from the Greek King,' who on parting with him bestows

United action became impossible, and most of the Byzantines went back to the capital; while Harald and his men started to attack the Moslem lands of North Africa, the richest country upon earth, as the Northmen believed. Here followed many years of war and plunder and adventure, and Harald gained 'exceeding much gold and all kind of precious goods.' But the whole of this wealth which he did not need for his own use, he sent back to Novgorod by the hands of trusty men, and stored it up in the keeping of his friend the Grand Duke Yaroslav. 'And after that he went to Sicily and took there various Burgs with cunning devices; and then fared he back to Constantinople with his host, and arrayed his journey for Jerusalem. All his wage-gold from the Greek king he left behind, and so did the Vaerings who went to the Holy City with him. Thus he came out to Jerusalem-land, and there all towns and castles, aye, and all the country, came unburnt and unwasted into the power of Harald. And he went to Jordan and bathed him there, as is the way of Palmers, and bestowed a great wealth on the Grave of the Lord, and on the Holy Cross, and on the other relics of Jerusalem-land. Moreover, he made safe the road to Jordan, and slew folk that plundered and wasted there.'[1]

But when he came back to Constantinople he heard that Magnus Olafson[2] had become King of Norway and Denmark. So he 'gave word' to the King of the Greeks that he would leave his service. Then Zoë charged Harald with

many splendid gifts, clothes of fur, a scarlet cape, a gilt saddle, and other things. (4) *Burnt Njal's Saga*, ch. lxxx., narrates the journey of Kolskegg through Russia to Miklagarth, his taking service under the Emperor, and his becoming Captain of the Vaering Guard.

[1] See chs. vi.-xii. of *Harald Hardrada's Saga*.

[2] His 'brother's' son; a mistake of the Saga for 'half-brother.'

keeping back some of the booty which belonged to the emperor; and on this charge the captain of the Varangian Guard was thrown into prison, and put in danger of his life. But with the aid of St. Olaf, or by his own skill and strength, he escaped from this peril, as from so many others before, and took a savage vengeance as he fled.

At the head of his men he forced his way into the sleeping chamber of the emperor and blinded him;[1] then, seizing the Princess Mary—whom he had fruitlessly wooed, and for whose sake Zoë had become his enemy—he sailed out into the Bosphorus with two galleys.[2] Skilfully 'leaping' the chain that guarded the entrance, he passed over the Euxine and the Sea of Azov,[3] and so returned, probably by way of Kiev and the Dniepr, to Novgorod and the Court of Yaroslav the Law-giver.

Here he took into his own keeping all the treasures which he had sent before from the South, 'and in that hoard was so much wealth that no man of the North had ever seen such in one man's owning.' For besides what he had won in battle, Harald had the good fortune to earn palace spoil three times in Constantinople, and 'that is the custom whereby, when the Greek King died,' his Vaering Guards might go over all the king's palace-hoards and take whatsoever they could carry off.

That winter Yaroslav gave Harald his daughter Elizabeth; and towards spring Hardrada and his wife journeyed from Novgorod to Ladoga, equipped a fleet, and sailed away from the East. First the adventurers turned aside into Sweden, where they met with one Svein, after-

---

[1] Chs. xiii. and xiv. of *Harald Hardrada's Saga*.
[2] Harald weighted his vessel, first behind, till his prow was half across the chain, and then before, so that the galley fell forwards into the sea on the other side, the crew rushing first to one end, then to the other.
[3] 'Ellipalta,' chs. xv., xvi. of *Harald Hardrada's Saga*.

wards famous as Svein Estrithson or Wolfson,[1] King of Denmark (1047-1074). Harald and he formed a league, plundered Zealand and other coasts of Denmark, and threatened Norway; Magnus in alarm called out his host against them; but before joining battle he offered Harald, as his kinsman, one-half of the kingdom, and a division of all the loose wealth they had between them. The offer was accepted, and so Hardrada came home to Norway as joint king. 'And to Magnus and his court men he gave wondrous things from the goods of the Greek King'— a mazer girt with silver, and a bowl of silver gilt filled with silver pennies, and two gold rings that together weighed a mark, and a cloak of brown purple lined with white skins— and other noble presents from that land 'where men say there are houses full of red gold.'[2]

And not long after this Magnus died, and Harald became sole king.[3] We have nothing to do here with his wars against Denmark and his old ally Svein Estrithson; but his friendship with the Icelanders is another matter. For 'great are the tales of King Harald that are set forth in those songs which Icelandmen brought to him or his sons, so firm a friend was he to all the folk of that land. Yea, once when there was a dearth in Iceland, Harald sent thither four ships with meal, and caused that meal to be sold right cheap; also he suffered the poor folk who could get victuals over sea to fare abroad; and he sent to Iceland a church bell,' to the same church for which St. Olaf had given[4] the wood. The king's Iceland friends, Thorleik

[1] See chs. xvii.-xviii. of *Harald Hardrada's Saga*, and the verses of Valgard of the Mead : 'Thou steer'st forth thy ship, O Harald, 'Neath fairest freight; thou sailest, With bottomless gold from the east land, From Holmgarth.' . . .

[2] All these, we may suppose, came from his southern treasures, though the skins might be Russian or Finnish.

[3] A.D. 1047.

[4] Cf. ch. xxxvi. of *Harald Hardrada's Saga*. A bell given by St. Olaf to a church in Iceland is also recorded.

the Fair, Haldor the son of Snorre, Wolf the son of Uspak, and the rest left deeply printed on the Icelandic poetry and chronicles the tradition of Hardrada's splendid personality. 'Masterful was he, and given to rule, a mighty warrior, and the boldest under weapons, more strong and deft in war than any other man; and so sage and cunning withal that, as all men said, no lord ever was in northern lands so deep-witted or so nimble of counsel.' But all men were not so contented with his reign of iron. Like the chief kings of Norway before him, Harald set himself to break down the power of the landowners. He cut off or drove into exile, one after another, all the nobles who opposed him; and the landsmen, says Hardrada's own Saga,[1] wanted nothing but a leader to rise in open revolt against him. Several of the exiles fled to Svein Estrithson in Denmark, and took service with him against the Vikings of the East Ways, the Wends, and the Courlanders, as well as against Norway; others took refuge in England or Normandy,[2] or went on Viking cruises in the Western Seas.

One curious incident is the intercourse between King Harald and his nephew Guthorm of Dublin, who often came over to see his uncle, but whose business was freebooting in the West, where he joined the Irish chieftain Margath[3] in ravaging the coasts of Wales, Cambria, and Cornwall.

After a long[4] but indecisive struggle of seventeen years with Svein, Hardrada made peace 'on these terms, that he should have Norway and Svein Denmark, unto the land-borders which of old had been between Norway and Denmark'; and it was now, in the ensuing time of rest,

---

[1] Chs. xxxvi. xlv. of *Harald Hardrada's Saga.*
[2] See chs. l., li., liii.-lv. of same.
[3] Otherwise Margad. Chs. lvi., lvii. of same.
[4] 1047-1064 A.D.

HARDRADA'S LATER ENTERPRISES

that Harald seems to have projected those Greenland, Vinland, and Polar journeys which made his name famous in the North as that of no other king had been.[1] But all these schemes were cut short by the fatal English venture of 1066.

Here after several triumphs Harald failed and perished; but in falling he brought down his foe with him. The two chief Norsemen of the world threw themselves on England in the same year and season; to crush the Northern invader the South had been left bare; and William of Normandy landed in the same days of early autumn which ended the struggle in Yorkshire.

---

[1] See Adam of Bremen, *Gesta Pontificum Hammaburgensis ecclesiae*, iv. 38 (247, Lappenberg). As to the Northmen's more distant wanderings, we may notice :—(1) About A.D. 1000, Gris Saemingson is recorded as the first Scandinavian merchant in Constantinople; cf. *Antiquités russes*, ii. 113; Heyd, *Commerce du Levant*, i. 74. (2) Before the conversion of the Northmen, Visby in Gothland was a great centre of commerce; and hoards of Anglo-Saxon and Arab coins, anterior to A.D. 1000, bear witness to this; cf. Bonnel, *Russisch-Livlandische Chronologie, Commentar*, p. 24; Heyd, *Commerce*, i. 63, 64. (3) Scandinavian, and especially Swedish, intercourse with Russia, particularly with Novgorod and its fairs, 860-1060, is shown by many details in Kunik, *Berufung der Schwedischen Rodsen*, ii. 131, etc.; Rafn, *Antiquités russes*, i. 295, 359, 432; ii. 119. (4) According to one tradition, the Goths of the Crimea originated in a ninth-century migration from Gothland, passing through 'Gardariki'; cf. Riant, *Expeditions des Scandinaves en Terre Sainte*, 64-5. (5) On the 'Majuj,' or Vikings, in Spain, their invasions of 844, 858-61, 966-71, and 1014, and embassy from Abderrahman II. to distant isle (Orkneys?) inhabited by these 'Magogs,' cf. Dozy, *Recherches*, ii. 252-340; Sebastian of Salamanca, ch. 26; Dudo of St. Quentin, in Duchesne, *Hist. Norm. Script*. 144 c.-152 A. In 844, besides attacking Brigantia Pharos (near Corunna), threatening Lisbon, and raiding Seville, the Northmen made descents on Africa, where a harbour near Arzilla was long known as 'the Majuj port.' Abderrahman II.'s envoy, Al Ghazal of Jaen, went from Silves by sea to Pirates' Court and returned *via* Compostella. In 858-61 the Vikings passed Straits of Gibraltar and went as far as 'Greece'; even before Charlemagne's death (814) they had appeared off the Narbonnese. In 1014, they apparently sacked Tuy, which some identify with St. Olaf's 'Gunnvaldsborg'; Olaf's 'Seliupool' has likewise been conjecturally fixed at estuary of Sil or Minho. (6) On early Norse descents in Erin, especially in A.D. 851, 852, 854, 859, 866, 869, 870, cf. *Annals of Ireland*, Dublin, 1860, pp. 159-163, 169, 175, 185, 189, 193.

CHAPTER III

PILGRIM TRAVEL

IN the first part of this survey of Mediaeval Geography we brought the record of pilgrim-travel to the close of the ninth century. The present chapter endeavours to supply an account of this religious exploration during the next 350 years, reaching to the end of the Crusading era, the fall of the Bagdad Caliphate, and the establishment of the empire of the Mongol Tartars as the dominant power in Asia and Eastern Europe. This long period may be subdivided into three; the pre-Crusading time, closing with the election of Gregory VII. to the Roman See; the time of the first three Crusades, closing with the Saracen recapture of Jerusalem in 1187; and the time of the later Crusading struggles, closing with the middle years of the thirteenth century, when the hopes of the Latin Kingdoms in the East were finally overthrown, and when new forces had worked a revolution in the relations of Europe with the Levant, Central Asia, and the Further East.

The first of these three epochs is naturally the least interesting. Beginning with the era of the collapse of the Frankish Monarchy of Charles the Great, it ends with the accession of that great Pontiff, under whose inspiration the Crusading movement finds its true commencement. With this movement opens a new era for Western pilgrims. Masters at length of the land so dear and so coveted, lords of the house where they had so long been servants, the Christians of the Latin world flocked in crowds to the recovered capital of the Holy Fields. On the other hand, in the two centuries before this revolution we have but a few records of Syrian pilgrimage, showing little activity in comparison of some earlier times, the age of Constantine, the age of Justinian, or the age of Charlemagne. And it would be unreasonable to expect otherwise. There is no darker century of Christian History than the tenth—at least in the lands west of the Adriatic; practical enterprise and scientific interest were never more nearly extinct in the races whose only bond of union was the Roman Christianity, now becoming so sharply antagonistic to the Churches of the East. Where men should have looked for a leader and a gospel, there was now but a very sorry ghost of the ancient Empire, sitting with crown and sceptre among the ruins of pagan power. For this was the age of the Pornocracy at Rome, and the Imperial Church was discredited and incapacitated for leadership by a succession of 'human monsters' (in the words of Baronius),[1] who neither took thought for decent morals nor good government, but only for the favour of Theodora, Marozia, and other 'patronesses,' for the enjoyment of such poor luxury as the age afforded, and for a complete immunity from the cares of administration. The Imperial Frankish State was a victim to the same effeminacy, and the same lack of governance; and the result to the Temporal power was a complete suspension for nearly a century of the Imperial name and fact in the West, the collapse of the first Romano-German

[1] 'Homines monstruosi, vita turpissimi, moribus perditissimi, usquequaque foedissimi.'

kingdom of Charles the Great, Charles the Bald, and Charles the Fat. From 888 to 962 this interregnum lasted, and when at last the vitality of the German people once more made head against internal discord and external invasion, the Empire which Otto I. restored was even in theory a narrower and less powerful body than the great State which, to all seeming, Charlemagne had so firmly welded together. The West Frank region, the France of the later Middle Age, was now for ever detached from the Eastern and more purely Germanic lands; but this detachment gave at first little promise of a strong and well-ordered state on the west of the Rhone, Saône, and Meuse. Nowhere in Christendom was the paralysis of the central power and the victory of political feudalism more complete. In Spain, at the close of the first millennium, the ruin of the Christian kingdoms seemed almost as absolute as in the eighth century. Barcelona and Compostella were both captured by the 'Hagib Almanzor,' the great Moslem general of the Caliph Hisham of Cordova (976-1002); Leon was overrun; the Christian advance in the central plateau was beaten back; and the Western Caliphate once more reached to the Pyrenees and the Asturias. In England, the development of Wessex, under King Ælfred and his first successors, gave better hopes of a united realm covering the main part of Britain, absorbing the settled Danes and Northmen, and bidding defiance to fresh Scandinavian attacks; but the last quarter of the tenth century blighted all this prospect. The new English Kingdom fell into hopeless decrepitude under Æthelred II.; and a similar decrepitude under Otto II. also threatened the revived German Empire at this very time. On every side the history of Latin Christendom in the tenth century is a sombre record.

The weakness, the comparative anarchy, and the

III.]   OUTLOOK AT CLOSE OF FIRST MILLENNIUM   115

ignorance of one civilisation is often in strong contrast with the firm rule and developed culture of others; and at this time Irish, Norse, Byzantine, and Moslem politics and society appear to absorb all the light that has passed away from the Latin world. The people, church, and schools of Ireland close the most brilliant period of their history just as the rest of the mediaeval Christian states are entering upon the Renaissance of the Crusading age. In the terrible pressure of Viking attacks a higher political life was developed, and while England was falling under Dane and Norseman, Ireland seemed to be reaching its unity as a great kingdom in the victory of Clontarf (1014). All through the dark age of the Imperial interregnum and the Saxon Dynasty the Irish school of art and literature, even of geography, was probably superior to any other in Western Europe; and here we may remember that no map of the earlier Middle Age has the accuracy of one attributed to an Irish scholar-monk.[1]

Still more remarkable is the position of the Byzantine Empire at this time (880-1025). In an age of general Catholic and Latin weakness and barbarism the Orthodox, Greek, or Eastern realm shows itself a better inheritor of the name and glory of Old Rome: its Church preserves a greater decency and displays a far larger measure of activity; its culture and literature suffer even less in the comparison. While Islam is crushing for the second time the Nazarenes of Spain, the Nazarenes of Nikephoros Phokas, John Tzimiskes and Basil II. are winning victories in the Levant, and lightening the hostile pressure upon Christendom at large. While the missions of the Latin Church are neglecting to carry forward the work of St. Ansgar in the North, or of Cyril and Methodius

[1] The 'Cottoniana.'

in the East, the Russian lands are won to the Orthodox Faith by the more persuasive influence and more gorgeous ceremonial of the New Rome. In contrast to the atrophy of all higher literature among the Franks of the tenth century, the writings of Constantine VII., *Born in the Purple*,[1] are worthy of more notice than they might claim at another time; for his works are typical of a school of culture which could hardly be found in Italy, France, Germany, or England in this age. They are typical also of a national and administrative activity which soon after recovered Crete and Cyprus, Antioch and Edessa, the suzerainty of Armenia and the Crimea, and the ancient Imperial paramountcy in all those Thracian lands which had been overrun by the migration of the Turkish Bulgarians from the Volga.

Once more the misery and poverty, the inaction and barbarism of Latin countries in the later ninth century, and the whole of the tenth, contrasts with an extraordinary activity of Northern and heathen peoples. The Scandinavian Vikings are not only the ravagers of Frankland and England; they are the superiors of Franks and English in the arts of life and in literary instinct, as much as in military skill; above all, their spirit of enterprise and movement marks them out as the leaders, if they were not to be the supplanters, of European civilisation.

Outside the limits of the Christian world, Islam claims in this period as much glory for its science and letters, as much praise for its brilliant and tolerant culture, though hardly as much attention for its universal ambitions, as in the first centuries of the Caliphate. But in spite of all appearances, the evidence of political, if not of social, decay is now becoming too strong to be ignored, and when the strength of Islam is regenerated for a time by Moors

[1] Porphyrogennetos.

and Berbers in the West, and by Turks in the East, it is a regeneration of arms and manhood, but not of science or of culture. Neither Seljuks nor Ottomans could found a second Bagdad; neither Almoravides nor Almohades could permanently restore the prosperity of Cordova and Kairwan. Yet, until its political ruin in the thirteenth century, the capital of Moslem Spain retained much of its intellectual glory; the name of Averroes was linked with the closing years of Mohammedan rule in the valley of the Guadalquivir; and even in its last and weakest time the Abbasside Caliphate maintained a centre of Moslem learning at the *Home of Peace* upon the Tigris.

From this survey we may gather that the Latin Geography of the later ninth, the tenth, and the early eleventh centuries, will prove to be a chronicle of little interest; on the other hand, this chronicle, the record of Western pilgrim-travel, is the necessary introduction to a decisive period; and however meagre in its interest or debased in its form, yet it is saved from oblivion by a certain rough abundance of material.

We must begin by briefly reviewing the somewhat arid though lengthy catalogue of journeys, religious, diplomatic and other, which alone furnish us with a sketch, one cannot say a picture, of the relations between the Holy Places of Syria and the nations of Western Europe, in the troublous times we have now reached. The Pontificate of John VIII. opens a new stage in these relations, about the year 880. For now we have, for the first time, a definite pronouncement from the Apostolic See on a point of high practical importance to the would-be Crusaders of that time which prepared men for the Crusades. All warriors killed in fighting against Pagans and Unbelievers—were they, whatever their sins, certain of Eternal Salvation? This question had been somewhat disputed, and the affirmative answer of John VIII. was a victory for the spirit of Islam within the bosom of Christendom. Before the rise of the Moslem danger, the balance of opinion in the Church would probably have been against such an unconditional absolution; but the seventh century had worked a change in the most peaceful of Catholics; and a repetition of seventh-century woes seemed again to threaten Christendom, as men saw the Karling empire dissolve. The papal pronouncement on this matter (c. 876-882) may be specially connected with the embassies of Theodosius, Patriarch of Jerusalem, and of Elias III., his successor, to the Church of Rome, the 'kings of the race of Charles, and the Clergy of the West' (A.D. 878, 881).[1] Of these missions the former made a long stay in Italy; as late as the 2nd of May, 879, the envoys are still with Pope John, who writes to Jerusalem excusing the delay, and promising

---

[1] On these, (1) 'John VIII. to the Bishops of Louis ["le Bègue" or "the Stammerer," King of France],' cf. Mansi, *Concilia*, xvii. 104; on (2) the 881 embassy of Elias III. to 'Charles the younger, Emperor, and to all bishops, princes, and nobles of the Kingdom of Gaul,' (April-October), cf. AA. SS., *i.e. Acta Sanctorum* (May iii., xlii.); *Spicilegium Acherianum*, ii. 370; on (3) that of Theodosius, cf. Jaffé, *Regesta Pontificum Romanorum*, 2462 (= Jaffé-Wattenbach, 3242; 'Theodosius, David, and Sabas, sent to John VIII. by Theodosius, Patriarch of Jerusalem, 878'); for John VIII.'s excuses on account of delay, cf. AA. SS. May vii. 699; Jaffé, 2462 (*Archives de l'Orient Latin*, i. 22-31.). Count Riant well contrasts the declaration of John VIII. with the letter of Adrian I. to Charles the Great (778) when the Frank King was starting on a Spanish campaign. In this no allusion whatever is made to this as a Holy War, though waged against Moslems. Leo IV., in 848, is the only Pontiff who in any measure anticipates John VIII. Calling the Franks to the help of Rome against the Saracens, he speaks of the certainty of Celestial Reward for any one killed in defending the capital of Christendom. But (1) the defence of Rome was a special matter and quite different from a general state of war against Saracens; (2) Leo's language is vague; (3) John VIII.'s pronouncement includes aggressive as well as defensive war. In 963, at the instigation of St. Polyeuctes, Nikephoros Phokas issued a similar edict, declaring that the soldiers killed in his Syrian wars were martyrs of the Church. (Cedrenus II., 368 (Bonn ed.); Zonaras, bk. xvi.).

in explanation a consignment of money and necessaries for the faithful of the Holy City. The Eastern Brethren must remain to take back these gifts; they only awaited the complete collection of the same. As to the second mission of Elias III. in 881 (entrusted to a pair of monks, named Gisbert and Reinard, who from their names must be Western and not Syrian ecclesiastics), this was avowedly for the collection of subsidies to free the Church of Jerusalem from various building debts. The credentials of this embassy, a letter from the Patriarch to Charles the Fat, last of the immediate successors of Charlemagne, gained some notoriety in the West; a century later it was known in England, and it is noticed of Asser the historian that he had read the document in question. About the same time as this appeal to the generosity of the Franks, and probably in the year 883, occurs the famous journey of the West Saxons Sighelm and Æthelstan. These 'gift-bearers,' probably priests, 'carried to Rome the alms which King Ælfred had vowed to send thither, and also to India, to St. Thomas and St. Bartholemew.' Between Rome and Malabar the mission probably visited Jerusalem, and the English king, in return for his offerings to these distant churches, received presents of silks, spices, gums, and Eastern shawls.[1]

All these journeys to and from the Levant have a certain similarity; they are all connected with the poverty, and so with the relief, of the subject Christians of the East; and they may all be paralleled by earlier missions. Pope John VIII. received the envoys of Theodosius and Elias III., and helped them to collect money in the West, just as earlier popes had done in past time: Martin I. in 649 and

[1] Cf. William of Malmesbury, *Gesta Regum Anglorum*, bk. ii. § 122; Florence of Worcester, *Chron.*, A.D. 883; Roger of Wendover, *Flores Hist.*, ed. Coxe, i. 154; Sighelm returned, and afterwards became Bishop of Sherborne.

652-3; Gregory the Great in 596, 600, and 603; Leo the Great in 453; Leo III. in 799, 800, 803, 807, and 809; Gregory IV. in 834 and 840. On all these occasions 'servants of God' from the Holy Land brought the letters, the appeals for money, and the requests for military or diplomatic aid, which the Eastern Christian poured out with ever increasing frequency after the Moslem conquest of Syria and Egypt. This was one side of the correspondence. The other was shown in the Papal embassies to Jerusalem, such as these of Julius I. in 342; of Anastasius I. in 400; of Innocent I. in 417; of Coelestine I. in 430; of Leo the Great in 453, 454, and 457; of Gregory the Great in 591, 597, 601, and 603; it is also manifested in the Frankish missions, of 765 from King Pepin, and of 797, 799, and 803 from Charles the Emperor. Of these last, the greater number (if not all) probably visited Jerusalem, but as a rule they seem to have been in the first place addressed to the Caliph at Bagdad. Thus they called out the return of courtesies from the Mussulman side, in the missions of 800, 802, and 807, from Harûn-al-Rashid,[1] just as the embassy of 825 from Lewis the Pious[2] had its response in that of 831 from Al-Mamûn. In the period we have now reached, this diplomatic intercourse is almost the only evidence of travel from

---

[1] The last of which is especially famous.

[2] Entrusted to the monk Raganarius. On the papal embassies to Syria (Jerusalem) above noticed, cf. Jaffé, *Regesta Pontificum Romanorum*, 81, 120, 156, 277-8, 291, 302, 728, 1127, 1396, 1515, 2462 (= Jaffé-Wattenbach, 282, 325, 373, etc.); for similar Roman letters and missions to Antioch, *ibid.* 102-7, 156, 169, 171, 993, 994, 1105, 1112, 1209, 2029, 2493, 3287, etc., extending from A.D. 415 to 1054. On the appeals to Rome *from* Syria, cf. *Acta Sanctorum*, May iii., i.-lxxii., *Tractatus praeliminaris de episcopis et patriarchis S^{ae} Hieros^{ae} Eccl^{ae}*; as well as the Appendix in the vol. May vii., 696-706, of the *Acta* (properly an addition to the vol. May iii.), on the Bishops and Patriarchs of Jerusalem; *Archives de l'Orient Latin*, i. 26-38, and especially 27-28; and Jaffé, *Reg. Pont. Rom.*, 291, 1110, 1396, 1515, 2462, etc.; Jaffé, *Monumenta Carolina*, 382, 383, 386.

Western Europe to Western Asia, and we are obliged therefore to notice in some details these meagre traces, which alone survive, of geographical movement in the tenth century.

About the year 900, the Patriarch Elias III. makes another appeal to the Christian West—and in this appeal he is joined by Pope Benedict IV. Their Holinesses address two encyclicals to the faithful; and both these letters introduce to the charity of good Christians a certain Malacena, Bishop of Amasia, and his mission of ransom. Some of his monks had been taken captive by the Turks of the Caspian, and he had started for Europe to collect the means for regaining their freedom. The rich and powerful Christendom of the West is again called to aid the elder sister of the East— on evil times now fallen, and evil tongues. And yet one thinks of Western Christendom, at this time, as in the darkest of its dark ages.

In 932 a letter from the 'Bishop of the Holy City' was read at a council in Erfurt; but this is the last trace for a time of the Franco-Syrian connection and of the quasi-protectorate of the Western Emperors over Christian interests in Palestine which had been marked by so many incidents:—by Charlemagne's buildings in Jerusalem, for the comfort of Latin pilgrims; by his acquisition of the Keys of the Holy Sepulchre as a friendly gift from the 'Just' Harûn; by the landed possession of the Syrian Church in France and Italy;[1] by the frequent embassies from Syria

---

[1] Thus the Church of Jerusalem had lands of its own at Neuvy and Mauriac, and apparently other real property held in common with the Abbey of Conques, cf. *Archives de l'Orient Latin*, i. 28; the Church of Sinai also had property in Normandy, cf. AA. SS., June i. 87-104, especially 89; Pertz, M.G. SS. (=*Monumenta Germaniae, Scriptores*), viii. 209, 210, 211, cf. AA. SS., June i. 87-104, on the life and travels of St. Symeon, the 'recluse of the Gate of Trier.' He is sent (c. A.D. 1034-35) from the Monastery of Mount Sinai into the West to collect moneys pro-

to Frankland and from Frankland to Syria; as well as by the alms sent yearly from the Karling Court to Jerusalem during a great part of the reigns of Pepin, of Charles, and of Louis the Pious.

Before the end of the ninth century this quasi-protectorate had passed from the Franks to the Byzantines; and during the next hundred years successive emperors of Constantinople asserted more and more closely their position as guardians of the subject Christians of the Levant. Thus Constantine VII.[1] in 947 sends gifts to the Patriarch of Jerusalem; and in April of the same year the clerk Niketas, in a letter to the Divine Emperor, inheritor of the name and virtues of Constantine the Great, describes the Easter Eve miracle of the Holy Fire, and the recent attempt of a 'cursed Emir' to abrogate this festival. Several matters of later date show the continuance of this Byzantine influence; the re-conquest of a great part of Syria by John Tzimiskes, a striking though temporary proof of the revived power of the Old Empire (968-975); the introductions which Fulk the Black of Anjou carried from Pope Sergius IV. to Basil II. (1011); and the treaty of 1021 with the Fatimite Caliph of Cairo, whereby the position of Constantinople as spokesman and champion of Eastern Christianity was fully recognised. Another proof of the same fact may be drawn from an interesting letter of Pope Victor II. to the Empress-Regent Theodora, in which

mised by Richard, Duke of the Normans. One Eberwin, returning from Palestine with a certain Richard, Abbot of Verdun, meets him at Antioch. Symeon tries to go with Eberwin and Richard overland, but is stopped at Belgrade; he returns to the sea, and so goes by Rome to 'Francia,' comes to Rouen, finds Duke Richard just dead, and stays with Abbot Richard and Eberwin. Poppo, Archbishop of Trier, soon after goes to Palestine and takes Symeon with him and back. Finally Symeon makes himself a cell at the *Porta Negra* of Trier, and dies in great sanctity.

[1] Porphyrogennetos or 'Porphyrogenitus.' For the interesting letter of Niketas, 'least among clerics,' cf. *Archives de l'Orient Latin*, i. 375-382.

the Pontiff complains of the exactions which the Greek guardians of the Holy Places levied upon Western pilgrims (1054-1057). A recent journey of Bishop Helinand of Laon was probably the immediate reason of this appeal; the story of his troubles in Jerusalem had turned back Lietbert of Cambrai after a toilsome pilgrimage as far as Laodicea; and loud and bitter were the complaints of Greek 'intolerance' now heard at Rome.

Once more, the well-known tract, *On the Situation of Jerusalem*,[1] which by some critics has been assigned to the epoch of the First Crusade, more probably belongs to the period of Byzantine dominion in Syria during the middle of the tenth century, and this period is also marked by various imperial letters to distant rulers,[2] expressive of the revived power of the Greek rulers in Western Asia.

Towards the close of the tenth century we have rather more frequent notices of Palestine pilgrimages from Western and Northern Europe. One St. John of Parma visits the Holy Sites six times, especially in 982. In 985, 986, and 988, the same journey is made by various monks from Monte Cassino, John of Beneventum, Leo, brother of Abbot Aligernus, and others, together with a certain Count Malefrid and his son. In 987 two Icelanders, Thorvald Kodranson and Stefnir Thorgilson, make their appearance at the Sepulchre of Christ—early results of that conversion of the Northmen whose consequences for Europe, for civilisation, and for Christendom were so deep and so far-reaching. Not without dramatic interest is the scene called up by this brief notice; the picturesque historian

---

[1] *Qualiter sita est civitas Hierusalem*.

[2] *E.g.* from John Tzimiskes to Leon, governor of Daron, probably genuine; and from the same to Aschod, King of Armenia, probably spurious. For Victor II.'s complaint, cf. Migne, *Patrologia Latina*, cxlix. 961-962; Jaffé, *Reg. Pont. Rom.*, no. 4015.

might well enlarge upon the meeting of the huge-limbed, fair-haired, blue-eyed sea-rovers from the North with those Christians, knights and merchants, priests and monks, whose most terrible enemies they had been, whose best allies they were yet to be.

About 990 Poppo, Abbot of Stavelot, and in 997 Frederic, Count of Verdun, each with a party of attendants, follow in the same 'holy road.' In 993[1] another noble, Hugh, Marquis of Tuscany, and his wife, Juliette, send[2] gifts of value to the Tomb of the Lord and to the Monastery of Sta. Maria Latina in Jerusalem; while about the same time (995) Orestes the Patriarch despatches an embassy to Pope John XV. (after the fashion of an earlier period), asking questions about various matters of ritual and discipline. A more doubtful record assigns to this epoch an urgent appeal from the 'immaculate Spouse of God' to the sympathy and succour of the West. This outcry against an even more unbearable Saracen oppression professed to come through the famous Gerbert, afterwards Pope Sylvester II.; but it is probably one of those rhetorical exercises to which the Crusading movement gave rise, and it may be referred to an epoch sixty or seventy years later than the great Pope who has so long stood sponsor for its eloquence. The story of Gerbert's pilgrimage need not perhaps be altogether discredited (c. A.D. 986); but his famous letter is surrounded by too many difficulties for modern enquiry to accept it, like Michaud,[3] as a 'precious monument' of the first pope who expressed the Crusading sentiment.[4]

---

[1] *Charta* of St. Victor of Marseilles, in Martène, *Amplissima Collectio*, i., 347-348.

[2] 29th October 993.

[3] Michaud, *Bibliothèque des Croisades*, ii., 491, ed. of 1829.

[4] Cf. *Archives de l'Orient Latin*, i., pp. 31-38, especially 34, note; 35, note.

With the eleventh century, as we have already noticed, a new age begins: old things are passing away; all things are becoming new. It is the same in travel as in everything else. Pilgrims are multiplied tenfold; and in the words of Ralph Glaber, 'first went the meaner folk, then men of middle rank, then many kings and counts, marquises and bishops; yea, and even women also, a thing unknown before.' Emmerad, a monk of Anzy;[1] Makko of Constance, about the year 1000, passing from West to East, cross the path, but without meeting the person, of Simeon the Armenian, who in the same year goes from East to West with the commendatory letters of the Patriarch Arsenius. We do not know if, at the Holy Sepulchre, they met with some of the many pilgrims from Georgia, who also entered 'Sacred Hierosolyma' this year. But this is only one occasion out of many when such a meeting between the furthest East and the furthest West of the Christian world may have taken place, or did actually happen by the side of the tomb of Christ. In 1002-3 an unnamed monk of Tours, a hermit from Bamberg, and the terrible *Black Count of Anjou, Fulk Nerra*, are the chief figures of our Travellers' Register. Fulk made in all three journeys to Syria; his death in 1040, at Metz, was the closing incident of the third and last of these expeditions. On his second visit (it is said) he had to buy admission to the Holy City for himself and his comrades. He was only allowed to approach the 'all holy grave of the Lord' on promising to insult the Cross; evading this ordeal, he contrived to bite off a piece of the rock with his 'teeth of iron,' and returned with this relic to his own country, where he

---

[1] Emmirad, Emirard or Emmerad, passes nearly seven years at Jerusalem according to AA. SS. April ii., 770. | The *Acta* for Jul. vi., 327, say of Simeon the Armenian that he also made the pilgrimage to Rome.

presented it in triumph to his Abbey of Beaulieu. Pilgrimage, like adversity, sometimes made strange companions; in 1003 we have the legend of one Guy, a hermit from Brabant, appearing at Jerusalem *at the same time* as Olaf Tryggveson, the first Christian King of Norway.[1] A whole flock of Scandinavian devotees were said to have followed soon after; and two French abbots, Gauzlin of Fleury in 1004, and Roger of Figeac in 1005, probably met some of these Northmen at the Holy Sepulchre. Many travellers perished on the way; Leger the deacon of Auxerre, who was 'given the sea for his grave'; Andrew the knight, buried at Jerusalem itself; Hictarius 'the blessed, taken from earth by the mercy of God,' before he had reached the tomb of Christ.[2]

A terrible event startled the pilgrim world in the first years of the eleventh century (1010). Hakim Biamrillah,[3] Fatimite Caliph, and temporal sovereign of Palestine, in that madness which the Druses recognise as a Divine

---

[1] This story is quite legendary. Olaf Tryggveson died in 1000 at the battle of Svold; but his people clung to a belief in his escape and continued life in exile, and the tale of his pilgrimage to Jerusalem formed a part of this patriotic myth.

[2] Cf. Migne, P.L., cxxxviii. 1229, 1232, 1252. On the other journeys of St. John of Parma, etc., etc., here referred to, cf. *Archives de l'Orient Latin*, i. 34, note; and AA. SS. Boll. May v. 179-180; [July iv. 281]; April ii. 770; July vi. 327; Sept. vi. 722; Sept. iv. 43; also Pertz, M.G. SS., vii. 63, 636-642; xi. 295; iv. 49; xx. 635; ix. 59; *Thorvald's Saga*, ch. x., ed. Arne Magnaeus, pp. 334-7; Constantine Porph., *De administrando imperio*, Bonn edition, p. 199; *Fornmanna Sögur*, iii. 32-34; x. 370; N. Arch. f. ä. d. G., iii. 353; Mabillon, Ann. O.S.B., iv. 170; Riant, *Expeditions . . . des Scandinaves en terre sainte*, 102-5, 117, 118; *Itin. Lat. Hieros.*, II. *series chronolog. peregrin.* A.D. 1064; *Recueil des historiens de la France*, x. 282; x. 15 (= Ralph Glaber, ii. 4, on Fulk). See also page 217 note.

[3] On Hakim, cf. what Nasir-i-Khusrau says at the end of his account of Syria. On the destruction and reconstruction of the Holy Sepulchre, cf. Ralph Glaber, *Chron.*, iii. ch. i.; Will. Tyre, i. ch. iv.; Cedrenus, ed. Bonn, pp. 515, 521: Zonaras, xvii.: Makrizi, in De Sacy, *Chrestomathie arabe*, i. 98.

Emanation,[1] destroyed the Christian buildings at the Holy Sepulchre; a few years later they were reconstructed (A.D. 1021); but a wild cry of vengeance was raised by many Catholics, and the Vandalism of Hakim unquestionably hastened the movement which, slowly gathering strength during the next eighty years (1020-1096), burst upon the Levant with the armed strength of the Crusaders. The treaty of 1020-1, which put the Syrian Christians under the special protection of the Byzantine Emperors, has been already noticed, and is significant of the increased prestige and power of Constantinople; but it is perhaps even more noteworthy, in view of future events, that so many Western and Latin sovereigns now began to take up again the part of Charlemagne in Palestine.

Thus the newly-converted Kings of Hungary, and especially the great St. Stephen, came forward as founders and benefactors, and more than one hospice for Roman pilgrims was due to their liberality.[2] Their example was followed in 1063-70, by the Republic of Amalfi, which shared so prominently with Pisa and Venice in maintaining the naval department of the Crusading armaments a generation later. The Norman Dukes also showed a keen interest in the Christian East; and Richard I., William Longsword, and Richard II. all sent messengers with alms and presents to

---

[1] The letters from the Jews of Western Europe inciting him to this action, which have passed into documentary history, were probably forgeries of a later date, invented to justify the anti-Semitism of the Crusading Era itself; cf. Adhemar of Chabanais, in Pertz, M.G. SS., iv., 137; viii. 210, 399; Ralph Glaber, iii. ch. vii.

[2] For Hungary, on joining the Latin communion, became a main route from Western Europe to Syria. 'Nearly all those who wished to go from Italy and Gaul to Jerusalem began to forsake the accustomed way by sea, and to make their passage through this king's country (Stephen's). He made the way very safe for all, and thus allured by his benevolence a countless multitude, both of noble and common people, to start for Jerusalem.' R. Glaber, *Chron.*, iii. c. 1 (M.G., SS., vii. 62).

the Grave of Jesus. An embassy of Richard II.'s in 1026 is specially recorded: on this occasion a namesake, the Abbot of Grace Dieu, made the pilgrimage on the Duke's behalf. Like Pope Gregory I., Richard II. also exchanged gifts and greetings with the monks of Mount Sinai. His successor, Robert the Magnificent, father of William the Conqueror, did even more. In 1035 he left his Duchy and went in person to Jerusalem; on his return he died at Nicaea, and was laid in the Church of the Panagia in that city.

Once more, the Lords of Aquitaine, perhaps the greatest Christian sovereigns in South-West Europe at this time, joined in the prevailing fashion. Duke William III., 'father of monks, founder of churches, friend of the Roman See,' made several pilgrimages both to Rome and to Compostella, where the Shrine of St. James was rising into fame. Unable himself to go as far as Syria, he sent there his trusty minister, William of Angoulême, with a great company of lords and prelates. This little army journeyed by the land route through Hungary, received the hospitality of St. Stephen, and enjoyed a prosperous and speedy 'accomplishment of vows.' Setting out in October, they were in Jerusalem by March, and home again at Angoulême in June.

Yet again, not a few eminent pilgrims came from England about this time, and especially during the reign of Edward the Confessor. Thus Svein or Swegen, the eldest son of Earl Godwine, and the scandal of his house, made his way to Jerusalem from Bruges at the time of the first Norman ascendency in England and consequent exile of the Godwine clan (1051-1052). His evil life had a good ending. He died upon the Sacred Road, at the city of Constantine, 'on Michael's Mass' (1053). Some years later, in 1058, Ealdred, Bishop of Worcester, more famous as Archbishop of

York [1] in the era of the Norman Conquest, went to Jerusalem 'with such splendour as none other had displayed before him; there he devoted himself to God; and a worthy gift he offered at the Lord's tomb, even a golden chalice of wondrous work.'

Lastly, among many instances of German pilgrimage at this time, let us take three famous journeys from the middle of the eleventh century. In 1038 Poppo, Archbishop of Trier, makes his way to Palestine along with the hermit Symeon, afterwards canonised by Pope Benedict IX. Again, in 1054, Fulcher, Arch-Chaplain of the Empire and Vidame of Arras, takes leave of his master, the Emperor Henry III., as he is on the point of setting out for the Holy Land with Lietbert, Bishop of Cambrai and Arras. This voyage was a sad failure; the pilgrims fell among thieves, were ransomed by the Catapan of Cyprus, and at Laodicea abandoned their further journey. For here they met with Helinand, Bishop of Laon, returning from Jerusalem, and he drew such a picture of the troubles that awaited them in the Holy City, that they decided to tempt God no further.[2] Once more, in 1064, Siegfried, Archbishop of Mainz, and Günther, Bishop of Bamberg, with two other prelates, the Bishops of Utrecht and Ratisbon, conduct an army of 7000

---

[1] He acted at the Coronation of William the Norman as chief Prelate (1066), Stigand of Canterbury not being recognised by the Conqueror.

[2] Riant suggests that Victor's appeal to the Empress Theodora (see pp. 122-3) was the result of a meeting at Cologne between the Pope and the Bishop of Cambrai. Lietbert had just returned from Laodicea with the account of Helinand's miserable experience in Palestine; Victor had gone to Germany to gather aid against the Saracens of the Mediterranean; we may notice that his complaint to the Byzantine Government is written from Cologne (at the beginning of December 1056). With this we may connect and contrast the alleged Firman of the Caliph Mozaffer in favour of the Christian Holy Places (1023). The last, however, was rejected as spurious by Heyd, in the original (German) edition of the *Commerce du Levant*, i., 116; cf. *Archives de l'Orient Latin*, i. 52; Boré, *Lieux Saints*, p. 5.

pilgrims to Syria. Their numbers and their adventures have led many chroniclers to call their journey a Crusade. From their losses it might well have been an armed attack on Islam, for of the seven thousand only two[1] returned. Attacked by Saracen robbers near Jaffa, and 'hunted for their rich apparel,' the party fled to a fortress, where they held out three days,[2] and then yielded, more from panic than real necessity. Offering their money for their life, they admitted the Moslem leaders within the wall. Seventeen Saracens entered, and one who 'took the headship of them' singled out the Bishop of Bamberg as his special victim. Unrolling his turban, he flung it round the Prelate's neck with a joyful cry—'Thou and all thou hast are mine.' But the fighting bishops of the West were different from the broken-spirited Christians of Syria. The Frankish Churchman felled his captor with a blow of his fist; the rest of the company, regaining courage, seized and bound all the other brigands who had come into their power, then, closing the gates against the hordes outside, they held the fort till the Lord of Ramleh came to their relief.[3] One other thing may be recorded of this journey. From letters of the Provost of Passau we see how Bishop Günther, of Bamberg, had arranged for his journey through the neighbour-

[1] Thousand.

[2] Cf. Marianus Scottus, *Chronicle*, A.D. 1064: the pilgrims when attacked by the Arabs, fled to a 'castellum vacuum *Cavar Salim* nomine' (*i.e.* probably Kafar Sallam, abandoned in the eleventh century by its inhabitants; cf. Schefer's *Nasir-i-Khusrau*, p. 63). On Siegfried's journey, cf. Eccard, *Corp. hist. med. aev.* ii. 122-3; Sudendorf, *Registrum*, ii., No. 12, pp. 14-15; *Ibid.* iii. No. 16, p. 28; Jaffé, *Mon. Bamb.* No. 28, p. 54, No. 29, pp. 56, 57; *Archives de l'Orient Latin*, i. 53-56. On Fulcher or Foucher of Arras, cf. Sudendorf, *Registrum*, iii. No. 6, pp. 12-13.

[3] The Latin accounts do not distinguish enough between the Saracen brigands (Bedouins, etc.) and the settled inhabitants and garrisons, who generally treated visitors well, for the sake of commerce; *e.g.* in this case, where the Egyptian garrison at Jerusalem rescues the prisoners at Ramleh.

ing districts, and how elaborate such arrangements could be, even in 1064. In the spring of 1065 Günther writes to his Diocesans, describing to them the course of his journey down to his arrival at Laodicea, on the homeward way; but soon after this he died without reaching Germany.

With the accession of Hildebrand to the Roman Pontificate as Gregory VII., we enter upon a new phase of development. For this daring statesman first definitely raised the question of a Crusade or armed expedition, under the sanction of the Christian Church, to recover the Holy Places of Palestine from the Saracens. The Byzantine reconquest had been only temporary, and the advance of the Seljuk Turks had driven back Eastern Christendom into narrower limits than ever before. Nothing but internal trouble, and particularly the quarrel with the Empire, seems to have taken from Gregory VII., and given to Urban II., the actual preaching and direction of the expedition which was planned for 1075 or 1076, but did not start till 1097, and was indefinitely postponed, in the opinion of most men, by the outbreak of the struggle with Henry IV. and the incident of Canossa. In the letters and appeals of Hildebrand we may see, moreover, the natural development of the policy of a Holy War, entailing salvation on all who took part therein. The close of the eleventh century realises the aspirations of Leo IV. and John VIII.[1] Now, also, we may see a natural result of the conversion of the Hungarians and the Northmen; events which reopened to Western Europe the long-interrupted land routes along the Danube Valley, which furnished more direct means of access to the Levant, and which supplied a new energy to Christendom. Once more, the end of the First Millennium and the excite-

[1] John VIII. to the Bishops of Louis [the Stammerer], cf. p. 118,

ment it had caused may not have been wholly without its influence on the Crusading movement. Many had expected the end of the world with the year 1000; the fatal era had come and gone without result; but the excitement thus aroused found a sedative in the passionate devotion of pilgrimage and an outlet in the fiercer delights of a Crusade.

It is therefore in the general awakening and expansive vigour[1] of the Christendom of Hildebrand's time that we must look for the primary cause of the Crusades. The outrages of Hakim, and the tyranny of the new Seljuk rule at the Holy Places, were but secondary causes; the victories of the Spanish Christians, and their renewed peril from the Almoravide invasion, only contributed what we may call details of encouragement or stimulus; the root of the whole movement lay in that new life stirring within the Christian people, even before the Council of Clermont, and finding more and more active expression in pilgrim travel, in merchant enterprise, in revived scientific enquiry, in an unappeasable and unmistakable determination to burst the bonds which Islam and heathendom had wound so tightly around Latin civilisation.

But, granting this, the secondary causes are none the less essential. Because history has been often too much occupied with the more obvious and more personal springs of action, yet in this, as in other epochs, it remains true in the nature of things, that historical writing must always deal more fully with the clear-cut incidents of secondary importance than with the vaguer and more general, even if more vital, conditions of national or racial fitness. We must recognise that the imminent danger of the Byzantine

---

[1] In 1087 Bari Mariners are said to have stolen the body of St. Nicolas from Myra or Myrrha, in Lycia; cf. the story of the Venetians in the ninth century carrying off St. Mark's body from Alexandria,

Empire from the Seljuk Turks, the cruel treatment of pilgrims at the Holy Places, the preaching of Peter the Hermit, the critical events in Spain, the writings and personal leadership of Hildebrand and Urban II., would have been useless without the revived strength and ambition of the Christian peoples ; we may admit that this revived strength and ambition would perhaps have found their outlet in a similar movement without any of these particular incitements ; but, after all, as events shaped themselves, it was the sting of these incitements which immediately set in motion the hitherto sluggish mass of Western Christendom and united its long-scattered forces. We must therefore look one moment at those events, or groups of incidents, which had so profound an effect on the Christian world of the eleventh century.

First, in the Levant itself, the rise of the Turks had certainly produced before A.D. 1080 a deep, dangerous, and irritating change in the position and prospects of Oriental Christendom, and so, in a measure, of all Christians. As early as the time of Justinian the formidable power of those Tartar tribes had caused Byzantine statesmen to look beyond the Oxus for a counterpoise to their Persian enemies. But it was in the tenth century that the Turks first began to play a leading part in the politics of the nearer East. As the Caliphate of Bagdad decayed, Turkish soldiers were more and more enlisted in its armies, and Turkish commanders put in charge of the palace and the capital, until the Pontiff of Islam became a tool in their hands. The action of Al Radi (A.D. 940), in creating the office of Emir al Omra, or Military Vizier, for the chiefs of these Turkish bands, was a practical abdication of the remnants of Temporal Power on the part of the Bagdad Caliphate, and the state of things thus produced was in many ways similar to that of the old Frankish monarchy in the time of the later Merwings and Charles Martel. Finally, a new immigration from Central Asia completed the change. In 1055, the Seljuk Turks, after cutting short the dominions of the Ghaznevides and humbling the successor of Mahmud, pressed into the valley of the Tigris, and formally received from the reigning Abbasside all the secular power of the Commander of the Faithful. The new champions of Islam soon cut short that Byzantine revival which in the tenth and early eleventh centuries had restored so much of the monarchy of Justinian.[1] In 1065-1068 Armenia and Georgia fell away to the Seljuk Empire; in 1071 a decisive struggle between Turks and Romans took place at Manzikert. Here were lost, at one blow, all the acquisitions of the last century beyond Mount Taurus; and this disaster was followed by the inroads of the Turkish cavalry to the shores of the Archipelago, the Hellespont, and the Bosphorus. Nor were these inroads transitory. It was now that the upland of Asia Minor permanently became a Turkish possession and a new home of the Turkish race. This was a heavier loss than Eastern Christendom had suffered since the first conquests of the Saracens; and the Byzantine Empire was now reduced in Asia to a narrow strip of coast, and was forced to recognise by treaty the permanent cession of Galatia, Phrygia, and the other provinces of the interior. The Turkish outposts established at Nicaea made the Greeks tremble for the safety of Constantinople itself; any year might see a further Turkish advance into Europe; and what the Ottomans achieved in the fourteenth and

---

[1] *E.g.* Crete, Cyprus, most of Northern Syria, including Antioch, and reaching as far South as Tripoli, Northern Mesopotamia, and a suzerainty over Armenia, Georgia, Albania, the other regions of what is now the Trans-Caucasus Province of Russia and the South of the Crimea.

fifteenth centuries was threatened as no vague contingency at the close of the eleventh.

Such was the political problem facing Gregory VII. when he looked eastwards; and the most sluggish of the statesmen in Western Europe might well think it time to stir very actively against the impending approach of Islam to the Adriatic and the Valley of the Danube.

Secondly, the Seljuk Turks, in becoming masters of the Abbasside dominions, acquired the control of Palestine. The Holy Sepulchre in Jerusalem was now guarded by a band of Turcomans; and their treatment of all Christians, both residents and visitors, was harsh and oppressive. Pilgrims returned to the West with terrible stories of insult and cruelty; and the nations which had tamely suffered the sacrilege of Hakim, rose as one man to avenge the wrongs proclaimed by Peter of Amiens.

Thirdly, the political dangers and popular resentments of the East were reflected in the furthest West. After many changes, it became clear, as the eleventh century drew on, that Spanish Islam was declining. Before his death in 1002, the Vizier Almanzor had restored the Mohammedan sway from Catalonia to Galicia; but in the next generation the Christian Kingdoms of Northern Spain had already recovered, and more than recovered, their position as it had stood before the accession of Abderrahman III. Meanwhile in 1031, the Western Emirate, or Caliphate,[1] came with startling suddenness to an end, after a few years of violent anarchy, and its dominions fell apart into various principalities. The result of this was soon apparent. More and more of the Central Highland was won back by the kings of Leon and Castile; the Valley of the Douro and the western coast down to Coimbra, fell again into Christian

[1] 'Caliphate' in name, only from Abderrahman III.

hands; and in 1084 Alfonso VI. recaptured the old capital of Toledo, and became master of the Upper Tagus, the natural heart of the peninsula. He was checked by a new and unexpected foe. In their despair the Spanish Moslems called in the aid of that religious and political movement in North Africa, which is generally known from its 'Marabût,' or Prophet, as the 'Almoravide' Revival. The Moorish enthusiasts responded to the appeal, and in 1086 their leader Yusûf Ibn Tashfin crossed over into Andalusia and defeated Alfonso in the battle of Zalacca. Succeeding waves of Berber invasion held Spanish Christendom at bay during most of the twelfth century; and after the disaster of 1086 a certain alarm was even felt lest Islam should again advance to the Bay of Biscay and the Narbonnese; but with the aid of volunteers from the other kingdoms of the West the new African peril was stubbornly faced and finally repelled. As originally conceived by Gregory VII. and his successors, it is probable that the Crusading impulse was directed, in part at least, to the more complete overthrow of Mohammedan power in the peninsula; but the embittered character of the eastern pilgrim-grievances, and the extreme danger of the Byzantine world, finally directed the entire stress of the new Christian armaments upon the Levant. It is clear, however, that Leon and Castile, Aragon and Portugal, shared at any rate in the benefits of the diversion created by the Syrian wars, and in the support thus incidentally afforded to Christian politics everywhere; while several incidents of the second and third Crusades showed very effectively the strength and closeness of the military alliance of the Catholic nations. Lisbon and the Algarve were both reconquered with the aid of men from Germany, France, England, and the Netherlands on their way to Palestine; and the close of the Crusading age saw the Latin warriors triumphant in Spain, however unsuccessful in the East.

Lastly, without attaching too much weight to the Voltairean innuendo that the Popes and the Clerical Order conceived and pressed on the religious wars with the object of killing off the dangerous manhood of the Western kingdoms, it is undeniable that we must give to the Clerical Order the credit of organising and directing the great Christian *Jihad*. The Church supplied the controlling intelligence; its exchequer supplied the funds for much that was undertaken; and, without the ecclesiastical inspiration, the Crusades, though perhaps in some form inevitable, would have had a very different character. At the same time, it must be recognised that this clerical inspiration gave the European nations of the Middle Age their first glimpses of essential unity; and that, along with some political reverses, the new federation of the Western World under papal leadership won successes of the first order in commercial and social progress, and certain negative advantages of no small moment in the military struggle. For even if the Holy Land was not finally recovered, nor the union of Eastern and Western Churches accomplished, yet Islam was checked for a time in the Levant, and permanently thrown back in Spain; the incursions of the Turco-Tartar hordes into Europe by way of Constantinople were postponed; and the land-travels of men who started from the Latin Kingdoms of the East led to a decisive and abiding extension of knowledge and civilisation. Thus the indirect, unexpected, and sometimes unwelcome results of the mediaeval duel of Cross and Crescent proved a rich compensation for the barren and abortive character of the central struggle in Palestine: European life was not impoverished but enriched by the religious wars; and the only doubt must be whether it was necessary through such tribulation to enter into the brighter age of the great discoveries.

Among the pilgrim travellers who may be traced between the accession of Gregory VII. and the first Crusade, perhaps the most important is the Anonymous devotee [1] who has left us a brief note-book made up partly of his own observations, partly of extracts from much older records. The earlier sections of this tract bear a close resemblance to passages of the fourth-century Bordeaux Pilgrim, but otherwise these jottings are almost without interest. The 'chamber covered with a single stone,' where Solomon wrote the Book of Wisdom, is a repetition from the most ancient surviving type of pilgrim-itinerary;[2] the water-pot of Mount Sinai, which incessantly ran with oil, may be compared with a similar story in Thietmar;[3] and in the confusion of 'Abraham's Castle' at Hebron with a similar fortress at Tekoa, there is only the repetition of what is often met with elsewhere. It is, perhaps, more curious that this work places Isaac's Mount of Sacrifice in the neighbourhood of the Patriarch's Castle, a more southerly location than is given by any other pilgrim.

With this little tract we leave the earlier Middle Ages and come to the transition period of the Holy Wars. One of the last aggressions of the New Turkish Islam was the reconquest of Antioch in 1084, after 116 years of Byzantine rule; but in November, 1095, the Council of Clermont decided that the Will of God[4] was the recovery of Jerusalem and all Syria: in the summer of

---

[1] Usually known as 'Innominatus I.,' the first of a long series of unnamed pilgrim travellers of this period, see pp. 203-7.

[2] *E.g.* the Bordeaux Pilgrim of 333.

[3] A.D. 1217.

[4] *Deus vult.*

1096 the main host of the Crusaders set out in five divisions; Antioch was won in the beginning of June 1098; and on the 14th of July, 1099, the great object of the enterprise was achieved, and the Cross was planted on the walls of Sion. The victory was followed by its natural results, and prominent among these was the arrival of pilgrims who possessed the leisure and the ability to do what had not been attempted since Bernard the Wise gave men a full description in the Latin tongue of all the Holy Places.

The earliest of this new group of writers came from a remote country of the West, the most recent conquest of the Norman Dukes of Rouen. This was the Englishman Saewulf or Saewlf, a native of Worcester, certainly not the first pilgrim traveller who followed in the wake of the first Crusade, but the first who has left us the narrative of his journey. He was evidently an 'Anglo-Saxon,' not a Norman or French follower of the invaders of 1066; and he pursued the despised calling of a trader. At least so we may fairly gather from William of Malmesbury's account. The latter tells us of one Saewulf, a merchant, who was in the habit of confessing to Bishop Wulfstan of Worcester (1062-1096), the last survivor of the Old English Episcopate. Wulfstan, himself a man of saintly life, so much respected that the Norman policy of 'No more Saxon Bishops' made an exception of his case, often urged the monastic vows upon Saewulf, whose morals, he thought, needed stricter discipline But his penitent clung to the freedom of his lay estate, and the Bishop fell back on a prophecy, that the sinner would in the future eagerly seek for and embrace that calling which he now refused when offered. William the chronicler lived to see the fulfilment of this prediction, when in his old age Saewulf became a monk in the Abbey of Malmesbury.[1]

It was probably in a fit of penitence that our merchant undertook his pilgrimage, sometime before his profession as a monk, in the years 1102 and 1103. We can only date the journey from the internal evidence of the narrative, but this is circumstantial enough for our purpose, and we may conclude with tolerable certainty that Saewulf's outward journey was in the summer of 1102, and his return in the summer of 1103. We may even go further, with the help of a fresh study of the text, and say that he started eastwards from Monopoli in Apulia on Sunday, the 13th of July, 1102, and that he began his return journey from Jaffa on the 17th of May, 1103.

The following details support this conclusion. Saewulf tells us that Baldwin, 'the Flower of Kings,' possessed Jaffa, Haifa, and other places on the Syrian Coast at the time of his visit; now Baldwin was elected in succession to Godfrey of Bouillon on Christmas Day, 1100. Again Tortosa was then in the possession of 'Duke' Raymond; from other sources we know that the Count of Toulouse took this city on the 12th of March, 1102, almost three years before his death, which occurred on the 28th of February, 1105. Once more, our pilgrim implies, though in a rather vague and negative manner, that Acre was still in the hands of the Saracens; and this state of things lasted till the 15th of May, 1104. Fourthly, Saewulf left Monopoli on the 'third day before the Ides of July,' the Feast of St. Mildred, an 'Egyptian' or unlucky day for sailors, and a Sunday;

---

[1] Cf. Will. Malm., *Gest. Pontif.*, p. 282; T. Wright, *Biog. Brit. Lit.*, Anglo-Norman period, p. 38. In the one existing MS. of Saewulf, this narrative is bound up with seven other fragments of ancient works. Avezac suggests that 'Saewulf' is only a nickname from frequent journeys, like 'Sea-dog,' but this is rather improbable.

even without the definite statement first quoted, we could fix the date here to the 13th of July 1102, which answers all the other requirements, and which, with the 22nd of July,[1] was notoriously ill-omened, according to the mediaeval calendars. Lastly, the traveller's embarkation at Jaffa, 'on the Day of Pentecost,' after the completion of his journey in the Holy Land, points to the Whitsunday of 1103 (17th May), as in 1104 the festival occurred on the 5th of June, three weeks after the fall of Acre, an event which he would scarcely have omitted.

Besides these allusions, hasty and slight enough, though not without their importance, there is little reference in Saewulf to the events of his time or to the facts of general interest in geography. He stops at Cephalonia to tell us that there Robert Guiscard died;[2] the great Norman conqueror was too recent and too mighty a figure in Mediterranean history for even a pilgrim to leave unnoticed; and it is possible that Saewulf's account of Constantinople contained more notices of contemporary men and matters. But both the opening and the close of the narrative appear to have been lost.

As to Saewulf's outward journey, we may notice that he does not go by way of Egypt, for the most obvious reasons. Egypt was a Moslem country, whose people, except for special circumstances, would naturally make common cause with their brethren of Syria against the invading Christians. In earlier times the route through Alexandria had been common enough, but now Christian pilgrims followed in the wake of their victorious armies and fleets along the northern shores, or through the northern waters, of the Mediterranean. Our present narrative shows us the traveller following the most adven-

[1] When Saewulf left 'Brandia' or Brindisi.   [2] July, 1085.

turous route, not troubling to take the safer and longer way by Constantinople, but trusting to the new-grown Christian power and trade at sea, and pressing directly forward from Italy to the Ionian islands, from Negropont to Rhodes, and from Rhodes to Palestine. Greece, on the other hand, he crosses by land, probably in preference to the troublesome circuit of the Morea; but the rest of his journey is on shipboard. His whole record demonstrates how great was the recent increase of European influence in the Levant, alike in war, in commerce, and in pilgrimage. Nowhere is this more clearly marked than in his picture of the great fleet at Jaffa, laden with palmers and merchandise, which was destroyed by a tempest in that perilous and unsheltered roadstead before his eyes.

Even the small amount of land travel performed by Saewulf was against his inclination; 'conscious of his unworthiness,' he is uncertain whether it was his sins, or the badness of his ship, that hindered him from taking without intermission, the 'direct course of the open sea.'

Of the first part of his journey—presumably from England to South Italy—he says nothing, and professes only to begin with a mention of the different islands at which he touched. But his narrative really commences with a list of the chief pilgrim ports—Bari, Barletta,[1] Siponte or Manfredonia, Trani, and Otranto, the 'entrance-harbour' of Apulia, then and later the favoured rival of Brindisi for the narrow passage from Italy to Albania. Saewulf himself did not make use of any of these; he took ship at the little harbour of Monopoli, about twenty miles from Bari, on Sunday, the Feast of St. Mildred (13th July, 1102). 'But,' he laments, 'as we set out at an unlucky hour, we had not proceeded more than three

[1] Saewulf's 'Barlo' or 'Barlum.'

miles, when the violence of the waves wrecked our vessel.' Refitting at Brindisi, and once more putting to sea with strange persistence on an ill-omened day, the pilgrim had a stormy passage to the isthmus of Corinth, by way of Corfu (24th July), Cephalonia (1st August), and Patras, famous for the martyrdom of St. Andrew. In this 'island' the relics of the apostle had long been kept before their translation to Constantinople; here Saewulf probably refers to the coast town near the entrance of the Corinthian Gulf; but, like other mediaeval writers, he employs the term of 'isle' in a generous sense for any seaboard place, without much attention to the precisions of geographical terminology.

In Corinth, where he 'suffered many hardships,' the traveller begins to notice St. Paul's journeys and their coincidences with his own wanderings, illustrating these by occasional explanations of singular perversity. In the Gulf of Corinth the sea route ended for a time,—after Saewulf had crossed to 'Hosta' or Livadostro, whence the party made their way, 'some on foot, some on asses,' over the Boeotian neck of land to the Strait of Euboea and the port of Negropont (23rd August). On this march they passed by the city of Thebes, or 'Stivas,' then celebrated as the seat of the largest Jewish colony in the Byzantine Empire. The voyage from Negropont brought Saewulf into the Ægaean, and he touched at a number of islands and harbours in his very devious course upon these waters—Petali or Spili, near Marathon; Andros, noted for its silk, samit, and 'sindals' (thin shavings of gold and silver, from which the precious tissues were made); Ancho, or Cos, where Hippocrates was born (for so we may correct the pilgrim's 'Galen'); Naxos, Patmos, and Rhodes. He does not appear to have visited Athens, 'where Paul preached,' and where a miraculous lamp still attested the sanctity of the place, nor the 'notable island' of Crete, nor Ephesus, 'where St. John entered the sepulchre living'; but he certainly spent some time at various points on the eastern side of the Archipelago, including Cnidus,[1] Syra, Samos, Scio, and Mitylene. At Rhodes he stays a little to tell us of its vanished marvel, the 'idol called Colossus,' one of the seven wonders of the world. This was 125 feet high; but in spite of its bulk 'the Persians destroyed it, with nearly all the land of Romania, when on their way to Spain. These were the Colossians to whom St. Paul wrote.' This ingenious confusion occurs in several of the later pilgrims, who, like Saewulf, mix up Colossae and Phrygia with the 'idol of Rhodes,' and name the people of the country after their great monument. But here the Persians are also confounded with the early Arabic Saracens, and the Levant conquests of the latter appear as part of the Moslem advance through North Africa into the West of Europe. Saewulf's Romania is, of course, the Roman Empire[2] of the East, and nothing else; it is an official and perfectly correct designation; but a very slight knowledge of Mohammedan history would have reminded the traveller that, when the Arabs conquered Rhodes, their capitals were at Mecca and Damascus, that Persia was then in no way the central country of Islam, and that a century was yet to elapse before the foundation of Bagdad.[3] Once more, the effects of the great earthquake in the third century B.C., and the damage wrought by

---

[1] Lido.
[2] From his later use of the term Romania, it may be that here also only the Asiatic Provinces are intended.
[3] A.D. 651, Conquest of Rhodes; A.D. 762, Foundation of Bagdad. However, the Persians of Chosroes II. captured Rhodes for a moment in 616. This may partly account for Saewulf's language here.

subsequent calamities, are, of course, forgotten; and the ruin of the Colossus is imputed wholly to the infidels who carted away for building purposes the remains of the fallen statue in the seventh century after Christ, under the first Ommeyad Caliph Moawiyah.

From Rhodes Saewulf made his way to Patara in Lycia, the home of St. Nicolas, and one of the resting-places of the English pilgrim Willibald 'during the rigour of the winter' of 722. Not far off was the town of Myra, where the Wonder-worker had his bishopric and became famous throughout the Orient of the fourth century. Here the saint died and was buried in the year 342; but his relics had been translated to Bari shortly before the time of Saewulf (A.D. 1087), perhaps to escape desecration at the hands of the advancing Turks. Now, at the opening of the twelfth century, Myra was an important harbour, the anchorage-in-chief for the shipping of the 'Adriatic' or Eastern Mediterranean, just as Constantinople was for the marine of the Ægaean Sea. Here, again, the English traveller reproduces a common usage of his time; exactly the same meaning is attached to the 'Adriatic' Sea by Matthew Paris and many other writers of the central mediaeval period. Near Patara Saewulf describes an entirely desolate town called 'Mogronissi,[1] or the Long Island of St. Mary,' which, from its churches and other buildings, appeared to have been inhabited by the Christians after they had been driven by the Turks from 'Alexandria,' or Alexandretta, at the head of the Gulf of Scanderoon. This Long Island was perhaps the modern Kakava, on the west side of which Avezac has pointed out the ruins of a church and town; Saewulf's remark is one more evidence of the changes wrought by the recent advance of Islam

[1] Makro-nesos.

under the devastating leadership of the Seljuk Turks. After worshipping at the Sepulchre of St. Nicolas, the pilgrim sailed to the 'Island called Xindacopo,[1] or *Sixty Oars*, on account of the force of the sea'; this is apparently the ancient Chelidonia at the south end of Finica Bay, and from this point Saewulf took to the open sea, and crossed the broad part of the 'Adriatic' to Paffus, Paphos or Baffo,[2] in the island of Cyprus. Another long voyage out of sight of land brought the wanderer, after seven days, to the coast of Palestine; one night the stormy weather forced his ship almost back to Cyprus; but at last the sunrise of the eighth day displayed the shore of the port of Jaffa, and with extraordinary and unexpected joy he landed in Syria on Sunday, the 12th of October, 1102. His journey from Monopoli had consumed no less than thirteen weeks, and all the time he had 'dwelt on the waves of the sea, or in islands or deserted cots and sheds,' hoping in vain for the hospitality of the suspicious and unfriendly Greeks.

On the advice of one who spoke, as Saewulf afterwards believed, by divine inspiration, the pilgrim went on shore at once, and thus escaped a terrible visitation. Next morning, as he returned from church, he heard the roaring of the sea and the shouts of the people. Dragged along with the crowd to the shore, he saw waves swelling higher than mountains, and countless bodies of the drowned scattered over the beach, while fragments of ships were floating on every side. The rage of the sea and the crashing of the vessels drowned the clamour of the spectators. 'Our own ship, being very large and strong, and others laden with

---

[1] Apparently a corruption of ἐξή-κοντα κωπαί shortened to 'ξοντα κωπαί.

[2] Here Saewulf makes a short excursion into the History of the Early Church, the Mission of Barnabas, and the life of St. Peter, confusing Acts xiii. 2-4 with Acts xv. 22.

corn and merchandise, or with pilgrims going and returning, still held by their anchors; but how they were tossed by the waves, how the crews were filled with terror, how they cast overboard their merchandise. What eye could refrain from tears?' Before long the ships were driven from their anchors by the violence of the waves, which threw them, now up aloft, now down below, until they were hurled upon the rocks, and there beaten to and fro till they were dashed all to pieces. 'For the violence of the wind would not allow them to put out to sea, and the nature of the coast did not permit them to put in to shore. Among the sailors and pilgrims who had lost all hope of safety some remained on the ships; others laid hold of the masts or beams of wood; many, in a state of stupor, were drowned without any attempt to save themselves. Some (incredible as this may seem), before my very eyes, had their heads knocked off by the rafters to which they clung for safety; while others were swept out to sea on the beams which they hoped would wash them to land. Even those who could swim had not strength to battle with the waves, and very few who thus trusted their own power reached the shore alive. Thus, out of thirty very large ships, of which some were (as they call them) *Dromunds*, some *Gulafres*, and some *Cats*, scarce seven were saved. Of men and women there perished more than a thousand.'

All the chief kinds of twelfth-century vessel were here represented, and all alike suffered severely, an incidental penalty of the new spirit of naval enterprise which was drawing the marine of the Italian and Provençal cities, with such force and suddenness, to the new homes of Christian power in the Levant. The Byzantine Emperor,[1] at this very time was famous for his navy of *Dromunds* or light

[1] Alexius Comnenus.

galleys; these ships appear to have been biremes, each tier of oars having twenty-five benches, and each bench two rowers; while the *Cats* of the narrative have been conjectured to resemble a modern Norwegian collier with a narrow stern, projecting quarters, and deep waist. With the misfortunes and terrors of Saewulf's companions we may contrast the good luck and daring valour of Godric the Pirate, a native of the same distant land as the Worcester pilgrim, who, a few years later, fought his way through the Saracen fleet, outside the same perilous roadstead, with a spear-shaft for his banner.

From Jaffa Saewulf went up to Jerusalem by a road which he, like Daniel of Kiev, found 'mountainous, rough, and dangerous on account of the Saracens, who lay in wait in the caves of the hills to surprise the Christians.' These brigands were swift and subtle in their movements: at one moment you saw them on every side; at another they were nowhere to be seen. On this road many travellers perished, poor and weak, rich and strong alike; many cut off by the Saracens, more by heat and thirst; many from want of drink, more from too much drinking. The Kingdom of Jerusalem was now only in its third year, and Acre and Ascalon (or Askelon) were still Moslem citadels. It was little wonder, then, that roads were still unsafe. Entering the Holy City from the west, under the citadel of King David, by the Gate of David (or of Jaffa), Saewulf first visited the Holy Sepulchre 'built by the Archbishop Maximus, with the patronage of the Emperor Constantine, after the discovery of the Lord's Cross.' Like the Russian Daniel, he notices that the church was open to the sky; it was situated, he avers, on the slope of Mount Sion.

This was the third Christian building which had marked the great scenes of the Crucifixion and Resurrection; the first, that of Helena or of Constantine, had been destroyed by Chosroes and his Persians, the 'Pagans' of Saewulf, in 614; the second had been demolished by Hakim Biamrillah in 1010; the third, standing in the traveller's day, was an erection of the years 1020-1026. Saewulf makes no clear distinction between these, and seems to think that what he visited was the original foundation of the fourth century; both here and elsewhere he was apparently at the mercy of the Syrian or 'Assyrian' Christians who undertook the duty or the pleasure of guiding or misguiding the ignorant but eager and credulous pilgrims from the West.

In Jerusalem Saewulf displays a good deal of learning, often without any close relation to fact. Titus and Vespasian destroyed Jerusalem to revenge Christ. Hadrian first extended the city westwards to the Tower of David. Justinian was said by some to have rebuilt the town and the Temple of the Lord, but this was not true, according to the witness of the *Assyrians*. The ordinary relics and holy sites are elaborately described by our pilgrim. He quotes the sentences[1] of Augustine on the question of Adam's burial, not at Golgotha, but at Hebron; and he locates Abraham's sacrifice of Isaac, not on Moriah, but on Calvary. One geographical monument is especially noticed.

At the 'head' of the Church of the Holy Sepulchre, in the outside wall, and not far from the place of Calvary, was the spot called *Compas*, which our Lord Himself signified and measured with His own hands as the *middle of the world*, according to the Psalmist's words: *For God is my King of old, working salvation in the midst of the earth.*[2]

---

[1] Not in the standard collection of these sentences by Prosper of Aquitaine. This story of Saewulf's is perhaps derived from James of Edessa, who talks about Adam's bones being distributed among the sons of Noah, and of Shem burying the head in Judaea.

[2] Ps. lxxiii. 12.

Thus we find the same legends supported by the same texts in the sixth, the seventh, and the twelfth centuries; and, from Saewulf's time, the *centre of the world* at Jerusalem, as a definite object which could be seen and touched, became more and more a fixed article of common belief; most of the anonymous pilgrims of the next centuries reproduce the story in exact verbal agreement; and with the Byzantine-Latin sketch of 1110 the same feature begins to appear in mediaeval maps. Reverence for tradition allied itself with convenience; and thus *Compas*, late into the fifteenth century, was retained as the favourite umbilical spot.

In the Temple area Saewulf professes to recognise the Beautiful Gate and the site of Bethel.[1] The Holy of Holies, he learnt, was in the hollow underneath the Rock of Sacrifice; there, too, the Lord, wearied with the insolence of the Jews, was accustomed to hide and to repose Himself; and there were to be seen the footsteps of the Son of God.

As to the Golden Gate on the east side of the temple area, Saewulf relates the story of Heraclius' entering here, with the Lord's Cross, on his victorious return from Persia, when 'the stones fell down and closed the passage, until, humbling himself at the admonition of an angel, he descended from his horse, and so the entrance was opened unto him.'

Near the Holy Sepulchre was the famous monastery of St. John the Baptist, the centre and home of the Order of the Hospital which drew its first origin from a small foundation of some pious merchants of Amalfi in the middle of the eleventh century (A.D. 1048), and assumed its military character (in 1118) shortly after the visit of Saewulf. In this memorable year the Order of the Temple also took

---

[1] We may contrast Saewulf's uncritical talk about Bethel with the accurate remarks of John of Würzburg.

definite shape; but the name of the greater society does not occur in our present narrative; and this omission, coupled with the reference to the earlier and less celebrated Hospital, in its primitive form, is another confirmation of our traveller's date, at the very commencement of the twelfth century.

The name of Ælia (or 'Helias') for Jerusalem, we are told, was interpreted by some as meaning 'House of God'; this Hebrew fable is accompanied by some Christian legends hardly less wild, such as that of St. John as the first holder of the 'Pontificate' of Jerusalem, or of St. Saba[1] as one of the 'Seventy-two disciples of the Lord.' The 'Chapel of Galilee' on the summit of Mount Olivet, the 'place of Galilee' on Mount Sion, the 'great city[2] of Galilee' by Mount Tabor, Nazareth on the Lake of Tiberias, the Garden of Abraham in Jericho, the Bethlehem cistern wherein the star[3] of the Magi fell (according to the report of some), are balanced, as instances of pious fable, by Saewulf's matter-of-fact relations about the tombs of Jerome, Paula and Eustochium, and the St. Saba monastery with its 300 inmates. Most of these religious men had been slain by the Saracens a short time before Saewulf's visit, and the survivors had fled to a convent of the same name (also noticed by Daniel of Kiev) within the walls of Jerusalem, near the Tower of David and the Jaffa Gate.

Saewulf's description of the sites of Hebron is especially detailed and interesting; he enlarges on the ancient workmanship of the patriarchal tombs and the smell of the precious aromatics in the same, the strong castle that

---

[1] St. Saba died 5th December, 532, aged 94.
[2] Cana (?)
[3] In Willibald the figure of the star is reflected in a well, c. 12. The Saewulfian form of the story occurs also in Mandeville.

surrounded the sacred monuments, and the humbler burial-place of Joseph at the extremity of the fortress.

In a digression upon the Jordan the pilgrim gives us a characteristic piece of his geographical phraseology: 'On this side of the river Judaea extends to the Adriatic Sea, to the port of Joppa; on the other side is Arabia, hostile to Christians and hateful to all who worship God, but possessing the mount whence Elijah was carried into Heaven in a fiery chariot.' Our pilgrim seems to have visited the Ghor in the Jericho region, after which he went North to Sichem, Haifa, Nazareth, and Cana of Galilee, where nothing was left standing except the 'Monastery called after Architriclinius.' Here again Saewulf preserves a characteristic mediaeval corruption, which, copied by hosts of later writers, appears (among others) in several of the Anonymous pilgrims and in John of Würzburg; and thus the 'Ruler of the Feast,' named after his title [1] becomes the patron saint of a celebrated convent and an important figure in the literature of mediaeval Christian devotion. Mount Tabor, 'covered in a marvellous manner with grass and flowers,' rose in the middle of the green plain of Galilee, overtopping all the mountains which surrounded it. On the summit Saewulf notices the three tabernacles or monasteries described by earlier visitors; and from this point the narrative now takes us *six* miles to the Sea of Galilee, and so, after some wanderings, to Mount Libanus. At the foot of this the Jordan boiled out in a very rapid stream from the two fountains of Jor and Dan, and far into its last home in the Dead Sea the milky waters of the river, whiter than any other, could be distinguished by their colour.

And so we come to the return journey. 'Having worshipped to the best of our power,' concludes Saewulf, 'at

---

[1] As given in the Gospel of St. John iii. 8, 9.

all the holy places in Jerusalem and round about, we took ship at Joppa, on the day of Pentecost' (1103); but fearing to meet with Saracen rovers in the open sea, the ship's captain sailed close along the shore, past various port-towns, some yet unconquered, some newly fallen into the hands of the Franks. Among the latter were Azotum[1] or Arsuf, Caesarea, Cayphas or Haifa, and Tortosa; among the former Tyre, Sidon, Acre, Byblus or Jebeil, Beyrout, Antaradus, Tripolis, Laodicea or Latakiyeh, and Gabala or Gibel, 'where are the mountains of Gilboa.' But in spite of their caution the travellers did not quite escape. Between Acre and Haifa, under the lee of Mount Carmel, they sighted a large Saracen fleet, led by the Admiral of Tyre and Sidon, who was 'carrying an army to Babylonia to assist the Chaldaeans in making war on Jerusalem.' For the Fatimites of Cairo were now sending relief to their brethren in Ascalon, hard pressed by King Baldwin; just as on the Northern frontier the Moslems were recovering from their defeats, attacking Edessa, and besieging the Norman Bohemond in his own conquest of Antioch. With his usual recklessness Saewulf confuses the mediaeval Babylon of Egypt with the Scriptural Babylon of the Chaldees, and the enterprises of Moslems in the twelfth century with the expeditions of various heathen enemies of the Chosen People before the Christian Era. All was one to him.

But in the present crisis matters at first were threatening; the Christians could only muster three vessels; and two of these, full of unwarlike and timid palmers, incontinently made off to Caesarea, disappearing with unpleasant speed round the great headland of Carmel; but their flight did not dismay the men of the *Dromund*, in which Saewulf

[1] The ancient Apollonia, north of Jaffa, not to be confused with Azotus or Ashdod to the south of Jaffa.

had trusted himself. In all they numbered two hundred warriors, and they were determined not to yield without a struggle. The ship's stern-castle was garrisoned and everything made ready for so stout a defence, that the infidels went on their way. It was not the first or the last time that a brave countenance proved the remedy in a desperate case.

This was on the Wednesday after Pentecost; and eight days later, striking across the open sea to Cyprus, the wanderers arrived at St. Andrew's Haven, near the northeast extremity of the island. Thence they reached the mainland of Asia Minor, and coasted along to Rhodes, passing Antiochetta, Myra,[1] Satalia, and many other ports, and narrowly escaping capture by pirates and shipwreck by sudden tempests. Hiring a smaller vessel at Rhodes, on the 23rd of June, that they might proceed more rapidly, they hastened on to Constantinople, touching at Stromlo, the ancient Astypalaea, 'a fair city entirely wasted by the Turks,' Scio, where they changed their bark again, and Tenit or Tenedos, near which, on the coast of 'Romania,' was the ancient and renowned city of Troy. Saewulf learnt from his Greek informants that the ruins of this place were still visible over a space of many miles; but he did not land here, any more than at the 'great town of Smyrna,' which he also passed at a distance, and briefly noticed in his narrative, now shrunken to a series of very meagre jottings. Coasting along the Troad, Saewulf entered the narrow sea which the Ancient World knew as the Hellespont, and the Middle Ages as the Strait of St. George. In the pilgrim's language, it divided the two lands of Romania and Macedonia, and through the narrow passage he sailed towards Byzantium, 'having Greece on the right and Macedonia on the

---

[1] *Stamirra* here. This, like *Astamirle*, was a regular mediaeval corruption of Myra or Myrrha in Lycia.

left.' The new meaning here given to 'Greece,' as if synonymous with Mysia, is even more strange than the arbitrary limitation of the terms Romania and Macedonia, the one to Asiatic, the other to the European, provinces of the Byzantine Empire. In the Dardanelles the traveller notices the 'keys' of Constantinople, *St. Phemius*[1] and *Samthe*, perhaps the ancient Eleonta and Xanthos, so close together, though on opposite sides of the Strait, that two or three bow shots would reach across. Here, then, the capital was almost impregnable from the sea. Saewulf passed the 'Arm of St. George' about Michaelmas time in 1103, and soon after this festival he arrived at Rodosto[2] with the wearisome experience of some four months' voyage from Jaffa.

Our pilgrim was now close to Constantinople, and before returning to his home he wished to perform his devotions in the capital of Eastern Christendom; but his narrative breaks off abruptly at the 'noble' place of Raclea or Heraclea, the modern Eregli, on the north shore of the sea of Marmora, 'whence, as the Greeks say, Helen was carried off by Paris Alexander.'

With Abbot Daniel of Kiev we come upon a new line of pilgrim travel. He is not the earliest Russian devotee whose journey to the Holy Land has been recorded. For among instances of this tendency we have some unnamed examples of 1022,[3] and the mention of St. Varlaam, Abbot[4] of the Lavra of Kiev, under the year 1062. But no record, journal, or note-book, kept by a Russian traveller of this age has yet been found.[5] Daniel, at the opening of the

[1] 'St. Phemius the bishop.' Apparently a mistake for the Virgin and Martyr Euphemia.

[2] 'Rothostoca' in Saewulf.

[3] Mentioned in the life of Theodosius of Kiev.

[4] 'Ηγουμένος—Abbot or Prior, which is also Daniel's title.

[5] Cf. H. M. Hagenmayer, *Ekkehardi uraugiensis abbatis 'Hierosolymita'*—Tübingen, 1877, pp. 360-362.

twelfth century, is the oldest Russian author, sacred or secular, who has described a journey from his country to any part of the outer world. More than that, he is among the very earliest names in Russian literature of any sort. Part of 'Nestor's' chronicle (A.D. 1066-1115) may be some ten or fifteen years older, or even more; but the father of Russian history is Daniel's only important predecessor. The present narrative was evidently much valued by our pilgrim's countrymen, for seventy-five manuscripts of it exist, the oldest being of A.D. 1475.

Daniel's pilgrim record may be fixed to the years 1106-7, with fair probability, from internal evidence. For one thing, he tells us that his journey was in the reign of the Grand Prince (Michael or) Mikhail Sviatopolk, son of Isiaslav and grandson of Yaroslav the Law-giver, who ruled at Kiev from 1093 to 1113. Again, he has a good deal to say of his intercourse with Baldwin, King of Jerusalem from 1100 to 1118, and of the peculiar kindness shown him by the Latin Prince. Thirdly, he notices that when he passed through the Holy Land, Acre belonged to the Franks. It was taken by them on the 26th of May, 1104, so Daniel's visit must have been subsequent to this and to the time of Saewulf's departure. Fourthly, he claims to have accompanied Baldwin on his expedition against Damascus; this appears to have been undertaken not earlier than 1106 or later than 1108. These are the chief marks of time. But it may also be remembered that Daniel, in his careful description of the miracle of the Holy Fire on Easter Eve, makes no allusion to any Latin Patriarch, but puts one of the Frankish bishops into the place which, as we know from other sources,[1] was usually assigned to his superior. This coincides with the well-attested fact that there was no

[1] Especially Fulcher of Chartres.

Latin Patriarch at the date of Easter, 1107; for the See was then vacant. Dagobert left the Holy City in 1103; and Ebremar, his substitute, also started for Rome towards the end of 1106. Once more, Daniel specially mentions the dangers from the Saracen outpost at Ascalon to Christian troops and travellers in Syria; and in this some have seen not only a general statement, but a particular reference to an attack of the Ascalon Moslems upon some passing Christians on the way between Jaffa and Jerusalem, in the year which saw the abortive commencement of the siege of Sidon (1107).[1] Our next pilgrim, King Sigurd of Norway, helped (as we shall see), to bring this enterprise to a successful end in 1110.

Daniel was the abbot,[2] or in modern Greek-Church phrase, the Archimandrite, of a Russian monastery, whose name and position are unknown. From his reference to the River Snov, a tributary of the Desna, which flows into the Dniepr by Kiev, we may conjecture that his home was in the Province of Tchernigov. Some have identified him with a certain Daniel, who was Bishop of Suriev in 1115, and died on 9th September 1122. In any case, he was closely connected with the Mother of Russian Cities, when at the height of her power and splendour, a good half-century before her first overthrow in 1169. The Eastern Slavs had been making steady progress since the migration of Rurik and his Scandinavians to Old Novgorod in the ninth century.[3] The reigns of Oleg, who established the 'Varangian' power in Kiev; of Igor and Sviatoslav, the great pagan Rurikides, feared even at the Court of the Eastern Empire; and of Vladimir the Saint and Yaroslav the

---

[1] Cf. William of Tyre, and *Des choses avenues en la terre d'Outremer*, xi. 4, Paris, 1879, vol. i. 384.

[2] Hegoumenos or 'Igumen.'
[3] Traditionally, A.D. 862.

Legislator, who brought their country within the pale of Christian civilisation, were glorious memories of Daniel's countrymen, even at this early time. Nor were these memories only of war and conquest, of victory over heathen enemies, or of daring raids upon the Imperial City on the Bosphorus. The Russia of the eleventh and twelfth centuries, though loosely knit in political organisation, was now a different thing from the barbarian country [1] described by Ibn Foslan and the other Arab explorers 200 years before. Jury trial, justices in assize, and other refinements of the higher mediæval society, it had developed in a manner closely similar to the lines of western progress. On the other hand, its Church and faith and architecture, its code of manners and morals, came to it from the Byzantines. When Abbot Daniel started on his journey, few could have suspected the weakness of the Russian States which seemed so flourishing, so continuous in the advance of their prosperity and power. The next generations were yet to show the fatal effects of a weak and doubtful centre, of divided sovereignty, of unchecked feudalism in government.

The field covered by Daniel's pilgrimage was much wider than that of most western devotees. Had he but recorded the early stages of his journey from the middle Dniepr, his work might perhaps have risen into the primary class of mediæval travel narratives. Even as it is, we must recognise the value of those glimpses of Eastern Europe which he gives us, those allusions to the pre-Mongolian principalities of Russia, whose life was so sternly cut short just a century after his time by the Tartar sword, but whose slavery proved the training for a second and greater time of national growth.

[1] Inhabited by people 'the most unwashen of all the men whom God has made' (Ibn Fozlan).

We must notice, moreover, in the present narrative, its fulness of detail, its evident good faith, its anxiety to verify by personal experience. Daniel is not without the credulity of the wonder-seeker, and he makes terrible slips at times; but wherever he has been himself his record is almost always a true one, and where he has not been, he frankly tells us that he only repeats the words of others. He makes no pretence to rhetoric. 'Forgive me,[1] my fathers, my brethren, and my lords,'—so he closes one of his last chapters,—'do not despise the ignorance in which I have been led to describe these holy places, in simple words, without the skill of letters. If I have written without learning, there is here at least no lie; for I described nothing that I did not see with my very eyes.'

Daniel's narrative shows us—if other evidence were lacking—that Palestine was far from settled in the early days of the Latin Kingdom. The roads from Jaffa to Jerusalem, from Bethlehem to Hebron, from Jerusalem to Jericho, were in constant danger from Saracen raiders. The garrison of Ascalon were especially feared in the neighbourhood of Lydda, in the wooded hill country near Solomon's Pools, and in the Wilderness of Judaea. In Central Palestine, Scythopolis or Bethshan was almost as great a centre for Moslem hostility. So the ways from Jerusalem to the Sea of Galilee, from Mount Tabor to Nazareth, and from one side of the Lebanon to the other, were all unsafe without an escort. Panthers and wild asses were noticed by Daniel in the plateau land to the west of the Dead Sea; lions in the almost tropical jungle of the Jordan Valley; and date-palms in the hot lowlands of Jericho and Bethshan.

The Russian pilgrim shows no sectarian feeling; no

[1] Ch. xcvi. in *Itinéraires russes, Société de l'Orient Latin*, 1889.

hostility is apparent—only once a shade of contempt[1]—in his language about the dominant Latin Church. On the other hand, his religion 'humbles his unworthiness in the dust.' We need not of course take too literally in his case, any more than in Bunyan's, the violent self-accusation of one who declares himself to be a scandal to the holy way, given up to every vice, and incapable of doing good. It is probable that our monk was led to undertake his journey, among other things,[2] by the news of the Christian conquest of Jerusalem; but he himself assigns his own sins as the sole cause of this act of penitence. Yet with a breadth of view which has often been an honour to the best men of the Orthodox Church, he rejoices to think how many holy souls, by practising good works, had reached the sacred land in spirit, and been acceptable to God, without once leaving home; while sinners there were ('of whom I am chief') who visited the holy places, and, swollen with pride, lost all the fruit of their labour.

Abbot Daniel was apparently entertained at the various Greek and Orthodox, even sometimes at the Latin Monasteries in Palestine; and in particular, he seems to have owed not a little to the guidance of an old monk from the Lavra of St. Saba in the Kedron Gorge. Naturally, his traditions are those of the Eastern Church;[3] his knowledge of the Bible and the Apocryphal Gospels is a knowledge of the Greek text, as we may notice in the case of the pseudo-Apostolic *Protevangelium of James*, from which he derives some of his sacred lore.

[1] In describing the service of the descent of the Holy Fire, 'The Latins began to mumble after their manner.'

[2] But we must remember that though records are scanty, it is clear that Russian Pilgrimage had begun to flow steadily to Syria before the first Crusade; and both in dress and manners the Eastern Slavic peoples, from their greater similarity to Oriental standards, may have found a better reception in the Levant than Latin travellers.

[3] The 'Assyrians' of Saewulf.

DANIEL'S NARRATIVE : ITS VALUE

From the present record, it would appear that Orthodox and Catholic, Greek and Latin, were now on friendly terms in Syria, as in other parts of the Levant; the same is attested by the historians of this period ; and, indeed, it is not till after the failure of the Council of Florence (1439-1443) that the Eastern and Western Communions frankly and entirely abandon the ground of common Christian feeling for that of sectarian hatred. As examples of this we may notice, among other points, the deference paid by the King of Jerusalem to the clergy and monks of St. Saba; the Orthodox control of the Church of the Holy Sepulchre, even its keys being in Greek hands; the superior honour allowed to the latter during some of the Eastern ceremonies when the 'Orthodox lamps' were placed on the sepulchre itself, while those of the Latins were only hung around ; and the union of the Franks with the Eastern Christians in the service for Holy Saturday, and in the cry of *Kyrie Eleison*. On the other hand, we have a somewhat different account from Fulcher of Chartres, Daniel's chief parallel, who was present on a famous Easter Eve, when the Holy Fire failed to descend. For, while confirming Daniel in almost all points, he says that in the Easter Eve service the Lections were first read by the Latins in Latin, and then by the Greeks in Greek; also that the Patriarch of the Latins opened the door of the sepulchre, whereas in Daniel this is done by one of the Greek bishops.

Along with many excellences Daniel has the defects of a scanty book- and place-knowledge. He puts Capernaum on the sea coast near Carmel. He identifies Lydda[1] with

---

[1] 'Formerly called Lydda, now Rambilieh' (ch. lxiii.). Daniel also identifies the Rama of Jeremiah, perhaps the modern Kuriet el'Enab, with Rama, 4 versts from Emmaus (cc. lxi. lxii.). The twelfth century seems to have recognised a Capernaum on the Mediterranean near Carmel as well as a Capharnaum on the Sea of Galilee. Daniel notices the latter in ch. lxxxiii.

Ramleh; Caesarea Philippi with the greater Caesarea on the coast; Samaria, Sebaste, or 'Sevastopol' with Nablûs or Shechem; Bethshan, the 'City of the Scythians,' with Bashan, the home of Og. On the Sea of Galilee, the rivers that flow into it, and the *town* of Decapolis, he has some very strange geography. Andrew and Peter are the 'sons of Zebedee.'[1] These and other surprises, especially the often extravagant measurements of distance, can all be paralleled from the narratives of Western pilgrims; they are perhaps sometimes due to corruptions of the text; and they usually relate to parts which the Archimandrite did not personally visit.

Abbot Daniel's account of the Holy Places, like Saewulf's, has the special interest of preserving some record of a state of things, and especially of buildings, which in great measure passed away under the Frank rule in Syria, as larger schemes of Western adaptation were carried out. Comparing his narrative with that of the Worcester pilgrim, we may say that it is in general fuller, as well as more accurate and more observant; several details[2] of interest in the Topography of Palestine, omitted by his predecessor, are supplied by him; while there is scarcely anything in Saewulf which is not also to be found in Daniel. His three excursions, to the Jordan and Dead Sea, to Bethlehem and Hebron, and to Damascus, were more extensive and more thorough than those of most other devotees; on the last-named journey he had the advantage of accompanying Baldwin and his troops, and thus of visiting places which no Christian could have seen without an escort. Lastly, his account of Jerusalem is remarkably clear, minute, and accurate, the result

---

[1] A confusion of Mark i. 16, 18, with Mark i. 19, 20.
[2] *E.g.* the Pit of Jeremiah.

of his long and studious sojourn at the *Metochion* of the
Community of St. Saba, near the Jaffa Gate.

On his outward way from Russia to Jerusalem Daniel's
record is somewhat meagre. Unfortunately, he begins with
Constantinople, so that we learn nothing about the first
part of his route. From the Bosphorus he goes over much
the same ground as Saewulf traverses on his return, by
Heraclea or Eregli, Gallipoli, and the north coast of the Sea
of Marmora, through the Dardanelles, and so into the
Mediterranean or 'Great Sea.' Here Russian pilgrims
turned to the left for Jerusalem, and to the right for the
Holy Mountain of Athos or for the city of Old Rome.
Near Heraclea a sacred oil rose from the depths of the sea,
in honour of the martyrs who had suffered thereabout.
With this story of Daniel's we may compare a remark of
Strabo on the discharge of 'asphalt' into the Ægaean
from Mount Athos, at the foot of which was another
Heraclea.[1]

Thence by way of Tenedos, Chios, and Ephesus (where
he tells us the usual wonders about the Tomb of St. John[2]
and the Seven Sleepers), to Patmos, the rich and populous
Isle of Cos, Telos or Nisyros, celebrated for its production
of sulphur or 'Herod's torment,' and Rhodes. In this
island Prince Oleg the Rurikide had spent two summers and
winters; this is abundantly confirmed from other sources;
it was in 1079, according to the Russian annals, that he was
made prisoner and carried to 'Greece.' From Rhodes Daniel,
like Saewulf, coasted along Lycia, famous for the Tomb of
St. Nicolas and the storax that was gathered upon the hills
near the coast. Cyprus was notable not only for the relics
of St. Epiphanius, but also for a miraculous cross of the

---

[1] Now Eraclitza; cf. Strabo, fragment 56.

[2] *E.g.* the holy dust rising yearly from this sepulchre.

Empress Helena, which, standing on a high mountain,[1] did not rest upon the ground but hung in space, supported by divine power. 'I, the unworthy, worshipped this holy and miraculous thing, and saw with my sinful eyes the divine favour resting on this place.' From Cyprus to Jaffa the pilgrim reckons 400 versts, or about 270 miles; from Constantinople to Rhodes it was 800 versts; and from Rhodes to Jaffa 800 more, making in all 1600 for the sea voyage to Palestine. These distances, like the 30 versts given for the way from Jaffa to Jerusalem, are fairly accurate; but with them we may contrast the 80 versts' interval between Gallipoli and Abydos, the 40 versts allowed for the length and breadth of Merom or Gennesareth, and the hardly less exaggerated dimensions of the Sea of Tiberias or Galilee, which is reckoned as 20 versts by 50.[2] In explanation or excuse we must not forget that the Roman Itineraries had now fallen out of use and memory, and that all measurements had to be worked out afresh. Ten versts from Jaffa brought Daniel to St. George of Lydda, where pilgrims often rested, 'but in great fear,' for the place was desolate, and unpleasantly near to the Saracen stronghold of Ascalon. On the foundation of the Moslem Ramleh in the early years of the eighth century,[3] the whole population of Lydda had been moved to the new town, and the ancient site had been a waste ever since; now, with the Crusading period, a partial revival of the old town began again.

Approaching Jerusalem, the traveller first saw the Tower of David, which then, as now, marked the entrance from

---

[1] Mount Troödos.

[2] All these are more than double the true figures; from Gallipoli to Abydos is about 30 versts; Merom is about 10 versts by 5 (7 by 3½ miles); the Sea of Galilee 18 versts by 10 (12½ miles by 7).

[3] Ramleh was founded by Suleiman Ibn Abd el Melek, A.D. 716; cf. Z.D.P.V. (= *Zeitschrift deutsch. Palästina-Vereins*), iv. 88.

Jaffa and the Mediterranean; and, reaching a plateau [1] which commanded the city about a verst from the walls, he dismounted and 'adored the holy resurrection with the sign of the cross.' Like almost all pilgrims, whether from West or North, Daniel entered by the Western or Jaffa Gate, otherwise known as the Gate of Benjamin, 'near to the house of the Psalmist.' At that time the 'Portal of King David,' as Saewulf calls it, belonged to a wall of narrower circuit than the present, and so lay somewhat to the east of the modern site. In exact agreement with his English predecessor, the Russian pilgrim notes that the Church of the Holy Sepulchre was open to the sky in the middle of the rotunda, and contained the Navel of the Earth.[2] Adjoining the umbilical spot, he adds, was the representation of Christ in mosaic, with the inscription: *The Sole of my Foot is the Measure*[3] *for the Heaven and the Earth*. To the 'Tower of David' he was admitted, though the 'lowliest of men,' with one companion, a certain Isdeslav; he gives an excellent description of this important fortress; and, like Bernard the Wise, he notices the neighbouring church of St. Mamilla, where so many martyrs had suffered in the Persian massacre of 614, 'when Heraclius was emperor.' A little further to the west of the city he also mentions an interesting link with the farthest west, an Iberian or Spanish Convent, which is probably the famous Monastery of the Cross.

Like St. Silvia of Aquitaine, Daniel shows an occasional reluctance to accept or even to repeat every sacred story;

[1] Perhaps an elevation on the way to Bethel, now called 'El Mesharif.'

[2] Some think the 'hell' described by Daniel near the Church of St. Stephen, which split asunder at the time of Christ's Crucifixion, is the mount above 'Jeremiah's Grotto,' just outside the present north wall.

[3] Daniel gives all his measurements of buildings in Russian Sagenes. A little later he quotes some striking passages from the Holy Week anthems of the Eastern Church.

the traces of the blood of Zacharias 'slain between the Temple and the Altar' was a favourite pilgrim wonder; but Daniel will not admit that these traces were any longer visible. 'Formerly they were, but not now.' Many devotees also had discoursed on the marvellous remains of the ancient temple. These had utterly disappeared, says Daniel; only some foundations were left. The present church was built by a 'Saracen chief named Amor.' In spite of the corruption of Omar's[1] name, it is sufficiently remarkable to find a Christian pilgrim so far informed about Moslem sovereigns and buildings.

We may suppose it was during the latter part of his stay in Palestine that the Russian Archimandrite passed sixteen months[2] in the Jerusalem House, Metochion, or Filiale, of the Lavra of St. Saba. This daughter-settlement, a stone's-throw only from the Tower of David, was now occupied, and apparently even crowded, by the monks who had just escaped from the destruction of the Mother-House and the slaughter of their brethren in the Kedron Valley. It is also noticed by Saewulf and by John of Würzburg; and it affords a good example of the constitution of the typical Lavra of the Eastern Church. Here monks were lodged in separate cells, though under a common superior; and it was easy to form out-settlements under this system, which afforded a sort of compromise between the original Hermit- or Anchorite-life of the earliest monks, and the community-life of later times and of Western pattern. Whether before, or after, the sack and massacre which had desolated the original House in the wilderness of Judaea, Abbot Daniel paid a visit also to this celebrated spot; and his account of the hanging monastery may be verified in all its details—the cells of the Lavra attached to the rock, 'like stars in the firmament,' and

[1] Cf. William of Tyre, i. 2; viii. 3, for the Omar tradition.   [2] Ch. i.

DANIEL IN PALESTINE

the dry torrent bed 'of utmost depth' beneath, shut in between high walls of rock, from which the habitations of the monks projected like the nests of birds. One of the brethren, an aged man, 'well versed in Scripture,' became the friend and guide of the Orthodox visitor, and under his direction Daniel journeyed to the Sea of Tiberias, to Tabor, to Nazareth, to Hebron, and to the Jordan. Another holy man of Daniel's acquaintance, an anchorite, 'terrifying of aspect, austere, and old,' lived as a Stylite or Pillar-saint in a deep cavern on the south side of the Mount of Olives, by the 'tomb[1] of St. Pelagia the courtezan'; and this is probably the same ascetic whom the Frank army consulted at the siege of Jerusalem in 1099.[2] From one or other of these informants the Russian pilgrim may have learnt the time-honoured and absurd traditions which placed Mount Hermon close to Jerusalem, and maintained the literal and annual turning back or retrocession of the stream of Jordan by a fanciful interpretation of certain passages in the Psalms.[3] But it was not from any 'Assyrian' Christian or native dragoman that he derived his suggestive comparison of the Sacred River with an object near his home and familiar to himself. Never before or after in travel literature, we may conjecture, was the Jordan likened to the Snov in Little Russia. Yet 'in every respect they were alike, both in size and in depth, and in their winding and rapid course.' Also the width of the Jordan at the 'Place of Baptism,' near Jericho, was the same as that of the Snov at its mouth; the numerous creeks of the two streams, and their great sheets of stagnant water, offered other points of likeness.

[1] This is very frequently noticed by pilgrims., *e.g.* by nearly all the *Innominati* or Anonymous Pilgrims.

[2] According to Albert of Aix.

[3] Cf. Ps. xlii. 8; cxiv. 5; Antoninus Martyr, ch. xi.

Daniel was not able to visit the Sea of Sodom, partly for fear of 'miscreant Saracens,' partly from the dissuasions of the faithful, who told him of the fetid smell rising from the infernal lake, which none could approach without torment. Speaking, therefore, only from hearsay, he naturally repeats the fashionable fables: 'The Sea of Sodom exhales a burning and malignant breath, which lays waste all the neighbourhood; it is dead and contains no living creature; and if the Jordan carries any fish into it, they perish immediately and cannot live one hour. A black pitch rises from the bottom and covers the banks, and the foul vapours are like burning sulphur; the torments of hell lie under the sea.'

Beyond Jordan, adds Daniel, inverting the sense of a famous text, was the land of Zebulon and Naphthali. Here many of the western pilgrims, especially in the fourth, fifth, and sixth centuries, had placed the Hill of Hermon; but the Slavonic visitor, with a certain independence not uncommon even in his traditionalism, locates it on the west or opposite bank of the river.[1] Hard by was the place where, at the baptism of Jesus, the Jordan turned back and the Sea of Sodom fled, 'seeing the Divinity in the midst of the waters;' in the same neighbourhood was Mount Gabaon, over which the sun stood still for half a day, so that Joshua might triumph when he fought against *Og, King of Bashan.* Along with this we may take some other distortions; the death of Absalom in a wood near *Hebron;* the house of the

---

[1] Basing this on an interpretation of Psalm xlii. 8, Daniel also describes (ch. xxxiv.) the Epiphany baptism of Catechumens in the Jordan at midnight 'like Christ Himself'; in chs. lv lvi., he seems to place the sites of both *Sodom* and *Zoar* west of the Dead Sea. The place of the sacrifice of Abraham at Hebron was one verst from the double cavern of Macpelah and also one verst from the valley of Greznova, which was two versts from Sigor or Zoar; the picture of the B.V.M., noticed in the Church of St Gerasinus near Jericho, is now in the Patriarchal Church of St. Constantine at Jerusalem.

Shunamite in the Saracen village of *Jericho;* and the healing of the two blind men at *Bethshan* or *Bashan.* The fertility of the Hebron region arouses Daniel's admiration; but the town he describes, like Saewulf, as now a mere waste. Here both these Christian travellers may be usefully compared with Benjamin of Tudela; but it is not likely that either of them was able to anticipate the Jewish traveller, and to enter those caves at Machpelah, which were only made accessible in 1119.[1] In the wooded hills near Mamre, where 'God paved the soil with white marble like the pavement of a church,' Daniel was in great danger from the Moslem brigands of the highlands and the 'miscreants' of Ascalon; but he came through in safety, with a good and numerous company, under the escort of a Saracen chief, and saw with his own eyes the stone column of Lot's wife near Segor, and many other places of 'indescribable sanctity,' on the west of the Dead Sea.[2]

After enumerating and describing all the places in and around Jerusalem,[3] the Russian pilgrim attempts a sort of Itinerary of Northern and Central Palestine; and here it is he makes some of his strangest slips. Caesarea *Philippi,* on the coast of Palestine, thirty versts from Jaffa, was but eight versts from Capernaum, which itself was not far from the Great Sea or Mediterranean. With this we may compare Benjamin of Tudela, who puts Capernaum four parasangs from the Kishon, and six from Caesarea; probably both travellers intended by the term some place on the

---

[1] Cf. Comte Riant, *Archives de l'Orient Latin,* iii. 411-421.

[2] Among these was the field near Bethlehem, whence the angel carried away the prophet Habakkuk to Babylon and back again in the same day and hour. Cf. Daniel, ch. lvi.; Lievin, *Guide Indicateur,* 331. This place is now shown near Mar Elias, between Jerusalem and Bethlehem.

[3] Among other peculiarities here, we may notice Daniel's 'Anatolia,' apparently meant for 'Araunah': like his 'Aphrodisius' for 'Epaphroditus,'

Syrian coast near Mount Carmel.[1] However this may be, here Antichrist was one day to manifest himself, and for this reason the Latins had abandoned the town.

Six versts from Capernaum brought one to Mount Carmel, where Elias slew the priests of *Babel*. Thence to Acre, now occupied by the Franks, and the coast towns to the north, 'philosophic' Beyrout, Great Antioch, Byblus, Tripolis, Latakiyeh, Little Antioch in Cilicia, and Satalia or Adalia. Most of these places Daniel passed without landing;[2] he probably took ship at Beyrout, and returned to Constantinople by the same way that he had come. But all this was after his tour in Galilee, a province of which he gives a detailed account. It lay, he believed, to the east or 'summer sunrising' of Jerusalem; Daniel's fondness for fixing his places by the position of the sun at the winter or summer solstice is a reminiscence of the days when clocks and watches did not exist, and when 'pointer-stones were set[3] up to mark the sun's furthest deviation north and south.' From the Holy City to Tiberias the pilgrim travelled in the company of King Baldwin, who was now 'making a road' against Damascus and the sources of the Jordan. Thus he passed in safety the Mountains of Gilboa,[4] on which no dew ever fell, and by way of the land of Samaria,[5] a district

---

[1] Athlît and Khd el Kéniseh have both been suggested. The latter is apparently the Capernaum of William of Tyre, Geoffrey de Vinsauf, and James de Vitry. See Guérin, *Samarie*, ii. 274-282, and Sir C. Wilson in the P.P.T. Soc. Edn. of Daniel.

[2] Sir C. Wilson thinks Daniel travelled by land either to Beyrout or to Sueidiah, the port of Antioch, and then embarked for Constantinople. At Beyrout Daniel also mentions the famous Eikon which shed blood and water on being pierced with a lance by blasphemous Jews.

[3] Cf. Sir C. Wilson in P.P.T. Soc. Edn. of Daniel.

[4] Gilboa was only 14 versts, 9½ miles, from Jerusalem, according to Daniel. He is thinking, perhaps, of Lubbân, between Jerusalem and Nablûs.

[5] Daniel repeats his confusion of Samaria with Nablûs in ch. xcv.

so fertile as to be the Granary of Jerusalem,[1] descended into the great trough of the Ghor down to the Sea of Galilee. The Jordan poured out of this lake, he tells us, in two *branches*,[2] one called Jor and the other Dan (here is an improvement on the older stories); and near this exit Prince Baldwin dined with all his troops. From this point the Crusading army parted company with their guest, crossing the Jordan at the old bridge, just below the lake; while Abbot Daniel pushed on to Tiberias, Tabor, and Nazareth, escaping from the perils of lions, and the still more dangerous, powerful, and impious Saracens of the Bethshan country. 'Unworthy sinner that I am, God granted me to see all this land of Galilee upon which I never hoped to set my eyes, all the holy places upon which Christ, our God, set His feet.' Truthfully and fully, declares the pilgrim, has he described all these places; some others had not been able properly to explore all they would fain tell of, and so had been led into error; others again, had sought refuge in lies and fables. But the company of Baldwin's troops and the guidance of the old Sabaite monk, who had passed thirty years in Galilee, saved the writer from such mischance. He was able even to make trial of the fish of the lake—especially a kind of carp, of taste excelling all other fish. The Lake of Merom he only saw from a distance, and he describes it very loosely, under the name of 'Gennesareth,' as *larger* in area than the sea of Tiberias, with which it was linked by a river, clearly a different stream from the Jordan of his

---

[1] At Bethshan, Daniel locates the stories of the Tribute Money, the Stater in the Fish's Mouth, and the healing of the Two Blind Men. See Matthew xvii, 24-27; xxii. 17-21; xx. 30-34.

[2] Instead of the usual 'fountains'; the two branches which formed the Jordan *below* the Sea of Galilee may be our Jordan and Yermak (Hieromax), or two arms of the Jordan itself, which then perhaps encircled Kerak of Galilee (Sennabrin). It is evident that in Daniel's mind the Jordan began at the Sea of Galilee.

geography. In this region was the great *town* of Decapolis, and towards the east Mount Lebanon, infested by Saracens, and therefore unapproachable, but from which (it was said) there poured down twelve large streams, six towards the east, and six towards the south; the latter fell into Gennesareth, the others flowed towards Antioch the Great; and the country between them was that Mesopotamia, or *Land between the rivers*, wherein Abraham's 'Charran' was to be found.[1] Lebanon was out of Daniel's reach; but Tabor, a 'marvellous work of God,' he ascended with six hours of 'toilsome labour.' His description is photographic. 'It is the highest of neighbouring mountains, and is perfectly isolated, rising in majesty from the middle of the plain like a round hay-cock, formed by art, and of a great circuit withal.' Here[2] took place not only the Transfiguration, but the meeting of Melchisedek and Abraham. In Nazareth, which had now passed into the hands of the Crusaders, the Orthodox traveller was received by the Latin bishop; and from this point, or rather from Cana of Galilee, he retraced his steps to Jerusalem in the company of a large caravan, by way of Acre, Haifa, Caesarea, and 'Samaria.'[3] The narrative ends with an elaborate account of the Easter services in the Holy City, with some words of praise and thanksgiving on the accomplishment of the traveller's pious hopes, and with a commemoration of those countrymen of his whose names he inscribed for the prayers of the faithful at the Lavra of St. Saba.

Desirous of witnessing the descent of the Heavenly

---

[1] All the mythical matter just preceding, like much in Saewulf, was probably derived from that licensed monstrosity, the 'Palestine Guide.'

[2] Or close by, between Nazareth and Tabor. 'Melchisedek often comes there [still] to celebrate the liturgy,' adds Daniel, ch. lxxxviii.

[3] Nablûs or Shechem.

Fire,[1] 'when it comes down invisibly from Heaven and lights all the lamps in the tomb of our Lord,' Daniel obtained an audience of Prince Baldwin. 'On the Wednesday in Holy Week, at the first hour of the day, wicked and unworthy as I was, I presented myself before the Prince and saluted him. Perceiving my insignificance, he affectionately called me to him, and said: "What will'st thou, Russian Abbot?" For he knew and loved me, being himself a man of little pride and of great goodness and humility. And I said: "My Prince and my Lord, I beg thee, for the sake of God and of the Russian Princes, permit me to place my lamp on the Holy Sepulchre in the name of the whole Russian land."' Permission was granted with special graciousness, and Daniel records with triumph that his own lamp, being inside the Holy Grave, caught fire, but none of the Frank lamps which hung above the Tomb were so fortunate. On Baldwin's entrance into the Church on Holy Saturday 'about the seventh hour,' Daniel walked by his side with the Abbot of St. Saba, and close by, among the throng of worshippers, were various friends and companions from Kiev and Novgorod—Isiaslav Ivanovitch, Gorodislav Mikhailovitch, the two Kashkitch, and others. 'Thanks be unto the mercy of God,' concludes Daniel, 'who has permitted my unworthiness to inscribe the names of the Russian Princes in the Lavra of St. Saba—to wit Mikhail Sviatopolk Isiaslavovitch,[2] in whose reign I made this pilgrimage, Vassili Vladimirovitch, David Sviatoslavitch, Mikhail, Oleg, and Pancrati Sviatoslavitch, and Gleb of Minsk. Their names and those of their wives and children, and of the Russian bishops, abbots, and boyars, together with

---

[1] On the Holy Fire, cf. Tobler, *Golgotha*, 460-483.

[2] He was the grandson of Yaroslav 'the Great,' Grand Prince of Kiev, famous for his Black Sea expeditions against Constantinople (1016-1054).

all faithful Christians, I did not forget to commemorate at each of the Holy Places.'

On his homeward journey, the Slav pilgrim was (like Saewulf), attacked by pirates near Patara in Lycia—unlike Saewulf, he did not wholly escape; but though plundered by the four galleys of these brigands, he saved his life, and returned at last to the city of Constantine and Holy Russia. At the end of his wanderings he returns thanks to God for his singular good fortune. Brigands, wild-beasts, and disease had all threatened, but none had been able to touch his life. Like an eagle taking its flight, he had been sustained by the Divine Power; like a stag, he had gone boldly forward, without fatigue and without loitering. To his own countrymen and to modern students there is another interest in the pilgrimage of Abbot Daniel; for it is an expression in written form of one of the chief elements of Slavonic life; it is among the earliest landmarks in the history of a great people; and it illustrates the truth that (with races as with individuals) internal progress is constantly associated with external activities, with an irresistible tendency to move out into the world, to travel, to explore, perhaps to colonise and to conquer.

From the time of the conversion of the Northmen a constant stream of Scandinavian pilgrims had flowed to the Levant. In the first quarter of the eleventh century, Denmark, Sweden, and Norway had all alike thrown off their national and official heathendom, though a local and unrecognised paganism long survived in country districts; and it was not long before some of the greatest Northmen are found upon the pilgrim way. Three chief routes were followed by Scandinavians bound for the Holy Places of Syria—the Eastern, the Western, and the Roman. The first of these, also called the Vaerings' Path, was historically the earliest and led through the Slavo-Scandinavian principalities of Russia to the Black Sea, the Bosphorus, and the Levant. The second, or long sea-way, led round the coasts of Northern and Western Europe and through the Straits of Gibraltar. The third, or overland track, conducted the Northman from any of the North Sea ports to Italy and the Threshold of the Apostles, leaving him to proceed eastward from one of the havens of 'Pouille.' We have already seen how Thorir Hund, 'according to the talk of many,' went off to the Holy Land after the battle of Stiklestad; and how Skopti Ogmundson, in the reign of King Magnus Barefoot, sailed with five ships along Flanders and France and Spain into the Mediterranean. Some said [1] that Skopti was the first Northman to sail through the Straits of Gibraltar;[2] in any case, he and his sons died on the journey, Skopti himself in Rome, others of his company in Sicily, others again in Palestine or Constantinople. But some of Skopti's followers made their way home again and brought with them wonderful stories of the splendid or sacred countries of the South; and herewith they roused the sons[3] of King Magnus among others to imitate their example. Magnus himself had been famous for his voyages and conquests in the British Seas and Isles; his children, like most men in Norway, had the same old yearning for glory and novelty and adventure; and, where Harald Hadrada had gone, his successors were eager to follow. As in Harald's time, Northmen could still win fabulous wealth in the service of the Byzantines; and whatever might be the fate of men in this world, they were certain of the next, if they journeyed in his steps to the Holy Places.

[1] So *Magnus Barefoot's Saga*, ch. xxii.; cf. *Sigurd's Saga*, ch. i.
[2] After the almost mythical Norva.
[3] But see p. 23, note.

And other weighty reasons induced the sons of Magnus Barefoot to take up the part of pilgrim warriors. The Latin Conquest of Jerusalem had created new centres of attraction for the enterprise, the commerce, the valour, and the religion of every Christian country; and Northmen might hope to carve out for themselves a kingdom in the Levant as good as the new principalities of Syria. In the Western Mediterranean such a kingdom had rewarded the daring and the persistence of a few Norman adventurers; and, beside the examples of Robert Guiscard and Roger of Sicily, there were the still greater achievements of William the Bastard in Valland and in England, to say nothing of the older feats of Rurik and of Rolf.

It was four years after the defeat and death of King Magnus in Ireland (1103) that Sigurd, his son, began his famous journey to Jerusalem (1107). First of all, he sailed to England with sixty ships, and, arriving there about the close of autumn, passed the winter at the Court of Henry the Norman.[1] In the spring of 1108 he set out again, coasted along the West of France, and came about the time of harvest to the 'James-land' of Galicia, already famous from its Shrine of Santiago at Compostella. Here the Norsemen spent the next winter, and here they soon fell to disputing, and then to fighting, with the 'Earl' or Ruler of the country, over their food supply, 'for it was a poor barren land.' To replenish their larder they plundered one of the Earl's castles, and sailed down along the west shore-land of Spain in search of better quarters. Near the mouth of the Tagus Sigurd met with a fleet of 'heathen Vikings,' and took eight of their galleys;[2] perhaps the reference here is rather to a Moslem than to a Norse pirate-squadron, for Sigurd's host had now arrived on

[1] Ch. iii. of *Sigurd's Saga*.   [2] Ch. iv. of *Sigurd's Saga*.

a field where the struggle between Christendom and Islam was proceeding very briskly. The successes of Alfonso VI. of Leon and Castile had been checked in Central Spain by the arrival of the Almoravides and the battle of Zalacca (1086); but he continued to win ground along the Western coast till 1093, even pushing up to the Tagus and storming Lisbon. In the next year, however, the Christian advance was forced back on this side also, Lisbon was retaken, and Alfonso summoned the chivalry of Christendom to his aid. Among the free lances who responded to the call was one Henry of Burgundy, an ancestor of kings, though never crowned himself; he showed himself a valuable and faithful ally, and he was rewarded with a partnership. In 1095 Alfonso conferred on Henry that county of Portugal which, like the Austrian Duchy, began as a military frontier against the heathen world, and which proved so useful and so important in checking the renewed attack of Islam on the West. The creation of this county or military district was especially to guard Oporto, 'the port' of Galicia; it had been effected only thirteen years before Sigurd's arrival at the mouth of the Douro; and it may be viewed as a second and more successful essay in a scheme of frontier defence, which in 1093 had set up a Lisbon County for Sueiro Mendes, and which, though defeated on the Tagus, was still maintaining the line of the Mondego at the time of Sigurd's visit.

The Northern Crusaders, like their followers in 1147 and 1189, rendered yeoman service to the struggling Christians of Spain under the Norse king's leadership (1108-9). The Castle of Cintra, near Lisbon, was a centre of Moslem raids, and Sigurd began by capturing it with all its stores and putting its defenders to the sword.[1] Next he attacked the

[1] Ch. iv. *Sigurd's Saga*, of which here quotes the court poet, Halldor Gabbler: 'No man he spared who would not take The Christian faith for Jesus' sake.'

great city of Lisbon, which even then, perhaps, contained 200,000 inhabitants, and which the Saga describes as half Christian and half heathen, marking the dividing line between the rival faiths and politics of Spain.[1] Since the year 716, it had been under Moslem rule, but for three brief intervals,—in 792,[2] 851, and 1093,—it had fallen back under Leon or Castile; now, in 1109, though fiercely attacked by Sigurd from the sea, and by 'Portugal' from the land, it was too strong to be taken, and its fate was deferred for almost forty years, when the same kind of combination won the decisive victory of 1147. The Norse fleet, though perforce content with an empty victory and some plunder outside Lisbon, did better work at Alcaçer[3] do Sal (a little further to the south), which fell into their hands after another battle.

From this point King Sigurd made straight for Norva Sound, and so entered the Middle Sea in the teeth of a great pirate host, which vainly tried to bar his passage;[4] here again the 'Vikings' of the Saga may be translated as Saracen rovers, or even perhaps as a more national force equipped by the Moslems of Andalûs and Barbary against their new enemy. Soon the Mussulmans had to meet an attack from a fresh quarter. Not satisfied with their Crusade in the west of Spain, the Northmen now fall upon the east coast and the Balearics. Here, where the ships of Moors or 'Bluemen' had so long found a refuge, a storehouse, and a fortress, the Christian warriors chased the Saracens into the hills, penned them up in caverns, and harried, slew, and burnt wide over Minorca and Iviça and Forminterra. In the last and smallest of these islands a great throng of the Unbelievers

---

[1] Ch. v. of *Sigurd's Saga*.
[2] On this occasion it was retained by the Christians till 812.
[3] 'Alcasse' in the *Saga of King Sigurd*, ch. v. The African Alcaçers, Algeciras, and the 'Moorish palace near Seville, called Alcasir,' have also been conjectured.
[4] Ch. vi. of *Sigurd's Saga*.

took shelter in a cave, whose entrance was fortified with two stone walls, an outer and an inner, exceedingly difficult to capture, or even to approach. For 'this cave lay on a cliff, very sheer and steep, but with a rock hanging over it from above. And the heathen kept guard well, and were not afraid of the Northmen, but defied and taunted them as cowards. Then Sigurd took a pair of ships'-boats and drew them up the rock above the mouth of the cave; and he had thick ropes fastened around the stem and stern and hull of each. Now in the boats went as many men as could find room, and they were lowered by the ropes to the mouth of the cave;' and so the heathen were driven with slings and missiles from their defence. Then Sigurd went up the rock with his host, and forced his way into the den, but the heathen fled behind their inner rampart. 'And upon this the king brought great trees, and made a pile of wood at the mouth of the cavern, and set fire to it; and thus he slew or burned all who were therein.'[1] It was a foretaste of the permanent Christian conquest under James of Aragon a century later.

Next spring the Northmen came to Sicily[2] and stayed there a long time with their kinsman, Duke Roger II. The vanity or ignorance of the Saga-writer here inserts a patriotic tale of Sigurd's condescension and generosity. Roger had bidden his visitor to a feast and served him with his own hand. But after six days of good cheer, the guest took his host by the hand 'and led him up to the high seat and gave him the name and right of King; but before there had been only

---

[1] As the skald Thorarin Curtfell (='Stuttfeld') sang:— 'The king's men up the mountain-side Drag two boats from the ocean's tide, The two boats lay Like hill wolves gray, Now o'er the rocks on ropes they're swinging Well-manned, and death to Bluemen bringing, They hang before The robbers door,' ch. vi. of *Sigurd's Saga*.

[2] Ch. viii. of *Sigurd's Saga*.

Earls in Sicily.' On the other hand, the Great Counts of the Norman Dynasty dated their kingly title from the papal grant of 1129; and if Roger, one of the most powerful and civilised of Christian princes, had wished for any such investiture by a temporal lord, he might have had it from the Eastern or Western Emperor. In any case, it is clear that if Sigurd really made any pretence of crowning the future patron of Edrisi and conqueror of Tunis, then a boy of thirteen years, it was only treated as a piece of pageantry by the Court of Sicily.

In the summer, the Norse leader went on his way over the 'Greek' Sea to Palestine,[1] landed at Acre (or at Beyrout,)[2] and went up to Jerusalem, where King Baldwin greeted him well. The Latin Prince knew how to make the most of so useful an ally in the critical struggle on the frontiers of Christendom, and he showed his guest all the honour in his power, even riding down with him to the Jordan and back again. 'And he made him a goodly banquet and gave him many relics, yea, a splinter out of the Holy Cross.' Yet the Norse king had not quite forgotten the heathen traditions of his people. For when he came to the sacred river, as he boasted on his return, he swam over the stream, and there, in a copse that was on the bank, he tied a knot and spoke over it a spell, and till that knot was loosed he thought his luck would not depart from him. But at the same time he promised to spread the faith in Norway to the best of his power, to pay his tithes, and to found an Archbishop's See, if that were possible. Still better, he showed himself for-

---

[1] Ch. x. of *Sigurd's Saga*.

[2] According to William of Tyre, Sigurd landed at Beyrout. The best text of *K. Sigurd's Saga*, chs. x.-xi. does not give the well-known fables of the clothes at Jerusalem, or of the golden horse-shoes at Constantinople, which are inserted to prove that no riches could dazzle the Northmen, or find them unprepared with an equal display of power and grandeur.

ward to take his share in the Holy War; with his fleet he helped in the capture of Sidon[1] (December, 1110); and all that autumn and winter Northmen and Franks made common cause against the Saracens. But at the beginning of the next year (1111), Sigurd left Palestine and sailed off to Cyprus and Constantinople.[2] At Angel-Ness, perhaps the mouth of the little river Ægos within the Dardanelles, he stayed a fortnight, waiting for a side wind which would fill his sails to the best advantage, and so set off his state entry into the Golden Horn. At last the right breeze came, and he coasted along towards the Great City; even the Norsemen, who had been so far and seen so much, were amazed at the wealth and population of the country along the Sea of Marmora and the Bosphorus. For 'over all that land there were castles and towns and villages, one following upon the other without any interval.' The full-spread sails of the Crusading fleet stood so close, one beside the other, that 'it looked as if it were all one wall' or one ship; and the whole of the people of the region came out to see the passing of King Sigurd. In honour of his visitor Alexius Comnenus opened the Golden Gate, through which the sovereign only was wont to ride, after a long absence, or a victory; silk stuffs were spread over all the streets of the city, from the Golden Gate to 'Laktiarn' or Blachernae, the quarter of the emperor's finest palaces; and the games were played in the Hippodrome for the entertainment of the guests. The love of sport was strong in all Norsemen, and nothing in their long journey was more delightful to Sigurd's warriors than the races and contests of the Arena at Constantinople. Within 'a high wall built round about a field, like a round bare Thing-Place, were earthen banks, and there men sat while the games were held in the open field.' Byzantium was

---

[1] Ch. xi. of *Sigurd's Saga*.   [2] Chs. xi., xii. of *Sigurd's Saga*.

then full of the masterpieces of ancient sculpture, so many of which perished in the sacks of 1204 and 1453, and the Scandinavian visitors, carrying their own traditions with them everywhere, looked on these as 'records of the Asfolk and the Giukings and the Volsungs'—figures worked in copper and metal with so much deftness, that they seemed to be living creatures, and really present at the games.[1] As to the sports themselves, they were so artfully managed that men seemed to be riding in the air; and 'shooting fires' were also displayed, together with the delights of harp-playing and song and all kinds of music.

At the end of these spectacles the Norse leader prepared for his return overland.[2] As to his ships, he made a present of them to the emperor, who gave him in exchange horses for his men and a guide through all the Byzantine lands. And so pleased were the Northmen with the city of Constantine and its ruler and its people, that many of Sigurd's army stayed behind and entered the Imperial Service. The rest with their king marched home through Bulgaria and Hungary, 'Pannonia' and Bavaria, into Suabia, where 'Lothaire, the Kaiser of Rome,' came out to meet and salute the returning Crusader. At midsummer the host reached Sleswick; and so, through 'Heathby' and Denmark, they came to their own land. 'And that was the talk of men, that never had there been a more splendid journey out of Norway than was the journey of King Sigurd.'[3]

---

[1] The Norsemen perhaps mistook the Greek sculptures for statues of their own heroes; or, rather, gave their own names to any 'heroic' images they saw, ch. xii. of *Sigurd's Saga*. The 'shooting fires' were probably fire-works; less probably, Greek-fire.

[2] Ch. xiv. of *Sigurd's Saga*.

[3] After his return, Sigurd contracted a Russian alliance, and married the grand-daughter of Valdimar of Holm-garth (Vladimir of Novgorod), ch. xxi. of *Sigurd's Saga*.

While Sigurd of Norway was still upon his travels, Adelard, or Athelard, of Bath, started upon a journey to much the same quarter of the world with very different objects. As a man of science rather than a devotee, a warrior, or a merchant, his visit to the Levant appears mainly as one of literary interest. 'He sought out the causes of all things and the mysteries of nature;' and it was with a rich spoil of 'letters,' or of manuscripts, that he returned to Europe to translate one of the chief works of early Saracen astronomy, the *Kharizmian Tables*, which had been compiled in the opening years of the ninth century under the direction of the Caliph Al-Mamûn.

After Sigurd and Adelard, the Catholic pilgrims of this period, down to the fall of the Kingdom of Jerusalem, are of less interest and importance. The number of these devotees who journey to Syria from the various distant countries of Western Europe is immense, but among them there are few who have left anything worthy of notice. In the pre-Crusading Era there was some reason to attempt an enumeration of the religious travellers who, in the absence of other material, often represented (as in the seventh or tenth centuries) nearly all that is known of geographical movement in Christendom. But, as the Dark Ages pass into the Crusading time, we are no longer left so poorly off. From the middle of the eleventh century mediaeval Christendom rapidly develops a civilisation in which every department of human activity and interest is gradually included. Among the rest, the problems of the world's shape, the positions of various countries, their inhabitants and products, occupy an ever-increasing share of attention, and both from the practical and the theoretical side the field of geography is explored afresh. Pilgrim-travel thus falls into a very secondary position, and a minute attention to the details of its later history would be entirely out of place; not merely because it would absorb a large amount of the attention that should be given to more important developments, but also because it is no longer possible to find the spirit of the genuine traveller so active among the visitors to the Holy Places. Almost all the later pilgrims (from the time of Sigurd) are of an inferior order to the best of the earlier wanderers among the sacred fields of Palestine. After the first quarter of the twelfth century there is hardly any one who can be compared with Arculf, Willibald, Bernard, Saewulf, or Daniel; in fact, men of this type now attempt something beyond mere pilgrim-travel, and those who go on the old rounds show less and less originality, power of observation, or reliable knowledge, with each succeeding century. As the type degenerates, its further decline is hastened by the compilation of standard guide-books, which may be faintly described as legendary and inaccurate, and from which the later pilgrim narratives blindly copy, to the ever more entire exclusion of anything independent or scholarly. Two of these hand-books, known as the *Old* and the *New Compendium*, are the source of most of the tracts on the Holy Road which have been left us, under various names from the time of the second Crusade to the close of the Middle Ages.

Over a material so poor in quality, so dependent in character, it is not worth while to spend much time; but we may briefly notice a few comparatively valuable specimens of this literature, such as the compilations of the so-called Fetellus about A.D. 1150, the records of John of Würzburg and of Theoderich about 1160-70, the pamphlet of the Byzantine Johannes Phokas of about 1185, and the notes of a group of Anonymous pilgrim travellers who journeyed to the Levant at various times between 1115 and 1187.

But if the Christian pilgrim-travel of the later twelfth century is of small moment, the Itinerary of a Jewish fellow wanderer is among the most valuable of the geographical works of the Middle Ages. The *Reports* of Rabbi Benjamin of Tudela have usually been treated as a pilgrim narrative, and classed with other Syrian travels of the same period. But this arrangement is impossible, from any but a mechanical point of view. Benjamin is a religious Jew who visits the old homes of his people, and so describes most of the sights noticed by the Christian devotees in the Levant; but this is only a small part of the truth. Taken as a whole, the Rabbi's outlook is not that of the religious wayfarer pure and simple. He is first of all a merchant and a collector of statistics for the use of other Hebrew merchants and patriots; his visit to Palestine is only incidental to a much more extensive tour; and he really forms a link between the older and narrower religious travel and such far-reaching and mainly secular enterprises as the missions of Carpini, Rubruquis, and the Polos, to Central and Further Asia. Besides this, the difference of creed prevents us from grouping Benjamin with the Christian pilgrim literature, probably as strange to him as he to it, and based upon entirely different interests and objects.

In the remainder of this chapter, therefore, we must finish the summary of Catholic and Orthodox Christian pilgrimage in the two sub-divisions that remain to be considered;—from the complete establishment of the Latin Kingdom to the fall of Jerusalem before the arms of Saladin;—and from the disasters of 1187 to the middle of the thirteenth century and the close of the Crusading Age. After this, Benjamin of Tudela and some other Jewish travellers must be treated by themselves, as an appendix to the more limited records of the older pilgrim wanderings, and an introduction to the wider history of diplomatic and commercial intercourse between Europe and Asia. That commercial life which is at the root of so much in both mediaeval and modern expansion, that element from which almost all progress of the pre-scientific kind takes its origin, is in Benjamin to a very marked degree, and clearly separates him, as it separates the Polos, from the purely theological travellers and geographers.

The short pilgrim-narrative which passes under the name of Fetellus or Fretellus, Archdeacon of Antioch, about A.D. 1200, is probably at least half a century earlier in its original and longer form, and it may perhaps be put back even to a time before the year 1150.[1] Thus from the context we see that it was written before the building of the choir of the Holy Sepulchre in the Crusaders' Church, and soon after the foundation of the Order of the Templars in 1118. The material of our present tract is of slight value, for it is badly arranged, full of legendary gossip, and without any literary form; it illustrates, however, an important class of pilgrim literature. This class may be defined as that of the impersonal guide-book, usually anonymous, brief, and arid in character, and in every way contrasting with the other type of personal record guaranteed by the writer's name, and giving a first-hand account of actual experience. Examples of the latter are of course to be found in the chief pilgrim narratives hitherto noticed, such as those of Saewulf, Daniel, and Sigurd; of the less valuable John of Würzburg and John Phokas a little later; and of Arculf, Bernard, and Willibald in the earlier

[1] Except for the reference in ch. xxxiii. to the commencement of the great Crusading Church at Tyre, which belongs to a later period of the twelfth century, at least in its complete stage.

Middle Ages. On the other hand, we have instances of the guide-book class from the time of the Bordeaux Itinerary; and here, in the twelfth century, we may sub-divide the same under two varieties, a Latin and a French, the latter represented by the *City of Jerusalem*, the former by the so-called Fetellus, or, more exactly, by the anonymous writer whom Fetellus re-edited.

In his description of Jerusalem the author alludes to a hospice for poor and infirm persons near the Holy Sepulchre; the same establishment is mentioned by Saewulf; and it was perhaps a descendant of the old house of Charles the Great, described by Bernard the Wise. The Dead Sea, over which no bird could fly, is here given the unusual name of 'River of the Devil'; and the destruction of the Cities of the Plain is described as an act due to the instigation of Satan. The Asphaltic Lake in which ancient ruins were clearly to be seen, and in whose neighbourhood wine and water alike became brackish, produced not only bitumen and gem-like rock-salt, but also alum and mill-stones of peculiar excellence. Hebron was the scene of the Creation, as Tabor was of the earliest tithe-paying. For, like Daniel and others, it is at Tabor that the narrative locates the meeting of Abraham and Melchisedek, and the payment of tenths by the conqueror to the priest of the Most High God. The sacred land of Arabia, according to Fetellus, was at first a solitude and a horror, but was irrigated, and so made fertile, by Moses. Mount Sinai in this country, the ascent of which was by 3500 steps cut in the rock, still retained some evidence of its special and terrible character; smoke and flashes of fire continually issued from it; and on Sabbath days there appeared upon the crest a heavenly light, sometimes lambent and like a fleece to look at, sometimes in the manner of lightning and with the noise of thunder. The monks and hermits who now lived there had the repute of extraordinary sanctity, and men knew and revered them from the furthest borders of Persia to those of Ethiopia. As to the latter, the legend of Moses' conquest of the same is here repeated; while Mount Hor or Petra is described with a special enthusiasm for the beauty of its scenery, and the death of Aaron on this mountain becomes in the pilgrim the story of a translation into Heaven.

Malbech or Baalbek, founded by Solomon, who called it the 'House of the Forest of Lebanon,' is the repetition of a local tradition[1] which still survives; like John of Würzburg and Daniel of Kiev, Fetellus, in giving the ordinary names of Jor and Dan, apparently makes the latter answer to the Hieromax, while the former or true Jordan is described with fair accuracy, 'becoming a lake near Baneas,' and afterwards passing from the *marsh* of Gennesareth or Merom into the Sea of Galilee. It is probably from the strange philology of the *Old Compendium* that our present summary copies the 'Hebrew' name of Aulon for the Jordan depression; from Moslem sources is, perhaps, derived the strange explanation of the sweetness of the Sea of Galilee, dependent on its receiving from outside all that would pollute other waters.

At Samaria this narrative places the burial and original tomb of St. John the Baptist, here agreeing with a very old local tradition noticed by St. Jerome; the Samaritan story of the offering of Isaac on Mount Gerizim is also reproduced; and we cannot help supposing that the compiler must have visited this region and been in special relations with some of the Samaritan community.

Along with many wild statements of distances there are equally wild identifications of places, persons, and historical events; thus Antioch was the same as Riblah or Reblata

[1] It occurs also in Benjamin of Tudela.

near Baalbek;[1] Dor was but another name for Caesarea, as Emmaus for Eleutheropolis; the temple hill was the Bethel of Jacob's vision; Tyre was founded by Phoenicians from the Red Sea; the army of Sennacherib was destroyed between Bethlehem and Jerusalem; and the Holy City was founded by Melchisedek, who was identical with Shem the Patriarch. Copying evidently from the same source as John of Würzburg and the Anonymous pseudo-Beda, Fetellus calls the Mosque of Omar the building of a certain Amir of *Memphis;* and close to this, he adds, were the dwellings of the Templars, the new Christian soldiers who guarded Jerusalem, and had special charge of the Noble Sanctuary, using the vaults of Solomon's palace for their stables.

At Acre, as the present guide-book truly remarks, there was now constantly arriving a great number of ships from the sea coast under Christian rule, between Ascalon and Mount Taurus; and hither the needful supplies of Asia flowed from Africa and Europe. This item of fact forms a welcome change to the legends with which Fetellus ends his Compendium—stories of the cave near Jerusalem to which a lion conveyed in one night the bodies of 12,000 martyrs; of the miracles of St. Saba when he entertained Thomas of Madaba in the country of the Arabs; of the miraculous Eikon at Beyrout, mentioned by so many pilgrims; and of the peculiar physical construction of the crocodiles at Caesarea, who lacked an anus and could only move their upper jaws.

If Fetellus supplies us with a fair specimen of the

---

[1] Haifa, also, in F.'s mind, was no other than Porphyrium (North of Sidon). This mistake is apparently from the *Old Compendium.* Diospolis, rightly identified with Lydda, is humorously explained as meaning 'Double City.'

anonymous compilation which served as a hand-book to twelfth-century pilgrims, John of Würzburg is an evidence that the spirit of the explorer and observer was still active among some of these devotees, even after the firm establishment of the Latin Principalities of the Orient. The German cleric who now comes forward with a singularly vivid and interesting narrative (soon after the second Crusade to all seeming had established the Christian power in Syria on a firmer basis than ever), may well compare in some respects with such a traveller as Saewulf. For though the younger man has left us a shorter and slighter record, he has given a more distinct impression of his personality and his feelings; and if we are to take him as the teacher and leader of Theoderich, his place is still more important in this obscure branch of the history of travel.

John, a priest of the Church of Würzburg, and afterwards bishop of that city, according to one authority,[1] seems to have visited Jerusalem between 1160 and 1170; and here on St. James' Day, the 25th of July, he was present [2] at a festival in the Church of St. Anne. Some [3] believe that John's description of the Holy Land was written soon after the year 1200; but in any case his visit must have taken place before Saladin's recapture of Jerusalem in 1187. It is also probable that his visit occurred just before the Byzantine restoration of the Holy Sepulchre buildings which Theoderich witnessed. The two pilgrims were probably well acquainted with one another, for John dedicates his work to a Dietrich, who seems to be no other than Theoderich.

Our present narrative is nominally limited to a description

---

[1] The manuscript at Tegernsee, which declares itself to be the property of the Monastery of St. Quirinus (at Tegernsee), and to contain a description of Palestine by the Lord John, Bishop of Würzburg. His name, however, is not in the registers of the bishops of this See.

[2] Ch. xvi.

[3] *E.g.* Fabricius.

of Jerusalem and the immediate neighbourhood; but this limitation is ignored in practice; and a detailed account of the sacred places of Galilee is also given, in which the compiler has apparently used an older and now lost guide-book. On the other hand, while John of Würzburg distinctly alludes to the famous record of Bede and Arculf, he does not quote the words or phrases of that narrative, nor does he follow its arrangement of the subject-matter. But in many places his expressions are identical with those of Theoderich; and, while it is possible that herein the latter may have simply copied from one whom he regarded as his teacher, there are various indications that both pilgrims drew from the standard travel-manual commonly called the *Old Compendium*.

'John, who by the grace of God is that which he is in the Church of Würzburg,' after offering his description of the Holy Land, with all good wishes for a 'portion in the Heavenly Jerusalem,' to his friend and follower Dietrich, begins his itinerary at Nazareth, 'the chief town of Galilee,' and thence proceeds to Mount Tabor, Little Hermon, and various other places of Galilee and Central Palestine. So at last he brings the reader through Samaria and Sichem up to Jerusalem, the 'glorious metropolis' of Judaea. 'According to philosophers,' the Holy City was placed in the middle of the world; and here among other wonders was that Temple of which Pharaoh Necho had been the destroyer, and Nebuchadnezzar the despoiler. Elsewhere, the pilgrim seems to believe that the building of Solomon was identical with that in which Christ was presented. Its restoration in the Christian period had been effected by Helena, mother of Constantine, by Heraclius, by Justinian, and by some Emperor of 'Memphis' in Egypt—thus John mentions various traditions without indicating any preference. The last name is evidently the confusion of a guide-book in common use, which informed the traveller that a Saracen chief (elsewhere termed the '*Admiral* Memphis' or 'Nymphis') rebuilt 'this Bethel' in honour of 'God most High,[1] because to Him all languages joined in rendering service.' Incidentally, John alludes to an interesting but entirely fabulous journey of the Great King Charles to this place, where an angel from Heaven presented him with a very holy relic,[2] which was brought into the West, and had a remarkable history. Charlemagne placed it in his capital of Aachen or Aix-la-Chapelle 'in Gaul'; Charles the Bald translated it to the Church of Charroux in Poitou; and in John's time it was still preserved there. Here we see how the story had grown of the Frank emperor's share in the history of Jerusalem, and how his real interest in Christian monuments and his authentic provision for the comfort of Christian wayfarers had been worked up into a series of sacred wanderings parallel to those of the mediaevalised Alexander.

Among other details we may notice that the 'Altar which stood in the open air' on the Temple area had been turned by the Saracens into a sun-dial; and John himself declares that many Moslems still came to pray there, as this dial conveniently pointed towards Mecca and the South. Again in the Palace of Solomon, at the south-east corner of the Haram area, were the stables of Solomon, of such size that 2000 horses or 1500 camels could be stalled therein; and near to this the Knights Templars had a great enclosure containing many spacious buildings, and a large new church not yet finished. John compares the Order of the Temple of God with its rival of the Hospital, much to the advantage of the latter, which

---

[1] *Allah kebir.*   [2] Ch. iii. of John of Würzburg.

gave ten times more in alms, and lay under no aspersion of treachery to the common cause like the Templars.[1] The inscriptions, hymns, and memorial verses (here recorded) in the Church of the Holy Sepulchre on Mount Sion and elsewhere, are a new feature in our Latin records of Palestine pilgrimage, and show how elaborately everything had been arranged under the Latin Kingdom for the convenience and instruction of Christian visitors; but the ornaments added by Manuel Comnenus[2] (which soon after figure so prominently) do not appear as yet. It is remarkable that John makes no mention of the Holy Fire described by Theoderich, and indeed by most pilgrims from the time of Bernard the Wise.[3] But in speaking of the Tomb of Godfrey of Bouillon, he launches out in a vigorous protest, on behalf of his German kinsmen, against the French claim to the leading or exclusive share in the victories of the Crusaders. 'For though Duke Godfrey is honoured for himself, yet the taking of the city is not credited to him and his Germans, although they had no small share in that exploit; but it is attributed to the French alone. And some dispraisers of our nation have even scratched out the epitaph on the famous Wigger,[4] because they could not deny that he was a German, and have written over it the epitaph of some French Knight or other. True it is that Godfrey and his brother Baldwin, who was king in Jerusalem after him, were men of our country. Yet since only a few of our people remained, and the rest in great haste and home-sickness returned,

---

[1] Alluding to the abortive Siege of Damascus in July 1148, when the Templars were accused of taking Saracen bribes; cf. John of W., ch. v.

[2] 1143-1180. His gifts to the Holy Sepulchre were probably after 1170.

[3] Cf. Theoderich, ch. vii.; also the reference of the almost contemporary Icelandic travellers (*Soc. de l'Or. Lat.*, and see Riant, *Scandinaves en terre Sainte*, 226-300; *Antiquités russes*, ii. 418, 422).

[4] Wicker of Suabia. This outburst is in ch. xiii.

the entire city has fallen into the hands of other nations [1] who took part in the Crusade, Frenchmen, Lorrainers, Normans, Provençals, Auvergnats, Italians, Spaniards, and Burgundians. And thus no part of Jerusalem, not even in the smallest street, was set apart for the Germans; [2] for they themselves took no care in the matter, as they had no intention of remaining; and so their names were never mentioned, and to the Franks alone was ascribed the glory. Yet this province would long ago have extended its boundaries beyond the Nile southward, and beyond Damascus northward, if herein were as great a number of Germans as of other nations.'

In spite of these complaints, however, John tells us a little later of a church newly built and called the House of the Germans, on the south side of the city, between Sion and Moriah; but on this (he adds, with a grievance even in his admission) hardly any except Germans bestowed the smallest benefaction. At some of the places near Jerusalem, the Würzburg traveller explains, describes, or comments in a rather unusual manner—as at Hebron, a city of *Saracenic* name, which in that language meant the City of Four, because four patriarchs were buried in the double cave of Machpelah; [3] at Tyre, where the Tomb of Origen is noticed; and in the Ghor, or Jordan rift-valley, which the writer defines, under the term of Avlon,[4] as stretching from

[1] Elsewhere in the concluding ch. (xxvii). John gives another list of nations and languages represented in Jerusalem, adding the names of Greeks, Bulgarians, Hungarians, Scots, English, Bretons, Ruthenians or Russians, Bohemians, Georgians, Armenians, Jacobites, Nestorians, Maronites, Syrians, Indians, Egyptians, Copts, and 'Capheturici.'

[2] But a few year later we hear definitely of a *Rue des Alemans* in Jerusalem.

[3] Adam, Abraham, Isaac and Jacob. The Bordeaux Pilgrim and many of the early Christian travellers give only three; but Adam is mentioned by St. Paula, and becomes gradually fixed in the tradition.

[4] John, like Fetellus and many others (*e.g.* some of the *Innominati*), copies the tradition of

Lebanon to the desert of Pharan. It is not likely that the pilgrim visited Mount Royal,[1] that 'royal mountain in Arabia which the Lord Baldwin, King of the Franks in Jerusalem, conquered and joined to that land for the Christians.' This fortress was east of the Jordan, between Kerak and Petra, far to the south-east of the Dead Sea; it was one of the greatest of the Christian fortresses in the East; and from its foundation in 1115 it kept open the route from Central Palestine to the Red Sea. The country of Idumaea or Edom, to which it really belonged, becomes, in John's hands, an extensive and peculiar region; for here he places the venerable metropolis of Damascus, Job's Uz, and the river Jabbok, as well as Mount Seir,[2] beneath which lay Damascus itself. In the same land was Mount Lebanon with its rivers, Orontes and Leontes (evidently intended by the traveller's 'Abana and Pharphar, rivers of Damascus'),[3] and Antioch, where for seven years 'St. Peter wore the pontifical tiara.' Such confusions are common enough among all the travellers of this time, when unable to verify from personal experience; and they do not weaken the impression given us by other parts of this 'libel'—the impression of a distinct personality, keenly patriotic and observant, and perhaps not without a certain just pride in his assurance to those who came after him, that he would not be envious of any one who 'wrote better about those matters' which he had attempted to describe.

Theoderich, as we have seen, is probably the Dietrich

the 'Hebrew' Aulon or Avlon; see chs. xx. and xxv. This Greek name was already in use for the Jordan valley and the Arabah in Jerome's time. Even the form of Ghor (*Gorius* in John) is not Hebrew but Arabic.

[1] Commonly known in Crusading History as *Le Crac de Montreal*, and called Shaubac by the Arabs.

[2] Mount Hermon, in this connection, is evidently meant.

[3] See ch. xxv. of John of Würzburg

mentioned by John of Würzburg as the follower to whom he dedicates his record. Possibly also he was the same as the Theoderich who became Bishop of Würzburg in 1223.[1] Like his master John, he was a German of the Rhine country; thus he compares the Church of the Holy Sepulchre with the sacred buildings at Aachen, and mentions how on Palm Sunday he and his friends buried, in the Potter's Field near Jerusalem, a companion named Adolf of Cologne.

Theoderich's narrative is quite as full, definite, and interesting as John's, and gives as great an impression of the intelligent eye-witness. Thus,[2] the people shouting their 'dex aide' and 'Saint Sépulchre' while waiting, 'not without tears,' for the descent of the Holy Fire; the stacking up of the pilgrim crosses on the top of the rock of Calvary and the bonfire made of them on Holy Saturday; the piling up of stones in the Valley of Hinnom by the too trustful pilgrims, who expected to sit upon them on the Day of Judgment;[3] the evidences of the power of the Templars and of the wealth and charity of the Hospitallers; the throng of ships in the dangerous harbour of Acre and the pilgrim's own buss among them; the Saracen habit of making a great noise about the simplest things, even the ploughing of a field; and the view from Quarantania, the so-called Mountain of the Temptation, over the plain of the Jordan Valley, covered with pilgrim figures; all these are touches which mark the observer and even the artist. Similar traces of reality, not without their picturesque side as well, are the Frank names of Belmont, Fontenoid, and Montjoye, which appear so typically in these pages by the side of Jerusalem and Hebron. Like

---

[1] This, however, is open to much doubt.
[2] Cf. Tobler.
[3] Theoderich, however, looks on this as a pious absurdity; 'the simple pilgrims delight themselves,' etc.

John of Würzburg, Theoderich gives us a picture of the Holy Land in the last days of Frankish rule, still dominated by Frankish customs and nomenclature, and covered by Frankish buildings.

Our pilgrim seems to have landed at Acre, to have gone thence direct to Jerusalem, to have visited Jericho and the Jordan, and to have returned by much the same road. The confusion of his account of the Sea of Gennesareth must be held as an objection to the theory of Tobler, that he made a special journey in Galilee, visiting Nazareth, Mount Tabor, and Tiberias. From one reference[1] it is clear that the writer was a priest; but the exact date of his journey can only be suggested approximately. Thus, in one place he refers to the execution of some monks by the fanatic[2] and 'sanguinary' Zenghi, the Father of Nûr-ed-din;[3] and elsewhere he names the patriarch Fulcher,[4] who held the See from 1146 to 1157. Once more, an inscription which he read on an altar[5] in the Temple of the Lord declared that the building of the same was finished in the fifty-third year after the taking of Jerusalem and in the sixty-third after the capture of Antioch; and this, though a confused reference, anyhow implies a time subsequent to 1150. Again he gives the date of 1161 to the Christian capture of Paneas,[6] Banias, or Belinas, near the sources of the Jordan; and in his description of the tombs of the kings of Jerusalem he brings us down to Amaury or Amalrich, in whose reign

---

[1] Ch. xxix.

[2] Theoderich, ch. xxx. Imad-ed-din Zenghi, the great leader, who first stemmed the Crusading conquests in the Levant, was usually called 'Sanguineus' by the Latin writers of the time; and this corruption of his name, as Gibbon says, afforded a comfortable allusion to his *sanguinary* character and end († 14th September 1146).

[3] Nur-ed-din Mahmud, † 15th May 1174.

[4] Theoderich, ch. xii. Exactly the same reference is given by John of Würzburg, ch. xiii. (except for the name of Fulcher).

[5] Theoderich, ch. xv.

[6] Theoderich, ch. xlv.

he certainly made his pilgrimage, and who died on the 11th of June, 1173. Lastly, he laments that he was unable to read the inscriptions on the arches in the Holy Sepulchre, as the colouring was so faded, except for the hymn *Christus Resurgens,* which was marked out in letters of gold, newly gilded, like the turret or cupola and the cross above the chapel.[1] All this John of Würzburg had seen in silver only; and the difference points to Theoderich's visit happening during the restorations of the Byzantine Emperor, Manuel Comnenus, which took place between 1169 and 1180. The two German travellers, however, use exactly the same language about the new church, not yet finished, which the Templars were building on Mount Moriah; on the other hand, John refers to another church 'now being built' over Jacob's Well at Shechem, while Theoderich speaks of it as complete. But this, again, we must balance by an exactly reverse statement as to the Church of the *Paternoster.* A new cistern, apparently the *Birket-es-Sultan,* is mentioned by Theoderich between Jerusalem and Bethlehem; and the discovery of this is first confirmed by other sources in 1176. From all this we may conclude that Theoderich's journey was between 1169 and 1173 (the death of Amalrich), and that it was only a little later than the visit of John of Würzburg. His wanderings in Palestine were evidently in the spring of the year; he saw the ripe barley in the Plain of Jericho on Monday in Holy Week; and on the Wednesday after Easter he was at Acre on his return journey.

In his description of Judaea Theoderich places Idumaea to the *north* of 'Jewry'; and considers the rich plain of the Jordan to extend as far as the Red Sea. From Mount Quarantania he beheld a most extensive view, reaching beyond Jericho and the 'Stream of Descent' to the Dead

[1] Theoderich, ch. v.

III.]   THEODERICH ; JOHANNES PHOKAS   199

Sea, Arabia, and even Egypt; and it was here that he saw more than sixty thousand pilgrims thronging over the flat plain to wash in the waters of the sacred river, almost all carrying candles, and visible also to the infidels who watched them from the Arab mountains on the other side of Jordan. In some respects Theoderich's Topography is better than the average; thus he carefully distinguishes Caesarea Philippi from Caesarea on the coast; and it is probably from an earlier 'authority,' such as the *Old Compendium*, that he derives his worst confusions,—Damascus standing on Mount Seir, the rivers 'Abana and Pharphar' flowing into the Mediterranean along the courses of Orontes and Leontes;[1] Mount Lebanon dividing Phoenicia from 'Idumaea';[2] and the Jordan passing through the Sea of Galilee between Bethsaida and Capernaum.

The pamphlet of Johannes Phokas is one of the last records of pilgrim-travel within the period of the Latin Rule in Jerusalem; and it is of special interest as representing Byzantine geography. The writer was a Cretan by birth, and the son of one Matthew Phokas, who afterwards became a monk and died in the Island of Patmos. In his early life Johannes served as a soldier in the army of Manuel Comnenus; but in the end he took the monastic vows, like his father, and it was as a monk that he visited the Holy Land in 1185.

In his *Brief Description of the Castles and Cities from Antioch to Jerusalem* he often mentions the Emperor Manuel; other personal reference is almost wanting in his work, except that near the Jordan he mentions a Spanish hermit who had formerly been a Stylite or Pillar-Saint near the Gulf of Adalia or Attalia; this man he had met in earlier days when marching with the Imperial forces. The author's style in this work is quite literary, and even rhe-

[1] Exactly the same in John of W.   [2] Perhaps for Ituraea.

torical, in marked contrast to the ordinary pilgrim traveller; he also shows a greater interest in classical antiquities and profane history and letters than most of his kind; and he furnishes a proof that mediaeval Constantinople, with all its faults, was more cultivated than most cities and countries of the West.

After a short preface, Phokas begins with a notice of Antioch; once it surpassed almost all the cities of the East; time and the hands of the Barbarians had extinguished its prosperity; but it still boasted of its lofty ramparts and towers. The famous hill Casius or *Caucasus*, whose height was so favourite a theme for ancient hyperbole, is less interesting to the traveller than the classical remains and the scenes of the exploits of various holy solitaries. Next to Antioch came Laodicea,[1] whose splendour had also been dimmed by time; then by way of Gabala, Gibel, or Jebeleh, the pilgrim moved on to Antaradus (Tartus or Tortosa) and Tripolis. In the interior of the country, among the mountains, Phokas describes the Chasysii or Assassins, a Saracen nation, but neither belonging to Islam or Christendom, and famous for their blind worship of their chief as God's Ambassador, and for their execution of his commands, even to the murder of the greatest princes. Mount Lebanon is treated with enthusiasm; its height, its clustering robes of snow, and its pine, cedar, cyprus, and fruit-bearing trees, all awake the visitor's admiration; while the harbours of Beyrout and Sidon remind him of ancient descriptions, and especially that of Achilles Tatius[2] in his *Leukippe*. The city of Tyre was still splendid, exceeding almost all the ports of Phoenicia; much larger than Tripolis, with a greater

---

[1] Ch. iii. of Johannes Phokas.
[2] Alexandrian rhetorician and novelist of about A.D. 450-510, author of the Greek Romance, *The Loves of Kleitophon and Leukippe*.

and finer harbour than Beyrout, and famous for a fountain of wonderful depth and volume. On all this coast, however, Ptolemais or Acre was the most populous town, where nearly all merchants and pilgrims landed, and where (from this very circumstance) prevailed a constantly-recurring pestilence and an irredeemable unhealthiness which made it highly dangerous to visit.[1]

Phokas next describes the Holy Places of Nazareth and its neighbourhood; then from Galilee he comes south to Nablûs and Jerusalem, travelling along a good road, all paved with stone, which ran the whole distance between Samaria and the Holy City.[2] Here the traveller, who often declines to repeat without examination the stories retailed to visitors, quotes Josephus[3] against the name *Tower of David*, as applied to the ancient fortress by the Jaffa Gate, and adds with cautious criticism that the existing tower of stone was perhaps only a substitute for the ancient one of marble. Passing on to other things, Phokas records how new ornamentation in gold had been executed for the Holy Sepulchre at the expense of the Emperor Manuel, 'my lord and master'; and like his Orthodox brother, Daniel of Kiev, describes the Kedron Convent of Mar Saba, over-hanging a terrific chasm, and adorned with marbles from the surrounding hills, where holy men, despising the world, endured the unbearable heat of this barren and savage gorge, and 'thus by means of quenchable fire extinguished the unquenchable.'[4] Still more wonderful was the Monastery of Khoziba, in the depths of the wilderness of Judaea, where the cells lay in the mouths of caves; where

[1] Chs. iv.-ix. of Johannes Phokas.
[2] Chs. x.-xiv. of Johannes Phokas.
[3] Who declared this Tower was built of polished white marble, 'while now it may be seen to be of common stone.'
[4] Among the saints buried here were the ancient *poets*, SS. Kosmas and John (of Damascus?).

the rock was heated by the sun till one could see tongues of flame bursting forth from it; and where the only drinking water of the miserable ascetics received the full heat of the noonday glare. Yet the monks of the Dead Sea coast did not pass their whole life in devotion; they parcelled out the land, planted trees and crops, and grew vines, building towers for the better care of their fields;[1] but the old Roman roads had fallen into decay, and the outlines of their stone pavements could only be faintly traced. Within two bow-shots of Jordan, the most holy of rivers, was another monastery which Manuel had restored; and among other pious works of the same benefactor in this neighbourhood Phokas reckons the monastery of Mar Elias, between Bethlehem and Jerusalem, and the great Church at Bethlehem, where the portrait of the Byzantine Emperor had been placed in the part reserved for the Latin rite.[2] In the vicinity of the sacred stream lived a Stylite Saint from the far West, 'a tall old Spaniard, pleasing and admirable, adorned with a species of Divine grace,' and full of wonderful tales of miraculous support from the devout assistance of a lion. Still nearer to Jerusalem were other traces of Spanish devotion;—a holy solitary from this country, tenanting the Pillar or *Castle* of Absalom, and a monastery close by the walls of the city, full of monks from the Peninsula, while among the inmates of St. Saba's Convent was another Iberian. Phokas gives no explanation of this remarkable prominence of Spanish devotees in the Holy Land during these last days of the Latin Kingdom; but at Mount Carmel, we may remember, he also mentions a hermit from Calabria in South Italy; and recollecting that our

---

[1] Chs. xix., xx. of Johannes Phokas.

[2] Phokas (ch. xxvii.) seems to attribute the original building of the latter to Manuel; but this would be almost incredible carelessness.

writer was a Byzantine, we may see from this one narrative how great a meeting-place for Christians of distant lands was the Syria of the Latin domination.

Among the pilgrim-travellers of the Crusading age, we may also reckon several Anonymous writers who describe the visits they paid to Syria before the triumph of Saladin and the misbelievers. The exact date of most of their journeys is uncertain; but the First of the *Innominati*, who is also the earliest, cannot well be put later than 1098,[1] and thus slightly precedes the conquest of Jerusalem by the Europeans under Godfrey of Bouillon. Next, perhaps, we have the Fourth of the group (in the ordinary numbering), who may be fixed to some point within the first quarter of the twelfth century. Neither of these offers any features of marked interest,—except that the last-named calculates his distances in German miles, and so appears to belong to a nationality which, as John of Würzburg bitterly remarks, was more prominent in the first days of the Latin Kingdom than in the later times of Levantine history. The Third place probably belongs to the Seventh of this category, whom some would place about the close of the reign of Manuel Comnenus (1180), but who need not be later than the middle of the century (1140-50). This tract is of even less value than the preceding; but the next of the series, commonly classified as the Sixth, and otherwise known as the compilation of the Pseudo-Baeda, occupies a much higher place. The author was an Englishman; his manual was compiled between 1150 and 1170, as far as can be judged; and while he shows certain resemblances with Fetellus, John of Würzburg, and Daniel of Kiev, his notes are probably based, for the

[1] See p. 138 of this vol.

most part, on the *Old Compendium,* which all the later pilgrims tend to reproduce with increasing servility. His English speech appears from his play on the words 'Desert of Sin' and his explanation of the same, on the *lucus a non lucendo* principle, as the 'holy waste.' Besides a short account of the chief places of the Holy Land, he gives us many of the stations of the route of the Israelites from Egypt to Canaan. To him, as to so many others of his religion, Hebron represented a prophetic revelation of the doctrine of the Trinity, for here Abraham 'saw Three and worshipped One.' Montreal Castle, the work of that 'brave lion Baldwin'; Tyre, recently taken by the Crusaders, with the help of the Venetian Navy; Damascus *built* by Eliezer, the steward of Abraham; the streams of Jor and Dan uniting at the foot of the mountains of Gilboa; the 'Hebrew' names of *Ghor* or *Avalon* for the Jordan Valley; Jerusalem in the middle of the world; the Church on the site of the Temple, said to have been erected by the Saracen *Admiral*[1] Nymphis; these are the chief memoranda of this narrative, and in almost all of them we are reminded of the Teutonic descriptions of John and Theoderich. But the account here given of the successive translations of the head of St. John the Baptist to Alexandria, Constantinople, Poitou, and Maurienne is probably extracted from Gregory of Tours; and the other curiosities of this note-book are mostly due to the common source of so much error and absurdity, (alike in the Würzburg travellers and in the *Innominati*), that *Old Compendium,* which perhaps appeared in the third

---

[1] In the text 'a certain Ammiraldus *Nymphis*,' for which last *Memphis* has been conjectured without much improving the sense, but at any rate making this passage agree more nearly with John of Würzburg.

The title of Ammiraldus is, perhaps, a corruption of the Arabic 'Amir-al-Mumenin,' more usually appearing in Christian histories under the form 'Miramamolin.'

decade of the twelfth century, soon after the Concordat of Worms (c. A.D. 1128).

The Fifth place in order of date (c. 1170) is given by some[1] to the (so-called) 'Second' Anonymous, who bears a marked similarity to the Bordeaux Pilgrim, with some likenesses to Saewulf and Theoderich. There is no decisive ground, however, for placing him before the era of 1187, and he has been brought down to the very end of the twelfth century by other scholars.

Sixth in order comes an Anonymous tract traditionally numbered as Third, and perhaps belonging, like the former, to about 1170-75. It ranks next to the Pseudo-Baeda in value, and is the most picturesque and interesting of the guide-books of the *Innominati*. Starting from Brindisi, the author brings us to the city of 'Clarence' in the 'Isle of Romania,' in which land he found a hundred and twenty-three towns, good wine, and sweet air. The rich islands of Candia and Cyprus are next described, and especial praise is given to Baffo, its aromatics and jewels, clever workmen and pious embroiderers. Two other Cyprian towns, Limasol and Famagusta, are also commemorated; the first for its place in the annals of Chivalry, being the capital of both the great orders of the Temple and the Hospital; the second for the refuge it afforded to the Community of the Holy Ghost, often confused with the Hospital, but in reality quite distinct. Nothing else is known of the Hospitallers of Cyprus or of the miracles worked by St. Patrick in that Island; on the other hand, the writer's notice of the frequented sea-passage from the ports of Cyprus to those of Little Armenia (or Cilicia) is amply confirmed by other sources, and prevents us from dating his compilation too early in the twelfth century. From his constant and emphatic reference to

[1] *E.g.* Röhricht, *Bibliotheca Geographica Palaestinae*.

the wines, aromatics, fair women, and other natural or artificial riches of the Levant, we may suppose this pilgrim was not so stern an ascetic as some others of his class.

The traditional Eighth and Ninth of this series come Seventh and Eighth in time; only one, according to this arrangement, is of later date, the so-called Fifth; and both belong to the last years of Latin ascendency (c. 1175-85). The notes of the first-named contain memorial and dedicatory verses, after the manner of John of Würzburg, and place the Mountain of Little Hermon near Tabor, without the usual pilgrim-duplicate of the 'Hill of Rejoicing,' in the environs of Jerusalem.

The Fifth and last of these Anonymous travellers has left a narrative in two parts, both of which seem to have been written after the fall of Jerusalem in 1187, but describe a journey made just before that date. Like Abbot Daniel and Bishop Willibald, the author refers, though in a guarded manner, to the extraordinary tale of the well into which dropped the Star of Bethlehem; like Saewulf and others, he alludes to the circle in the Holy Sepulchre which the Lord declared to be the middle of the world. He also gives special attention to the subject of the European Settlements in Syria, such as those of the Genoese, Pisans, and Venetians, maritime allies of the Frank land-forces, cunning traders, and invincible at sea, who for their services had been freed from all tribute and toll, and lived under their own jurisdiction. Jealous and quarrelsome among themselves, they did much to mar the glory and happiness which their courage, skill, and wealth deserved. The chief orders of knighthood and the ecclesiastical divisions of the Latin Kingdom are also noticed, and we are told (by the way) that the Greek Bishop of Sinai was obedient to the Roman Patriarch at Jerusalem. Among other notable things recorded are the four annual changes of colour in the water of Jacob's well—from clear to muddy, from muddy to red, and from red to green; the fountain of Siloam flowing only *three* days a week; the apples of Adam plainly showing the marks of his teeth; the cotton which in Syria could be sown like wheat; and the balsam trees which would only give fruit to Christians. Like James de Vitry, the writer confuses Haifa and Porphyrium; like others, he disputes the vain story of rainless Gilboa; and like all mediaeval Latins, he finds in 'Mohammed' and 'Mohammedanism'[1] terms clearly expressive of idolatry, and useful as designations for the calves of Bethel.

Last of all, there are various scattered and mostly insignificant writings of the twelfth century, which either refer to the Syrian travels of their authors, or give some account of the country and its Topography from the experiences of other men, but which are of too slight a character for separate treatment. As examples of the former class, we may briefly refer to the works of Theotonius, prior of the Convent of the Holy Cross at Coimbra in Portugal, who twice visited Palestine about 1112-13; of Belardus of Esculo, a pilgrim of a few years later (c. 1118-20);[2] and of Burkhard or Gerard of Strassburg, whose itinerary is of 1170-75. To the latter subdivision (of stay-at-home writers) belong Dermot or Dermatius, an Irishman, who professed to have made a pilgrimage about 1115, but whose *Exhortation* is probably a rhetorical exercise; Achardus of Arroasia, who wrote on the Temple of Solomon about 1120;

---

[1] 'Machometh' and 'Machomeria.'

[2] The important itinerary of Nicholas Saemundarson, Abbot of the Benedictine House of Thingeyrar in Iceland (whose journey and description is perhaps connected with the Copenhagen copy of the famous Plan of Jerusalem or *Situs Hierusalem*), is treated elsewhere, pp. 217 note, 430 note.

Hugo of St. Victor, whose tract of 1135 on the Holy Places is an excerpt from Baeda; Peter, a deacon of Monte Cassino, author of a similar pamphlet in or about 1137; and Gerard of Nazareth, afterwards Bishop of Laodicea, the biographer of various saints resident in the Holy Land (c. 1160).[1] To these we may add the name of another Scandinavian devotee, Gissur Hallson, whose account of a visit to Jerusalem in 1150-52 has disappeared, and who must therefore be taken as typical of a third class, the pilgrims who travelled but have left no record.

After the fall of the Holy City into Saladin's hands the relations of the Christian world with Syria are completely changed, and Christian travel in the Levant begins to follow new lines. We have already noticed the beginnings of this tendency: even in the earlier twelfth century pilgrimage has ceased to attract the really enterprising spirits, who increasingly find their natural calling in the journeys of an expanding commerce, and leave the *Via Sacra* to a humbler class of wayfarer. Here we need only repeat that the last seventy years of the Crusading Age supply us with nothing of special interest in pilgrim-travel. In fact, there is only one first-hand description of the Holy Places in this time from a Latin source which is worthy of the name, or anything more than a series of allusions, travellers' tales, or citations from older writers. This is the old French pamphlet known as the *City of Jerusalem*, of about 1220. By the side of this, but much inferior to it, are certain passages in the works of historians and chroniclers, such as Ernoul and James de Vitry, which may be taken as representative of a large body of Crusading literature referring to the geography of the Levant. While, in con-

[1] As well as Othmar, another writer on the Holy Places, c. 1165.

clusion, there are certain minor notices with which we may continue and end in this period the list that has been sketched out for the preceding epochs. Among these slender records the *City of Jerusalem* naturally comes first.

This tract, though it professes to describe the condition of Jerusalem just before Saladin's reconquest, is undoubtedly of the thirteenth century; in some manuscripts it is accompanied by a second part, dealing with the other Holy Places of Syria. Generally speaking, the notes here given us resemble those of John of Würzburg and of Theoderich; they are somewhat less detailed, but they embrace a larger number of subjects. In the second part, the mention of Château Pèlerin shows that the date of writing must have been subsequent to 1218, when this Castle was erected by the Templars. Again, the Castle of St. Margaret, on Mount Carmel, is supposed to date only from 1209. Once more, the writer refers to St. Chariton as having been dead eight hundred years, and he died in 410. Thus the date of composition cannot well be earlier than about 1220. At the very opening we meet the startling statement that Jerusalem was no longer in the same place where it stood when Christ was on Earth; then it was on Mount Sion; but in the author's time only a few monks inhabited that hill, guarding the traditional place of the Last Supper. Next, we hear of the four Master Gates of the Holy Sepulchre; of the Church of St. James of Galicia; of a street devoted to the cooking of food for the pilgrims (and thence called Mal-quisinat, of the street of the Germans, noticed by John of Würzburg; and of the Exchange or Market of the Syrians and Latins. Besides these, we have the Pool of the Germans, perhaps connected with the establishment of the Teutonic Order, and the Abbey of the Georgians, founded by monks from that country of Avegia or Amazonia which mediaeval and classical legend placed in the neighbourhood of the Caspian.

The second part of this tract describes the ordinary route from Acre, Haifa, and Caesarea to Jaffa and Jerusalem; from Jerusalem to the Jordan; and from the Holy City to Samaria and Galilee—altogether giving a pretty complete list of the journeys usually performed by Syrian devotees, and furnishing a number of distance-reckonings which on the whole are extraordinarily accurate.

On the other hand, the miraculous and legendary element is very strong here, and the mediaeval readings of Bible history appear in their most complete distortions. As examples of this we have the Chapel of St. Cornelius, 'who was, after my Lord[1] St. Peter, Archbishop of Caesarea'; the cloak of St. James of Galicia; the print of the ten fingers of Christ upon the stone; and the miraculous virtues of the Tomb of St. Catherine, which many wild beasts on the mountains lived by licking. A peculiar story occurs about the Sea of Galilee. 'On this sweet water lake of Jordan, Christ ate with His Apostles after His resurrection; but He only ate the backs of the fish, which were as large as the roaches of the sweet waters of France, and all the rest He threw back into the water, when the fish at once moved again and swam away.' As to Tiberias, one of the stories from the Arabic Gospel of the Infancy re-appears in this tract; for here was the torch which the Jews threw at the Messiah when He showed them how to dye. In Tortosa we have a mention of a very ancient Church of the Theotokos, which the author supposes to have been the oldest in the world to receive this dedication; and, like James

[1] 'Monseigneur.'

de Vitry, he ascribes the building of the edifice to St. Peter.[1]

In the seventh, eighth, ninth, and tenth chapters of Ernoul's chronicle we have a description of various parts of Palestine outside Jerusalem. This chronicle is of the year 1231; it abounds in mistakes, and is of very small value. Beginning with a notice of the Jordan, 'where it rises, how it goes, and where it rests,' the troubadour-annalist repeats the old rabbinical stories of the fountains of Jor and Dan (or Dain), of the Ark of Noah and its building at Arka on Mount Lebanon, and of the mountain of salt on the shore of the Asphaltic Lake. The miracle of Cana is transferred to Tiberias, and the incident is distorted into the 'Wedding of Archedeclin' or Architriclin,—thus outdoing Saewulf and the *City of Jerusalem*. For here Architriclinus, the 'Ruler of the Feast,' appears as the bridegroom, and the narrative of St. John is still more completely corrupted than we find it in the Worcester Pilgrim. In mentioning the town of 'Crac' or Kerak, and its lordship to the east of the Jordan, Ernoul refers to the Monastery of St. Catherine on Mount Sinai, within the boundaries of the same lordship, an outpost of Christendom, on the borders of Islam, where thirteen monks lived in privation 'like Moses.' Near this was the Red Sea, through the midst of which ran the River of Paradise which Scripture called Pison or the Nile. In the neighbourhood of Jericho (or 'Jericop') 'walled with adamant,' Ernoul describes the snake-charming which he

---

[1] This still remains almost perfect, having been turned into a Mosque, but it was of course much later than tradition made it, and both the fifth-century Church of the B.V.M. at Ephesus and the fourth-century Chapel of the Virgin's Tomb at Jerusalem were older. Many other favourite legends of this neighbourhood re-appear in the *City of Jerusalem*; as, *e.g.* the miraculous picture at Beyrout, and that at Sardenai, near Damascus (*Notre Dame de la Roche*).

may have witnessed there; and as to the Samaritan community near Sebaste, he has a remarkable story of its origin. It was colonised, he tells us, by the branches of that race which were to be found in Alexandria and Damascus; and it is curious to notice that these Egyptian and Syrian 'churches' lasted to the seventeenth century. For the rest, Ernoul is meagre and unsatisfactory; his identifications of ancient sites are generally wrong; and his estimates of distance are reckless. Scarcely anywhere does he show evidence of personal travel and first-hand knowledge, and his account of Jerusalem is apparently derived almost verbatim from an earlier work.

Among the descriptions of Syria which are historical, antiquarian, or ecclesiastical rather than geographical, but which, being accompanied by some personal knowledge of the country, cannot be altogether neglected, that of James de Vitry is prominent. Having been appointed Bishop of Acre in 1217, the year of a futile Crusade under King Andrew of Hungary, James went with the Christian army to the siege of Damietta in 1218, took an active part in military operations both in Galilee and Egypt, and did not finally leave Palestine till 1227. Thus he spent nearly ten years of his life in the country whose history he attempts to narrate, and if he had been endowed with any geographical interest, he might have compiled both maps and descriptions far superior to anything then in Christian use, but his mind was the mind of a rhetorician and a wonder-seeker, and his hysterical style and marvellous tales are untempered by any good critical faculty.

His identifications of places are little better than those of the most casual visitor; thus we find here the old mistakes of Pelusium 'or' Belbeis, Beersheba 'or' Gibelin, and so forth; 'Hierapolis or Maubech' is his reading of 'Heliopolis or Baalbek'; the name of Ælia, given to Jerusalem, he derives from Ælius, a Roman quaestor, who rebuilt it after the destruction by Titus; Petra he places near Rabbath Ammon 'where Uriah was slain.' Whenever the Saracens got possession of the Holy City they set up the *image* of 'Machomet' in the Temple; the historian adds a short and less mythical account of the three great Orders of Chivalry closely connected with the Holy Sites.[1]

As to the maritime intercourse between Western Europe and the Levant, De Vitry is emphatic and suggestive. The skill of the Italian merchants, as seamen, might have been of great service to the Crusaders, if they had not been more inclined to fight with one another than with the infidels, and if they had not shared to the full the degeneration which had overtaken the Latins in Syria and destroyed the efficiency of the Temple and the Hospital. Further on, among his notice of various sects, De Vitry gives us a little incidental geography, as in the case of the Jacobites of Nubia, Æthiopia, and other countries 'as far as India'; of the Nestorians, living under that most puissant lord, Prester John; of the Georgians, among whom dwelt the Amazons; of the Mozarabic Christians of Africa and Spain; of the Assassins who were the chief sect of the *Essenes;* and of the Sadducean Jews living near the Caspian Mountains and the wall of Alexander, who in the time of Antichrist were to return to Palestine. Lastly, in his Topography of Syria, De Vitry includes a province between the Tigris and the Euphrates,

---

[1] Of these the Hospitallers are said by De Vitry to have arisen from the purely peaceful enterprise of some *Lombards,* especially of Amalfi, who brought merchandise to Syria, and by presents induced the lord of Egypt to let them build a Latin Church near the Holy Sepulchre. The St. John of the Hospital was, according to De Vitry, [not St. John the Baptist but] St. John the Charitable, of Cyprus, afterwards Patriarch of Alexandria; cf. *History of Jerusalem* (abbreviated), ch. lxiv.

and a considerable portion of the Arabian Desert, while among the three provinces that make up Palestine he reckons Philistia and its capital at Caesarea *Philippi*.[1]

There only remain certain scattered contributions to what may be called the Mediaeval Library of Syrian geography, and of these only a few record actual travel; the rest are nothing more than declamations, allegories, pious appeals, or topographical notes of a second-hand order. The former class feebly maintains an old tradition in the persons and writings of Antony of Novgorod (c. 1200), Wilbrand of Oldenburg (c. 1212), Thietmar or Thetmar (c. 1217), and Sabbas of Servia (c. 1225-1237).

Among these the 'great sinner' Antony, a poor counterpart of Abbot Daniel, represents the other main centre of Russian life, the northern and commercial capital in the South Baltic basin, of which he was archbishop or metropolitan. He passed through Constantinople on the eve of the Latin Conquest, and perhaps visited Palestine under the restored Moslem rule, shortly after the death of Saladin; but among the details of his journey there are scarcely any of general interest, and he has only recorded little more than his impressions of the sacred places of *Tsargrad*. Yet, in the allusions he occasionally makes to the religious intercourse between the Byzantine world and his own people, and to the condition of Byzantium itself in the last days of the Older Eastern Empire, he establishes his claim to a position of higher value than the average pilgrim wayfarer of this age. Thus, in the gold paten which the Princess Olga had caused to be made for the divine Liturgy in the Church of the Eternal Wisdom; in the 'model' eikon of the Slav saints, Boris and Gleb; and in the relics of the Russian priest Leon,

[1] Cf. *History of Jerusalem*, chs. lxxv.-lxxx., xcvi.

who had thrice made the journey to Jerusalem on foot, and of the Lady Xenia, the daughter of Bracislav—we have suggestions of a constant but little-known movement from North-Eastern Europe to Constantinople and Palestine. And the same suggestion lies in Antony's mention of the convent of Matchukov; of the Russian 'embolon' among the treasures of the Imperial City; and of the embassy under Tverdiatina Ostromiritza, Nedan, Domagir, Dmitri, and Negvar from Roman, Grand Prince of Vladimir, to the Emperor Alexius III. These envoys were present at a church council over which the Emperor[1] presided; and this council was probably that held in the May of 1199, at which the patriarch John Kamateros also assisted. It is to Antony's descriptions of the Church of St. Sophia and of the imperial 'palace of gold' that most of his readers would now turn with especial interest; but they would be disappointed if they hoped to find here anything more than catalogues of relics, (such as the shield of Constantine and the right hand of John Baptist), descriptions of 'holy and appalling' miracles, or details of religious services, such as that of Matins in the Patriarchal Church.

With the journey of the Archbishop of Novgorod that of another metropolitan naturally connects itself. In 1225-1230 Sabbas of Servia, afterwards one of the saints of the Eastern Church, made his way to Palestine, and compiled an account of his visit in the Serbian dialect of Slavonic; but no proper record has survived of the alleged travels of St. Euphrosyna, Princess of Polotsk, to the same country, at the end of the twelfth century; and these have accordingly been dismissed, without sufficient cause, as purely mythical. The itineraries of the Western pilgrims, Wilbrand of Oldenburg and Thietmar, are almost entirely repetitions of

[1] Angelus Comnenus, 1195-1203.

material we have already had to notice, especially from John of Würzburg and the *Old Compendium;* Wilbrand, however, has rather more to tell us about Asia Minor than the generality of palmers. Last among the first-hand descriptions of Palestine within this period, albeit of very doubtful character, is the supposed book of James Pantaleon, Patriarch of Jerusalem, commonly assigned to one of the last two decades of the Crusading age (1250-1270).

The latter class of these writings, those of an entirely derivative and untravelled authorship, include a tract by Peter of Blois (about 1190), on the *Length and Breadth of Palestine,* mostly composed of extracts from Jerome; the *Tripartite Relation* of the monk Haymar (about 1199), on the manners and resources[1] of the Hagarenes or Saracens, addressed to Innocent III.; another work of the same ecclesiastic, on the condition[2] of the Holy Land about the same date; a short anonymous description of Jerusalem; and a fragment of an Itinerary[3] from the last years of the twelfth century. Besides these, Radulphus Niger, as he tells us himself in his chronicle (c. 1200), composed a short account of the three principal pilgrim-routes to Jerusalem; Gervase of Tilbury, in his *Otia Imperialia,* of about 1211, makes frequent reference to the geography of the Levant; the same service is performed by Eustathius,[4] who may be reckoned among the latest and least of Greek geographers, in his *Parekbolai* of nearly the same date (c. 1212); while Roger of Wendover and Matthew Paris in their *Flores Historiarum* (1236) and *Chronica Majora* (1250), conclude the list of mediaeval annalists in this period who give us an allusive

---

[1] Vires.
[2] Status.
[3] In Libr. at Linc. Coll. Oxford.

[4] Cf. Müller, *Geographi Graeci Minores,* ii. 373-391.

treatment of Syrian matters and Syrian intercourse with the West. But Matthew Paris devoted more special attention to this subject; for he also composed a map of Palestine and a variety of notes about Palestine travel and topography, elsewhere noticed at length, and apparently based in part upon a little guide-book of about 1231, *Les Pèlerinages pour aller en Jérusalem*. It is more questionable whether Matthew's work was in any way related to other still more insignificant productions of this time, such as Philippe Mousquet's rhymed description of the Holy Land (c. A.D 1241), or the parallel sketch of Martinus Polonus (c. 1240-1245).[1]

[1] Other pilgrimages of the Scandinavians to the Holy Land and in a less degree to Rome and Compostella, offer many points of geographical interest. For want of space we can only note here the journeys, *e.g.* (1) of Alfvin Haraldson, grandson of Cnut the Great, c. A.D. 1060 ; (2) of Svein Nordbaggi, Bishop of Röskilde, c. 1086 ; (3) of Lagman Gudrödson, King of Man and the Isles, c. 1095 ; (4) of Eric the Good, King of Denmark, 1102-3 ; (5) of Svein and Eskill, brothers of Asker, first Archbishop of Lund, c. 1150-3 ; (6) of Eskill Kristiernson, second Archbishop of Lund, c. 1164-8 ; (7) of Andres Skialdabandr from Norway, c. 1229-1230. All the above were peaceful attempts, some unsuccessful, to visit Jerusalem, with or without a journey to Rome. Armed expeditions to Palestine from Scandinavia are instanced in the fleets of (1) Wimmer or Guinemer of Boulogne and his Danes and Frisians, 1096 ; (2) the Northmen of Man and the South Isles, 1189-91 ; (3) the Danes and Frisians who visit Compostella and ravage Moslem Spain, 1189-91 ; (4) the Norwegian Crusaders, Hreidar, Ulf Laufnaes, etc., 1190. The pilgrimage of Abbot Nicholas of Thingeyrar, 1151-4, is the best example of the Norse 'Rome route.' From the Eyder he traces an itinerary by Paderborn and Mainz to Basel, Vevey, Mount St. Bernard, Aosta, Pavia, and so, by Rome, to Bari and Monopoli. Cf. *Scriptores Rerum Danic.*, i. 379 ; iii. 338 ; *Scriptores Rerum Germanic.* xvii. 794-6 ; Dozy, *Recherches*, ii. 295, 329-340 ; Riant, *Expeditions* . . . *des Scandinaves en Terre Sainte* pp. 62-7, 69-86, 125-6, 129, 131-9, 144-215, 221, 225, 226-9, 230-9, 272-7, 290-305, 312-338, 342-3, 348-353, 354, 365.

CHAPTER IV

BENJAMIN OF TUDELA AND OTHER JEWISH TRAVELLERS

A NEW chapter of mediaeval travel begins with Rabbi Benjamin of Tudela (c. 1159-1173). His *Records* are the earliest important contributions of the Hebrew race to geography; they also mark a distinct advance in the movement from west to east, whose outlook now began to reach beyond the Euphrates to Central and Further Asia. Before the time of Benjamin the Jews had often made extensive journeys, being for various reasons especially useful as envoys, negotiators, commercial travellers, and spies; but their geography, written and traditional, seems to have remained at the rudimentary stage of the Old Testament, until in the twelfth century their learned men first condescended to study earth-knowledge as a science, or at least as a body of fact. At first this new development was chiefly noticeable in the more careful and systematic visitation and description of the old home-lands of the Hebrew people, and above all of Syria itself; and here, as elsewhere, it was apparently through the Crusading movement that the change was brought about. For Jews as well as Christians availed themselves of the fresh opportunities of travel to the Holy Places of both religions; and they also attempted to discover and make known to their brethren in Europe, more perfectly than before, the condition of the congregations of the Levant, and the position and state of the sepulchres and memorials of the great Hebrews of former days.

Long before the twelfth century, however, a certain movement to and fro between distant Hebrew communities is faintly shown by scanty records, and it will be convenient in this place to treat these indications as introductory to the central figure (and narrative) of Rabbi Benjamin, concluding this chapter with a short account of Petachia of Ratisbon and the other Israelites who contributed in theory or practice to the progress of geography before the middle of the thirteenth century.

The Talmud of Palestine or of Jerusalem, which in its present form is of the later fourth century, contains some topographical allusions, not only to regions within the borders of Syria, but also to Egypt, Mesopotamia, and Arabia. The sixth century Babylonian Talmud has a more extended geographical outlook; notices of Media, Persia (or the Land of the Magi), and Cappadocia are here to be found along with remarks on the general structure of the world, the depths of the sea, the mountains of darkness, the seventy nations, and the mysterious river Sambation. The journey of Isaac the Jew, in the embassy sent by Charlemagne to the Caliph Harun al Rashid (801, 802), was a proof of the extreme value of the Hebrews as intermediaries in mediaeval intercourse; it was Isaac who furnished the report of this mission to the Frank Emperor; and it would seem that the connection of the Frankish and Babylonian synagogues was revived by this journey, for it is from about A.D. 850 that we find the name of France mentioned in the decisions of the Eastern Rabbis. Several other Jewish travels of the ninth century are recorded. Thus about 820 Jacob ben Sheara was sent to India by a Moslem prince to procure certain astronomical works; again, some fifty years later, another Hebrew wanderer, whose name is appropriated by the half-fabulous relation of Eldad the Danite, paid a visit to the Jews of Arabia (c. A.D. 870). Saadia Gaon, of the Fayûm in Egypt, who died A.D. 942, and Nathan, a Babylonian, who flourished in the middle of the tenth century, have only left some cursory geographical references to distant lands, such as India; but the son of Saadia, one Rabbi Dosa, brings us into connection with a far more important person, Chisdai or Chasdai ben Isaac, the physician and minister of Abderrahman III., Caliph of Cordova. This Hebrew statesman busied himself in collecting information upon the state of the Jews in all the countries that sent embassies to Cordova; and in pursuit of this object he wrote (among others) to Joseph, Prince of the Khazars, giving him an account of Andalusia, and asking in return for some news of the state of the Jewish Kingdom in South-Eastern Europe. This letter was brought from the Guadalquivir to the Volga by three Jewish messengers, Saul, Joseph, and Jacob ben Eliezer (c. A.D. 959). It was in answer to a similar enquiry about the Jews of Egypt that Chasdai received a similar report from Rabbi Dosa; and we may perhaps consider that this correspondence was identical in object with the travels of Benjamin of Tudela and Petachia of Ratisbon, only pursuing a different method to arrive at the same end. That end was clearly the establishment of a better understanding between the various communities of Jews scattered over the world; for this understanding, in favourable times and circumstances, might develop into a political unity, and thus the Jews might again play a prominent part in history; even without this, the financial schemes and fraternal charities of the dispersed children of Abraham could not fail to derive the greatest benefits from the maintenance and development of such a correspondence as this of Chasdai. The very survival, indeed, of the Hebrew race and religion was obviously in danger from a complete interruption of the communications between the often distant synagogues; and it was therefore natural that extensive travel should be more in fashion among the Hebrews, to whom it was so necessary, than among those Christians, whose ideal for many centuries was rather one of self-contained exclusiveness within the sacred limits of the Greek or Roman civilisation. Many of these Jewish wanderings have passed unrecorded; others are only commemorated by a bare allusion; others, again, survive in a fragmentary state. But from what remains to us it is clear that throughout the earlier Middle Ages, as in later and more civilised times, the Jews of the most remote countries had a system of correspondence, by letters and messengers, which was surprisingly complete, probably superior to any telegraphy in use among the Christians of the Dark Ages, and responsible for a steady and remarkable growth of Jewish influence.

In this connection we may notice that the tenth century gave a certain strange and novel promise of the restoration of Israel. The conversion of the Khazars, or at least of the ruling clans of this people, was not the only or the earliest incident of the kind, but it was apparently the most important acquisition of mediaeval Judaism; and it offered some hopes of providing a new and powerful centre for the revival of Jewish nationality and ambitions. Among older instances of successful proselytism, the leading place perhaps belongs to that Hebrew dynasty and kingdom in Yemen, or Homeritis, which, from about B.C. 120 till the first quarter of the sixth century after Christ, maintained the law of Moses as a privileged creed in South-Western

Arabia. It was at last overthrown by the Christians of Abyssinia (A.D. 522), fifty years before the birth of Mohammed, after a savage persecution of the Nazarenes, to which the Koran apparently alludes;[1] but down to the time of Mohammed's final success various Hebrew communities, scattered over other parts of the Arabian peninsula, maintained themselves in considerable strength. With the triumph of Islam the Prophet's fatherland was closed to all except his followers; and obstinate Jews were expelled without compunction; but in the next century they found a new home on the north of the Black Sea. About A.D. 740 a certain Bulan, King of the Khazars, determined to abandon the older heathendom of his people. From the sixth century this nation, usually considered as Turkish, had been rising in importance and civilisation; they had gradually given up their nomadic habits, had adopted a settled life, and had even developed a considerable commerce in fish, furs, and slaves;[2] their chief town, Amil, or Bilangiar, near the modern Astrakhan, commanded the mouth of the Volga, and claimed an important share in the trade of the Caspian; while other Khazar settlements stretched from the estuary of the 'Atil' to that of the Don. Another of their towns, Semender, is mentioned by Edrisi as near Amil; and their dominion seems to have extended from the Caucasus and the wall of Derbent in the South, to the latitude of Moscow in the North. According to one tradition, the ancient Persian ramparts in the Caucasus were erected by Chosroes the Just as a barrier against the heathen Khazars;

---

[1] Chap. lxxxv. *The Starry.* 'By the star-bespangled Heaven; By the predicted day; By the witness and the witnessed—Cursed were the contrivers of the pit of fire, when they sat over against it and were witnesses of what they inflicted on the believers . . . they hated them for their faith in God.'

[2] In exchange for money and other products of the South.

and beyond this rampart the early Arab travellers recognised the same people as masters of the land. Thus Sallam the Interpreter (in about 840) and Abul Hassan Ali, of Bagdad, surnamed Al Masudi (in about 930), visited and described Khazaria; so did Ibn Foslan in 921-2, with the special object of drawing it within the pale of Islam; nor were these the only efforts in this direction. A great number of the people did actually embrace the religion of Mohammed, and some of the early Caliphs relied upon their friendship and support against Constantinople. In the pursuit of their trade, however, the Khazars admitted merchants of all races and faiths with equal tolerance, and sometimes they appeared rather to incline to the Byzantines against the Saracens; but in the end they decided in favour of the third of those claimants who were pressing them so hard. Prince Bulan made careful enquiry, and found a brilliant solution of the difficulty. For the Christian admitted Judaism to be the second best form of belief, and the same was granted by the Moslem; from this it was clear that the best was really that which each disputant placed next to his own; and Bulan therefore decided in favour of the Jewish creed, just as Hellenic opinion after the Persian war decided in favour of Themistocles. After this, teachers of the law of Moses were brought into the country, and Judaism became a condition for the holder of the sovereign power; but otherwise the old toleration continued till the overthrow of the dynasty and the nation, about the year of Christ 1000. Chasdai heard of all this through the envoys of the Byzantine Empire at the Court of Cordova, and it was by the medium of a German Jew (among others) that he forwarded his famous epistle to King Joseph. The international character of the Hebrew race in the tenth century could not be better illustrated.

Josippon ben Gorion (otherwise Joseph Gorionides) the translator of Josephus; Sherira ben Khanina, the principal of the Hebrew Academy at Pumbeditha, A.D. 967-997; and Assaf, the mathematician, of c. 1050, all wrote upon geographical matters. With the name of Assaf are connected some stories of fabulous travel, but none of these are really important to the subject.[1] After the commencement of the Crusades, Nathan ben Jechiel of Rome (A.D. 1101) and Abraham ben Chija, an astronomer of Barcelona (1100-1130), continue the list of Jewish geographers— if such a name can be applied to writers mainly concerned with other interests, in whom geography is merely incidental and allusive. On the other hand, Abraham ben Meir ben Ezra, of Spain, who died at Rome in 1168, was a considerable traveller, and visited in the course of his wanderings many places of Italy, Provence, France, North Africa, Syria,[2] Mesopotamia,[2] and the Islands of the Mediterranean[2] and British Seas.[2] Unfortunately he left no proper account of these journeys. Again Yehuda ben Elia Hadasi, a Karaite Jew, lived and wrote in Constantinople in the middle of the twelfth century (c. A.D. 1147), and in the course of his works referred to various matters of physical geography and to the dwelling-places of different religious sects. In the same manner, Abraham Halevi ben Daud (or David), resident at Toledo in 1161, may be counted among the Hebrew writers who touch on things geographical; and the same may be said of the great Moses Maimonides (A.D. 1131-1204), who was also a practical traveller, but whose works have only the most meagre reference to the subject of earth-knowledge.

The change from these arid and scanty jottings to the

[1] For they only give scattered references.
[2] These are less certain than the visits to Italy, France, and North Africa, but, according to one tradition, he wrote two pamphlets in London.

Records of Rabbi Benjamin is a great and significant one; but it is not likely that he, any more than his predecessors, attracted the attention of many except his co-religionists. His work was so primarily concerned with the men of his own race, and the affairs and traditions of his own faith, that it would be impossible for it to win the interest of mediaeval Christendom. Thus the value of his narrative was scarcely recognised until religious prejudice ceased to govern the mind and literature of Europe. He visited the Jewish colonies, from Navarre to Bagdad, and described those beyond the Tigris as far as China; but he wrote for his own nation, and few others cared about him for many centuries.

The date of Benjamin's travels may be fixed from internal evidence within fairly narrow limits. Thus his own language makes it tolerably clear that he visited Rome after the year 1159; that he was in Constantinople during the reign of Manuel Comnenus, perhaps in the month of December, 1161; and that his account of Egypt must have been composed before 1171. It is also pretty evident that he came to Bagdad after a partial revival of the temporal power of the Abbasside Caliphs had begun with the vigorous action of Moktafi (A.D. 1150-1160). The date of the Rabbi's return, as given in his own preface, brings us to the year of Christ 1173; he appears to have been in Antioch immediately after the accession of Bohemund III. in 1163; and he evidently arrived in Sicily, on his way home, before the close of 1169, while Archbishop Stephen of Palermo was still governing the island, as Chancellor, during the minority of William II., the Good.

Benjamin's account of Bagdad and the Caliph is especially important in relation to what he says or implies about the state of the Seljuk power in Western Asia (c. A.D.

1150-1160). This Empire, after a short time of undivided strength, had now split up under three chief dynasties, and was practically separated into three independent kingdoms. Sultan Sanjar of Merv, a master-builder, a patron of art and letters, and a brave but unlucky warrior, ruled on the Murghab and the Oxus, controlling all the north-eastern provinces: Massûd, the 'Grand Sultan' or nominal head of all the Seljuk houses, claimed to rule at Bagdad, and to control the Caliphate and the central provinces; Sultan Zenghi of Mosul (with his sons Seïf-ed-din and Nur-ed-din) was the most important prince on the western frontiers of Asiatic Islam. But just before the era of Benjamin's visit to Mesopotamia, great changes took place in these short-lived kingdoms, through the deaths of both Massûd and Sanjar (1152-1153); the situation was made worse for the Seljuks by the fact of Sanjar's death following upon a ruinous defeat at the hands of the Ghozze Turks from beyond the Syr Daria; and the watchful Abbassides took full advantage of their opportunity. The Caliph Moktafi, who had been as submissive as his predecessors for the past sixteen years (1136-1152), now threw off the yoke and recovered a good deal of that imperial power which the weakness or necessity of Al Radi and others had forfeited in the earlier part of the tenth century. Moktafi's successor, Mostanshed or Mostanieh, maintained the position thus won; and it was probably under this, the second of the 'restored' Caliphs, that Rabbi Benjamin came to Bagdad. His language confirms what we know from other sources on a different point.

The family of Zenghi, from jealousy or policy, pursued an opposite course of action from Sanjar and Massûd, favouring that revival of the Caliphate, which the senior Seljuks endeavoured in every way to hinder; thus they gained many privileges and a sensible increase of authority and dominion from the favour of the Abbassides. For, however feeble the direct temporal sway of the Moslem pontiffs, they never ceased to exercise a considerable secular influence, like the Popes of Rome, through the medium of spiritual decrees, favours, or censures.

It is fairly certain that Benjamin was not only a Rabbi, but also a merchant; and that his object was as much to acquire and diffuse information about the commercial, as about the religious, state of his countrymen in distant lands. From Spain to Jerusalem his route seems accurately reflected in his narrative; but from Jerusalem to Damascus and Bagdad it is evident that he does not follow the same method; for here no direct course is indicated, but only a series of wanderings, backwards and forwards, and from side to side. It is doubtful how far east he penetrated, but it can hardly have been much further than Bagdad, where he must have resided for a good space of time. Here it was that he probably compiled the latter part of his *Records*, which he divides under 'things seen' and 'things heard.' The almost complete absence of the smaller places and personal names so fully recorded hitherto, as well as the comparatively vague, unhistorical, and unscientific character of this trans-Tigris section, allow of scarcely a doubt as to the second-hand or traditional method now used to supplement the first-hand observation of the earlier narrative. This earlier narrative may be sub-divided in two parts, in one of which 'things seen' are described, in due order of time, along a direct and clearly indicated route; whereas in the other, beginning from Jerusalem, although Benjamin still treats exclusively of matters within his own knowledge, no proper sequence is observed.

Lastly, it seems probable that much of the time consumed by the Rabbi's journey was spent in five places—

firstly, in Constantinople; secondly, in the 'City of David'; thirdly, in the capital of the Abbassides and of the Jewish Princes of the Captivity; fourthly, in the great port of Alexandria; and last, in Sicily, then perhaps the most civilised of Christian countries: that part of the *Records* which goes beyond the experience of the traveller himself was mainly compiled, there is reason to believe, in one or other of these centres of mediaeval life and thought. With Benjamin's information usually agrees that furnished by the contemporary Arabic geographers and historians; and it has been proved by the researches of Zunz and others that the names of Hebrew magnates, rabbis, and merchants, given by our wayfarer in so many cities from Saragossa to Bagdad, were really borne by prominent Jews of the twelfth century.

Benjamin's *Records* open with a preface which tells how the son of Jonah, of Tudela, in the Kingdom of Navarre, travelled through many distant countries; how, when he returned, he brought this report of the same to the country of Castile, in the year 933 (A.D. 1173); how he took down in writing, at each place, what he saw or what was told him by men of integrity, whose names were known in Spain; how the names of such reliable informants were mentioned by the author under their respective abodes; how the said Benjamin was admitted to be a man of wisdom and possessed of deep knowledge; and how, after strict enquiry, he was also found to be a true witness and his words altogether to be approved as credible.[1]

This being assured, the reader might proceed with confidence in the steps of the Rabbi. The 'Reports' which follow have perhaps been abridged from a longer original; but the

[1] This preface is of practically the same age as the narrative it introduces, but by another hand. Cf. Asher, *Benjamin*, II., 1.

FROM SARAGOSSA TO MARSEILLES

abridgment has not, in any case, been made with the ruthlessness from which Petachia has suffered.

Starting at Saragossa and passing through Tortosa, Benjamin came down the Ebro to Tarragona, famous for cyclopean and pelasgic buildings unique in Spain; these had been much damaged by the Saracens in A.D. 719, on the overthrow of the Gothic and Christian kingdom; but in 1038 the Archbishop of Toledo had undertaken a partial restoration, and thus the traveller's attention may have been especially drawn to works which the present age has again neglected.

Barcelona was a small port, but already frequented by merchants from various cities and lands—Genoa, Pisa, Sicily, Greece, and the more distant harbours of Palestine, as well as Alexandria in Egypt. Thence the route lay through Gerona, Narbonne, famous for its university and legal school, Beziers, and Har Gaash or Montpellier, a great centre of trade, ranking next to Marseilles in this part of the world, where Benjamin notices the presence of merchants from Portugal or 'Algarve,' Lombardy and its havens of Genoa and Pisa, Egypt and Palestine, England and Francia, Greece and the Roman Empire. The Rabbi next visited Nogres or Bourg de St. Gilles on the banks of the Rhone, a famous place of Christian pilgrimage, and the home of a prominent Jew, the steward of the leading Christian prince of the Languedoc, Raymond V. of Toulouse. This fortunate and powerful Hebrew, Abba Mari, was representative of the extraordinary prosperity of his race in Provence and the Narbonnese, where Jews might be counted by hundreds in every large city, and where alike in commercial and intellectual matters they wielded an influence of strong and subtle potency. Their connection with the heretical movements, even then (c. 1160) agitating the South of France, has often been asserted; it is easy to prove in one sense, difficult in another; probably their presence and the favour they enjoyed in the dominions of the House of Toulouse was a sign rather than a cause of the impending trouble; as yet it was not thought unbecoming to employ Jews in positions of high trust about the Papal Court itself.

At Marseilles, Benjamin took ship for Genoa, and about four days brought him from the ancient Phokaean city, still celebrated for its commerce and learning, to the leading Christian port of the Western Mediterranean. Genoa, 'mistress of the sea,' was now at war with Pisa; but the rivals had many points in common. Both were fortified; both were often disturbed by civil strife, when fighting raged from house to house; both possessed a brave and daring populace, which would not suffer the rule of any king or prince, but entrusted power to senators chosen by the popular vote; lastly, both had to lament a scanty Semitic element.

At the close of the thirteenth century a later Jewish traveller reported an almost total absence of his nation on all this coast, from Provence to Rome; Benjamin only speaks of two Israelites in Genoa, twenty in Pisa; but the former were visitors from the distant Ceuta on the Straits of Gibraltar; and the Tuscany of this period, like the Scotland of more recent time, seems to have offered but a poor opening for Hebrew industry.

From Pisa and its 10,000 embattled houses, covering a vast area, the Rabbi moves on to Lucca and Rome, the 'metropolis' of Christendom. Even here Jews[1] were

---

[1] Milman, *History of the Jews*, bk. xxvii. vol. iii. pp. 325-336, edition of 1863, maintains the theory that the Jews were better treated in Italy than in any other Christian country, partly because here they monopolised wealth and trade far less than elsewhere. The reference of the text makes it probable that Benjamin's visit to Rome was be-

found, some of them being in the service of the Pope; one was the steward of the household of Alexander III. and minister of his private property. Thus the same pontiff was the patron of 'Rabbi Daniel and Rabbi Jechiel' and of St. Thomas of Canterbury. Few better illustrations could be desired of the tolerance and liberality which is sometimes to be found even in the most exclusive circles of mediaeval society.

In Rome Benjamin has many singular legends to tell us, chiefly taken from the writings of Joseph ben Gorion, and occasionally resembling similar tales in Christian writers of which one in William of Malmesbury may be taken as an example.[1]

Beginning with a fairly matter-of-fact account of the 'large place of worship called St. Peter's, on the site of the wide-spreading palace of Julius Caesar,' the visitor goes on to describe the eighty halls of the eighty eminent kings who were all called Emperors, from Tarquin to Pepin, the father of Charles, who first conquered Spain from the Saracens Among these was the palace of Titus, who was rejected by three hundred senators for having wasted three years over the conquest of Jerusalem, which, according to their will, he should have accomplished in two years; here also was to be seen the hall of King Galba, containing three hundred and sixty windows, one for each day in the year, and having a circumference of nearly three miles. Besides these wonders Benjamin describes a cave underground, containing a king[2] of Rome upon his throne, with his queen and about one hundred nobles of his Court, all embalmed and in good preservation. To the Jewish visitor an object of special interest was St.

tween 1159 and 1161, or between 1165 and 1167; for Alexander III., though he reigned 1159-1181, was away from the city 1161-1165 and 1167-1177, through fear of Frederic Barbarossa.

[1] *Gesta Regum*, ii., §§ 205, 206.
[2] 'The King' is Benjamin's expression; which, he leaves uncertain.

John Lateran;[1] for there were two copper pillars made by Solomon, whose name was engraved upon each. The Jews of Rome related that every year, about the 9th of Ab, the time of the destruction of both Temples, these pillars sweated so much that the water ran down from them. There also was the cave in which Titus, the son of Vespasian, hid the vessels of the Temple which he brought from Jerusalem; and in another cave on the banks of the Tiber were the sepulchres of the ten martyrs, or ancient teachers of the Mishna, who suffered in the period between Vespasian and Hadrian. Opposite St. John Lateran there were statues of Samson (with a lance of stone in his hand); of Absalom, and of Constantine, who built the city 'which is called after his name,' the last an equestrian figure in copper gilt. Benjamin's Absalom may be difficult to recognise, but his Samson is obviously a Hercules. We may remember that as the Jewish traveller puts his own national legends to fit classical objects in Rome, so King Sigurd's Norsemen in Constantinople recognise in the figures of the Hippodrome their own heroic Asers, Volsungers, and Giukungers.

The next stage in the journey brings us through Capua to 'Pozzuolo or Sorrento,'[2] 'a large city built by Tsintsan Hadareser, who fled in fear of King David.' This town had been inundated by the sea, and the streets and towers of the submerged dwellings were still to be seen. It is now well known that on this coast various old Roman villas are covered by the sea, and this naturally gave rise to stories of submerged cities. Here, also, were hot springs producing the petroleum used by physicians; and hot baths, a cure, or at least a relief, for almost every disease, and

---

[1] *In Porta Latina*, Benjamin calls it.
[2] From Joseph Gorionides, bk. i. ch. 3, who also speaks of petroleum here. Cf. Asher's *Benjamin*, ii. 27.

frequented in summer by crowds of sick folk from the whole of *Lombardy*. Thence fifteen miles along a causeway under the mountains (built by Romulus, the founder of Rome, who feared David King of Israel and Joab his general) brought the traveller to Naples and Salerno, where was the chief medical school of Latin Christendom,[1] and where no less than six hundred Jews resided. For the mediaeval world, whatever it might say against the Hebrew race, generally admitted their pre-eminence in medicine. Close by was Amalfi, where most people were busied in trade, not tilling the ground, but buying everything for money, though it is hardly true, as Benjamin says, that 'no one dared make war upon them,' for the city had been sacked by the Pisans in 1135; yet it existed as a Republic till 1310, and had its consuls at Naples till 1190; Edrisi,[2] about 1150, calls it a flourishing city; and even in its decay it was still among the foremost ports of Italy. Next Benjamin passes into Apulia, 'the Pul of Scripture'; and here were the harbours of Trani, a favourite place for pilgrim-embarkation to Jerusalem; of Bari,[3] which had been destroyed by 'William of Sicily,' or rather by the Byzantines of Manuel Comnenus during the reign of William the Bad; of Taranto and Brindisi; and of Otranto, where the Rabbi seems to have taken ship for Corfu.

This Island,[4] long part of the Norman Kingdom of

---

[1] Cf. Ordericus Vitalis, *Hist. Eccl.* II. iii. 11; Gibbon, ch. lvi. (and Bury's notes, vol. vi. pp. 189, 190, of his edition) on the doctors of this school, who were already celebrated in the tenth century and from whose work we possess fragments as early as the eleventh. Cf. Asher's *Benjamin*, ii. 28.

[2] Edrisi, ii. 258. Cf. Asher's *Benjamin*, ii., 30-1. Here, in contrast to Salerno, were only twenty Jews.

[3] Benjamin calls it St. Nicholas de Bari, after the Church and Priory built in 1098 and endowed by the Norman Roger, Great Duke of Apulia, Calabria, and Sicily, and successor of Robert Guiscard.

[4] Here, alas, there was only one Jew.

Sicily and South Italy, had been reconquered in 1149 by the Emperor Manuel during that last flicker of Byzantine energy which marks his reign; but the traveller seems ignorant of this change in its fortunes, and marks the confines of the 'Kingdom of Greece' at Arta and Patras.[1]

Of the Peloponnese Benjamin tells us little, merely noticing Lepanto and Corinth; but the great colony of two thousand Jewish dyers and silk-workers in Thebes arrests his attention, while the learning of the Rabbis in the same colony excites his admiration.[2] No scholars like them were to be found in the whole Greek Empire, save at Constantinople; the Hebrew element formed a tenth of the entire city of 'Stivas'; and among their other trades or handicrafts that of St. Paul was professed by many. Hard by, on Mount Parnassus, was another settlement of Israelites, two hundred strong, engaged on the unusual and uncongenial pursuit of agriculture.

From Thebes to the large city of Negropont, a resort of merchants from all parts; from Negropont, by way of Vlachia or Wallachia, to the great commercial harbour of Armiro on the Gulf of Volo, frequented by Venetians, Pisans, and Genoese; and from Armiro to Saloniki, Dmitrizi near the Strymon, and Christopoli opposite Thasos, Benjamin makes his way to Constantinople. He evidently passed along the great coast road on the northern shore of the Archipelago, and so perhaps entered the city of Constantine by land, unless (like many) he took ship again at Christopoli.[3]

[1] Benjamin's story of the origin of Patras is copied from Joseph Gorionides, bk. ii. ch. 23.

[2] Cf. Gibbon, ch. liii. pp. 71, 72, of vol. vi. in Bury's edition.

[3] Benjamin also passed (a) *Jabustrisa*, probably a Slav colony, dating from the great Slav immigration into Greece in the early Middle Ages. Asher believes that it was a Wallachian town; and the Vlachi or Wallachians, an atheist people nimble as deer—great brigands, independent of all, and dreaded by all, not professing Christianity and having many Jewish names,—are men-

No mediaeval traveller has given a better account of the Queen of Cities, and here at any rate the original text of the narrative has probably escaped abridgment. The circumference of this 'Metropolis of the Greeks and residence of King Manuel' was then a matter of eighteen miles. One half of the city was bounded by the Continent, one half by the sea; and of this sea two arms here met together, the one a branch or outlet of the *Russian* waters,[1] the other of the *Spanish*. Great stir and bustle prevailed at Constantinople, because of the meeting therein of so many merchants, who came from all parts of the world both by land and sea, from Babylon and Mesopotamia, from Media and Persia, from Egypt and Palestine, as well as from Russia, Hungary, 'Budia' or Bulgaria, Lombardy, Spain, and the land of the Petchinegs.[2] Bagdad alone equalled it as an emporium. Moreover, at Constantinople was the house of prayer called Santa Sophia, the metropolitan seat of the Pope of the Greeks, who were at variance with the Pope of Rome. All the other places of worship in the whole world did not equal this in riches. It had pillars and lamps of silver and gold, and altars as many as the days in the year. It is not surprising if the Rabbi trips among all these details of Christian ritual; but a large number

tioned by the Rabbi immediately afterwards; (β) *Rabenica* or *Ravenique*, which seems now impossible to fix, though noticed by several mediaeval writers, *e.g.* Henri de Valenciennes, *Chronique*, ed. Buchon, p. 259; (γ) *Gardiki* or *Cardiki*, a little town on the Gulf of Volo and the seat of a bishop; and (δ) *Bissina*, mentioned by other mediaeval writers as Vissina, Vessina, and Bezena. *Saloniki* in Benjamin's figures contains more Jews (500) than any other town in Greece, Thebes excepted. 'Mitrizzi' or Dmitrizi, near the ancient Amphipolis, is another Slav town. *Drama* is the *Dramine* of Villehardouin: Nikephoros Gregoras also calls it *Drama*; it was near the ancient Philippi.

[1] The Black Sea is probably so called because the Russian piratical dashes across it had now become a famous tradition.

[2] This people, after 1122, when the Byzantines decisively repulsed their raids, had become more peaceful, and taken to agriculture and trade.

of altars has never been the practice of the Eastern Church.

In the Hippodrome, near the wall of the palace, the imperial sports were held; and here the birthday of Jesus the Nazarene was yearly celebrated with unequalled games. Representatives of every nation under Heaven might then be seen gathered together to witness the amazing feats of jugglery and the fights of animals which were exhibited in that place.

Benjamin describes like an eye-witness, and it has been already suggested that he was present at the festivities of the Christmastide of 1161, when Manuel Comnenus [1] married Maria, daughter of the Prince of Antioch. Likewise we may well believe that the Jewish visitor saw the interior of the Palace of Blachernae,[2] its pillars and walls covered with pure gold, its pictures of ancient battles and of the victories of Manuel himself, its throne of gold adorned with jewels, and its hanging crown suspended over the imperial seat and blazing with precious stones so brilliant that the room needed no other light.

The tribute brought to Constantinople every year from all parts of Greece, in silks [3] and purple cloths and gold, filled many towers. The city alone paid up to 20,000 florins a day, from the rents of hostelries and bazaars, and from the gate and harbour duties paid by merchants entering the

---

[1] 1143-1180.

[2] On the Palace of Blachernae, cf. Niketas, *In Manuele*; Kinnamos, v., 3; Gibbon, ch. liii. (vi., 75-8, of Bury's edition. Also see Bury's additional note 9 to vol. ii. p. 546); Le Beau, 88, 38, who collects the details from Niketas and Kinnamos; Ducange, *Constantinopolis Christiana*, especially bk. ii., ch. iv., pp. 113-123; Petrus Gyllius, *De Constantinopolcos topographia*, 1632; Jules Labarte, *Le Palais impérial*, etc., 1861; and D. Th. Bielaiev in his (Russian) essay on the Palace, 1891.

[3] Gibbon, ch. liii., quotes Falcandus, the Sicilian historian (c. A.D. 1190), on the Greek manufacture of silk. Falcandus describes all the various methods of working the silk, and what he says is a good illustration of Benjamin's language here.

town. Many rich Greeks dwelt in the country, in appearance like princes, dressed in silk and gold and jewels, and riding upon horses. Not less rich was the land itself, producing bread, meat, wine, and every delicacy; and here also were learned men,[1] well skilled in the Greek sciences, and living in peace and comfort, every man under his vine and fig-tree.

This brings Benjamin to the one shadow which he throws over his glowing picture of the Byzantine State. The Greeks had every luxury and every grace, but nothing could give them the spirit of men. Like women they were fit for no warlike enterprise themselves, but hired soldiers of all nations (whom they called *Barbarians*) to carry on their wars with the Sultan of the Turks.[2] No Jews dwelt in the city with the Christians; they were forced to reside beyond that arm of the sea where they were shut in by the 'Channel of Sophia,' and they could reach the central part of the town by water only.[3] Nor could any Jew ride upon a horse, save one, the Imperial Physician, by whose influence the Hebrews enjoyed many relaxations of oppression, even if no abatement of the popular hatred. For Jews were still often beaten in the streets, and tanners took special delight in pouring the water of their tanneries upon passing Israelites, 'who being thus defiled become objects of contempt to the

---

[1] Asher, *Benjamin* ii. 51, refers (following Gibbon, ch. liii.) to the Byzantine scholars and writers, Anna Komnena, the Empress Eudocia, Stobaeus, Suidas, Eustathius, Tzetzes, etc., in illustration of this remark of Benjamin's, which probably has especial reference to Jewish scholars.

[2] Thogarmim. Among the soldiers whom the Greeks hired were (1) the Vaerings or Varangians, mostly Scandinavians and English, but also sometimes including various Slavonic or Slavo-Norse bands; (2) these Slavo-Norsemen reckoned separately, viz. Russians, etc.; and (3) Bulgarians, Alans, Khazars, and other Turco-Tartars.

[3] The Jewish quarter, called 'Stanor' (Stenon), was beyond the Galata Tower, near the entrance of the Port of the Golden Horn, and on the Bosphorus itself, cf. Villehardouin, 153; Asher, *Benjamin*, ii., 53.

Greeks.' The Hebrew trade was largely in silk and cloth, and the place of their colony, 2500 strong, was called Pera.[1]

From the Bosphorus, through Rodosto,[2] Gallipoli, Mitylene, Chios,[3] Samos, and Rhodes, the next stage of the *Records* brings us to Cyprus, where Benjamin was shocked by the atrocities of certain heretic Jews, 'Cyprians and Epicureans,' who profaned the eve of the Sabbath and hallowed that of the Sunday.[4] Two days' journey from Cyprus was the frontier of Armenia the Little, and the realm of Thoros,[5] King of the Mountains (of Cilicia), whose rule extended to the land of the Turks. From this it is clear that the Rabbi passed by this coast before 1167, when Thoros died, at peace and in terms of vassalage with the Emperor Manuel, whom he had so long resisted. Benjamin evidently crossed from Cyprus to the Cilician or Armenian coast at Corycus or Korghos, the Kirkes of Edrisi; and he passed out of the dominions of the Byzantines or 'Javanites' at Malmistras or Mopsuestia, which he strangely identifies with Tarsus.[6]

A little beyond this border-line was Great Antioch, strongly fortified, but overlooked by a very high mountain, and possessed by Prince Bohemund[7] the Poitevin, surnamed the Stammerer, who had succeeded to the principality of

---

[1] By the Greeks themselves it was generally reckoned a part of Galata.

[2] Rhadesta in Ptolemy, formerly Bisanthe; the modern name first occurs in Benjamin, next in Villehardouin.

[3] Then, as now, famous for its Mastic, cf. Edrisi, ii. 127, who makes Samos also a centre of this trade.

[4] So they were not Karaites, anyhow, as Rapaport points out.

[5] The town of Dhuchia was on Thoro frontier. Benjamin's phrase apparently refers to his past life as a rebel. On Kirkes, cf. Edrisi, ii. 130.

[6] The two places, of course, were thoroughly distinct, and about 45 miles from one another; both had been reconquered by Manuel Comnenus in 1155.

[7] Bohemund III., Prince of Antioch, 1163-1200. On Antioch, cf. William of Tyre, iv. 9, 10.

Antioch shortly before. Thence by way of Latakiyeh [1] and under Mount Lebanon, where Benjamin stops to tell us about the Assassins of this region, who did not believe in the tenets of Mohammed, but in those of one whom they thought 'like unto the prophet Kharmath.' [2] This chief, their Lord and Master in life and death, was called the Old Man; [3] his dwelling was in the city or castle of Cadmus,[4] which was the Kedemoth of Scripture in the land of Sihon; and his followers were the dread of all men, for they would kill even a king whom their chief had once devoted to death. They were now at war with the Frank Christians, and especially with the Count of Tripoli; [5] but they were just as hostile to many of the Moslems. The name of these *Assassins* is no doubt from the drug *Hashish;* and their society, which somewhat resembled the Templars as a union of war and religion, was formed in Persia (c. A.D. 1090) by Hassan [6] Ibn Sabah, at the Castle of Alamut, in the province of Rudbar. Radically, it was a branch of the Shiites or supporters of Ali,

---

[1] Lega, the ancient Laodicea, founded by Seleucus Nikator in honour of his mother, cf. Edrisi, ii. 131.

[2] Founder of the 'Carmathians,' who held a doctrine of the transmigration of the souls of their founders. In this the Assassins do not seem to have agreed.

[3] A translation of *Shaykh* [*-al-Hashishin*]; also called the Shaykh or Chieftain ('Old Man' *i.e.* 'Elder) of the mountain, *i.e.* Lebanon.

[4] Cf. Asher, *Benjamin*, ii. 66, on the 'castle of Cadmus in the country of the Anziery.'

[5] The Count of Tripoli in Benjamin's time was Raymond II., also Count of Toulouse and St. Gilles. In 1187 he bequeathed the county of Tripoli to his godson Raymond, son of Bohemund III., thus uniting the counties of Antioch and Tripoli from the accession of Raymond's brother, Bohemund IV.

[6] A minority derive the name of 'Assassin' from this Hassan. Cf., besides *Rashid - ed - din*, (1) A. Jourdain, *Dynastie des Ismaeliens*, and *Notices et extraits*, ix. 143, etc. (2) De Sacy, *Chrestomathie arabe*, i. 130; *Mémoire sur les Assassins*, 1809; and *Réligion des Druzes*, 1838. (3) Quatremère's *Notices sur les Assassins ou Ismaeliens* (*Mines de l'Orient*), iv. 339. (4) Hammer, *Geschichte der Assassiner*, 1818; E. trans., 1835. (5) Wilken, *Geschichte der Kreuzzüge*, ii. 240. (6) Ritter, *Erdkunde*, viii. 577, etc. (7) Falconet, in *Acad. Inscr.*, xvii. 127-170; (8) Guyard, in *Journ. Asiat.*, 1877.

and so in alliance with the Fatimites of Egypt. The title of *Ismaelites* or *Ismaelians*, also applied to them, was derived from Ismael, seventh Imâm in the line of Ali, a descendant of whom became founder of the Fatimites (A.D. 909). Properly, the Assassins seem to have been themselves a branch of the Ismaelian secret societies, who under a cloak of extreme religiousness taught absolute atheism, lawlessness, and licence, and hence were called *Mulehet* or heretics; Alamut, their stronghold in the Lebanon, was destroyed by Sultan Bibars in 1270. Jebail or Byblus, Benjamin's next resting-place, was governed at this time by seven Genoese chiefs, among whom was one Julian (otherwise Hugo, otherwise William), of the great family of the Embriaci; to the same house also belonged the Admiral of the Fleet which in 1109 took Byblus from the Saracens.[1] This victory was a leading incident of the Crusades, and the conqueror became feudal lord of the city on condition of paying a certain annual sum to the state of Genoa. In Benjamin's time the Government of Byblus seems to have been carried on by a committee of seven, six nominated by Genoa, while the seventh was always an Embriaco, who acted as President of Council. Two days' journey from Byblus was Beyrout; a little further Saida or Sidon;[2] and within twenty miles of the last lived the Druses. This strange sect confessed no religion; their dwellings were on the summits of the mountains and on the ridges of rocks; and they were not subject to any prince. Mount Hermon was their boundary. Their life was vicious, and in their folly they believed in

---

[1] Cf. William of Tyre, xi. 9, who says that Hugo Embriaco was governing Byblus when he wrote (about A.D. 1180), 'being a grandson of the Hugo who conquered it,' but all other historians call the conqueror Willelmus.

[2] On Beyrout and Sidon, cf. Edrisi, i. 354-356. On Benjamin's story of the temple and idol of the Ammonites at Jebail, cf. Strabo, xvi., ii. 18.

the transmigration of souls; but they were unequalled mountain-climbers and lived on friendly terms with the Jews.

Benjamin was perhaps aware that their belief was derived from the Carmathians, and that, according to some, the Druses were originally known as Karmath; like the Assassins and Fatimites, they were probably a branch of the Ali schism and so 'Ismaelians'; but the mad Fatimite Caliph Hakim became their especial hero and prophet. All these secret societies of early Islam adopted allegorical interpretations of the Koran, and so broke every law of Mohammed with untroubled conscience.

In Palestine itself Benjamin's descriptions are usually much more accurate than those of the Christian pilgrims; and in Jerusalem he shows to peculiar advantage, giving us many facts and little fiction. Thus he notices the ancient stones in the lower part of the city walls, the stables and hospitals of Solomon, the large prayer-house called 'Sepulchre' (containing 'the tomb of that man, visited by all pilgrims'), and the cupola, on the site of the Temple, built by, or bearing the name of, Omar. In the Holy City, small as it was, people of all tongues met together—Jacobites and Armenians, Greeks and Georgians, Franks and Jews; these last, two hundred in number, had command of the dyeing trade, and paid a yearly sum to the Christian king for the monopoly of this business. The four Gates of Jerusalem, the Tower of David, the position of the Templars, and the character of their Order, the isolated situation of Mount Sion, the sepulchral monuments[1] and high mountains in the neighbourhood, are all well, though briefly, described by the Rabbi; he is less satisfactory in some of the more out-lying parts of his own sacred country. Caesarea

[1] He tells us a long story about the miracles that had defeated a recent attempt to enter the sepulchres of the kings of Judah.

on the coast he identifies with Gath of the Philistines; Haifa, the ancient Ephah, with Gath Hachepher; St. George with Luz; Tebnin, near Tiberias, with Thimnatha in Judaea; and Capernaum with Kephar Thanchum, as well as with Meon, the abode of Nabal the Carmelite. This last confusion, arising from the two *Carmels*, is found in many other writers; and the same may be said of Benjamin's tale of the Salt Pillar of Lot's wife—'incessantly licked by sheep, it always grows again, and remains as large as ever'—a method of treatment which we have already had in the sixth-century pilgrim, Antoninus Martyr.

With all this we are fortunately able to contrast many excellent and reliable descriptions and references; thus an admirable account is here given of the Samaritans of Gerizim, of the Patriarchal Tombs at Hebron, and of the neighbourhood of the Sea of Galilee, as well as of the cities of Acre and *New* Tyre. The last-named, a beautiful town, with an unrivalled port, was still famous for its glass and purple dye;[1] both of these manufactures were largely in the hands of Jews, who, moreover, possessed a great ship-owning interest. The harbour was guarded by two towers, and a chain drawn across the entrance of the haven every night; from the walls of the city the curious might see the remains of the ancient or Crowning Tyre, 'whose merchants were princes and whose traffickers the honourable of the earth,'[2] but which now lay, with all its towers, markets, streets, and halls, at the bottom of the sea. Acre, the frontier town of Palestine, was the place where most pilgrims disembarked—at least, of those who took the maritime routes. When Benjamin visited the country, it was still in Christian hands, and was rightly considered the

---

[1] Cf. Edrisi, i. 349.   [2] Isaiah xxiii. 8.

key of the Syrian coast and the most capacious and convenient of Syrian harbours.

But it is perhaps at Hebron that the author's Palestine narrative is most valuable. He carefully distinguishes between the old town on the hill, long since ruined, and the newer city in the field of Machpelah. Here was the Church of St. Abraham, which, under the Mohammedan rule, had been a *synagogue* (or mosque); and in this building supposititious tombs of the patriarchs had been erected. But if an additional fee were given to the keeper of the cave, the real graves might be seen; an iron door opened, which itself dated from the time of the Hebrew forefathers; and, with a burning candle in his hand, the visitor could descend into a first and second cave, neither of which held anything, and so arrive at the third, where the genuine tombs of Abraham, Isaac, Jacob, and their wives lay beneath an ever-burning lamp. All around were tubs or casks full of the bones of Jews who had been laid to rest in this most holy place, which (outside Jerusalem) had but one rival, Nablûs[1] and Mount Gerizim. This also was visited by Benjamin; but here he found no Jews and only a hundred Samaritans, observers of the Mosaic rite, pure and simple, and called Cutheans. Among their priests were descendants of Aaron; the whole people claimed to be of the tribe of Ephraim; and they possessed the bones of Joseph. In their *synagogue* on Mount Gerizim they offered sacrifices and burnt-offerings, and this, they pretended, was the Holy Temple, where sacrifices alone might be offered. But in spite of their unquestionable advantages, they were not full or true Hebrews; three of the letters of the alphabet were wanting in their pronuncia-

---

[1] Benjamin accurately identifies Nablûs and Shechem, this being just one of the points where a Jew would go right; cf. Edrisi, i. 339.

tion; and from the lack of one their glory was taken away, from the absence of the second they were denied the gift of piety, and from the loss of the third it followed that they had no humility. All their ceremonial care against defilement, their changes of garments and frequent washings, did not avail them.

In the vicinity of the Sea of Galilee the *Records* notice a pretended sepulchre of Samuel, apparently forgetting the plain statement of the Old Testament; loosely assume Paneas or Belinas to be the ancient Dan; and, though not giving the ordinary tradition of the fountains Jor and Dan, only substitute another legend of more elaborate inaccuracy. The Jordan, according to this story, issued from a cave; three miles beyond, the sacred river united its waters with that of the *Moabite* Arnon; in front of the cave were vestiges of the image and altar of Mikha, an idol of the children of Dan; while close by was the site of Jeroboam's shrine of the Golden Calf, at the northern confines of Israel, towards the 'uttermost sea.'[1] So much for legend. With this Benjamin, after his manner, proceeds to mingle some other and more useful, because contemporary, notices.

In Tiberias there lived a Jewish astrologer or astronomer named Abraham, a typical figure in an age when learned Hebrews so often acted as scientific advisers to the great men of the earth, Alfonso X. of Castile, Sultan Seïf-ed-din of Mosul,[2] the early Caliphs, or the French kings of the twelfth century.

Benjamin of Tudela differs widely from Petachia of Ratisbon in his reckoning of 'two hundred' Jews as resident in the Holy City; and remembering that the

[1] The Mediterranean, as in Deut. xi. 24, which Benjamin is probably quoting.

[2] His Hebrew astronomer was named Joseph; cf. also Salomon of Mesopotamia in the ninth (?) century, at Nineveh; R. Isaac, son of Baruch, A.D. 1080; and another R. Isaac, who flourished in France, c. 1150. See Asher, *Benjamin*, ii. 104.

Crusaders killed most of 'Christ's enemies' when they took Jerusalem in 1099, and barred it against those enemies as long and as much as possible, the estimate of 'one' Hebrew, given by the German visitor, is certainly more plausible. But there is a striking agreement between the present *Records* and other writings of this time, on the question of the fertility of Judaea; near Bethlehem the country then abounded with rivulets, wells, and springs; and only a land of considerable resources could have sustained the numerous Christian garrisons which were camped upon its soil. One of these fortresses, to which is significantly given the Spanish name of *Toron de los Caballeros*, and which is placed at Shunem, was probably that now called the Frank Mountain, to the south-east of the birthplace of Jesus; another was at Ibelin (the 'Yabneh of Antiquity'), where the great Crusading house of Balian had their home, and from whence they controlled their fiefs of Jaffa and Ascalon.

It is curious that our *Records*, which aim at preserving a memorial even of the smallest Jewish communities, and which mention the obscure and decayed Ashdod or Palmis, o rarely noticed in the Middle Ages, are wholly silent as to Safed, near the Lake of Chinnereth or Merom. For here was the principal Hebrew settlement in Palestine; here in later time, and probably at the era of Benjamin's visit, was a university; and near here the traveller must have passed, if we may judge from his description of Merom or Maron, and its cave, with the sepulchres of Hillel, Shammai, and other noted teachers—a place of Israelitish pilgrimage within sight of the 'City set on a hill.'[1]

Damascus, lately seized by Sultan Nur-ed-din, 'King of the Turks,' son and successor of Zenghi the old Atabeg

[1] *I.e.* Safed, which Petachia also omits, cf. Asher, *Benjamin*, ii. 107-8.

of Mosul, and head of the Western branch of the Seljuk race, was admired by the Hebrew wayfarer in common with almost every visitor and in nearly the same language. Its surroundings, within a circuit of fifteen miles, presented gardens and orchards unequalled in the world; so Moslems termed the Ghutah of Damascus, along with Samarcand and the neighbourhood of Basrah or Bassora, one of the three earthly Paradises. All this beauty was the result of careful irrigation; for the waters of the Abana and Pharpar (the *Amana* and *Parpar* of Benjamin) flowing from Mount Hermon, on which the city *leant*, were carried in pipes through the streets, markets, and houses of the oasis.

A great trade was maintained here between merchants of all countries; among these were numbered three thousand Jews, two hundred Karaites, and four hundred Samaritans; here was also a seat of Hebrew learning, perhaps a survival from the days when the Ommeyad Caliphs reigned at Damascus, and impartially encouraged every kind of science, even when professed by Christian or Rabbinical doctors. The most notable of their buildings, the great mosque or 'synagogue' of Walid, is described with curious awe and wonder by the traveller. It was formerly the palace of Ben-Hadad, as men said, and one of its walls was framed of glass, constructed by enchantment, and provided with as many openings as there were days in the year. Into these the sun threw its light in regular succession, and as each of the openings was divided into twelve degrees, according to the hours of the day, even the lesser parts of time could be known. It is singular that Benjamin should speak with bated breath of a building in glass which was only the most splendid of many similar works in Moslem countries or in China; but he came, not from the highly civilised Spain of Andalusia, but from the upland of the

Pyrenaean region, where such refinements were unknown, or known merely as the triumphs of wizardry.

In Damascus, moreover, lived the President of the University of Palestine; until the Crusades this seat of learning had been (for some time) in Jerusalem; the Frank conquest extinguished it; but in Benjamin's day it was revived in a new home. On this point the *Records* are confirmed by Moses Maimonides[1] and by Petachia, who also agree in the statement that the presiding teachers of Syrian Judaism were nominated by the Prince of the Captivity at Bagdad.

Half a day's journey from Damascus brought the author to Baalbek, not often visited or mentioned by mediaeval travellers, and identified in one Hebrew tradition (here reproduced) with Baalath in the Valley of Lebanon, built by Solomon for Pharaoh's daughter. The stones of this 'palace' were of enormous size, Benjamin truly remarks; some measured twenty spans by twelve; no binding material held them together; and the whole thing was so marvellous that some believed it could only have been erected with the help of Asmodai. Tadmor in the desert, the classical Palmyra, was also built by Solomon with equally gigantic stones; it was walled, and contained two thousand warlike Jews, living far from any other habitation, and engaged in constant hostilities with the Christians of Northern Syria and with the subjects of Nur-ed-din.

It is not clear how the author made his way onwards from Damascus to Aleppo and the Euphrates; for, though he mentions Tadmor, Hamath on the Orontes, Hazor, and Bales or Pethor on the Euphrates at the edge of the desert, it is by no means certain that he passed through all these places; his statements of distance are very loose, and the

[1] *Commentary on the Mishna,* A.D. 1167.

order of the localities mentioned is not always that of a possible itinerary; but of course the lapse of time and the confusion of his notes may account for many inaccuracies. He alludes to an earthquake which had recently laid Hamath in ruins (A.D. 1157); and at Bales or Barbarissus he records the tradition of Balaam's Tower ('may the name of the wicked rot'), which the enemy of Israel built 'in accordance with the hours of the day.' In Aleppo he dwells on the scarcity of water, whereas Edrisi[1] praises it for the abundance of its supply; all the Arabic authorities, however, agree with Benjamin that this town was the residence of Nur-ed-din, and strongly fortified by him (1149-1173). On the whole, it seems probable that the Rabbi struck the Euphrates at Barbarissus and followed the course of the great river through Davana or Kalat Jiaber (the 'fort Jiaber' of the Crusaders),[2] to Rakka; then, drawn aside like Marco Polo by the commercial renown of Mosul, we may suppose that he proceeded due east from the borders of Mesopotamia through Nisibis, by a route sometimes ascribed to Alexander the Great. Jiaber, he tells us, had been retained by the Arabs when the Turkish invasions swept over this country, and dispersed the children of the desert in their ancient wastes; Rakka, the Callinicum of the Greeks, the scene of Al Bateni's astronomical observations (c. 900) in an old palace of Harûn-al-Rashid, was also on the frontier of the Thogarmim,[3] and rejoiced in a colony of seven hundred Jews and a synagogue built by Ezra the scribe, 'when he returned from Jerusalem to Babylon.' Another of Ezra's buildings was at Charrhae or Haran, the place of the defeat of Crassus and the call of Abraham; this spot was

---

[1] Edrisi, ii. 136.

[2] Thirty-five miles below Bir or Birrah, an important river station, where the Euphrates proper began, according to one view.

[3] The Turks, cf. Lebrecht in Asher, *Benjamin*, ii. 318-392.

reverenced 'even' by Moslems, who suffered no one to erect anything where tradition placed the House of the Common Forefather, that 'true Moslem,' 'sound in faith,' preacher of the Divine Unity, and teacher of the Resurrection, who, according to Arabic belief, was himself (with Ishmael) the builder of the Kaaba at Mecca, and specially called to be an Imâm or Minister of God to all mankind.[1]

A third synagogue, ascribed to Ezra, is noticed by Benjamin at or near Bezebde, or Jezirah ben Omar, an island in the Tigris, and a famous market for the trade between Mosul and Armenia, where the Rabbi locates a mosque of Omar constructed from the Ark of Noah. Four miles from the river at this place was the Mount Ararat on which the Ark had rested; thus the 'Sublime Mountain' is calmly transferred from the other side of Kurdistan. In the same confused manner Benjamin alludes to the great Hebrew academy at Pumbeditha in Nehardea, which he identifies with El Jubar on the Euphrates as the centre of a large colony of Jews and the home of many eminent scholars.

From Haran the narrative brings us to Nisibis, a 'large city plentifully watered,' and celebrated for the long struggles waged over it by Rome and Persia; here was a considerable settlement[2] of Jews; and from this meeting-place of commercial and military highways we move onward to the Tigris, reaching the Arrowy Stream a little north of Mosul. In the neighbourhood of the 'Muslin' town, on the other side of the river, and united by a bridge with the

---

[1] For these titles of Abraham, as given in the Koran, and all the other details of the Moslem view, cf. chs. ii., *The Cow*, verses 118-128, 130, 134, 260, 262; also lxxxvii., *The Most High*, verse 19; iii., *The Family of Imran*, verses 60, 89; vi., *Cattle*, verses 74-84, 162; xvi., *The Bee*, verse 121; xxix., *The Spider*, verses 15-30; xliii., *Ornaments of Gold*, verses 25-27.

[2] One thousand.

mediaeval city, was Nineveh, the true Asshur of the Old Testament, though Benjamin transfers this name to the site of Mosul. The Assyrian capital lay in ruins, but its site was covered with small towns and villages: hereabouts were no fewer than seven thousand Jews, and among them a prince of the House of David and a certain Rabbi Joseph, the court-astronomer of Seïf-ed-din, brother of Nur-ed-din of Damascus.

Benjamin was now on the 'confines of Persia,' but he stops here to mention Carchemish on the Euphrates, Rehoboth on the same river, and a number of towns and Hebrew sepulchres of note in the neighbourhood of Pumbeditha, before commencing his minute and interesting account of Bagdad, the 'metropolis of the Emir-al-Mumemin al Abassi,' or Abbasside Caliph. This potentate, acknowledged by all Saracenic kings, and holding a dignity over them just as the Pope of Rome enjoyed over the Christians, was very friendly towards the Jews, and many of his officers were of that nation; he understood all languages, was versed in the law of Moses, and read and wrote the Hebrew tongue. He enjoyed nothing but what he earned by the labour of his own hands; and therefore manufactured coverlets which he stamped with his seal, and which his officers sold in the public market; these articles were bought by the nobles of the land, and from their produce his necessaries were provided. His character was that of a trustworthy and kind-hearted man, but he was generally invisible. Pilgrims who came to Bagdad from distant lands, on their way to Mecca, often asked to see the 'brightness of his face,' but were only allowed to behold and kiss one end of his garment. All the brothers and relatives of the Caliph were then confined within their palaces and chained in iron, because of a recent rebellion.

Only once a year did the Caliph leave his palace, at the Feast of Ramadan, when he was escorted by princes of Arabia, Persia, and even of Tibet, a land distant three months' journey from Arabia. The reigning Pontiff, among other buildings, had erected houses, streets, and hostelries for the sick poor, and sixty medical warehouses were provided with spices and other necessaries from his stores. Out of pure charity he had also arranged for the confinement and chaining in irons of all the insane, who were particularly numerous during the hot season.

The 'Prince of the Captivity' (as the Jews called him), the 'noble descendant of David' (in the Moslem phrase), was treated with great honour. All the Caliph's subjects had to rise in his presence and salute him with respect, under a penalty of a hundred stripes. On his visits to the sovereign he was escorted by horsemen who cleared the road, himself also riding on the noble animal so often forbidden to the Jews, and wearing a turban on his head covered with a white silk cloth and surmounted by a diadem. His authority is defined by our traveller as extending over Syria, Mesopotamia, Persia, Khorasan, Sheba, 'which is Yemen,' Diarbekr, Armenia, the Land of Kota near Mount Ararat, and the Land of the Alans, which was shut in by mountains, and had no outlet except by the iron gates made by Alexander. Nor was this all. His sway was also acknowledged by the Jews in Sikbia and the provinces of the Turkomans, as far as the 'Aspisian' Mountains; in the country of the Georgians 'unto the river Oxus (these are the Girgasim of Scripture and believe in Christianity)'; and even in Tibet and India. This Jewish High Priest also possessed hostelries, gardens, and orchards in Babylonia, and extensive landed property. His 'cathedral,' the 'metropolitan synagogue,' was ornamented with pillars of rich marble, plated with gold and silver; it was but one of twenty-eight prayer-houses in the city of Bagdad, which also contained ten Hebrew colleges, but only one thousand Jews.

So much for the leading personages of Bagdad. Coming to other and lesser matters, Rabbi Benjamin has to tell us of the Hebrew professors and doctors of the city, and of the Caliph's palace and park, three miles in extent, filled with all kinds of beasts and trees, and containing a pond or lake supplied with water from the Tigris. He also dilates upon the palm-trees, gardens, and orchards, the philosophy, science, and magic of the Mohammedan capital, as well as upon the trade, flowing in from all countries, which made it the equal, and more than the equal, of Constantinople. We cannot but suspect an error in the traveller's figures, when, after all this exposition of the glories of an unparalleled 'metropolis,' he assigns a space of three miles as the circuit of the same, and puts the number of its Hebrew citizens at a bare seventieth of the Jewish Community at Kufa. It has already been noticed that Benjamin's visit was probably in or about the year of Christ, 1164, in the reign of Mostanieh-abul-Modhaffer, when Bagdad had to a great extent recovered from its political decay, and stood as high as ever in social and commercial importance. A century was yet to elapse before ruin came upon it with the descent of the Mongols (1258); and in the middle of the Crusading age it may well have contained half a million of human beings, a larger populace than could be found in any city of Europe.

After Bagdad, as all careful readers have noticed, Benjamin's narrative takes a different character. It is no longer, for the most part, a record of personal travel; it is rather an attempt to supplement the first part 'of things seen,' by a second, 'of things heard.' The writer now hastily surveys the world from China to Bagdad, and concludes with a few particulars about lands to the far south or north, like Arabia or Germany, as to which we cannot always assume any personal experience on his part.

This second portion of Benjamin's work, abounding in fables and pleasant stories, is not so characteristic as the former. For the compiler is more truly a representative, not of the geographical mythology of the twelfth century, but of the historical and scientific spirit of observation and enquiry which was slowly gaining upon the legendary temper of mind; and it will be enough to pass very briefly over the concessions here made to romance, noticing, however, the not infrequent passages where the author reverts to matter of fact.

Babel or Babylon, a mass of ruins extending over thirty miles, and containing the furnace of the three children, the tower[1] of the dispersed generation, and the palace of Nebuchadnezzar, is described, in the manner of sixth-century pilgrims, as unapproachable on account of the scorpions and serpents that infested the site. Various wild stories follow, not without some foundation in a long-past history, but tricked out with almost Mandevillian extravagance, on the Hebrews of Southern Arabia, who were Rechabites, and lived twenty-one days' journey through the desert of 'Yemen or Sheba.' These rigid Talmudists were a terror to their neighbours; they fasted all their lives except on Sabbaths and Holy days; and they were always dressed in black. Their capital was at the large city of Thema or Tehama; and their country extended sixteen days' journey towards the northern mountain range. Still more incredible particulars are related of various other tribes of independent and warlike

[1] Probably the *Birs Nimrâd*. Twenty thousand (?) Jews lived in the neighbourhood of Babel, according to Benjamin; cf. Asher, ii. 139-140.

Israelites in the heart of the sacred land of Islam, forbidden to (and guarded against) all but Moslems since the days of the Prophet. At Telmas were 100,000 similar, though not as perfect, Rechabites; in the notorious Chaibar (so long the obstacle to Mohammed's progress and the chief cause of his later hostility to Judaism), were 50,000 more; while in Thanaejm no less than 300,000 Hebrews lived in a town fifteen square miles in area. Uninhabited deserts separated these people from their neighbours.

In Khuzistan or Elam, Benjamin is mostly occupied with the tomb of Daniel at Susa,—just as at Napacha, near Babylon, he gives many details of the synagogue and sepulchre of Ezekiel. As usual, Moslems vied with Jews in paying honour to the ancient prophets of Israel. Noble Mohammedans resorted to Napacha to pray at Ezekiel's tomb,[1] which no one dared plunder or profane, even in time of war; here also Arabian merchants were to be found mingled with Hebrews, at the Festivals of the New Year and Atonement, when devotions were accompanied by trade, and the best was made of both worlds. The library attached to Ezekiel's shrine at this place, containing some manuscripts coeval with the first temple; the lamp that burnt day and night upon his grave, and had never been extinguished since the son of Buzi lighted it himself; and the neighbouring tombs of Ananias, Azarias, and Misael,—were all treated with profound respect by the Unitarians of the Later Revelation, who honoured this holy spot with the surprising title of an 'agreeable abode.'

Likewise at Susa Jewish gratitude remembered the

---

[1] Near Napacha, which perhaps is the Nachaba of Ptolemy. Benjamin also notices the tomb of Nahum at Ain Japhata, perhaps the site near Elkoth, East of the Tigris, where Col. Shiel was shown the sepulchre of the Prophet. (*Journal of the Geographical Society*, viii. 93).

honour lately paid to the sepulchre of Daniel by the leading Seljuk prince and 'Supreme Commander of Persia,' Sultan Sanjar Shah-ben-Shah, whose rule stretched from the Persian Gulf to Samarcand and even Tibet—a four months' journey, with the Shatt-el-Arab at one end, and at the other the mountains and forests where the musk-producing deer was found.

When this great Emperor once visited Shushan, he ordered the coffin of Daniel the prophet, enclosed in an outer shell of glass, to be hung from the middle of the bridge over the 'River of Susa' by chains of iron: and so it remained in the time of Rabbi Benjamin. For the greater honour of the illustrious dead, no one was suffered to fish in the river over a space of one mile on each side of the coffin.

From Shushan, whose exact position is still matter of controversy, Benjamin proceeds (in his enumeration of localities, if not in any real itinerary of his own) to the River Holwan and the district of Mulehet (near the south coast of the Caspian), possessed by a sect who did not believe in Islam, but lived on the summit of high mountains, paying obedience only to the Old Man in the land of the Assassins (of Lebanon). These mountains are evidently grouped by the Rabbi with the range of Khaphton or Zagros, in which were scattered more than a hundred congregations of dispersed Jews, descendants of the captives of Shalmaneser, like the Hebrews of Arabia already noticed. In connection with the Jews of Amaria (reckoned among the congregations of Mount Khaphton) the *Records* here insert the long and wonderful story of David-el-roy, a pseudo-Messiah, whose feats of magic were performed through a knowledge of the Secret or Unrevealed Name of God. Armed with this mystery, it was easy for him to cross rivers on his outspread shawl, and to appear and disappear at will.

At Hamadan or Ecbatana Benjamin notices the tombs

of Mordecai and Esther; at Ispahan was the royal seat of Persia; at Shushan the palace of Ahasuerus and tomb of Daniel, so highly honoured by the greatest of the Seljuk Princes.[1] Then, from Shiraz or Fars, near the Persian Gulf, seven days' journey only would conduct the traveller to 'Giva' on the Oxus, close to the northern frontier of Sultan Sanjar; the distance is but a wild guess; yet, however the route may be shortened in the Rabbi's figures, all that he says about Khiva, its extensive market, its traders of all lands and tongues, and its flat surroundings, is plausible enough. The fifty thousand Jews of Samarcand seem more open to question; but the future capital of Timur, before the advent of the Mongols, must have contained a great populace, if we are to put any faith at all in the Arab geographers and historians who from the days of Ibn

[1] The position of the ruins of Susa or Shushan has been much disputed, but it may now be fixed at (a) Susan, 35 Kil. 35 of Dizful in Khuzistan, close to the intersection of N. Lat. 32, E. Long. 48, rather than at (β) Shuster on the River Karun. The tradition of the tomb of David is very old; it is to be found, e.g. in Aasim of Kufa (c. A.D. 735). As to various minor localities of Persia referred to in Benjamin, it may be noticed here, e.g. that the Rabbi's notes on the River Samarra and the sepulchre of Ezra show how the land, especially in the Shatt-el-Arab, must have considerably altered. Perhaps even as late as the twelfth century the Tigris had a separate channel to the Gulf. For Benjamin's City of Dabaristan or Tabaristan (cf. Edrisi, ii. 180), Asher, ii. 168, suggests Farahabad, the capital of Tabaristan province. The Rabbi's 'Sura,' to the north of Bagdad, was for eight hundred years the seat of a famous Hebrew University; his 'mountains of Kazvin' are probably the Elburz range (Damavand, etc.); his 'Amaria' was perhaps in the Holwan region, between Bagdad and Kazvin, where Jewish tradition is still very strong and traces of Jewish blood very common. The Jews of Mulehet, according to Benjamin, united with the Assassins in their raids; yet they obeyed the Prince of the Captivity at Bagdad, — a curious inconsistency. On the Rabbi's 'land of Kuth' (or Kotha) in North Persia, cf. 2 Kings xvii. 24-30; Josephus, *Antiquities*, ix. 14, 3. On Hillah and Babel, cf. Edrisi, ii. 160. On Mosul, cf. Edrisi, ii. 148; Petachia, 171 (Wagenseil). On Napacha and the tombs of Ezekiel and Nahum, so carefully described by Niebuhr, cf. Petachia, 197 (Wagenseil), where many details are added. See also Asher, ii., 130-2, 136-143, 155-162.

Haukal reserve their choicest superlatives for the Sogd. The Hebrews of Northern Persia, as we have seen, sometimes combined with the Assassins; they also allied themselves on occasion with the 'Ghozzes,' Ogûz, or 'infidel' Turks,[1] who adored the wind and lived in the desert, ate no bread or cooked meat, but endured a diet of raw flesh, and being destitute of noses, breathed only through two small holes. These monsters had invaded Iran 'about eighteen years before'; probably Benjamin here refers to their great triumph over Sanjar (1153-1156), when so many cities, from Merv to Rai, fell before them.

On the Persian Gulf a considerable trade was still maintained from Arabia, Yemen, India, and the islands of the ocean, but there is no mention here, as in some earlier writers, of the junks of China.

From the island-mart of Kish,[2] ten days brought the voyager to a famous bed of pearls[3] at El Katif, and seven more landed the sailor at Khulam[4] in Malabar, on the

---

[1] Cf. Edrisi, i. 181; ii. 208-9. The Ogouzes or Ogûz in the twelfth century moved across the Oxus to the east of Balkh; for a time they paid tribute to Sultan Sanjar; but in 1153 they revolted, and in 1156 they defeated and captured Sanjar, after a reign of twenty years in Khorasan. After this the Ogûz stormed and sacked Merv, Nishapur, and many other places.

[2] Kish (otherwise Kaish, Keis, Qäs Küs, or Keïsh) was the greatest port of the Lower Persian Gulf, between the eleventh and the fourteenth centuries; its prosperity comes between that of Sirâf and Ormuz. Perhaps it is the Kataia of Nearchus' Journal; it is mentioned by M. Polo; by Edrisi and Abulfeda, who both apparently copy an earlier writer; and by Kazwini. Cf. Asher, ii. 175-8.

[3] Edrisi, i. 377, gives the same account as Benjamin of the origin of pearls. El Katif is the Kotaif of Ibn Batuta, and perhaps corresponds to the ancient Gerra. Cf. Edrisi, i. 371; Asher, ii. 178-9; it lay on the Arabian shore of the Persian Gulf, and had an important trade.

[4] Otherwise Coulam, Culam, Culan, Quilon, etc; Edrisi, i. 176-7, says the King of Kulam adored Buddha —that is, at the end of the eleventh century or beginning of the twelfth. In the fourteenth century Ibn Batuta vaguely calls the same potentate an 'infidel,' but perhaps Benjamin's statement may refer to the Parsis, who had fled for refuge to Malabar from Moslem persecution. See Ouseley's *Travels*, vols. i. and iii., and K. Ritter's *Erdkunde*, ii. 58;

confines of the Sun-Worshippers, children of Kush, all black and given to astrology, but trustworthy in trade. Here the heat was so great that all business was transacted at night. Secretaries of the king reported upon the lading of every ship; and after this report had been made, goods might be left about anywhere. If any articles were lost, they were almost always brought to one of the royal officers specially appointed to receive such things. Benjamin describes the black and white pepper,[1] the cinnamon, ginger, and other spices of this land, where the dead were not buried but embalmed, and he reprobates the enchantments of the people, whereby the images of the sun revolved at dawn with wondrous noise:—'this their way is their folly.' Far beyond, a journey of two-and-twenty days, lay Khandy or Ceylon,[2] inhabited by adorers of fire, among whom were twenty-three thousand Jews. The worship of these necromancers, their suttee, their Moloch-like ritual, and what some have thought to be their doctrine of transmigration, are, perhaps, the grounds on which Benjamin identifies them with the Druses of Syria.[3]

v. (i.) 594; v. 615; M. Polo, iii. 25; Barbosa, 223 (Quatremère's Extracts), pp. 157, 172, 173, of Hakl. Soc.'s edition of the *Description of East African and Malabar Coasts*. On the white and black Jews of Malabar, reported by R. Benjamin, cf Buchanan, *Christian Researches;* Asher, ii. 183-5; Ritter, *Erdkunde*, v. 595, etc. The white Jews of this colony claim to have settled in Malabar A.D. 231.

[1] Noticed by every subsequent traveller of importance in these regions, especially Ibn Batuta. For 'cinnamon' some read 'sugar-cane.'

[2] Cf. also the forms 'Cingala' and 'Gingaleh'; see Ritter, *Erdkunde*, vi. 44-62, who refers to Aristotle's famous comparison of the positions of Taprobane and Britain (*De Mundo*, ch. iii.) Benjamin's remarks on the Jews of Ceylon suggest an intercourse with Bagdad, *via* Kish — a mere trifle to the intercourse between the Moslems of Ceylon and Spain.

[3] Their priests were unequalled necromancers. In their houses of prayer were ditches in which large fires burnt; through these they passed their children, and herein they threw their dead. The great of this land sometimes burnt themselves alive, all applauding and saying, 'Oh, happy are ye, and well shall it be with you.' While the devotee was burning, all showed the greatest joy, and played upon instruments. A few days after, the spirit of the dead

From Ceylon the passage to China was made in forty days. The Rabbi is one of the earliest Europeans to give to the great country of the Far East its modern name,[1] but he only says that it lay to the eastward, and that, according to some, the star Orion dominated the ungovernable sea of Nikpha which bounded it. From the Arabian travellers or story-tellers our present narrative borrows the story of the ship-wrecked mariners saving themselves in the hides of oxen and by the help of giant birds.[2] At this time, as before, there were certainly Jews in China; and it may have been this circumstance which led Benjamin to mention the country; but he says nothing of these Hebrew colonies; all he records about the Indian Ocean and the Far East is probably from the hearsay of merchants and others whom he met in Bagdad and at the head of the Persian Gulf.

Aden,[3] 'the Eden of Scripture,' in 'Continental India,'

(or, as Benjamin thinks, the devil in his image) would appear to his family, and give orders for the discharge of his debts. On the Druses of Ceylon, cf. *Anciennes Rélations*, 85, 165. On the Jews of Ceylon, cf. Asher, ii. 188, and Edrisi, i. 72, who gives them a position of great importance, even politically. The name 'Cingala' or 'Gingaleh' is perhaps connected with the tradition of an old Chinese colony near Point de Galle. Evidently Benjamin conceives Gingaleh as different from Khandy.

[1] From the dynasty of the Kin; like 'Cathay,' from the previous dynasty of the 'Khitai' or 'Kitai.'

[2] Large eagles called griffins, who, taking them for cattle, carried them to their nests to consume at leisure. The man then cut open the hide and escaped. All this is probably extracted by Benjamin from the standard book of travellers' tales to which Sinbad owes so much; cf. Edrisi, i. 96-7.

[3] 'Continental' or 'Middle' India, 'called Aden, and in Scripture Eden in Thelasar.' There is a striking parallel here with M. Polo, who makes Abyssinia, *plus* the West Arabian shore, equivalent to 'Middle,' 'Second,' or 'Continental' India; cf. the Persian name of 'Black Indians' for Abyssinians. Benjamin's Khulan, mentioned immediately after Gingaleh, is perhaps Socotra, or some point near the entrance of the Persian Gulf. After it the *Records* name Sebid (also in Edrisi, i. 49), a very important market, one hundred and thirty-two miles from Sana'a in Yemen. For 'Eden in Thelasar,' cf. 2 Kings xix. 12; Isaiah xxxvii. 12; the only Thelas(s)ar known is in Mesopotamia.

famous for its mountains and its independent Jews, was at war with the Christian Kingdom of Nubia.[1] This reference (as far as the Hebrews are concerned) seems like a confusion with an earlier time, before the rise of the Prophet Mohammed, when Jewish kings reigned in Yemen and fought with the Christians of Abyssinia. In the twelfth century Nubia was slowly becoming Islamised; but this process was not completed till the time of Albuquerque; and the Rabbi's notice of Christianity in that land is sound enough.[2] Thence he brings us through the Desert of Sheba to Assuan, on the banks of the Nile or Pison, which here came down from the country of the Blacks, inhabited by people who often resembled the beasts, ate herbs, and were without the notions of other men. From Assuan the *Records* transport one to Khalua, whence caravans crossed the Desert of Sahara in fifty days 'even unto Zavilah,[3] the Havilah of Scripture, which is in the land of Ghanah.'[4] This caravan trade was often endangered by sand-storms, but its commerce was extensive; iron, copper, fruit, pulse, and salt were offered against gold and precious stones. The whole

---

[1] On Christian Nubia, cf. Masudi, i. 17-23, in Quatremère's Extracts. On Assuan, the ancient Syene, so important in early geographical calculations, cf. Ibn Haukal, Edrisi, and Masudi, ii. 4, 9 (Quatremère).

[2] Governed by a 'Sultan-al-Habesh,' by whom perhaps we must understand Prester John of Abyssinia. The Nile, in Benjamin, is also called a 'Sea.' Masudi speaks of the Nubian Apes, ii. 30 (Quatremère), and perhaps is here misunderstood by Benjamin. Cf. Asher, ii. 192-5.

[3] Zavilah, Zuila, or Zuela, is in Edrisi, i. 258-9, and is said by him to be remarkable for the splendour of its bazaars, streets, and buildings.

Khalua or Alua is the Ghalua of Edrisi, i. 33, and is mentioned by several Arabic writers as a place of Jacobite Christians and a starting-point for Sahara caravans. Ghanah was doubtless the country of the modern Jenné on the Upper Niger; cf. Azurara, *Discovery and Conquest of Guinea*, Hakl. Soc. Edition, pp. xlviii.-l. of vol. ii.

[4] The town of Kuts, mentioned immediately after as having thirty thousand Jews, and lying on the frontiers of Egypt, is identified by some with Thebes, by others with Apollinopolis Parva; cf. Makrizi, i. 194 (Quatremère); Asher, *Benjamin*, ii. 196.

land (or collection of lands) here referred to lay to the west of Abyssinia.

Benjamin's visit to Cairo, the 'Metropolis of the Arabs of the Sect of Ali,' must have been shortly before the overthrow of the Fatimite Dynasty in 1171; but he says nothing about Egyptian politics, though he describes[1] the overflowing of the Nile, the cause of its rise from heavy rains in Abyssinia, and the Nilometer which marked the stages of that rise. The granaries of Joseph in Old Cairo, Misraim, or Memphis, Benjamin does not identify (like so many Christian travellers) with the Pyramids, which are separately described, as constructed with wondrous magic and unequalled in the world. The Rabbi concludes his Egyptian notes with an interesting picture of Alexandria. This great city had now been a Moslem town for five hundred years; but its commerce, as a market for all nations, was still immense. It[2] was built upon arches by Alexander, the Macedonian, and in the outskirts of the city was the school of Aristotle the Preceptor. Its long straight streets were sometimes a mile long. The port was partly formed by a pier extending a mile into the sea; and here was the High Tower, Lighthouse, or Minar, on the summit of which stood the famous mirror of glass, wherein one

---

[1] It is curious that Benjamin does not mention the 'Canal of Joseph' here, noticed by Edrisi, i. 308-9. The Rabbi's identification of Pithom with the Fayûm is doubtful. Benjamin's Tsoan or Zoan, is apparently a suburb of Cairo near the Mokattam Hills. On the Nilometer, cf. Edrisi, i. 310-2: and Ritter, *Erdkunde*, i. 835. On the Delta mouths, see Edrisi, i. 313, who makes only four, two being natural branches, the others canals. Benjamin's Old Cairo or Memphis is apparently Fostat; cf. Edrisi, i. 301-2. The *Records*, in their treatment of Egyptian places, do not show any regular itinerary, but merely enumerate the Jewish settlements. The Jews of Egypt were apparently not subject to the 'Prince of the Captivity,' owing to Fatimite dread of Bagdad influence. Cf. Asher, ii. 196-211.

[2] Cf. William of Tyre, xix. 24; Edrisi, i. 297-8; Asher, ii. 211-216.

could perceive the approach of hostile ships at a distance of fifty days. At last a Greek named Theodore, of great cunning, arrived in the port, bringing tribute from his king, who was then subject to Egypt. Under the pretence of good fellowship he made the keeper of the lighthouse dead-drunk, broke the mirror, and escaped.[1] The Christians then resisted the Egyptians with better success, and took Crete and Cyprus from them. Without its mirror the Pharos could still be seen at the distance of a hundred miles, and its light at night was a guide to all mariners. From all parts of Christendom merchants resorted to Alexandria,—out of Tuscany, Lombardy, Apulia, Amalfi, Sicily, Valencia, Rakuvia, Spain, Catalonia, Roussillon, Germany, Saxony, Denmark, England, Flanders, Hainault, Normandy, France, Poitou, Anjou, Burgundy, Mediana, Provence, Genoa, Pisa, Gascony, Aragon, and Navarre.[2]

[1] Another version, given by Langlès, tells how a Greek spy persuaded Walid, sixth of the Ommeyads, to let him dig under the Pharos for treasure. With this excuse he destroyed both Pharos and mirror. This Greek spy was in the service of the Emperor Justinian II., or Anastasius II.; he pretended to be a deserter and to embrace Islam, and so won the confidence of Walid. According to the Arabs, the observatory of Alexandria was in the Pharos.

Cf. the pilgrim Arculf, who tells us how St. Mark's body then lay (A.D. 690) in a church in Alexandria; also Bernard the Wise, who describes how the Venetians, some time before his visit in 867, had carried off the relics of this saint. The story given above seems to have been one of those which has almost gone the round of the old world, and has survived in Indian, Mohammedan, and European forms. T. Wright refers to the Old English poem of the Seven Sages as a parallel.

[2] The list of the twenty-eight Christian states, whose people traded to Alexandria, is unfortunately uncertain here and there. For 'Valencia' Asher suggests 'Florence,' as Valencia in 1170 was not Christian at all; for 'Rakuvia' Asher reads 'Ragusa'; for 'Catalonia' Sprengel suggests 'Coralita' or 'Sardinia.' 'Roussillon' is a natural correction for the reading of the text, which is not 'Russia' in the MSS. and first editions, though (as being nearer to this form) it has been so altered. The 'German' Traders are, no doubt, from Lubeck and other Hanse towns. 'Mediana' is perhaps 'Maine.' For 'England' Zunz would read 'Galicia.' Cf. Asher's *Benjamin*, ii. 218-220.

Mohammedan traders came thither from Andalusia, Algarve, Africa, and Arabia, as well as from the countries towards India—Abyssinia, Nubia, Yemen, Mesopotamia, and Syria. From India they imported all sorts of spices, which were bought by Christian merchants. The city was full of bustle, and every community had its own fonteccho there. It was probably from the lists of these fontecchi,[1] from the consulates (in modern language) of the various nations, that Benjamin compiled this catalogue of the traders of Alexandria.

Leaving Egypt, a land unequalled upon earth for its cultivation, the traveller briefly notices Mount Sinai (on the top of which stood a Church of Syrian monks) and the arm of the Red Sea,[2] one day's journey from the Holy Mountain, which itself was an off-shoot from the Indian Ocean. For himself, he seems to have returned to Europe by way of Damietta[3] and Messina,[4] where then assembled most of the pilgrims for Jerusalem. Benjamin's note-book concludes with a glowing description of the splendours of

[1] These fontecchi ('pundak' in Benjamin, from the Greek πανδοκεῖα) were great compounds or enclosures containing the establishments of foreign traders, and including—(1) their shops and warehouses; (2) baths, taverns, and bakehouses; (3) a church or chapel; (4) a market-place where men could carry on trade, even if debarred from the Moslem city outside. They were therefore more than 'hostelries for sea-captains,' as Asher calls them.

[2] Ailah or Elim belonged, says Benjamin, to the Bedawin Arabs; cf. Edrisi, i. 322. On Mount Sinai, cf. Edrisi, i. 332, and Ibn Haukal, who both speak of the Christian House of Religion here. On Benjamin's place-names in this last section

Asher suggests that the author is here copying an older pilgrim-route from Egypt to Mount Sinai, and thence back to Damietta, whence the voyage to Palestine was often undertaken.

[3] The 'Tennis' or 'Khanes' of Benjamin (near Alexandria, on Lake Menzaleh, which sometimes was also called 'Tennis') is perhaps the Ahnas of Edrisi, i. 128. Here Asher thinks the Rabbi may have embarked for Sicily, but Alexandria is more likely. Damietta was probably his first place of call after embarkation.

[4] Cf. Edrisi, ii. 81. Benjamin's name of 'Lunir' for the Straits of Messina is peculiar. Cf. Asher, *Benjamin*, ii. 224.

Sicily,[1] and a few remarks upon Germany, 'the country of Bohemia called Prague,' France,[2] and Russia. This last was very extensive, mountainous,[3] and full of forests, and from the gates of Prague to those of Kiev one travelled over the lands of Slavonia,[4] where intense cold kept all men within their houses in winter, except for the chase of sables or white squirrels.[5]

Rabbi Moses Petachia, a contemporary of Benjamin of Tudela, visited the Levant within a very short time of his predecessor. For Benjamin names one Daniel as the Prince of the Captivity at the time of his stay in Bagdad; while Petachia says this same Daniel had been dead a year

---

[1] On Palermo cf. Edrisi, ii. 76-8. Benjamin's remark, 'Palermo is the Viceroy's seat,' probably refers to Stephen, Archbishop of Palermo, Governor and Viceroy of the Norman Kingdom. This was Stephen of Rotrou, son of the Count of Perche; he was brought to Palermo and made chancellor (c. A.D. 1166), during the minority of William II. of Sicily, by the Queen Mother Gentiles, who acted as Regent, and her uncle the Archbishop of Rouen.

Sicilian coral, referred to by Benjamin as a 'stone' found near Trapani, is also mentioned by Edrisi, i. 266-7, and modern books; some of it is valuable in commerce. Here is another proof of Benjamin's calling as a merchant. Cf. Asher, ii. 224-5.

[2] Benjamin's 'Mount Maurienne' was of course in the county so called, then including nearly all modern Savoy. The name is said to signify 'Moors-land,' and to come from the Arab invasion of the ninth century. A count of Maurienne and Turin joined in the second Crusade.

[3] The best explanation of this phrase seems to be the Carpathians.

[4] Called Khna'an or Canaan by the Jews because of the prevalence of slavery in Slav lands, even in Bohemia (cf. Genesis ix. 25, 'cursed be Canaan, a servant of servants shall he be' . . .). Hence the belief of some Jews that the Slavs were descendants of the Canaanites. Among Jewish references to Slav lands we have those of Petachia (c. A.D. 1180), of Chasdai, the famous Physician and Minister of Abder-rahman III. of Cordova (c. 970), and of Eliezer Ben Nathan (c. 1100). The last-named seems to have visited Russia, and refers to the images (pictures or eikons) of saints on the doors and walls of the houses. It is curious that Moses Maimonides uses the term Khna'ani or Canaanites for the 'Syrian' or Arabic invaders of the Magreb or North-West Africa.

[5] 'Vaiverges,' the Polish 'Wieworka.'

before his arrival in the Moslem capital. But we must remember that the weaknesses and inaccuracies which are sometimes to be found even in Benjamin appear much more prominently in Petachia, who is in every way an inferior writer and observer. So it may be well not to rely too strictly on calculations depending upon the minute accuracy of one of his remarks. Thus, while he agrees, as above noticed, with Benjamin in his mention of the chief Rabbi Daniel, he gives a different name [1] for the Father of the same chief Rabbi.

As with the writer, so with the text. If there is ground for thinking that Benjamin's record has been abridged by an editor, in the case of Petachia we know that what has come down to us is only a selection. Large portions of what appear to have been highly interesting matter have been omitted by Rabbi Yehuda the Pious, who acted as Petachia's literary mouth-piece, and who frankly refused to write down many of the details that were given him.[2] This Rabbi Yehuda, like Petachia himself, was a native of Regensburg or Ratisbon in Bavaria, on the Upper Danube; he was related to several famous Jewish doctors [3] of the later twelfth and earlier thirteenth centuries; and to him, as a man of ripe judgment, the traveller submitted this report. The result is an arid and disappointing summary of a journey that deserved a better fate.

Petachia, after returning from his travels, if not before, seems to have left the 'Jewish Athens' on the Danube, and removed to Prague [4] in Bohemia. The date of his

---

[1] Chasdai in Benjamin, Shelomoh in Petachia.

[2] Cf. p. 13 of the London edition of 1856: 'R. Yehuda would not write this down.'

[3] For instance, his brothers were R. Isaac Hallaban (the White) Ben Jacob, and R. Nachman of Ratisbon. Yehuda died A.D. 1217 or thereabouts.

[4] So Zunz. The first edition of Petachia was published in this city A.D. 1595, with a German translation.

journey must apparently be fixed to a time before Saladin's reconquest of Jerusalem and the Holy Land[1] (1187); on the other hand, his narrative is clearly a little later than Rabbi Benjamin's; we may, perhaps, assume that it lies between the years 1180 and 1186.

Petachia's outward route, compared with Benjamin's, is somewhat like that of Carpini in relation to his successor Rubruquis. On one side we have a northern overland journey traversing many of the regions of Eastern Europe; on the other side is the Southern Mediterranean or maritime way more usually followed, at least by the literary travellers, from Western Christendom to the Levant. It is very doubtful, however, whether this can be said of the mercantile movements which, after all, must have employed a far larger number of the journeying classes.

Petachia[2] set out from Prague, travelled through Poland, and so came to Kiev on the Dniepr, even then, in its decay, the capital of the Russian lands. From Russia the Rabbi came in six days to the lower course of the Dniepr, where it flowed through the land of Kedar, the country of the Petchinegs and Komans. Through this land of Kedar, the Ukraine of the Slavs, he next passed, and he describes it faithfully enough, like some of the more trustworthy of subsequent wanderers in the Tartar steppes. The people of this country had no ships or boats, but crossed the rivers on rafts of horse hide, which they tied to the tails of their living

---

[1] This appears from Petachia's language, but with less clearness than one could wish.

[2] In Wagenseil's Altdorf edition of 1687-1691 occurs the title, 'Circuit of the Rabbi Petachia of Ratisbon, brother of R. Isaac the White, Author of the Tosephoth, and of the Rabbi Nachman of Ratisbon. And R. Petachia went round all the countries as far as the river Sambation.' Here we follow the divisions and pages of the London edition of 1856, by Dr A. Benisch and W. F. Ainsworth.

horses.[1] These last swam across and towed their masters after them. The inhabitants of Kedar [2] ate no bread, but only rice and cheese, millet boiled in milk, and raw or half-cooked flesh. In this land strangers travelled under escort; the savage natives drank from skulls, and from vessels of copper, which were made to resemble skulls; their favourite oath was by the drinking of blood. As a nation they lived in tents, were excellent archers, and very far-sighted, perceiving objects at more than a day's distance from them; their country, which it took Petachia sixteen days to cross, was wonderfully level; and their government was in the hands of noble families who lived unhampered by kingship. So far, allowing for certain exaggerations of detail, the Rabbi is in complete agreement with all careful observers of the Turco-Tartar races from the time of Justinian; but when he adds that the folk of Kedar had beautiful eyes, because they ate no salt, and were favoured with the scent of fragrant plants, it is more difficult to follow him.

A gulf of the sea ran in between Kedar and Mesech [3] between the Ukraine and these shore-lands of the Crimea and Sea of Azov, to which the name of Khazaria clung centuries after the Khazar dominion had passed away. Through Khazaria the Rabbi travelled about eight days; at the end of this country, seventeen rivers [4] which surrounded it came together and flowed into 'the Sea,' whether the Black Sea or the Sea of Azov is not quite clear. On the shallow waters of the latter, in

---

[1] Cf. Xenophon, *Anabasis*, v. 10; and the *Tarikh-el-Kamil* on the passage of the Oxus by the Mongols in A.D. 1219. . . .

[2] In Kedar were no true Jews, but a number of heretical Sectaries, who claimed membership with the chosen people.

[3] The land of the Khazars.

[4] The 'end of Khazaria' should bring one at least to the Don, but this remark on the seventeen watercourses, requiring a river with many branches, does not suit the lower Tanais, Dniepr, Volga, or any other stream of this part.

their furthest western reaches, often known to ancient geographers as the Putrid or Stagnant Sea, to more modern times as the Sea of Sivash, Petachia is unusually detailed. If the wind blew from this fetid marsh, it was always fatal, he declares, to passers-by. All this exaggeration is to heighten the contrast with the sweet waters of the Black Sea on the other side of Meshech or Khazaria. The Rabbi's terms of Kedar and Meshech, apparently used as generic expressions for distant and barbarous countries, are not in agreement with the ordinary usage of Jewish writers, to whom Kedar is properly Arabia. As to the Khazars, so famous in earlier times as the one important addition which the Jewish Faith ever gained in Europe, they had ceased to rule this country for more than one hundred years when Petachia wrote; first the Petchinegs, then the Komans, had supplanted them in their Kingdom of the Steppes, as they, in their turn, were to be supplanted by the Mongols. In Petachia's time the old connection of the Crimea with Constantinople had been strengthened by Manuel Comnenus, and the people of the Peninsula professed a direct allegiance to the Greek Emperors. To the Jewish traveller the Karaite or heretic Jews were naturally of special interest. These Sectarians, who rejected the Talmud and all tradition, holding only by Scripture, had been long established in the Crimea, and their principal stronghold was near the modern Baktchi-Sarai. It is not unnatural to suppose that they were in part, if not entirely, descendants of the Judaising Khazars.

The next stage in his journey Petachia calls the *Land of Togarma*[1] which adjoined the *Country of Ararat*;[2] and in

[1] Togarma, Petachia says, was tributary to the King of Greece. Cf. Benjamin of Tudela's 'Thogarma' and 'Land of the Thogar-mim' in the sense of 'Turks' country.'

[2] Petachia seems to allude to the two Peaks of Ararat in a later

IV.] PETACHIA'S ROUTE; KHAZARIA TO BABYLON 269

these two expressions Georgia and Armenia are evidently intended.[1] Beyond this region was the Land of Islam. Herein he travelled in eight days to Nisibis and the neighbourhood of Diarbekr, having the range of Ararat[2] on his right; and at the end of these mountains he seems to have gone eastwards from Nisibis to Mosul or Nineveh on the Tigris, where the whole land was black like pitch. The famous city which had once stood here, and the forests that had covered this region, were now no more; they had been blotted out like the Cities of the Plain.

At Mosul Petachia fell ill, and from his language here we may suppose that he was a rich man, and had some fear of his death being hastened by the natural desire of the Sultan to seize half his property, as the law allowed at the death of a travelling Hebrew. Determined to disappoint the grasping Moslems, the Rabbi recovered, and crossed the Tigris on a raft of reeds 'dressed in his most beautiful clothes,' perhaps in a sort of bravado.

From Nineveh[3] the traveller floated down the current of the Arrowy Tigris to Babel, making the distance in fifteen days. It is confusing enough to find him using

passage (p. 49 of Eng. Ed.), but he declares the Ark was no longer there.

[1] Both Georgians and Armenians considered themselves descendants of the Thogarma of Genesis x. 3, whom some Jewish writers, e.g. Benjamin, evidently think the ancestors of the Turks (Thogarmim). There were but few Jews in the large cities of this country.

[2] As in other writers, Ararat is here used in two senses;—first, the Peak itself, the Turkish Kus-dagh; and secondly, the range extending from the neighbourhood of Erivan and the River Aras, down south-west to Cilicia and the Mediterranean, i.e. the Taurus. Near the modern Diarbekr Petachia only mentions Khosen-Kapha, perhaps the Saphê of Ptolemy, the Supha of Plutarch, the Turkish Hisu-Kapha. Nisibis was an old Jewish centre of learning, of which Ezra was the traditional founder; and there was a large Hebrew colony in Petachia's days. Cf. Ritter, Erdkunde, xi. 558, who quotes a record of 30,000 Jewish families living in and near Nisibis.

[3] Here Petachia was much interested in an elephant that executed criminals.

almost indiscriminately the three senses of the Rabbinical Babel (as employed for Mesopotamia, in a wider and a narrower sense, as well as for a small district on the East of the Euphrates, round Pumbeditha), but here the second sense appears dominant; for on reaching Babel the writer goes in one day to Bagdad, the Metropolis of Islam, and the seat of the Caliph, the 'Great King who ruled over nations.' Petachia probably visited the city shortly after Benjamin, and in the same period of the Caliphate. He describes[1] it as very large, more than one day across, and three days in circuit; for the rest, he agrees closely with Benjamin, especially as to the position and importance[2] of the Jewish colony and their Chief, the Head of the Captivity.

From Bagdad, Petachia travelled to Nehardea,[3] Shushan, Mella,[4] and the graves of various Jewish Saints and Prophets. Hereabouts he saw a remarkable animal, the flying camel, which could go a mile in a second; and in the same region he collected many Jewish legends and miracle-stories, especially of the tombs of Ezra and Ezekiel. Here, also, his geography becomes more than ever confused; from the neighbourhood of Bagdad to Mount Sinai was all one mountain range, as men reported.

The Jews of Babel, Persia, and Kush (Æthiopia or

---

[1] One detail he adds to Benjamin, on the Gates of Bagdad, made of polished copper, so bright that horses were frightened by their glitter, etc.

[2] He also gives the same extravagantly low number of 1000 Hebrews (with 2000 students in the Jewish academy) in a city, which at this time perhaps contained 500,000 people. Petachia's Caliph was perhaps either—(1) Mostadhi Billah, 1170-1179, or (2) one of his two successors. Carmoly suggests Nasir-li-din Allah (Abul Abbas Ahmed VII, A.D. 1180-1225).

[3] Often mentioned in the Talmud and Josephus, a stronghold of Babylonian Jews corresponding to the modern hamlets of Werdi and Irzah.

[4] Perhaps the modern Mella, near Birrah, on the Euphrates, although this is much disputed. The tomb of Daniel at Shushan or Susa is described in very similar language by Benjamin, Petachia, and the Arab geographers, e.g. Ibn Haukal.

Abyssinia),[1] were numerous, but often suffered from oppression, as in 'Greece.' The Land of Israel was still subject to the Christians (so Petachia hints rather than asserts), and the Jews who lived there obeyed the spiritual commands of the Head of the Academy at Bagdad.

Since he had left Bohemia, the Rabbi had journeyed steadily towards the East, until his arrival at the 'Home of Peace'; now he began to turn westward. He does not himself remark upon what now appears the most curious feature of his journey, his outward route by the northern and eastern shores of the Black Sea, and consequent arrival in Western Asia from a side almost unattempted by other mediaeval travellers. Nor does he give any complete itinerary of his return. He is content to furnish us with ampler details of Palestine.

From Bagdad to the site of old Babylon, from Babylon back to Mosul and Nisibis, thence to Haran, Aleppo, and Damascus—so Petachia made his way into the Holy Land of his race. He probably did not visit the river Sambation, which his narrative begins by fixing as the goal of his journey; he only repeats the Rabbinical fables about this marvellous stream, which stopped dead on the Sabbath,[2] and which lay forty days' journey beyond the tomb of Ezekiel, near the Mountains of Darkness, where the lost Ten Tribes were imprisoned, awaiting the Coming of the Messiah.

In the Cities of the Caliph the traveller saw the envoys of the king of Meshech,[3] whose realm extended as far as

---

[1] The regular Hebrew use of Kush (cf. Genesis x. 6; 1 Chron. i. 8, 9; Isaiah, xi. 11) might imply either (1) Æthiopia or (2) Arabia. Petachia could hardly be ignorant that for five centuries and a half Jews had been excluded from the Holy Land of Islam; but cf. Benjamin on 'Continental India, called Aden' and its 'independent Jews.' Taking him in the Abyssinian sense, Petachia's notice is interesting, and agrees with other information.

[2] Cf. the other name 'Sabbation.'

[3] 'Apparently here used for Magog,' Benisch conjectures.

the aforesaid Mountains of Darkness, beyond which lived the sons of Jonadab Ben Rechab. Here[1] we seem to have a confused reminiscence of an earlier time; for the marvellous conversion of the Seven Kings, and the visits of Rabbis even from Egypt to instruct the people of Meshech in the law of Moses, can hardly refer to anything else than the Hebrew Kingdom of the Steppes.

The heat of Babylon was such that to Petachia it seemed a different world; from his description of the Jewish Service here we might suppose that a knowledge of the Ancient Hebrew music was still preserved. Some of the marvels which our traveller notices at the tombs of the Prophets may be frankly accepted; for the supernatural lights that appeared here were no doubt due in his time as well as ours to springs of naphtha. Petachia's appetite for the miraculous is stronger than Benjamin's, but his marvels are of the same kind.

When he visited the 'goodly city' of Damascus, it had already passed under the hand of the king of Egypt, the great Saladin, and this fate was soon to overtake the rest of Syria. East of the Jordan, in the land of Sihon, king of the Amorites, and of Og, king of Bashan, where no grass or plant could grow, was the grave of Shem, which Petachia saw with his own eyes, and thus could verify to be eighty cubits (or forty yards) in length. Thence he moves onwards to Tiberias, Acre, and Jerusalem, through a land teeming with marvels. In the Holy City there was only one Jew residing, and he paid a heavy tax to the Christian king.

Outside the walls, and standing upon Mount Olivet, was a palace built by the Ishmaelites, while Jerusalem was still in Moslem hands. The Salt Sea of Sodom and Gomorrah, in the midst of an eternal desolation, was no longer marked by

[1] Understanding by Meshech the Land of the Khazars

the Salt Pillar of Lot's wife, as many had supposed; and the circuit of the whole land of Israel, which fame swelled to so vast a size, could be made in three days. At Hebron the abridgment has preserved rather more of the original narrative. The keeper of the Cave of Machpelah, it may fairly be supposed, had no authority to admit a Jew to the earliest home of his race, whence Christians or Moslems have so long claimed to exclude him; but a gold piece gained entry for Petachia. The Rabbi, however, was not satisfied; he became convinced that the real tombs of the Patriarchs were still behind the bolts and bars of an inner wall; and with another gold piece he procured admission within the cave itself, where he saw iron doors such as no man could make, through which blew winds of unearthly power.[1]

After Benjamin of Tudela and Petachia of Ratisbon, there are several indications of continued Jewish travel to Syria and other parts of the Levant; few of these are of importance; but one at least is worthy of a place in the class headed by the Rabbi of Navarre, if we look either at the extent of these travels, the variety of the same, or the intellectual ability and linguistic skill of the wanderer.

This was Jehuda Charisi ben Salomo, who about A.D. 1216 journeyed by way of Egypt, through Alexandria and Fostat, to Judaea and Jerusalem. In Palestine he also visited Gaza, Ascalon, Acre, and Safed; Damascus, Hems or Emesa, Hamath, and Aleppo in Northern Syria; Edessa, Harran, and Rakka on the middle course of the Euphrates; Mosul, Bagdad, and the Shatt-el-Arab on the Tigris and in the basin of the Persian Gulf; in whose neighbourhood he

---

[1] Petachia's description of the Machpelah Cave ranks with that of Benjamin of Tudela as the most detailed and interesting in mediaeval travel from Europe.

also saw the tombs of Ezra and Ezekiel. He returned to Moorish Spain, his native country, by way of Greece, passing through Thebes, so famous as a Jewish centre, among other places. In Christian Spain he made extensive journeys, as well as in France; in the former he acquainted himself with Toledo, Calatayud, Lerida, and Barcelona; in the latter, with Narbonne, Beaucaire, and Marseilles, from which last he sailed to Alexandria. Setting out on his journey in about 1216, he ended his wanderings in 1218 or 1219,[1] and among his other achievements he translated one of the most celebrated and delicate of Arabic literary treasures, the *Assemblies* of Al Hariri.

Of other Hebrew travellers or geographers in this age we need only notice the anonymous pilgrim, who about 1188 writes a letter detailing the miseries of the Jews in Jerusalem; Samuel ben Simson, who (some twenty years later) makes a report on his travels among the Sepulchres of the Fathers in the Promised Land (c. 1210); Hillel 'of Palestine,' whose similar work, of almost the same date, is only known by a fragmentary quotation (c. 1212); and Jacob Antoli of Provence, who does not appear to have left his home on any more distant wanderings than to Naples, where he died in 1231. His chief work lay in the translation of Ptolemy's *Almagest* (from the Arabic of Averroes), and of the *Celestial Movements* of Al Ferghani; to the latter he added a geographical chapter, on the position of various towns; and his version was the basis of the standard Latin imprint of the great Arabic astronomer.

[1] He died c. 1255. On the Hebrew travellers and geographers, before and after Benjamin of Tudela and Petachia of Ratisbon, cf. Zunz, *Essay on the Geographical Literature of the Jews*, at the end of Asher's edition of Benjamin, vol. ii., pp. 230-314.

CHAPTER V

DIPLOMATIC AND MISSIONARY TRAVEL—CARPINI, RUBRUQUIS, ETC.

THE Conquests of the Mongols, and especially their two great attacks on Eastern Europe in 1220-22 and 1238-39, excited the attention of Christendom, not indeed enough to still internal strife, but enough to call out a number of writers, travellers, envoys, and statesmen who made the Tartars their special object of study. Horror was blended with approbation; these new Huns were as revolting as the men of Attila, but they promised to be very useful in breaking down the power of Islam. Thus Alberic Trois-Fontaines, Roger Bacon, Matthew Paris, Vincent of Beauvais, and the Pontiffs Gregory IX. and Innocent IV., with many others, show a considerable hesitation and confusion of mind in face of the new problem which had so suddenly arisen. Alberic records with triumph in 1222 how the Mongols retreated on hearing of the fall of Damietta into Christian hands (1219); but in 1239 he describes them sympathetically.[1] Matthew Paris tells how in 1238 the fear of the Mongols made the people of Gothland and Friesland keep away from the Yarmouth herring fishery;[2] but elsewhere he

---

[1] Referring (*Cronicon*, A.D. 1222) to John 'de Palatio' Carpini, probably our John de Plano Carpini ('Pian de Carpine') for his information. See p. 391.

[2] M. Paris, *Cronica Majora*, under A.D. 1238; Rolls Series Edition, vol. iii. pp. 488-489.

narrates the cold reception given to the 'Ismaelian' or 'Assassin' Ambassadors at the Court of Henry III., in their appeal for aid against the Tartar hosts of Hulagu Khan (1238); and he even seems to credit the story of Mongol descent from the Lost Ten Tribes, in spite of the cruelty and barbarism of their warfare. Roger Bacon, on the contrary, appears to prefer an identification with Gog-Magog and the armies of Antichrist, who had broken out of their Caucasian or Arctic prison, and come forth to desolate the world.

After the battle of Lignitz (9th April, 1241) and the ravaging of Silesia, Poland, Hungary, and Moravia, Frederic II. urged common action upon Henry III. of England and the other princes of Christendom; and this advocacy was alone enough to make the Papacy lukewarm in its support of the same object. Thus, although Innocent IV. writes with many brave words to the Archbishop of Aquileia, and Gregory IX. addresses consolations to the Queen of Georgia and the King of Hungary; although a papal encyclical is issued for a Crusade against the heathen enemy; and although passionate exhortations are addressed to Christendom to 'cease quarrelling, if one would not cease living,' little was really attempted, and absolutely nothing was done, in a military sense. At the Council at Lyons, however, in 1245, two diplomatic missions were organised, as embassies from the Pope to the Grand Khan of the Mongols; and on the 9th of March the Pontifical commission was granted to the envoys. Of these two missions, the first was to take a northern route by way of Poland and Russia; the second was to proceed along a more southerly course, through Asia Minor and Armenia. The former or Northern embassy was committed to John de Plano Carpini, a Franciscan of a House near Perugia and Provincial of his Order at Cologne; the latter or Southern legation was put under another Franciscan, a Portuguese named Lorenzo. Carpini started on the 16th of April, 1245; delivered his letters to Kuyuk Khan near Karakorum; and in the autumn of 1247 reappeared at Lyons and handed Kuyuk's answer to the Pope.

On this most important journey (which there is a certain regrettable tendency to depreciate) John de Plano went out by way of Bohemia, Poland, and the Ukraine to Batu's camp upon the Volga, and returned by the same way; his chief companion was one Benedict the Pole, a Brother of his own Order, who acted as his interpreter in the earlier part of the journey.

As to Lorenzo of Portugal, we hear nothing more about his mission, but two years later we meet with him as a Papal Legate in Asia Minor. Apparently his enterprise was merged in, or superseded by, the new embassy of 1247. This was addressed especially to the Mongol General Baitu, commanding in Armenia; it was likewise undertaken by commission from the Pope; and it was placed in the charge of another Friar, Brother Ascelin or Anselm. Simon of St. Quentin accompanied Ascelin, as Benedict the Pole accompanied Carpini, and an abstract of Simon's account, but probably a mere fraction of the whole, is preserved by Vincent of Beauvais. Ascelin was coldly received by the Mongols, and did not return till 1250. The staff of this mission was comparatively large, but as in some other enterprises, the more pretentious schemes proved the least successful. According to some writers the Dominican Andrew of Longumeau, or Lonciumel, was also associated with Ascelin, in the earlier part of his journey; he is not mentioned by Simon; but the Friars Alexander, Alberic, and Guiscard of Cremona (the last of whom joined as an interpreter at Tiflis), were certainly in the party.

A Frenchman was also at the head of the next important legation from Christendom to the Mongols; for even before the return of Ascelin another mission had been sent to Central Asia by the Most Christian king, the great St. Louis. It is not unlikely that Carpini and Benedict, on their return from the East, were presented to Louis IX., as it is more than probable that they came into contact with his servant William de Rubruquis at the French Court. When the king went to the Levant in 1248, he took Rubruquis with him; and in the autumn of this year Friar William was present at the reception of a Mongol embassy in Cyprus by his sovereign. This embassy was from Ilchikadai, the Tartar General commanding in Persia, who offered St. Louis an alliance against the Moslems; it was headed by one David, who spread wild stories of the conversion of the Nomade Court; and King Louis promptly sent his answer by the hands of Friar Andrew of Longumeau in the February of 1249. Andrew reached Kuyuk's 'Horde' on the Imil, but he found the Great Khan had just died, and the Regent Ogul Gaimish received him as if he were bringing tribute from the King of France, and sent an insulting answer. Both the self-styled legation of David and the action of Ogul Gaimish were afterwards repudiated in the strongest terms by Mangu Khan.

In 1251 Brother Andrew returned from the Mongol Court, with some messengers of the Empress Regent and various supposed evidences of the existence of the true faith among the Mongols; and it was on the strength of these, especially in relation to Sartach, the son of Batu, occupying an important command on the north of the Black Sea, that the French king now despatched William de Rubruquis on a fresh mission.

But while the king was still at Caesarea, busy in fortifications against the Saracens, there arrived at his Court, from the Latin Emperor of Constantinople, one Philip de Toucy. This Philip had gone not long before on a mission from the Latin Emperor Baldwin II. to the Komans, and on this journey he was probably the companion of one Baldwin of Hainault, who married a Koman princess and made his way to the Mongol headquarters in North Central Asia. Philip de Toucy told King Louis the story of his journey; and this story made Rubruquis decide to travel through the Koman country in preference to the Armenian route. His embarkation probably took place at Acre; his meeting with his comrade Benedict of Cremona may have been in Constantinople; and from his mistakes in referring to the Dynasty of Trebizond, it would appear that he had quitted the Court of King Louis before the Comnenian embassy arrived at Sidon in 1252 on a visit to the French king.

Of these various legations only two have left us a detailed account of their journey; but the narratives of Carpini and Rubruquis, from their fulness, their accuracy, and their high intelligence, are an ample compensation for the loss of others; and among all the materials for the history of exploration in the Middle Ages there are none, before the time of Marco Polo, which occupy so prominent and so indispensable a place.

And first of all, as to the earlier work. Carpini does not begin with the narrative of his travels properly speaking,—with his itinerary or geographical report,—nor does he, like Rubruquis, interweave this with records of a sociological character. On the contrary, he leaves to the very end the task of telling us about his journey, and gives the earlier, and by far the greater, portion of his book to a minute account of the manners, customs, and history of the 'Mongols whom we call Tartars.' It was in this sense, no doubt, that he understood his task. We must not forget his position; it was the same as that of the other envoys; all of them alike were commissioned for definite purposes, they did not go to please themselves.

Our traveller, who afterwards became Archbishop of Antivari (1248),[1] and was now sent as Legate of the Apostolic See to the Tartars and the other nations of the East (1245), made a journey of sixteen months (February 1246 to June 1247) among the Oriental peoples, from the privations of which he died soon after his return. Everything, he declares, which was described by him, was either the result of personal experience or of the information gained from trustworthy Christian captives of the Tartars. The Pope ordered him to observe with care; and both he and Brother Benedict the Pole, his companion and interpreter, obeyed this order to the letter. So Carpini asseverates, and he further warns his readers not to disbelieve what was new and startling in his narrative merely because it was fresh to them. The Tartars, he considers, had become by the increase of their power and dominion, a pressing danger to the Christian Church, and in spite of all difficulties he was anxious to see that danger in its own home, and to forewarn his own people, by adequate information, of the true nature of the peril which hung over them.

In the course of his wanderings Friar John evidently acquired a more bitter hatred of the Mongols than was the case with Rubruquis; but their general characterisation of the Tartars is strikingly similar, and this close agreement is a confirmation of the trustworthiness of both narratives.

---

[1] Reckoned in Albania till 1878, when annexed by Montenegro; it lies between the Lake of Skutari and the sea; and now, as in the thirteenth century, is the seat of a Roman Archbishop; it was under the Venetians from 1450 to 1571. As the nearest East Adriatic harbour to Bari, it had a special interest for Italians.

The report of Carpini's mission opens with an account of the position, extent, and nature of the Tartar[1] country. It lay in that part where the east was (as it were) joined to the north, and it was bounded on the east by Kitaia, Cathaia, or North China, by the Solangi of Korea and Manchuria, and by the aboriginal tribes of the Amur basin. To the south was the land of the Saracens, and to the southwest that of the Uigurs;[2] west lay the Naimans; north was the all-encircling Ocean. The nature of the soil of Tartary was sometimes a plain of sandy clay, sometimes a region of high mountains. Here and there were woodlands, but often were great stretches without any trees at all. Not a hundredth part was really fruitful; the oases had been created by irrigation from the rivers, which, after all, were very scarce. Hence there was also a noteworthy absence of towns. Karakorum, indeed, near the seat of the Mongol Court at *Syra Orda*, was reputed fairly large;[3] but Carpini, though within half-a-day's journey of this settlement, never saw it. Viewed as a whole, 'Tartary' was only good for pasturage.

The climate of this vast region was unequal, extreme, or, as we should now say, 'continental.' Snow-storms, hailstorms, and thunder-storms were of frequent occurrence, and the winds were of terrific strength, driving the dust with blinding force. Summer was the rainy season, if so it could be called, for the rain was hardly enough to freshen the thin grass. The changes of temperature were sudden and most trying; and with such a climate it was little wonder that Tartary was poorer and more wretched (for all its vastness) than could be expressed in words.

[1] In the narrower sense of Mongol.
[2] *Huiuri*; Uigurs and Naimans both seem to have belonged to the Turco-Tartar stock. The latter were outside (to north and north-east) of the pre-Mongol empire of the Kara-Khitai, the former were mostly within the same.
[3] An illusion dispelled by Rubruquis. Cf. p. 361.

In appearance the Tartars were different from all other men—their faces broad, their cheek-bones projecting, their noses small and flat, their eyes tiny, and their eyelids drawn far up. Almost all had slender waists, small feet, moderate stature (inclining to shortness), scanty beards, shaven crowns. Their back hair they grew long like women.

Their wives were 'many as they would,' and they allowed marriage between all except the nearest relations. Their dress was the same for both sexes—tunics of buckram, velvet, or silk stuffs, cut open from top to bottom, and doubled over the breast. Married women, however, had various peculiarities of apparel; but the unmarried were habited almost exactly like men.

Their dwellings were round tents of osier work, with a circular hole in the top, admitting light and letting out the smoke. A fire always burned in the middle of the hut, whose size varied according to the owner's dignity. These homes were moved freely from place to place—even in war—sometimes being taken to pieces and packed, sometimes transported whole on bullock waggons, but never left behind.[1] Their chief riches lay in their cattle—camels, oxen, sheep, goats, and horses, of which last they had a greater number, Carpini imagined, than all other nations put together. As to their religion, they believed in one God, but had no liturgy or ritual; and a reverence for certain idols of felt and silk, which they believed to be the guardians of their cattle, was almost their only trace of a cult, except for a vague honour paid to the Southern Quarter of the Heavens as especially noble.

The ceremonial of passing foreign visitors between two fires was carefully observed, and very lately a distinguished stranger (or rather vassal) had fallen a victim to Mongol

---

[1] So Herodotus, i. 216; iv. 46, 106, 121, and Aeschylus, *Prometheus Bound*, vv. 709-722.

etiquette. Michael, a son of the Prince of Tchernigov in Russia, had come to the Court of Batu on the Volga; he had been passed between the two fires to purify him and avert the evil eye; and he had then been ordered to bow towards the South in reverence to Ghenghiz Khan. On his persistent refusal, Batu sent one of his peers, the son of Prince Yaroslav of Vladimir, to threaten him; and as he would not yield, he was kicked in the stomach by one of the conqueror's 'satellites' till he expired. With him suffered one of his attendants, a certain Feodor, who urged his master to be firm, and win the crown of martyrdom.

Carpini's description of this function, as it was performed upon himself, may be compared with that given by the Byzantine Zemarchus on his embassy to Dizabul the Turk in 569. Two fires were lighted, and two spears were placed alongside. A cord was stretched along the top of the spears, and some shreds of buckram were tied on to this cord. Under it, and between the fires, passed those who were to be purified. At the same time, two women standing by threw water about and made incantations. When any one in Tartary had been killed by a thunderbolt, all the people of the district were forced to undergo this purification.

The Tartars also adored the great powers of nature, the sun and moon, water, earth, and fire; for his own part, the traveller seemed to think they might soon make important changes in their religion; but the future conversion of so many of their tribes to Islam does not appear to have been clearly foreseen by him.

The Mongols were extremely jealous of their horses, as the mainstay of their military power. While Carpini was in their country, Andrew, Prince of Tchernigov, was executed on an unproved charge of conveying these animals out of Tartary and selling them elsewhere. His younger brother came with the wife of the murdered man to Batu to beg for the safe continuance of their land-holdings; and with circumstances of peculiar barbarity the two were forced into an unwilling, and (as they believed) incestuous, union.

Though without any law, properly speaking, these Tartars had certain traditions and prejudices which took the place of a judicial system. Thus it was regarded as wicked to touch fire, or even approach it, with a knife; to flick arrows with a whip; to break one bone with another; to beat a horse with its bridle; or to kill young birds. But there was no prejudice, on the other hand, against robbery, torture, the pitiless slaughter of their helpless victims, or fornication. They had no idea of future rewards and punishments, but only of a life after death where the present conditions would be exactly reproduced.[1]

Their fondness for divinations and incantations was remarkable, and the answers of their devils they believed to be the oracles of God. All their enterprises were begun at the New or Full Moon, to which they did homage as to their Great Khan or Emperor.

By fire they thought all things were purified, and they showed it the greatest reverence. Those who were present at a deathbed could not go to Court again till the New Moon. A spear shrouded in black felt at the door of a tent signified a recent death. A table with food and drink was buried with the departed chief, who was also provided with a mare and foal, and a stallion with bridle and saddle, as his provision for another life. Gold and silver were also sometimes interred. The burial car and hut of the deceased were usually destroyed, and his name[2] none might

---

[1] Many curious similarities here with American—Indian beliefs.

[2] Carpini quotes the practice of Okkodai Khan himself, the first successor of Ghenghiz, in some of these particulars.

mention to the third generation. Two great cemeteries existed in Tartary—one for their magnates, the other for their warriors who were killed in Hungary; and both these were specially guarded and rigidly protected against intrusion. The traveller himself nearly suffered death for accidentally entering the latter. The Tartars, proceeds Carpini, had some good qualities to balance their barbarism, cruelty, pride, avarice, and filth. They were marvellously obedient to their lords; though the brigands of the world, as regarded all other peoples, they were scrupulously honest among themselves; haughty and arrogant to every foreigner, they were familiar, friendly, and respectful to one another. In all enterprises their patience, endurance, and cheerfulness were worthy of admiration. Their women had (unusual) chastity; so at least Carpini, unlike others, was inclined to think; though he allowed that their talk was foul to a degree. Of their darker vices he says nothing.[1] Though often drunk, they were never quarrelsome in their cups.

On the other hand, their pride was disgusting. The traveller himself experienced this; he was also a witness of their insulting behaviour towards Yaroslav of Vladimir, titular chief[2] of the Russian princes, towards a son of the King and Queen of Georgia, towards the Prince of the Solangi (of Korea and Manchuria), and towards many great Sultans of Islam. The deceitfulness of the Tartars in their dealings with other nations was inexpressible; bland at first, they usually sought to circumvent with cunning, and if this did not succeed, they would bite like scorpions. They were

---

[1] Cf. Döllinger, *Heidenthum u. Christenthum*, bk. ix. section i., subsection ii., quoting Sylv. de Sacy, *Journal des Savants*, June 1829, p. 331, on the Uzbeg descendants of the 'conquering hordes' of Ghenghiz and Timûr.

[2] After the earlier fall of Kiev 1169, and still more after the Mongol sack of 1240, Vladimir became the chief of the Russian principalities, in rivalry with Suzdal.

abominably dirty and foul in their habits; avaricious, insolent, and untiring beggars, when they had got hold of a wealthy stranger; and yet mean and niggardly hosts withal. At times, when pressed by hunger, as in the siege of a certain city in Cathay, these Tartars would eat human flesh; but in general they could live on incredibly little, and there was nothing too loathsome for them to devour. They were so stingy that they would not even throw bones to their dogs till they had sucked out the marrow.

Their favourite drink was mare's milk—Kosmos or Kumiss—with a kind of home-brewed beer. A gruel of millet and water was also much in use among them, and on a cup or two of this they would go for the rest of the day.

For adultery, theft, and murder among their own tribes they inflicted death; traitors were scourged, even the greatest by the humblest; for there was little difference in honour between one Tartar and another; in fact, they were a very pure democracy, tempered only by military organisation. From this last came their excellent archery and horsemanship, the result of incessant practice from babyhood. Even the women rode, and often shot, admirably, using short stirrups like the men. But nearly all their time the females were employed in working at the clothing, furniture, and leathern goods of their lords. The peaceful occupation of the men was herding, and little else. They made use of a breed of swift camels, both wild and tame (ancestors of the present stock of Central Asia), whose speed astonished the Friar, and has since become proverbial.[1]

Carpini next takes us through the history, ethnology, and geography of the Tartar races. He tells us that the land of

[1] Especially since the discoveries of Prejevalski. Cf. the stories of Petachia on the wild camel.

Mongal or *Mongol*, 'in the Eastern parts' of the world, had four chief divisions. These were the countries of the Yeka, or Great Mongols; of the Su, or Riverine Mongols, 'from a stream in that land called Tartar;'[1] and of the Mekrit and Merkit Mongols, who inhabited the Baïkal basin.

All these tribes were essentially one; but, to speak precisely, Ghenghiz Khan belonged to the Yeka. He first conquered the other Mongols and then the Kara-Khitai, Uigurs, and Naimans. After the subjugation of the semi-civilised and partly Nestorian 'Huiur'[2] the victors adopted the script of the conquered, and pushed on to the subjugation of the Komans, of the Kirghiz Steppes, and of South Russia, on one side, and of the Kitai, Khitai, or people of North China, on the other. In the last-named region there was fabulous wealth; silver was used, when stones were scarce, for missiles, and the civilisation, literature, and religion of the people was a thing to study. 'For it is said they have the Old and New Testaments, the Lives of the Fathers, houses like churches, stated times of prayer, saints and hermits.' They worshipped one God, honoured Christianity and its Scriptures, gave alms, believed in Eternal Life, and though baptizing little, were of a gentle and humane disposition. In appearance, they were somewhat like the Mongols,

---

[1] Cf. M. Paris, *Cronica Majora*, A.D. 1234 'dicti Tartari a Tar flumine . . . habentes ducem nomine Caan.' Avezac divides the Tartars into three main groups—(1) Tunguses to the east and north-east; (2) Mongols in the middle; (3) Turks in the west.

The Mongols' original home was in the Mountains of Burqân Qâldun and the country between the Baïkal and the Upper Amur. The Empire of the Kara-Khitai (arising out of the expulsion of the Kitai or Khitai Dynasty from North China in the twelfth century) roughly answered to the modern Eastern or Chinese Turkestan, the Russian Province of Semiretchensk, and the basin of the Chu; the Naimans occupied the modern north-west Mongolia. Omyl, Imil, or Emil is perhaps the modern Chuguchak, near Lake Ala-Kul, in the extreme north-west corner of 'Mongolia,' and just inside the Russian frontier.

[2] Uigurs; mainly within the Kara-Khitai realm.

beardless and broad in face; their language was quite peculiar. They were unsurpassed artificers, and their land was rich in corn, wine, gold, silver, and silk.

Carpini adds some rather legendary details of the Mongol conquest of 'Lesser India' and of the 'Black Saracens called Æthiops'; of the Tartar repulse from 'Great India' by the Christian warriors of Prester John, aided by Greek Fire; and of the monstrous races of men-women and dog-shaped folk reported from distant parts of Asia. Similar fables he repeats about the Adamant Mountains[1] in the Caucasus or 'Caspian' range; about the cloud-concealed natives long imprisoned in the bowels of the hills; and about the Troglodytic races, who fought the Tartars by underground passages, and lived in terror of the mysterious and fatal sound which accompanied the rising of the sun.

Ghenghiz Khan, proceeds Carpini more justly, was not only a conqueror, but an organiser and lawgiver. He ordained that his successors must be elected by a proper Kuriltai or Meeting of Chiefs; that any pretender to the throne not so elected was to suffer death; that the conquest of the whole world was to be steadily pursued; and that their army was to have a definite organisation by tens, hundreds, thousands, and tens of thousands. Further, the Great Khan divided his empire into provinces, and fixed the location of each general or prince in command; these princes fixed the station of the chief officers; the chief officers did the same with their inferiors; and so on. Throughout all the Tartar posts reigned implicit obedience, in death or life, in peace or war.

The Great Khan enjoyed many special privileges—as to his *harem*; as to his posts and messengers (who everywhere and at all times could requisition horses and provisions);

[1] Discovered, he says, by Ghenghiz himself.

and, in general, as to the property of his subjects throughout his dominions. For, in the last resort, everything belonged to the successor of Ghenghiz. One of the most valuable of his perquisites was the tribute of brood-mares paid to him by all the chiefs.

On his accession the new Khan Okkodai followed his father's counsel by attacking the Saracen Komans and the Mussulmans of Khwarezm[1] or Khiva, Turcomania, and Eastern Persia; the work of Ghenghiz was vigorously carried forward; and his children soon captured the famous merchant cities of Barchin, Iamkint, Ornas, and others in the Valley of the Jaxartes or Syr Daria.

Here, as elsewhere, the old life and prosperity of Central Asia was blotted out. The Pagan Turks were next conquered, and then the Russian principalities, where the slaughter was terrible, especially at the 'metropolis' of Kiev. In his journey through this country Carpini passed by countless skulls and bones, and saw for himself that the once great and populous Mother of Russian cities had scarce two hundred houses left. Hence the devastation spread to Poland and Hungary; to the land of the heathen and Moslem Finns and Turks in the Middle Volga basin, the Mordvinians, the Bulgarians,[2] and the Bashkirs;

---

[1] The Khwarezm Shah is the 'Altisoldanus' of Carpini. This Kingdom, centring in Khorasm(ia) or Khiva on the Lower Oxus, at the opening of the thirteenth century, had extinguished the Seljuk dominion except in Asia Minor, and extended over all modern Persia, Afghanistan, Baluchistan, most of Russian Central Asia, and much of Trans-Caucasia.

[2] Carpini evidently considers, and rightly, that Great Bulgaria, near the modern Kazan, was the Bulgars' mother-country; his 'Great Hungary' was the land of the Bashkirs. He evidently knows little or nothing about the Parossits, who were said to live by the smell of cooked meat. But he is quite correct in thinking that the Samoyedes' tents and clothes were of beasts' skins. As to the dog-faced men, and the story that they spoke two words in human fashion and barked for the third, Carpini is careful to mention this only as a common tale.

and even to the half-fabulous Parossits, the Samoyedes, and the dog-faced, ox-footed monsters of the far North.

In or near Armenia the conquering Mongols found, as some asserted, a still more wonderful race—one-footed and one-armed, with a hand in the middle of their breast, but possessed of matchless speed. These folk, when tired of hopping on their single feet, whirled themselves round and round.[1] Armenia, Georgia, and the Seljuk Sultanate of Rum or Asia Minor, whose capital was at Iconium or Koniyeh, were all overrun and placed under tribute, as was the land of the Bagdad Caliph. This great potentate was now bound to go humbly to the Mongol Court and pay his respects every year.

Except in the case of the rout of the entire army, those Tartars who fled in battle, or failed to follow their leaders, were put to death. In every company of ten, each was responsible for the rescue of any messmate taken prisoner. Their weapons were bows, arrows, lances, and hand-axes, with leathern or iron shields, breast-plates, and greaves. The horses also had body armour. The military ruses and ambushes of the Mongols were infinite; their scouting system most efficient; their rafts (for crossing rivers) ingenious beyond what one might have expected in so rude a people. Their simulated retreats, their dummy warriors dressed to represent real men and horses, their distant and 'back-shooting' archery, their tactics in drawing on, entrapping, surrounding, or pursuing the foe—all these well deserved Carpini's clear description and enthusiastic praise. Not less notable were their siege tactics; mining and flooding, battering machines, Greek Fire, and other devices,[2] if unsuccessful,

---

[1] Here the believing spirit might gratefully recognise the Cyclopodes of Isidore.

[2] *E.g. Arvinam hominum liquefactam*, poured in torrents upon the besieged.

they supplemented by the most rigid blockade. Nowhere had their perfidy and cruelty been more shown than at the surrender of towns, where no treaty bound them; and they only admitted in fact, whatever they might pretend in theory, an unconditional surrender. Some nations talked of having peace and treaties with them, but that was absurd, for they really made peace with no one, except on absolute submission. Once their vassals, men were forced to fight for them, and all promises of good treatment became empty words. Thus, while Carpini was in Russia, he witnessed the merciless exactions of a Saracen envoy of Kuyuk and Batu, who was levying a poll tax[1] and impressing troops for his masters.

Their vassal princes, forced to make the terrible journey to the Great Khan's Horde or Court in Mongolia itself, were lucky if they escaped with insults. Sometimes they were poisoned, sometimes executed on false charges, their sons and daughters retained as hostages in perpetual captivity. The traveller saw this treatment inflicted on Yaroslav of Russia, as well as on a prince of the Alans and a prince of the Solangi. The property inherited by such a hostage on the death of his father was rarely suffered to come into his possession.

In all states laid under tribute, a Mongol Prefect (or 'Baschat') was the real ruler, and in such states numberless extortions were practised by individual Tartars. A city in Russia had been lately destroyed for opposing the orders of a Prefect of this kind. All lawsuits between the tributary princes had to be decided at the Great Khan's Horde. A recent case of this was the well-known success

---

[1] Viz. on each household one white bear's skin, one black beaver, one black 'zabulun' or zibelin, one black iltis, one black fox. Slavery was the only alternative to payment.

of Prince David of Georgia, a son of adultery, who, by appeal to the Mongol Court at Syra Orda, had ousted Melic, the rightful heir.

Carpini adds a list of more than forty peoples or countries[1] who had fallen victims to the Mongol armies, and of some others which up to that time had manfully and with success resisted them. Unquestionably these scourges of Heaven intended, if possible, to conquer the entire world, literally and absolutely, according to the will of Ghenghiz and the style of the Great Khan's seal: *'Kuyuk the strength of God: God in Heaven and Kuyuk Khan on earth; the seal of the Lord of all men.'* Already they had conquered the whole earth, outside Christendom, and Christians might be sure that their time was coming. Even while Carpini was in Mongolia, Kuyuk raised his banner against the Church of God and the Roman Empire. There could be no mistake what was involved in the Tartar Conquest—it was unheard-

[1] These were—(1) the Khitai, Kitai, or inhabitants of North China; (2) the Naiman(s), or people of north-west Mongolia; (3) Solangi of Korea and Manchuria; (4) Sarihuiur; (5) Bashkirs, Bashkirts, or 'Bascarts of Great Hungary'; (6) Parossits; (7) Cassi; (8) Jacobitae, Jacobites, or heretical Christians of various Asiatic countries; (9) Mordui or Mordvinians of the Middle Volga; (10) Samogedae or Samoyedes of the Petchora country; (11) Kara-khitai or 'Black Chinese' of Turkestan; (12) Komans; (13) Tumat; (14) Kergis or Kirghiz, extending from the Angara to the Altai; (15) Casmir or Kashmir; (16) Saracens of Persia, etc.; (17) Alans or Assi of Modern South Russia; (18) Georgians or Obesi of Tiflis, etc.; (19) Nestorians; (20) Turks; (21) Perses or Persians; (22) Voyrat; (23) Kara-nitai; (24) Huyur, Huiuri, or Uigurs; (25) Bisermins or Moslems of Turkestan and North Persia; (26) Turcomans, of the Akhal Tekke region, etc.; (27) Bileri, or 'Great Bulgarians' of the middle Volga, the later Kazan and its neighbourhood; (28) Cangitae or Canglae; (29) Khazars or Gazari of Modern South-East Russia; (30) Tarci; (31) Su Mongol; (32) Merki(t) Mongol; (33) Mekrit Mongol; (34) Corola; (35) Comuci; (36) Burithabet; (37) Comani (? re); (38) Brutachi (Jews); (39) India Minor or Æthiopia; (40) Circassians; (41) Rutheni or Russians; (42) Baldach or Bagdad; (43) Sarti or Sarts of Western Turkestan.

Carpini's prophecies were not wholly unfounded; it was the conquest of China and Persia that most effectually diverted the Tartar deluge from Europe in the generation succeeding his journey of 1245-6-7.

of misery, starvation, and insult, such as was faintly tasted by the wretched envoys who had gone from Europe to the Mongol Courts.

Carpini ends this part of his work by prophesying that Livonia and Prussia will be the first attacked; by warning every Christian king and nation to make common cause against 'these devils'; by promising success to a vigorous and united resistance; and by detailing various suggestions as to the arms, discipline, fortifications, and military ruses most useful in combating them. He then passes to the record of his journey, perhaps the longest continuous piece of land-travel accomplished by any European up to this time.

Carpini's route was by the Northern Way from Latin Europe to the Mongol Headquarters. After quitting Lyons, on the 16th of April 1245, he first tells us of his arrival at the Court of Bohemia. Here the King, Wenceslas I., an old friend and patron of the traveller's, advised him about the next stages of his journey; recommended him to go forward by way of Poland and Russia; and promised the assistance of various people of importance in the former country, relations of his own, whose aid would be effective in all the countries of the Eastern Slavs.[1] Furnished with letters and an escort, as well as fresh supplies of money, the friar now pushed on from Bohemia to Silesia, where he was received by Duke Boleslav, a nephew of Wenceslas, and himself a personal acquaintance of John de Plano's. This prince, like his uncle, forwarded the mission in every way, welcomed the legates at his Court in Lignitz, and helped them on to the borders of Conrad, Duke of Lenczy, his immediate

---

[1] Cf. pp. 733, 734 of Carpini's text, Avezac's edition, in vol. iv. of the Paris Geographical Society's *Recueil de voyages et de mémoires;* all references are to this edition.

neighbour on the south-east, then resident in Cracow.[1] There, Carpini tells us—apparently intending his readers to understand the palace, castle, or dwelling in which this Duke Conrad greeted the embassy—a providential meeting occurred. 'Vassilko,' Duke of 'Russia,' had just arrived, and from him the Latin envoys were able to gain much valuable information about the Tartars. His brother Daniel had just applied for, and obtained, a safe conduct for a journey to the camp of Batu;[2] and he warned the new-comers to be under no illusions. Any one who wished to travel among the Tartars must have rich presents, and must expect to be persecuted with demands; the 'give, give' of the horse-leech was the Nomade method of dealing with all visitors; no business could be done, and no respectful treatment could be secured, without compliance in this custom.

The Russian princes here noticed were both on Carpini's further route between Cracow and the Volga. 'Vassilko,' otherwise Vassili or Basil, 'Duke' of Vladimir and Volhynia, (1214-1271), and Danil or Daniel, 'Duke' of Galicia (1228-1264), had first struggled bravely against the Mongol deluge, and then bowed to the inevitable, hoping by diplomacy to recover a little of what they had lost in war. Daniel especially was famous for his valour at the fatal battle of the Kalka (13th May 1223), and for his later negotiations with the Papacy. Herein he followed the example of his father, Roman of Vladimir, with a similar result. For as he wished for military aid against the Tartar rather than any ecclesiastical benefits, and as Innocent IV. had only a kingly title and spiritual favours, but no troops, to bestow, the life faded out of the hopes which Carpini had excited or afforded,

---

[1] Pages 734-6 of Carpini's text in the *Recueil*.
[2] Pages 735-6 of Carpini.

and the promises which he had lavished, in his eagerness to see the Russians march under the Papal Banner. However futile in result, it was probably this matter which mainly determined Friar John to take the northern overland route, in preference to the Mediteranean and Black Sea voyage enterprised by Rubruquis; but neither at this nor at any subsequent time was a permanent understanding effected between the Eastern Slavs and the Church of Rome.

In deference to Vassilko's advice, the Latin envoys, before leaving Cracow, purchased or procured beaver skins and other furs, current as money in some of the regions of their future wanderings, and essential as aids to diplomacy among the Mongols;[1] under the escort of Duke Basil they passed on into Red Russia; and here their host, in deference to a suggestion from his visitors, summoned the bishops of his jurisdiction to hear the letters of the Pope and confer upon the question of reunion. The Apostolic Briefs very strongly urged the Separatists to return to the One Fold; and Carpini added arguments of his own, until he thought his hearers were almost persuaded; but the end was, as usual, evasion and postponement. The Russians would give an answer when Daniel came back from Batu's Camp; as to what their Roman friend had said, it was too important to be decided off-hand.

With a 'Ruthenian' guide Carpini now pushed on to Kiev.[2] In his company were two other friars who had joined him on the way—Brother Benedict in Poland,[3] and Brother Stephen in Bohemia; from the former of these we have a summary of the journey which in a few points is a useful supplement to John de Plano's own record. Arriving

---

[1] Page 735 of Carpini.
[2] Page 736 of Carpini.
[3] Benedict's additional narrative of the Carpini mission is on pp. 774-779 of the text in the *Recueil* of the Paris Geographical Society, vol. iv. edition Avezac. See pages 390-1, note of this vol.

on the Dniepr after a trying march, the legates found the whole country wasted and desolate, emptied of its men (whom the Mongols had killed or enslaved), and overrun by Lithuanian brigands. Only the 'metropolis' itself, the 'Court of the Golden Heads,' so terribly sacked in 1240, seemed to show a certain, though attenuated, survival of settled life; and here, on the edge of the wilderness, where no more towns could be looked for, Friar John took counsel with a Tartar 'millenarius,' or chief-of-brigade, and received an emphatic warning about his further journey. Only with Tartar horses could he hope to accomplish the enormous stages which lay before him, 'for they alone could find grass under the snow, or live, as animals must in Tartary, without hay or straw.'[1] This advice appeared sound enough; for the Europeans were beginning to discover that only Tartar bodies could stand the fatigue and hunger of the route. Already, at Danilovka in the Ukraine, Carpini had almost died from the privations which ultimately killed him, and had made them carry him in a cart through the deep snow and intense cold, 'so as not to hinder the business of Christendom'; but he was partially restored by the halt at Kiev; and on the 4th of February, 1246, he was able to start again.[2] His course now lay down the Dniepr, and leaving the heights from which the half-ruined city of Saints[3] looked over the great stream, he made his way to Kanov, or Kaniev, one hundred versts nearer the sea, where the country was under the immediate rule of the Tartars. Still following the course of the river to the south-east, Friar John soon came to another hamlet, where ruled a certain Alan, named Micheas, who represented the Mongol chiefs in this part. He was a man full of all malice and iniquity, who had before now sent to Kiev some of his body-guard to induce the envoys to come to him on

[1] Pages 736-737 of Carpini. [2] Page 737 of Carpini. [3] See page 391.

the way, falsely pretending that these were the officers of Kurancha or Corenza, his superior officer. His sole motive, of course, was to extort presents, and having entrapped the strangers, he squeezed them unmercifully; as a small compensation, he put them forward on their journey to the next stage, where Carpini places the first encampment of the Tartars, and where the visitors painfully endured the extortions and insolence of these barbarians and their maddening curiosity.[1] The utter indifference of the Nomades to Papal Legates, Papal Letters, or even the Papal Name, pained and astonished the writer, who ineffectually declared to a scornful and suspicious auditory the earnest wish of the Roman Pontiff for peace between Christians and Tartars, for the conversion of the heathen, and for the cessation of the Mongol attacks upon Poles, Hungarians, Moravians, and other peoples of Western Europe. The result of this oration was a renewed demand for presents, and the loan of a guide and pack-horses to the encampment of Kurancha.[2] This personage was the highest in power of any of the Tartar officers whom the envoys had encountered; he was in supreme command of the forces on the western frontier, and had under him sixty thousand men, wherewith to repel any sudden attack from the side of Europe; but he was subordinate to Batu, whose only superior was the Grand Khan himself. He would not allow the new-comers to pitch their tents close to him, and sent his slave-stewards to find out what presents he might expect. Like Micheas, his demands rose with compliance, but at last he appeared to be satisfied, and Carpini was admitted to his presence. At the entrance the friar was instructed to bend the left knee three times, to avoid touching the threshold with his foot, and to explain the object of his journey before the Mongol general and the

[1] Page 738 of Carpini.   [2] Page 739 of Carpini.

other nobles who had assembled to hear his declaration. Kneeling, he presented the Papal Letters, but the interpreter from Kiev proved almost useless, and as no one else more competent appeared, the Briefs had to remain unexplained and untranslated.[1] The Legate, however, contrived to express his wants, and with fresh guides and horses he resumed his journey for the camp of Batu on the 26th of February, 1246. Kurancha's 'Horde' was probably in the neighbourhood of the present Ekaterinoslav, and we know from Carpini himself that its station was on the western side of the Dniepr. All this time, therefore, the mission had been following the 'Borysthenes,' in its south-easterly trend as far as the celebrated rapids; now the route changed to a direction almost due east, and the travellers struck across to the head of the Sea of Azov, and crossed the Don near its mouth, somewhere about the site of the present Rostov. Thence to Batu's Court, 'on the border of the Koman country,' the way was east-north-east until the Lower Volga, about one hundred miles above Astrakhan, was reached.[2]

Going as fast as the horses could trot, and changing them three or four times a day, the envoys rode on from morn to night (and often through a great part of the night itself), till Wednesday in Holy Week, the 4th of April. Meanwhile they were traversing the country of the Komans, the steppe-zone between the Ukraine and the Aral Sea, watered by four great rivers, to the first three of which Carpini gives the Slavonic[3] names of 'Neper,' Don, and Volga. In this point he shows an advanced accuracy; even Rubruquis, who is so much fuller and so often more precise, is here wanting; but Friar John had the advantage of Russian guides, while his

---

[1] Pages 740-742 of Carpini.
[2] Pages 742-744 of Carpini.
[3] 'Neper' is in the 'Anglo-Saxon' or 'Cotton' Map of c. 990. This is the earliest West European notice of any of these names.

great rival, from choice or necessity, is here dependent on his classical knowledge, and so can supply nothing but an echo of the old confused nomenclature. The new terms not only bring us in touch with the modern world, but bring us into an atmosphere of comparative certainty and precision.[1]

First of all, the traveller approaching from the west crossed the Borysthenes of the ancients, the Neper of the men of Kiev, on whose banks were two large Mongol bivouacs, that of Kurancha to the west and that of Mauci to the east. The latter was the second son of Chagatai Khan, and so a grandson of Ghenghiz himself; and Carpini had good cause to remember his name, for he caused the embassy more grief and trouble than any other of the Tartar magnates. Second of the great prairie-streams was the Tanais or Don, the accepted boundary, from the early days of Greek geography, between Europe and Asia, and along this now roamed a Nomade prince named Catan (perhaps the *Scatai* of Rubruquis), who had married Batu's sister. Third, still further to the east, came the Volga (the Rha of Ptolemy, the Atil, Athil, Itil, Etil, or Edil of most mediaeval geographers, the Northern Tigris of Roger Bacon), whose vast basin was the special domain of Batu, and the centre of the Mongol power in the West. Fourth, and last, was the Jagac, Yaik, or Ural, on each bank of which a *millenarius*, or *mingatan*, followed a definite beat up and down the stream. This indeed was the practice of all Tartar officers who had to guard the courses of these rivers; in winter they usually moved down to the sea, while in summer they kept more towards the north and the 'mountains'; perhaps by this last phrase we may understand a supposed extension of the Urals to the west, till they sank down into the great plain of the German and Baltic lands. It is Rubruquis, and not Carpini, who makes

[1] Page 743 of Carpini.

the Maeotid *marshes* of the North (transferred by him to the neighbourhood of the Baltic) to be the spring and source of these Euxine and Caspian water-courses.[1]

As to the sea which receives them, John de Plano vaguely imagines it to be all one and the same 'Sea of Greece,' or 'Greater Sea,' from which flowed off the Bosphorus, or Strait of St. George, to the city of Constantine; along the shore of this sea, as on the waters of the Dniepr, the envoys went perilously over the ice 'many days in many places'; and into it flowed, according to our author, Volga and Ural as well as Don and Dneipr. Thus at last, after uncounted hardships, the mission arrived at Batu's Horde, which then seems to have been encamped very near the mouth of the Atil, and on the left or eastern bank of the same, perhaps about the site of the modern Krasnoi Yar. Here they had to endure the Nomade ceremony of purification, and to pass between two fires; they were not allowed to pitch their tents within a league of the Tartars; and they were closely questioned by Eldegai, the 'procurator' or 'proctor' of Batu, as to their ability to make a suitable present, as to the motives of their journey, and as to their credentials. All having been explained, or taken as capable of honest explanation, the envoys were admitted to their first audience, and allowed to present their letters, to make a formal declaration of their purposes, and to ask for interpreters. This request was granted on Good Friday, the 6th of April, 1246; translations of the 'commendatory' documents were made into Russian, Arabic, and Mongol; and a copy of the last was offered to Batu, who read and noted it carefully, but neglected to pay any attention to the bodily wants of his unlucky visitors. Since their arrival they had only been able to procure a little millet in a bowl, and

[1] Carpini, p. 743; cf. Rubruquis, p. 250.

this proved but a foretaste of the sufferings that awaited them.[1]

The ceremonial of Batu's Court, as described by Carpini, was a good example of the rude but martial splendour that prevailed in the Tartar camps, and in a measure survived until the extinction of Tartar independence. Like the Great Khan himself, the Conqueror of Russia and of Hungary surrounded his presence with apparitors, 'proctors,' door-keepers, chamberlains, and all the state of Nomade royalty; his throne was raised above all else, even the seats of his sons and brothers; to these last a bench in the middle of the tent was allotted. The rest of the people sat behind them on the ground, men to the right, women to the left. Among Batu's possessions were some large and handsome linen pavilions, trophies of the Hungarian campaign, which the Christian visitors could not have inspected without bitter recollections of the past and some uneasiness for the future. No one was permitted to enter the presence of Batu, unless he were an attendant or specially summoned; the envoys, on being introduced, did reverence, and were then placed upon the left of the throne, the inferior or women's side, which on their return from the Great Khan was exchanged for the right or more honourable quarter. Near the door of the audience chamber was the usual table with gold and silver goblets of Kumiss and other drinks, and whenever the Prince put the cup to his lips there was a burst of song and playing on guitars. If Batu rode in public, an awning or canopy was carried over his head, according to the etiquette observed with all the members of the House of Ghenghiz. The great general was kind to his own men, but they held him in the utmost awe. To his enemies he was most cruel and dangerous, for in him were united craft of a high order

[1] Pages 743-745.

and long experience in war. To the Latin visitors he showed himself, if not a generous host, at least an affable and easy gaoler, and on the day following the presentation of the Papal Letters, Carpini was ordered to proceed to the Court of Kuyuk with the guides he had obtained at Kurancha's Horde (Easter Eve, 7th April). Some of the mission were to turn back and return to Europe with the news of what had been accomplished up to that time; among these was probably Stephen of Bohemia, who had been one of the chief supports of Friar John in the earlier stages of the journey. It does not appear whether the Russian interpreter from Suzdal, who had served Carpini with unusual efficiency at Batu's camp, went any further with him. In any case, none of those who now started on their return were able to pass [1] the line guarded by Mauci until the chief of the embassy reappeared, on his way back from Mongolia; they passed a wretched time of expectation and anxiety; and as they gradually abandoned hope of their more advanced companions, lost in the deserts of Central Asia, they came also to despair of their own deliverance.[2]

On Easter Day (8th April 1246)[3] John de Plano, with Benedict the Pole and the rest of his company, started on the last and most arduous section of the route, lying between the Volga and the Selenga; 'and most tearfully we set out, not knowing whether it was for life or for death.' The travellers were all so reduced by their privations that hardly a man could ride. During the whole of Lent their food had been nothing but a porridge of millet and water, and their drink snow melted in the kettle.[4]

At this point Carpini attempts a general survey of the

---

[1] After recrossing the Don.
[2] Carpini, pp. 745-747.
[3] According to Benedict, Easter

[3] Tuesday, 10th April.
[4] Carpini, p. 747.

regions adjoining the line of march which he had followed up to now, and these countries and their inhabitants he classifies according to *climate* or latitude. On the whole, the picture he thus gives us is clear and accurate, and speaks well for his power of combining, sifting, and presenting a mass of detail full of difficulty and bewilderment. Beginning with Komania, which occupied the entire steppe or prairie-zone to the north of the Euxine, the author places behind this belt of land three rows of nations, gradually nearing the Pole. First, reckoning from east to west, was the group made up of Russians, Mordvins, Bilers of Great Bulgaria, and Bascarts or Bashkirs, all inhabiting what answers roughly to the Black-Earth zone immediately north of the steppe region. Next came the Parossits and Samoyedes dwelling in the dense forest region to the north of the fertile corn-land and sparser wood-country of the south; while north of these, again, were the dog-faced men on the 'desert shores' of the Ocean, wandering over the open *tundra*, in the neighbourhood of, and even beyond, the Arctic circle.[1]

To the south of the Koman country the Friar places, apparently reckoning from east to west, the Caucasian and Crimean tribes, and the nations of the Northern Black Sea littoral—Alans, Circassians, Gazars or Khazars, and Greeks, as well as Iberians, Georgians, Armenians, and Turks. To these well-known names he adds the Zicci, a branch of the Circassians or Cherkesses, in the Western Caucasus; the Cachs or Kakhetians of the rich vine-growing country east of Tiflis; and the Brutachi, who were said to be Jews and shaved their heads. Rubruquis and Benjamin of Tudela also confirm this very ancient and interesting tradition of a Caucasian Israel, referred to by so many Jewish writers, and

[1] Carpini, pp. 747-748.

perhaps derived from the forcible transplantings of Assyrian conquest. Western Europe was divided from Komania by the intervening lands of the Hungarians, and of certain Russians (the Ruthenians of Galicia are evidently intended); it was a very long country, measured as it lay between the rising and setting sun; and the envoys' journey through it was most wearisome. From the beginning of Lent till the eighth day after Easter, they rode on at high speed, changing horses five or even seven times a day, except in those waste regions where such frequent relief was impossible, and where stronger horses had to be taken for longer periods. All the natives had been either killed, enslaved, or expelled by the Tartars, and the same was the case in the next land traversed by the mission. This was the country of the Kangitae,[1] often reckoned as a part of Komania, where the same nomade life and the same Pagan religion was to be found, where a terrible scarcity of water prevailed, and where the wilderness, without tents or houses, was only marked by the human skulls and bones, scattered upon the ground like cattle-dung. Through this desolation (where some of the escort of the Russian Prince Yaroslav had lately missed their way and died of thirst) Carpini travelled from the 16th of April to the middle of May, when he entered a land of 'Bisermins,' or Moslems. Here the Koman language was still prevalent, but the people held the religion of the Saracens. Their difference of faith had not saved them from a share in the common ruin; everywhere the visitors passed by desolated cities, towns, and villages.[2]

Friar John had now crossed the vast plains of what is called by modern geography the Ural-Caspian depression, from the basin of the Volga to that of the Syr Daria, and from the region of Pagan tent-dwellers to the country of the

[1] 'Kanglae' of Rubruquis.   [2] Carpini, pp. 748-749, 750.

settled Mohammedan principalities of Central Asia; but he was not able to learn the name of the great river along which his route here lay; and he only records a few names of cities and sovereigns. The country used to have a lord who was called the High or Supreme Sultan;[1] but he was killed by the Tartars with all his house, and the writer could not even discover his name. In his place the Mongol chiefs, Buri (or Burin) and Kadan, sons of the same mother,[2] ruled the country from their camps near the borders of the province— which borders the Friar does not say. He adds, however, more definitely, that on the northern frontier, towards the land of the Kara-Khitai and the ocean, was the station of the Tartar Prince Sitan, a brother of Batu.

From the middle of May to the middle of June, from Ascension Day till a week before the Feast of St. John the Baptist, Carpini pressed on through this Saracen land, along the valley of the lower and middle Jaxartes, past the famous but wasted towns of Iamckint, Barchin, Ornas (or Otrar?),[3] and many others. After the election of Okkodai, the irresistible Batu had conquered all these places, then belonging to the Empire of Khwarezm or Khiva, the domain of the High Sultan. Iamckint and Barchin soon fell, but the siege of Ornas was long and troublesome, and the city was only taken by the heroic expedient of turning the course of the river Syr, and flooding the town with all its houses and its people. Before this time Ornas had been very rich and populous, and the seat of a great market to which Christians resorted from Khazaria and the countries of the Slavs, but which was dominated by the Saracens as masters of the

---

[1] 'Altisoldanus,' evidently the Khwarezm Shah, or ruler of Khiva. (Carpini, p. 750).

[2] A wife of Okkodai, according to Rashid-ed-din; of Chagatai, according to Carpini (p. 666). In either case, Kadan and Buri (as Rubruquis writes the name) were grandsons of Ghenghiz.

[3] See p. 391.

city. The great river that flowed by the town passed on to Iamckint, and fell into a sea (the Aral), after traversing the country of the 'Bisermins'; thus Ornas was a sort of inland port,[1] controlling the most northerly of the chief trade routes that crossed Central Asia.

Carpini was now emerging from the boundless level of the steppe regions; for this country, reaching on the south towards Jerusalem and Bagdad, had very high mountains, the western outliers of the Thian Shan; and these hilly districts continued to check the rapidity of the envoy's progress, till he again descended into the prairie region, on the other side of the great Asian dividing wall. The next stage of the journey was through the country of the Kara-Khitai. Here the Mongols had rebuilt an old town called Omyl or Imil; here their Emperor had fixed one of his residences; and here the visitors were entertained by the representative of the Great Khan, all the nobles of the town clapping hands before their guests as they drank. Imil was part of the special domain of the new Tartar Emperor, Kuyuk;[2] it had been founded by the Kara-Khitai more than a century before;[3] its destruction was the work of Ghenghiz; and its restoration was due to Okkodai. Hard by was a lake of moderate size, along which the mission journeyed for several days, and which was probably (although Carpini neglected to ask the name) the famous basin of the Ala-Kul, just north of Kulja, on the other side of the Zungarian Ala Tau. On the shore of this lake was a little hill, whence issued in winter such storms of wind that one could hardly pass, and that not without great danger. The surface of the water was studded with islands, and many small streams fringed with woodland flowed into this lofty mountain basin, celebrated in modern times as the only

---

[1] *Unde est quasi portus.*   [2] Rubruquis' *Keu* Khan.   [3] About 1125 A.D.

prominent lake in Higher Asia which is not drying up, but deepening and expanding.[1]

To the south of this region, as Carpini heard, was a great desert (probably the Takla-Makan section of the Gobi), in which lived wild men, destitute of speech, or even of joints in their limbs, so that when they fell down they could not rise of themselves. They were clothed in felt of camels' hair, and built shelters against the wind of the same substance; if wounded with arrows, they staunched the blood with grass, and fled swiftly before their pursuers.[2]

Passing one of the Great Khan's Hordes or Residences, near which they rested a day in the midst of unusual plenty, the Legation now entered the country of the Pagan Naimans,[3] a cold mountain region where heavy snow and biting winds greeted the strangers even at midsummer (28th and 29th June). Like the Tartars, these Naimans, as well as their neighbours, the Kara-Khitai, were purely nomadic, making no attempts at agriculture, but dwelling in tents and living on their flocks. Now they had been ruined and nearly exterminated by the Mongols, and their extensive country, through which Carpini travelled many days, was as desolate as the steppe regions to the west. Finally, the envoys, leaving behind them the last of the barriers which lay between them and their goal, entered Mongolia, and on the 22nd of July arrived at the court of Kuyuk, the Great Khan Designate, for whose election the *Kuriltai*, or National Assembly of the Mongol Notables, had just gathered. Okkodai, the first successor of Ghenghiz, died in 1241, five years before; but a long delay in the appointment of a new emperor was a common

---

[1] Carpini, pp. 750-752: *Nouveau Dictionnaire de Géographie Universelle* [Vivien St. Martin and L. Rousselet] art. *Ala-kul.*

[2] Carpini, p. 648.

[3] Some, though only a minority, of the Naimans seem to have been Nestorian Christians.

incident enough among the Shepherd Kings, and it was no easy matter to convene the Nomade princes from the four corners of the Mongol Empire. Now that the electors had fairly met, Carpini's guides, fearful of being too late for the ceremony, hurried him on remorselessly. Going as fast as the horses could trot, rising at dawn, and travelling till night without a rest, he was often obliged to go supperless to sleep, taking in the morning 'that which should have been eaten overnight,' until the Imperial Horde was reached.[1]

The *Kuriltai* seems to have met, not where Rubruquis found the Mongol Headquarters, but at a place called Ormektua, half a day's journey from Karakorum. Both 'cities' were probably close to the head-waters of the Orkhon, about three hundred miles S.S.W. of the Baikal Lake, or Holy Sea. *Syra Orda*, as Carpini expressly tells us, was only the name of the Imperial tent, court, or palace.[2]

When the legates reached the Court, Kuyuk assigned them a tent and a fixed allowance of meat and drink, but would not admit them to an audience until the formalities of the election were over. Meantime translations of the Papal Letters and of the other credentials of the mission had been forwarded by Batu, and after five or six days Carpini and his party were received by the Empress-Mother Turakina, who had acted as Regent from the death of Okkodai, and whose tent was the meeting-place of the electors. Accompanied by a Tartar guard, the Latin visitors inspected the splendid pavilion that had been set up for the *Kuriltai*—all of white velvet,[3] and large enough, as Carpini thought, for more than two thousand people. Around this was a wooden paling, ornamented with various designs, and pierced by two gates, one for the Great Khan and for him

---

[1] Carpini, pp. 752-753.
[2] 'Statio,' Carpini, p. 757.
[3] 'Purpura,' Carpini, p. 755.

alone, the other for all the privileged persons who were allowed within the enclosure. Here all the Tartar chiefs met together; outside the Court they rode in circles over hill and dale with their attendants; when coming to their Council they left their horses picketed about two bow-shots from the tent, but kept their arms. Their apparel was magnificent, and it varied with each day of the ceremonies; from white to red, from red to blue, from velvet to silk brocade. When Kuyuk visited the Court in person, red was the colour specially reserved to do him honour. The horses of the nobles had gold-plating on their bits, saddles, cruppers, and body-armour, and any of the common people who ventured near them, or came within a certain distance of the tent, were soundly beaten by the guards. Meantime, within the pavilion, the chiefs discussed the election, soberly till noon, more freely afterwards; for at mid-day they began drinking their mares' milk, and till evening they drank so deeply that it was a sight to see. As they became more and more genial with their liquor they called the Frankish strangers inside, offered them Kumiss, and, on their refusal, plied them with mead, vainly struggling to overcome the obstinate temperance of their visitors. Among all the princes or envoys who had come to the installation of Kuyuk, none (except the Russian Yaroslav) were treated with such honour as Carpini and his friends.[1] Yet the whole world, as it were, was represented at this remote Mongolian Horde; 'for there were more than four thousand ambassadors, some bringing tribute, some offering presents, as well as Sultans and chiefs who had come in person, and others again who were governors of provinces.' Among this crowd the most notable were Prince Yaroslav of Suzdal; two sons of the King of Georgia; several princes

[1] So Friar John himself declares, p. 757.

of the Khitai and Solanges,[1] from North China, Manchuria, and Korea; the emissary of the Caliph of Bagdad; and more than ten Moslem Sultans. From the Dniepr to the Ho-ang-ho, and from the Persian Gulf to the Arctic circle, all nations were eager to conciliate the favour of 'God in Heaven and Kuyuk Khan on Earth.'[2]

Carpini spent, as far as his recollection would serve, a good four weeks at or near the *Syra Orda*, and during this time he believed the election of Kuyuk took place, though with great secrecy, and without any formal proclamation. But, from the homage paid him, it was clear that the matter was settled; whenever Kuyuk came out of 'the tent,' they sang to him and inclined before him certain staves tipped with tufts of red wool, symbols of royalty which no other chief enjoyed, and which, under the name of *tughs*, were for many ages among the accompaniments of Tartar sovereignty in Asia.[3]

The new Emperor was, in the writer's judgment, about forty years old, perhaps forty-five, or even more; he was of medium stature, very reserved and dignified in manner, with a reputation for great sagacity and (as the Christians at his Court supposed) so favourable to the faith of Jesus, that he would certainly declare himself a convert at no great distance of time. Christians (probably Nestorians) were regularly maintained in his service and supplied from his table; a Christian chapel stood just in front of his tent, and here service was publicly performed—all which was peculiar to Kuyuk's household, and not permitted at the Court of any other of the Mongol princes.[4] Three or four leagues from the *Solemn Court*, Imperial Pavilion, or *Syra Orda*, was another camp or tent, called the *Golden Horde*, in a

---

[1] Otherwise 'Kitai' and 'Solangi.'
[2] Carpini, pp. 754-757.
[3] Carpini, p. 757.
[4] Carpini, pp. 765-766.

fine large plain near a river [1] flowing between mountains; here Kuyuk was to have been enthroned on the 15th of August, but the ceremony was deferred for ten days, on account of a storm of hail. This tent was a splendid piece of work; it rested on pillars plated with gold and studded with gold nails; the top and sides were covered with silk brocades;[2] and here at last, on the 24th of August, the enthronement of the new Khan took place. A great multitude came together, all standing with their faces towards the south, and constantly bending the knee to that quarter of the heavens. These observances continued for a long time, until the moment of the installation had come. Then every one went back into the tent, and placing Kuyuk on the imperial seat, knelt before him and did homage—first the princes and then the rest of the people,—the Friars alone abstaining from the customary genuflections, though afterwards they also paid the required marks of respect to the new ruler. The day ended with a gorging and drinking bout; carts of cooked meat were devoured with salted broth for sauce; mares' milk, mead, and other liquors were not spared; and that night saw only the beginning of a feast that lasted [3] in proportion to the magnitude of the event it marked.

Here, at the Court of the *Golden Horde*, Carpini and his friends, with the emissaries of the Solanges and many others, were admitted to the Imperial presence; Chingay,[4] the protonotary, wrote down the names of all the envoys to be received, and the names of those from whom they came, and repeated the same in a loud voice before the Great Khan and the assembled princes and nobles of the Court. The ambassadors were then searched, lest they should have

[1] Probably the Orkhon.
[2] *Baldakinis*.
[3] Seven days. Carpini, pp. 757-758.
[4] The Uigur prime minister of Okkodai, who had lost favour in the regency of Turakina. Carpini, pp. 758-759.

upon their persons any knives or other weapons; they were warned not to touch the threshold, to enter the presence by the left door, and to bow the knee four times before the throne. The presents offered to Kuyuk by various envoys were here on view in the *Golden Horde*—silks and samits, velvets and brocade, furs and gold embroidery, with a canopy or awning[1] covered with precious stones and designed for state occasions. From one governor of a province came a caravan of camels covered with brocade and a number of horses and mules in armour of leather and iron; on a hill near the Horde more than five hundred carts-full of gold, silver, and silk were divided between the Emperor and the Mongol princes, as accession-spoil; the Latin envoys alone had no gifts to offer. From the Khitai of North China came a wonderful tent of red velvet, fitted with a daïs for the Great Khan's throne, and a royal seat of ebony, marvellously carved, rounded behind, and approached by steps, with lavish adornments of gold, pearls, and precious stones. Benches were placed around the throne; those on the left were higher, and here sat the female courtiers, the harem of the Great Khan; on the right there were no raised seats, but the chiefs sat on seats of lesser height in the middle of the tent, their inferiors crowding behind them.[2]

While Carpini was at the Mongol Court, Prince Yaroslav of Suzdal died there—from poison, it was said, administered by the Empress-mother with her own hand, at a banquet she gave him. In any case, he fell ill after that same banquet, and within seven days he died, and every one knew how much the Tartars coveted his lands. To complete the work of robbery, the Empress summoned to court Alexander, son and heir of Yaroslav, afterwards famous and hallowed

[1] Or umbrella. Carpini, p. 760.   [2] Carpini, 758-761.

as Alexander Nevski; she was anxious, so ran the message, to give him his father's lands; but all believed that, if he came, it would only be to death or life-long captivity; fortunately for himself, he was too wary to put his head in the lioness' den.[1]

When the envoys of the Pope had been first presented to Kuyuk, he dismissed them for a time to the Court of this same treacherous Empress;[2] he did not wish to meet the western visitors just then, the Friar believed, for he had determined in two days' time to unfurl his standard against the whole of that Western World from which they came. The imperial hospitality was not too generous; for the month which the envoys spent at *Syra Orda* was a time of terrible privation; the allowance of food which was given to the four of them was barely sufficient for one; the market was too distant for them to buy anything; and the guests were in constant fear of the diminution of their supplies to starvation point. But for the help of a Russian,[3] named Kosmas, who, as a goldsmith, was in high favour with the Khan, the whole party might have died. This Kosmas had made the imperial throne above described, and the seal with which Kuyuk's answer to the Papal Letters was authenticated, and he showed the visitors the haughty inscription on the seal of the 'Emperor of all men,' which our author gives in full elsewhere.[4] From some Russians and Hungarians, who spoke Latin and French, and had come to Ormektua in the following of various princes and ambassadors, Carpini learned many details of the Khan's private life; but most of all he gathered from some Europeans, who had been in Mongolia,

---

[1] Carpini, 761-762. He did not visit Mongolia until Turakina had ceased to hold power, after the complete installation of Kuyuk.

[2] Carpini is here apparently referring again to his first days at Ormektua, before the election had taken place, p. 762.

[3] Ruthenian, p. 763.

[4] See p. 292 of this vol.

either as prisoners or artificers, for lengthy periods—ten, twenty, or thirty years.[1]

After the Latin strangers had been some time under the Emperor's charge (for their stay at the Court of the Empress-mother fortunately lasted but a few days), they underwent a visit from Chingay the protonotary. On behalf of the Khan, he required a written statement of the objects of the mission, as at Batu's Court; and in the same way, and almost in the same words, this statement was furnished once more; perhaps Kuyuk wished to see whether this would tally with the account given by Batu. After a few days the legates were summoned to the imperial presence, and required to state in words all that they wished to say. They did so; and various questions were now put to them, on behalf of the Khan, by Chingay, Kadak, the 'procurator' of the Empire, and other notaries and secretaries. Thus they were asked if any one at the Papal Court could interpret the Russian, Saracen, or Tartar writing; they replied that none of these were in use at Rome, and that the best course was to write in Tartar, and to give a translation of the same, through the interpreters, in Latin. Among these interpreters was one Temer, who had been a 'knight' or attendant of the unhappy Yaroslav, and had now entered the service of the Khan.[2]

The suggestion of the Friars was adopted, and on the 11th of November they were again summoned, and furnished with the Mongol answer to their letters, written in Latin and Arabic as well as 'Tartar.' This document was read out and translated word by word; as the Latin version was recited, the Friars were called on to explain each phrase,

[1] Carpini, 762-763.
[2] Carpini mentions three others of Yaroslav's suite, 'Dubarlaus,' clerk who had also entered the Khan's service, and two menials, Jacob and Michael; cf. pp. 763, 764, 771.

lest a mistake had been made in any word; finally the various draughts were compared with one another to guard against omissions in any copy. The legates were provided with Tartar guides for the return journey, and Kuyuk proposed to send back with them some ambassadors of his own to the States of Christian Europe and the Pope. He would not, however, demean himself by making the offer, but only gave a hint through the interpreters, who urged the Franks to ask the Khan to execute what was in his heart; the hint was not taken, as Carpini and his fellows dreaded nothing more, in the interests of Christendom, than such an embassy, which would be a mission of espionage, and not improbably an immediate cause of war. So the matter dropped; the Friars returned to their quarters, and two days after, on the 13th November, they obtained their formal dismissal, together with a missive from the Great Khan, signed and sealed with his own seal. The Empress-mother gave them some parting presents, a fox-skin gown with the fur outside, and a piece of velvet, but in this case she does not seem to have added her favourite gift of poison.[1]

The return journey was even more arduous than the way out, and the winter weather proved even more trying than the summer. From the 13th of November to the 9th of June Carpini struggled across the vast expanse between the Mongol Court and Kiev, resting usually (when rest was taken) in places on the open plain where one could scrape a hole with the feet, and pile up the snow around as a protection, poor at the best, for often one woke to find all covered with the drifts. It was on this homeward route that the mission passed through a Bisermin or Moslem town, called Lemfink, which was probably in the middle basin of the Syr Daria, and perhaps near the present Tchimkent; this

[1] Carpini, pp. 764-767.

is the only indication given us by Carpini's narrative of the course he was now following, but there is no reason to suppose that it was very different from that of his outward way.[1]

On the 9th of May Friar John reappeared at Batu's Camp upon the Volga, and reported to the Western Viceroy what the Khan had written in answer to the Papal Letters. Batu asked no further questions, but warned the envoys fully and truly to declare the whole counsel of Kuyuk to the Pope and the lords of Christendom. So saying, he dismissed them with safe conducts, and passed them on to his lieutenant, Mauci, who guarded (as we have seen) the left or eastern side of the Dniepr basin. Here they rescued those members of their party who had vainly essayed to return with news of the progress of the mission on its first arrival in Batu's Horde; these unfortunates had long abandoned all hope of Carpini's life and of their own freedom, and expected nothing but an early death after a few hard years of slavery. Still pushing on, the reunited friends passed the next barrier or cordon, that on the right or western bank of the Borysthenes, where 'Corenza' commanded. Again Kurancha begged hard for presents in return for the guides he furnished, but, 'as we had nothing,' says Carpini drily, 'we gave nothing.' In six days more, Kiev was reached (9th June); this last stage of the journey was marked by a meeting with the Russian Prince Roman on his way to the Tartars, and with two other nobles of the same race, a certain Prince 'Aloha,' and the 'Duke' of Tchernigov, who accompanied the mission on its return through Russia.[2]

On Carpini's arrival, the people of Kiev came out in a body to meet him and his friends, rejoicing over them as

[1] Carpini, p. 768.   [2] Carpini, pp. 768-769,

if they had risen from the dead ; the Princes Daniel and Vassilko entertained them at a conference on the question of the Reunion of the Churches; and for eight days their stay was prolonged, in the enjoyment of a hospitable welcome-home, which continued throughout all Russia, Poland, and Bohemia. The Friar concludes with a too promising declaration of the Eastern bishops, whose supposed wish 'to have the Pope for their special lord and father, and the Holy Roman Church for their lady and mistress,' was not destined to find much confirmation in subsequent events. It is perhaps more interesting to notice that Kiev was already showing some revival as a centre of trade ; Carpini now met with several merchants from Italy, who had come here by way of Constantinople, the Black Sea, and 'Tartary,' apparently in the period between his first visit and his return ; among others were Michael of Genoa, Manuel of Venice, and Nicolas of Pisa,[1] representatives of the three chief commercial cities of the Mediterranean.[2]

Carpini returned to the Papal Court in the summer of 1247,[3] and in the very next year a fresh attempt was made by a Christian sovereign to open friendly relations with the Tartars, and to hasten their supposed inclination towards the Catholic faith.

This was the mission of Andrew of Longumeau, who was sent out by King Louis IX. of France, while holding his Court[4] in Cyprus in 1248. The King's letters were addressed especially to Ilchikadai, the Mongol general commanding in Persia, and through him to the Great

[1] 'Pisanus,' perhaps one Nic. Pisano,—the Venetian family of the Pisani.

[2] Carpini, pp. 769-772.

[3] Soon after his return, he died of the privations he had undergone.

[4] During his absence from France, on the unlucky venture of the sixth Crusade.

Khan; and they formed an answer to the embassy of a certain David, who had been sent by Ilchikadai to St. Louis, in the earlier part of the same year (1248). David had reported that the mother of the Great Khan was a Christian; and that the Mongols were eager for an alliance with Christendom against Islam; he even reported a decisive conversion of the Great Khan himself and all his lords, at the Festival of the Epiphany, three years before. This story of David's was interpreted, and partly supported (according to one authority) by Brother Andrew of Longumeau, who, like Simon of St. Quentin, had been to the Tartars in the company of Friar Ascelin[1] and the Southern Papal mission of 1247, and had met this David at Kars, at the camp of the Mongol general, Baiju, commanding in Armenia.

Now Brother Andrew was put in charge of King Louis' return Embassy, and started from Cyprus for the Court of Kuyuk in the middle of February, 1248-9, and in the company of the aforesaid David, and of various priests, friars and laymen, one of whom is said to have been an Englishman.

The outward journey was probably made from Cyprus to the coast of Little Armenia or Cilicia; and from Little Armenia overland through Asia Minor, in a north-easterly direction, by way of Sivas and Erzerum, to Tiflis, the capital of Georgia. Hence we may suppose Andrew went to Tabriz, where he would find Ilchikadai, and perhaps part with David. Proceeding onward, the mission apparently skirted the southern shore of the Caspian, and by way of Urgenj or Khiva, Talas, and Tchimkent, pushed on to the Horde of Kuyuk on the Imil.[2] At Talas Andrew saw some Teutonic

---

[1] Ascelin himself did not return to the Papal Court till 1250.

[2] On the return journey, according to Rubruquis, the Embassy followed the eastern shore of the Caspian for a long distance.

or German captives among the Tartars, a striking proof of the far-reaching character of the Mongol raids; but he seems to have been unable to get speech with these prisoners. In its main object the mission was a failure; for, on arrival at the Imil Court, it transpired that Kuyuk Khan had lately died, and the visitors were received with haughty insolence by the Regent Mother, Ogul Gaimish, who treated them like tribute-bearers, and sent them back to King Louis with insulting letters.[1] On their return in 1251, they presented a somewhat fantastic report to King Louis, identifying the Mongols with Gog and Magog and the host of Antichrist, and giving some account of the great conqueror, Ghenghiz Khan, his wars and legislation, his encounter with Prester John, and his supposed conversion to Christianity. Among other evidences of this adoption of the True Faith by some of the Tartars, they noticed eight hundred chapels, or what they supposed were chapels, in one camp—all mounted on the waggons of the Nomades. They also reported positively as to the Christian belief of Sartach, the son of Batu. On their way to and from the Mongol Court, they passed many ruined cities and piles of human bones; and, like preceding and subsequent embassies, they witnessed and experienced the Tartar ritual of passing all strangers (and their presents) between two fires of purification.[2] Their journey lasted a year, and the average rate of their progress was ten leagues a day. Brother Andrew's location of the prison of the Gog-Magogs and companions of Antichrist, from which

---

[1] Which Joinville gives as follows:—' Bone chose est de pez... cete chose te mandons, nous pour toy aviser; car tu ne peus avoir pez se tu ne l'as à nous, et tel roy et tel ... et touz les avons mis à l'espée. Si te mandons qui tu non envoies tant de ton or et de ton argent chascun an, que tu nous retieignes à amis; et se tu ne le fais, nous destruirons toy et ta gent. ...

[2] This was done with especial care in the case of Friar Andrew's mission, as some of its presents were destined for the dead Kuyuk Khan.

the Mongols had escaped, among rocky mountains, in a sandy desert at the eastern extremity of the world, clearly refers to the Great Wall of China.

We know nothing more of the various attendants who are said to have accompanied Andrew—Jehan Goderiche, a priest and a member of the same Order, two other Friars, various clergy of Poissy, Gerbert de Sens, Herbert 'le Sommelier,' two clerks, two sergeants-at-arms, Robert 'the clerk,' John of Carcassonne, and one William; but it is clear that this Legation accomplished one of the most important and interesting of those diplomatic and missionary journeys, which form a supplement to the travels of Carpini and Rubruquis, and its route along the south and east coasts of the Caspian is unique in Christian mediaeval travel.

The ill-success of Brother Andrew's mission did not prevent King Louis from making another effort, during his stay in the Levant, to open intercourse with the Mongols. In 1252 he despatched a Franciscan, named William, famous as William de Rubruck or Rubruquis, from his native place of Rubruck in French Flanders, on a new mission. Carefully avoiding anything in the nature of an official embassy, the French sovereign furnished Rubruquis with letters to the Emperor of the Tartars, and to the Mongol Prince Sartach, occupying a district to the north of the Black Sea, whose mythical conversion to the Christian faith had been so constantly asserted. A more critical mind than St. Louis' might well have been deceived by the plenitude of lying rumour as to the Catholic inclinations of the Mongols. As Friar William only occupied a semi-private position, he had but a small supply of money, but he was provided with a richer stock of church vestments and books, Queen Margaret giving him an illuminated Psalter, and the King presenting him with a Bible, and perhaps with that fine Arabic manuscript which he mentions among the other treasures of his baggage. Rubruquis probably embarked at Acre in the spring of 1252, and in the company of that Philip de Toucy,[1] who, as we have seen, had given St. Louis a great amount of fresh intelligence about the Tartars. It was indeed his journey, his visit to the French camp, and his report of the travels and successes, marital and other, of Baldwin of Hainault, which immediately decided both the mission of Rubruquis itself, and the outward route of that mission. In Constantinople Brother William must have remained nearly a year, for it was not till the spring of 1253 that he left the Imperial City for Mongolia; and his travelling companion, Friar Bartholomew of Cremona,[2] probably joined him in Byzantium, from an earlier station in the Bithynian Nicaea. The mysterious personage so often mentioned in his narrative, 'Homo Dei Turgemanus,' the interpreter or dragoman of the party, may have come from Palestine or may have joined at a later time; there is nothing to show his origin or nationality; and it must remain uncertain whether he was properly a guide and 'turgiman,' named or nicknamed 'Homo Dei,' or whether he was 'Turgemanus, that Man of God,' monk, or hermit, who was willing to act the part of an interpreter for his Christian brethren. In any case, he did not prove himself an efficient helper. The other companions or servants[3] who followed Rubruquis to the Hordes of Batu and of Mangu have no clear individuality in his pages, though some of them may have helped to pass

---

[1] Son of Narjot (Joinville's Narjoe) de Toucy, formerly Regent of the Latin Empire of the East.

[2] Who stayed behind at Mangu's Court when Rubruquis returned.

[3] *E.g.* Gosset.

on the knowledge of Central Asia, now acquired, to later travellers.[1]

It was on the 7th of May, 1253, that Rubruquis set out on his journey from Constantinople, and sailed into the Black Sea, Pontus, or *Mare Maius*, whose length he estimates at fourteen hundred miles, with a breadth at the narrowest point of only three hundred,[2] between Sinope and the Crimea. Georgia and Constantinople, Turkey and the Land of the Khazars,[3] were in his mind the four chief regions of the Euxine shore-land, at the extremities of east and west, of south and north, respectively. The traveller seems to have made a straight course from the Bosphorus to Kherson or Sevastopol,[4] and then (without touching land) to have coasted along the Tauric Chersonese as far as Soldaia or Sudak, between the modern Livadia and Theodosia or Kaffa. This Sudak was then a port for all the maritime traffic between Turkey and Khazar-Land, between the north and south coasts of the Black Sea. It had not yet been eclipsed by Kaffa and the influence of the Genoese; and it formed (the author here invents a geography of his own) one extremity of the triangle formed by the Crimean or Khazarian Peninsula.[5] To the east of this the River Tanais[6] fell into the sea hard by the city of Matarcha or Matrica (in other

[1] For Rubruquis's own report attracted wonderfully little notice for many centuries.

[2] It is really seven hundred long, by four hundred broad. Cf. Rubruquis, pp. 213, 214, of text in vol. iv. of Paris Geographical Society's *Recueil de voyages et de mémoires*, to which all references are made here.

[3] Gazaria, Cassaria, Caesarea, *i.e.* Khazaria, are all names for the Crimea at this time, used by Rubruquis, p. 214, and all refer to the Khazar rule therein.

[4] The traditional scene of the martyrdom of St. Clement; cf. the sixth-century pilgrim, Theodosius, in *Itinera Terrae Sanctae*, Saec. IV.-XI. (i.), Tobler's edition, 1877; Soc. de l'Orient Latin; also Rubruquis, pp. 214-215.

[5] Surrounded by the sea on three sides, or 'pretty well triangular in shape,' Rubruquis, p. 214.

[6] Don; cf. Rubruquis, p. 215.

words, at the Strait of Kertch or Yenikalè); for the Sea of Azov was clearly not reckoned as anything more than a marsh created by the Tanais. No large vessels could enter this water, which over a length and breadth of *seven hundred miles*[1] never exceeded six paces in depths; small barques, however, came from Constantinople up the Tanais to buy fish. Beyond the estuary of this famous stream, for so many ages the traditional boundary of Europe and Asia, lay the Georgians or Iberians, and various other peoples,[2] who yielded no submission to the Tartars; to the south was Trebizond, with an imperial family related to the Greek Emperors of Nicaea, once lords of Constantinople; and in this kingdom, in all the lands to the north of the Black Sea, and even in many countries to the west of it,[3] the overlordship of the Mongols was unquestioned. As far as Slavonia all men yielded them tribute, and in this tribute was now included certain payments in a metal often exacted by pastoral conquerors from more settled races. For every household had to furnish an axe and a quantity of unwrought iron.

Rubruquis landed at Sudak on the 21st of May, and found himself already famous. Merchants from Constantinople had been spreading news of this 'envoy' from the Holy Land, and in spite of his wish to travel as a simple monk, without any formal character, he was obliged to admit his character as an ambassador, bringing letters to Prince Sartach from King Louis. During his short stay at Sudak, the traveller was lodged and entertained

[1] Modern Geography estimates it at 235 miles by 110.

[2] *E.g.* Zicci; cf. the Ζυγοί, Ζυγιοί, and Σιγυννοί of Strabo, xi., ii. 14. The Georgians or Iberi at this time were certainly under tribute to the Tartars. Cf. Rubruquis, p. 216, on the Zicci,

Suevi, and Hiberi, who 'do not obey the Tartars.'

[3] Such as ancient Thrace, the 'Wallachia' or 'Vlachia' and 'Lesser Bulgaria' of this time; cf. Rubruquis, p. 216.

by the Bishop, whose good report[1] of the Mongol chief was not confirmed by the Friar's own experience a little later. Misled by the advice of some Byzantine merchants, Rubruquis purchased and borrowed ox-waggons for his journey, in preference to horses; such carts the Russians used to carry furs; and in them the Franks now stored such poor offerings as they had brought to conciliate the Tartar magnates. The slowness of these *arbas* more than outweighed their advantages; with them one travelled twice as slow as in riding; but the writer could only learn by experience, and he set out from Sudak on the 1st of June with four waggons that had been bought, two that had been lent, five horses for the five members of the party, and two drivers or servants. The rank and file of the mission was composed of the dragoman, Brother Bartholemew of Cremona, one Gosset, who had special charge of the presents, and a boy called Nicholas, who had been redeemed from slavery at Constantinople by Friar William's compassion and the money of the French King.[2]

Along the coast of the Crimea, from Kherson to the mouth of the Tanais, the sea was faced by lofty promontories (the mountain range of the Tchatyr Dagh), and the shore was studded with villages. Between Kherson and Sudak alone the traveller counted forty of these hamlets, and nearly every one had its own language or dialect; not the least remarkable was the Teutonic of various Gothic settlers. To the north of the hilly region that faced the Black Sea was a flat country, full of springs and rivulets, and abounding in trees, and beyond this wooded zone, again, was the steppe or prairie, a mighty plain, that stretched five

---

[1] Professedly derived from a personal visit; Rubruquis, p. 217.
[2] Rubruquis, pp. 218-219.

days' journey to the isthmus which marked the beginning of the Tauric Chersonese. This narrow spit of land was intersected by a ditch, running from the Eastern to the Western Sea, from the Azov to the Euxine, and affording a means of defence against the outer world. In the neighbourhood of the isthmus were brine-springs, and from these the Mongol lords, Batu and Sartach, derived great wealth. For all the neighbouring Russian and Koman lands, and many of the Black Sea traders, drew their supply of salt from thence.[1]

Rubruquis now pauses to give an account, somewhat in the manner of Carpini, but less methodical and elaborate, of the manners and customs of the Tartars. It was on the third day after leaving Sudak that Friar William first encountered the masters of the great country he had begun to traverse; the shock of surprise which this meeting left upon his mind was like that on a man suddenly transported into another world or another age of history; but it was as much a part of his duty to his sovereign to describe the people, as the country, of the terrible and mysterious Mongol Empire.[2]

These Barbarians, who had suddenly grown to such power, and conquered all Scythia, from the Danube to the rising of the sun, were utterly different from the nations of Europe in their mode of life, which was absolutely nomadic. In no place did these Tartars make use of fixed dwellings; everywhere they shifted their tents according to the fancy of the moment, not knowing where their next habitation would be. Yet this wandering life had certain limitations. Every chief or captain, according to the number of men in his command, knew how far his pasturelands extended. In winter they usually moved down to

[1] Rubruquis, pp. 219 220.   [2] Rubruquis, pp. 220-238.

winter regions in the South; in summer up to cooler lands in the North. The waterless steppes were only grazed over when covered with snow, for this snow was a substitute for water.[1]

As to the homes of these Nomades, they were tents of felt, mounted on a framework of wicker or sticks, circular in shape, and contracting at the top to a little round hoop, above which a sort of collar projected, serving as a chimney. This felt was often coated with chalk, clay, or powdered bone; it was usually white, but sometimes black; and it was decorated, especially at the 'chimney' top and at the entrance, with embroidered designs in colour, sometimes depicting vines and other trees, sometimes birds and beasts. These huts were carried on carts, and measured up to thirty feet in diameter; as many as twenty-two oxen[2] were counted by the author, drawing one cart and tent; the axle of this cart was as large as the mast of a ship, and the driver stood at the door of the house, and thence managed his team. For the transport of their bedding and valuables, the Tartars made large chests of osier wood, which were carried separately from the tents and ox-waggons, on high carts drawn by camels. A chief or rich man often had one or two hundred such carts, and when he encamped, pitching his dwelling with the door to the south, he would station these carts in rows like walls on each side of the tent. In the case of a prince like Batu, there was not only the central hut, where he himself had his Court, but separate tents for each of his wives,[3] and a number of smaller dwellings for attendants and servants. For the wives of a chieftain, the place of honour in camping was to the west; inside the tent of the

---

[1] Rubruquis, p. 220.
[2] In two rows of eleven each.
[3] Numbering twenty-six in this case.

master, the women's side was on the east or left, the men's on the west or right, while his own seat or couch was on the north, and (when sitting) his face was always turned towards the south. Over his head was hung a felt image or puppet called his 'brother'; above the chief wife, or 'mistress,' was another doll, her 'brother'; and higher up, between the two dummies, was a third, considered to be the guardian of the whole dwelling. Other images of the same kind were hung at the entrance, one on the men's side with a mare's tit, one on the women's with a cow's udder;[1] at all festivals these idols were solemnly sprinkled, beginning with the 'brother' of the master, and this ceremony was followed by libations to the various quarters of the heavens. First, they made offering to the south, to do reverence to the element of fire; then to the east, to do honour to the air; next to the west, to show respect to the water; last to the north, in remembrance of the dead. Each libation was thrice repeated, and accompanied with genuflections. Before drinking, the Head of the House poured a little on the ground, or (if seated on horseback) on the horse's mane. At the entrance of the great tent stood a bench with a skin of milk, or other drink, and cups, and this was specially intended for the entertainment of guests.[2]

The chief Tartar liquors were Kumiss, *Kosmos*, or mares' milk in summer, and a *cervisia* or beer of rice, millet, and honey in winter; at their drinking bouts music played, and the revellers clapped their hands and danced to the sound of the guitar. Their food was indiscriminate and often revolting;[3] but their skill and care in making jars of ox-hide and shoes

---

[1] For men, among the Tartars, milked the mares, women the cows.

[2] Rubruquis, pp. 220-224; cf. also Carpini, 616, 618-620, 745.

[3] *E.g.* rats and mice; conies with long tails like cats and marmots are also mentioned among 'small animals good to eat.'

of horse-hide, in drying and mincing flesh, and in saving every particle of food, even to the bones, was admirable. The making of Kumiss was a great feature of Tartar life, and Friar William describes it minutely; its taste he compares to that of milk of almonds; its effect was to intoxicate weak heads and to make the inner man most joyful. Pure water was carefully avoided by the Nomades, but cows' milk, butter, and curds (of which they made a special preparation) were all used by them.[1]

In his journey Rubruquis saw no deer, and few hares, but many gazelles and wild asses like mules, to say nothing of the great-horned sheep or *argali*, which no other Christian traveller mentions till quite modern times. Hawks and falcons played an important part in Tartar life. With them the rich procured a large portion of their food; the poor made a living from their flocks of sheep, and the slaves had to be content with dirty water.[2]

As to clothes;—silks, gold stuffs, and cotton cloths were obtained from Cathay and other lands in the East, or from Persia and the South; furs came from Russia, Great Bulgaria, and the Bashkir and Kirghiz lands in the North. In winter the Tartars always wore at least two robes of fur, and often a third; the innermost, against the body, was of course the best and most valuable; the outer one would be of inferior fox, wolf, or even dog-skin. Trousers of fur were also worn, and the wealthy padded all their clothing with silk stuffing, wonderfully soft, light, and warm. For this silk the poor substituted cotton cloth or fine wool, picked out of the coarser, which was used for the covering of tents or baggage trunks, or for bedding. Tartar ropes were also of wool, with a third part of horse-hair intermixed, and

---

[1] Rubruquis, pp. 224-229; cf. also Carpini, 637, 638, 640, 671; Matthew Paris, iv. 386-389; Joinville, 147-148.
[2] Rubruquis, pp. 229, 230.

their felt saddle-cloths and rain cloaks were excellent inventions.[1]

The men's fashion of shaving the head in a square, leaving a tuft of hair falling down to the eyebrows and tresses at the side, which were plaited together as far as the ears; the dresses of the women, like those of the men, only somewhat longer; the peculiar head-gear of the married females, and especially of the wealthy among them; the hideous fashions of face-painting or smearing; and the marriage customs[2] obtaining in 'Scythia,' are described by Rubruquis in close agreement with Carpini, but less adequately and forcibly. He adds some curious details about the Tartar dread of thunder,[3] the separate occupations and household duties of men and women, and their mode of washing.[4]

The laws and criminal justice of the Mongols offered many points of resemblance to the Europe of that time. Torture was freely used to compel confession; 'grand larceny' was punished with death, small thefts with cruel beatings; no one could interfere in a duel; adulterers, sorcerers, wizards and witches (except the authorised Shamans of the Mongol State religion) were liable to the extreme penalty. But the slaves of a master were wholly at his disposal; all the females might be made his concubines; the *youngest* son's inheritance might include the harem of his father;[5] otherwise no widow married again. On the death of any member of a household, and especially if the departed were a 'Master,' or Head of a Family,[6] loud lamentation

---

[1] Rubruquis, pp. 230-232.
[2] *E.g.* marriage by capture.
[3] When, *e.g.*, they wrapped themselves in black felt, expelled all strangers from the house, and crouched under cover till the storm was over.
[4] Rubruquis, pp. 232-235.
[5] Except, of course, his own mother.
[6] So perhaps we should amend Rubruquis' language ('if any one dies,' etc.), which is confused and ambiguous here, p. 236.

was made, as in the more southern countries of the East; no taxes were paid for a year; and no other dwelling might be entered by the relatives for that time. In the case of the nobles, the place of burial was always concealed, except from those who watched the grave and prevented the approach of any stranger. The Komans usually raised a great mound over such a lord, and set up a statue or image of him, facing east, and holding a cup in his hand. Other sepulchral monuments were shaped like small pyramids; others, again, were like towers; others like courtyards covered with flat stones, round or square in shape, with four high vertical stones at the corners facing the four points of the compass; others were mere empty houses of stone. Horses were generally killed at the tomb, and their skins hung up facing the four quarters of the heaven; mares' milk was left for the dead man to drink and meat for him to eat. Scrupulous care was taken to prevent any one, except the soothsayers or Shamans, visiting the sick, for fear lest an evil spirit or deadly wind should come with those who entered.[1]

Such were the people among whom Rubruquis now found himself. However different in other respects from Europeans, these Tartars had as keen an eye to their own interests as the most civilised people; but as children of nature they showed their insolence, suspicion, inquisitiveness, and greed with even less reserve. A crowd of horsemen surrounded the new-comers, who were forced to wait under the shadow of their carts; seeing the provisions of the travellers, the 'escort' demanded a share, and after tossing off one flagon of wine, asked for another. No man, they said, would come into a house with only one foot. Baffled by Friar William's resolute economy, they examined him narrowly about his identity, his credentials, and the object

[1] Rubruquis, pp. 235-238.

of his journey; but for answer they obtained only a declaration that their visitor had understood Prince Sartach was a Christian, and was therefore bringing him letters from the King of France. Rubruquis had also a missive from the Latin Emperor of Constantinople to Scatai, who commanded under Sartach in the Crimea; and with the help of this he procured at last oxen, horses, and guides. But this bargain was not struck without a long and wearisome wrangle, the Tartars constantly renewing their demands for presents, not only of bread and wine, but of articles such as knives, gloves, purses, and belts. When refused they became abusive, calling the Friar an impostor; for, as he soon found, they considered themselves lords of the world, and thought there was nothing which any one had the right to deny them; as for gratitude, even with a view to future favours, they made no pretence to it; it was something that they abstained from actual violence, but their manner was such that the writer felt, at quitting them, as if he had escaped out of the midst of devils.[1]

On the next day Rubruquis continued his slow journey to the North, and soon came upon Scatai and his men, who formed the first Tartar cordon near this part of the Black Sea coast; but for two months, from the time they left Sudak until they came to Sartach, the Friar and his companions never once slept in house or tent, but always beneath their waggons or on the open ground. Nor did they once behold a city, town, or village, but only the gravemounds of the Komans.[2]

The Horde or Camp of Scatai was on the move when Rubruquis encountered it, and it seemed a veritable city of

[1] Rubruquis, pp. 238-239, 240.
[2] Rubruquis, p. 240. Soon after entering the Tartar country, Friar William had his first taste of mares' milk, which made him break out into a sweat with horror and surprise, though afterwards he came to think it very palatable.

tents and carts, herds and men, though the number of warriors in this 'station' was really below five hundred. Scraping together a few small presents, the visitors came into the presence of the Chief, who was seated on a couch, holding a lute in his hand, with his wife beside him, a hideous and almost noseless Tartar dame, her face besmeared with some black unguent. The letters of the Emperor of Constantinople, written in Greek, were presented; but as no one could interpret them, they were sent back to Soldaia for translation, and hence ensued four days' delay (5th June—8th June).[1] But the time was not wholly wasted, for Rubruquis met with some Christians[2] at this little Mongol Court, and made some progress with the conversion of an enquiring Saracen, who was finally repelled by a terrible dilemma. For all over Tartary it was held that no Christian could drink mares' milk. This was equivalent to renouncing the faith, and all converts must choose between Christianity and Kumiss.[3]

On Whitsunday or Pentecost (7th June), Scatai furnished the mission with guides, and they set out the following day 'due north,'[4] towards the isthmus of Perekop, rejoicing as if they had just passed through one of the gates of hell. Their so-called guides robbed them audaciously, and progress with the ox-waggons was very slow, but at last they reached the end of the province, where the two seas came close together, and the narrow tongue of land was crossed by the famous trench already noticed.[5] On the mainland, to the north, there was another Mongol camp, full of men as hideous as lepers, whose duty was to collect the tax from the salt lakes and springs of the Crimea.[6]

[1] Rubruquis, pp. 240-244.
[2] Alans, Hungarians, and Ruthenians, or Russians.
[3] Without which last, the Moslem plaintively declared, it was impossible to live in these deserts. Rubruquis, pp. 244-245.
[4] Perhaps rather N.N.W.
[5] Cf. Herodotus IV., 3.
[6] Rubruquis, pp. 245-246.

Thence 'due east,'[1] over the vast Continental steppe for ten days, to the next Tartar camp, the sea lying on the right or south all the way, while on the left or north was the almost waterless prairie, over thirty days' journey across, without forest, hill, or stone, but affording one of the finest of pasture-grounds. No long time ago the Komans[2] or Kipchak were masters of this country from the Danube to the Tanais,[3] a distance of two months' journey for a hard rider. Beyond the borders of Europe, moreover, all was theirs to the Etil,[4] a stretch of ten days' good travel; and in the north their land bordered upon the forests of Russia, which extended from Poland and Hungary to the Don. The whole of this region had been ravaged by the Tartars, and was still being ravaged every day. When the miserable Russians could pay no more tribute, their oppressors drove them off like sheep into the wilds, and set them to herd their cattle. Beyond the Russian woods lay Prussia, which had been lately conquered by the Teutonic Knights, and the traveller supposed they might easily win Russia also, if they would only take it in hand. For the good Friar somewhat hastily concludes that the Tartars would all fly into the deserts if they heard 'that great priest the Pope' was ready to make a crusade against them.[5]

However this might be, for the present there was nothing better to do than to push on to the east over the great plain, with only earth and sky in view, save for a glimpse now and again of the Sea of Azov on the right hand. Even the waste had its advantages; for there the travellers escaped the filthiness, insolence, and importunity of the Tartars, whose camps never supplied their guests with sufficient food, and whose chiefs always demanded presents, without any return.[6]

[1] North-east rather.
[2] Called Valans (Alans) by the Teutons, adds Rubruquis, p. 246.
[3] Don.
[4] Volga.
[5] Rubruquis, pp. 246-247.
[6] Rubruquis, pp. 247-248.

Soon after the middle of July [1] Rubruquis arrived at the Tanais or Don, probably not far from the modern Kalatch. At this stream, as men believed, was the boundary of Europe and Asia, just as the Nile was the border of Asia and Africa; and here, on the line of the direct track between Sarai and the Crimea, Batu and Sartach had established a ferry and a colony of Russian peasants to guard and manage it. This village lay on the east bank of the river; small boats were used for transport; and in order to carry over the waggons two of these punts had to be used, lashed together, each bearing one of the wheels. Another village, lower down the Tanais, marked the place for winter crossing, when the Tartar camps had all been moved southward; and below this again the river formed 'a great sea [2] of seven hundred miles,' before it reached the Pontus. Its course marked the Eastern border, not only of Europe in general, but of Russia in particular; its source was in the *Maeotid fens*, which extended to the ocean in the north.[3] Like its tributaries, it abounded in fish, and its west bank was forest-clad. Where the writer crossed the stream it was about as broad as the Seine at Paris; the same point marked the usual northern limit of the annual Tartar march, which now, at the end of July, and the season of rye harvest, began to turn south once more. In this region wheat could not thrive,[4] but millet abounded. The dress of the Russian peasants hereabout [5] was peculiar enough for special notice, especially the high-pointed felt caps of the men, resembling

---

[1] 'A few days before the Feast of St. Mary Magdalene' (22nd July), says Rubruquis, p. 249; he took from this time to 31st July to traverse the space between the Tanais and Sartach's camp, which was three days' journey from the Etil or Volga.

[2] The Azov Sea.

[3] Rubruquis here creates a new set of Maeotid Marshes in the north or north-west of Russia.

[4] Nowhere does it thrive better.

[5] In the ferry settlement.

German fashions, and the gowns of the women, trimmed from the feet to the knee with fur of vaire or minever.[1]

After some delay and great difficulty, Friar William obtained fresh horses and oxen, the want of which kept him three days at the Don; the next problem was to find the camp of Sartach in a country so destitute of people and of landmarks; but at last, on the 31st July, the mission arrived at the Horde of the 'Christian' prince, happy as shipwrecked mariners on reaching port. The region beyond (and to the north of) the Tanais, was truly beautiful, with rivers and forests, in whose dense woodland lived the Finnish tribes of *Moxel* and *Merdas*, of Cheremiss and Mordvin. Among these races the former were pure Pagans, the latter Saracens; they had no towns, but only little hamlets in the forest; in the past they had fought under the Mongol banners against the Germans, and now they hoped that these brave adversaries would come and rescue them from the Tartar yoke. The Moxel householders were given to hospitality, entertained passing merchants without stint, and were not over jealous of their wives; their furs were celebrated; and honey, wax, hawks, and herds of swine were among their other treasures.[2]

Beyond the country of the Moslem Merdas was the Etil, the Volga of Carpini and the Russians; Rubruquis had never set eyes upon so great a river; and he wondered 'where away in the north' such a mighty mass of water[3] could take its rise. He could only learn that it flowed immediately from Great Bulgaria, or the neighbourhood of the present Kazan, and passed southwards into a great lake (the Caspian Sea) which had a circumference of four months' journey, a mountainous shore on three sides, and

[1] Rubruquis, pp. 249-251.
[2] Rubruquis, pp. 251-252.
[3] Rubruquis, p. 259.

the steppe country to the north. Where he crossed these rivers, the Don and Volga, were not more than ten days' journey distant from each other, but in their lower course they diverged widely, the one flowing into the Pontus, the other into the aforesaid lake, which was known, among other names, as the *Sea of Etil*.[1]

Far away in the south, beyond 'this desert,' which Friar William was now traversing, lay the high mountains of the Circassians, where the Christian Alans still maintained their resistance to the Tartars; the Lesghian Saracens, likewise independent, lived on the shore of the Caspian; and beyond Lesghia were the iron gates of Alexander, at the pass of Derbent, through which the writer passed on his homeward way. For the present, he was travelling in a country where the Kipchak or Koman tribes had till lately been masters; where some of the greatest Mongol camps had been fixed since the invasion of Batu; and where strangers, bound for the Imperial Horde, had to obtain leave for their further journey.[2]

On the 2nd of August the mission was admitted to an audience of Sartach, at a point three days' march from the Volga. The letters of King Louis were produced, and as Arabic and Syriac translations of the same had not been forgotten, their tenor was easily made clear.[3] The effect was good; permission was given to proceed; and on the next day Rubruquis set out for the Horde of Sartach's overlord, the great Batu himself. Yet he did not pass this stage without much trouble, from the greed and

---

[1] Also the names of *Caspian Sea*, and *Sea of Sirsan* (? Shirvan) are given by Rubruquis, p. 265.

[2] Rubruquis, pp. 252-253. Carpini, pp. 659, 679.

[3] At the camp were some Armenian priests who knew Turkish and Arabic, and one at least who understood Syriac. This was a companion of that David, who in 1248 came to St. Louis in Cyprus, on the mission from Ilchikadai above referred to.

curiosity of Sartach's people, and especially of a Nestorian, named Koiak, who enjoyed great power in this 'Court,' and who carefully detained most of the books and vestments of his Christian brother, in charge for him against his return. Koiak had heard of the King of France from the Sieur Baldwin of Hainault,[1] and believed him to be the greatest lord among the Franks; but this did not hinder him from the unscrupulous appropriation of Frankish property under the guise of safe-keeping; and his final advice to the travellers ('be patient and humble,') was scarcely encouraging. One fond illusion was shattered at Sartach's camp; his visitors were warned not to call him a Christian any more;[2] he was a Mongol,[3] and nothing else. To the writer it even seemed as if he mocked at Christians.[4]

Three days after resuming the journey (5th-8th August), Rubruquis came to the passage of the Etil, the greatest of rivers, four times larger than the Seine, and very deep. This last portion of the road was infested by brigands, mostly escaped slaves of the Mongols, such as Russians, Hungarians, and Alans; and apart from this danger, there was the fear of perishing from hunger in the waste. Sartach had treated the strangers with even closer meanness than was usual among the stingy Mongols. Not once in the four days of their stay did he offer his guests any food, and as he gave them nothing 'for the road,' they were driven to live on the biscuit they had brought with them as a delicacy for the Tartar nobles.[5]

[1] Whose visit was probably after 1240, when he married a Koman princess. He seems to have been successful in reaching the Great Khan's Horde in Mongolia, perhaps after the visit of John de Plano Carpini.

[2] Nor would he be called a Tartar, but this distinction, however valid, was not much observed, even by the subjects of the Khans; cf. p. 263 of Rubruquis' text in the Paris *Recueil*, vol. iv.

[3] 'Moal,' in Rubruquis.

[4] Rubruquis, pp. 253-258, 259, 263.

[5] Rubruquis, pp. 258, 264, 265.

On the Volga, as on the Don, Friar William found a ferry at the point where the steppe track reached the stream; this ferry marked the extreme northern point of Batu's own summer march. Now, at the beginning of August, the Viceroy was retracing his steps and moving down south towards the Caspian Sea. This alteration of the Horde's movement brought Rubruquis near the mouth of the Volga, and in his later journey onward to Mongolia, he passed close by the northern side of the mysterious inland water. On his return he came along its western coast; Friar Andrew of Longumeau had already travelled by the southern and eastern shores; and so clearer notions of the Mediterranean of Central Asia began to obtain in Europe. It was not true, Friar William remarks with emphasis, that it was a gulf of the ocean. This opinion was supported by the authority of the great St. Isidore,[1] but it was baseless, all the same. To the south were the 'Caspian' mountains and Persia; to the *east* the mountains of the Assassins or of Mulehet; to the west the highlands of the Alans, Lesghians, and Georgians, and the Iron Gates of Derbent. Finally, to the north were the great plains once ruled by the Komans,[2] but now by the Mongols, and on this side the Etil flowed into the sea.[3]

Rubruquis seems to have reached the Volga not far from the present Saratov; five days to the northward was Great Bulgaria, the later Kazan;[4] and the Friar was perplexed to think what devil had carried the religion of Mohammed so far. But his immediate business was to find Batu, who

---

[1] One of Rubruquis' chief authorities, but used by him with discrimination.

[2] Especially those called Canglae, the Cangitae of John de Plano Carpini.

[3] Rubruquis, pp. 265, 266.

[4] Which he also estimates at thirty days' journey north of the Iron Gates of Derbent. All along this route there was no city, only some villages where the Etil fell into the sea.

had started on a slow southward course which would not be reversed till January. So he dropped down the Volga in a boat, until he overtook the Horde, lying like a great city stretched out for three or four leagues round about the Chieftain's dwelling. 'And as among the people of Israel each one knew on which side of the Tabernacle he had to pitch his tent, so here men knew on which side of the Court they must place themselves.' Only to the south of the central tent and its main entrance no one could plant himself.[1]

Here at last Friar William had an audience of the Mongol king-maker, who alone could settle the question of his further journey. It was a painful thought to the Frank stranger that he had lost the vestments wherein he might have come before the conqueror of Eastern Europe as befitted the representative of a great king. Friar John of *Polycarp* (as he terms Carpini) had never appeared in such humble guise, lest he should bring a slight upon the Pope who had sent him; Andrew of Longumeau had not passed this way on his journey to Kuyuk;[2] and Rubruquis, in his friar's dress, barefooted, with uncovered head, a 'spectacle even to himself,' could do little to maintain the dignity of his mission and of France. Warned to keep silence until invited to speak, he waited 'the space of a miserere' for the first word of the Mongol Prince. Batu was seated on a long divan, gilt, and raised upon a daïs of three steps, with one of his wives beside him; he looked intently upon his visitors, and to Friar William he seemed about the height of 'my Lord John of Beaumont.'[3] The Viceroy's face, 'covered with red spots,' was hardly a pleasant sight, but still more painful was his indifference to religion. For

[1] Rubruquis, pp. 266-267.
[2] Rubruquis' *Keu* Khan.
[3] A companion of St. Louis in the Holy Land.

when Rubruquis, receiving permission to address the throne, urged him on bended knees to seek the celestial goods of baptism, he quietly smiled, while his courtiers broke out into laughter and loud clapping of hands. Finally, after the usual enquiries as to the name, country, sovereign, and objects of his guests, he offered them Kumiss and dismissed them, at the same time announcing his decision as to their future progress. The party was to be divided; the two Friars, William and Bartholemew, with the dragoman,[1] were to go on to the Great Khan Mangu, and the clerk Gosset was to return to Sartach. But Gosset's superiors, in spite of this pronouncement, were detained in Batu's company for five weeks, and followed the slow wanderings of the Mongol encampment down the Volga. Their sufferings were pitiable. 'Sometimes my companion[2] would say to me, almost with tears in his eyes, It seems that I shall never get any food again.' For the market which followed the Horde was so far away that one could not reach it, 'having to-travel afoot from lack of horses.' After a time, the strangers were relieved by some Hungarian prisoners, 'who had once been clerks,' and by a Christian Koman who had been baptised by certain Franciscans in Hungary; from these good friends they obtained food and drink; and in return Rubruquis wrote out for them the *Hours* of the Virgin and the Office for the Dead, as he could not satisfy their request for books.[3]

Friar William reckoned Batu's company at five hundred heads of families, but not more; from day to day he saw him riding with all his Horde, and so could make a fair estimate; but he does not seem to have been admitted to another audience, though he was told of the great chief's

[1] 'Homo Dei Turgimanus.'
[2] Apparently Bartholemew of Cremona.
[3] Rubruquis, pp. 268-272.

constant enquiries about his visitors from the West. It was not until the 14th of September that arrangements were made for their departure. They were then furnished with a guide, who warned them with a contemptuous air of the length and cold of the journey ('four months in such weather, that stones and trees were often split with frost'), and threatened to abandon them on the way if they showed any weakness. Next day, the 15th of September, they set out, furnished with garments suitable for so extreme a climate—sheepskin coats and breeches, boots and hoods of Tartar fashion, felt stockings, and other necessaries.[1]

Their course was 'ever Eastward' till the Feast of All Saints (1st of November), and all the way they traversed the land once inhabited by the Kanglae, or Eastern Komans, having on the left or north the 'Great Bulgaria' of the Middle Volga, and on the right or south the Caspian Sea. Twelve days' journey from the camp of Batu, they came to the Jagac, Yaik, or Ural River, which flowed from the land of Pascatir (or the Bashkirs) in the North, and like the Etil, fell into the Caspian. The language of this people of Pascatir, Rubruquis believed, was the same as that of the Hungarians; they were a purely pastoral race, owning no fixed dwellings, and bordering on the west upon Great Bulgaria. The Bulgarians, indeed, had some small cities, but going eastward there were no more towns to be met with; all the tribes were shepherd and Nomade in their habits. From Pascatir (otherwise 'Old' or 'Great' Hungary) once went forth the Huns, 'who were afterwards the Hungarians,'[2] and ravaged as far as France and Egypt; with their fleet horses they crossed the barriers of Alexander in the Caucasus; and with them

---

[1] Rubruquis, pp. 273-274.
[2] 'Hence it is the same land as Great Hungary' (emendation for 'Bulgaria'), Rubruquis, p. 274.

went the Bulgars, the Vandals, and the Wallachians. The Bulgars of the Danube, near Constantinople, were of this stock. The Vlaks, Illaks, or Wallachians, lived originally near the Bashkirs, and from a mingling of these races came the people of the Land of Assan,[1] south of the Danube.[2]

The languages of the Russians, Poles, Bohemians, Vandals, and Slavons, adds Rubruquis with perfect correctness, were related; all these nations were supporters of the Huns (and Hungarians) in old days, and of the Tartars in later time. Here the Friar begins to wander a little, blaming the unlucky Slavs for the very thing they had struggled against so desperately; but he was naturally sure of his information, as he had it from those holy men,[3] the Dominican Missionaries, who had been among the Bashkirs before the advent of the Tartars.[4] Since then, 'Pascatir' had been conquered by the neighbouring (Saracen) Bulgars, and some of the people had become Moslems.[5]

'So we rode through the land of the Kanglae,' resumes the narrative, 'from the Feast of the Holy Cross to that of All Saints (15th September to 1st November), and nearly every day we went, as near as I reckon, about as far as from Paris to Orleans,[6] or sometimes more, according to our supply of horses. Sometimes we changed these two or three times a day; at other times, we went two or three days without

---

[1] This is the Southern Bulgaria here named after its famous Chief Asen.

[2] Rubruquis, pp. 274-275.

[3] The studied misrepresentation of things Slavonic in the West is largely based on old Roman Church prejudice against Eastern non-Papal Christianity.

[4] A mission we only know of from Rubruquis, except for a vague reference in Alberic Trois Fontaines, *Cronicon*, 564, under A.D. 1237. Four preaching friars, says Alberic, hearing of the intended Tartar attack upon Komania and Hungary, travelled for a hundred days as far as Old Hungary, and told on their return how the Tartars, as they found, had already invaded and subjugated Old Hungary.

[5] Rubruquis, pp. 275-276.

[6] Ninety-five kilometres.

any change. Out of twenty or thirty horses that we had (to choose from), we, as strangers, always got the worst. True, they gave me a strong horse, because of my weight, but I dared not ask if he rode easily or no, nor, indeed, could we venture to complain about anything. So we had to bear great hardships, sometimes riding two on one horse; hungry and thirsty, cold and weary, oftener than one could reckon.' Not until evening did they have proper food, a shoulder and ribs of mutton, and some meat broth, which the half-starved travellers found invigorating and even delicious; but, besides this, they had only some millet gruel at the beginning of the day. On Fridays they had to break their fast, for the choice was mutton or nothing. At times this meat had to be eaten half cooked, or nearly raw, for the only fuel was the dung of animals, a few briars, or, still more rarely, a little brushwood from the banks of streams. At first their guide was at no pains to conceal his contempt for his moneyless comrades, but he gradually came to know and like them better, and then would often take them 'to pray; at the camp of some rich Mongol close to their route. Many of these wealthy Nomades were descendants of Ghenghiz Khan, 'scattered all over this vast sea-like plain'; they knew little of any world but their own prairies; and yet, like other Tartars, they catechised their visitor with a host of impertinent questions:—Whether the Pope were really five hundred years old or no; whether the Franks had many cattle, sheep, and horses; what the ocean sea was like, about which their visitor talked so much; and how it could possibly be endless, without shores.[1]

On the 31st of October the mission somewhat altered its route; from due east to south-east, as Rubruquis describes it. Thirty-four days had elapsed since he crossed the Ural

[1] Rubruquis, pp. 276-278.

River, probably below Uralsk, and we may fairly suppose that he had long passed the 'huge wide lagoon' of the Sea of Aral,[1] and had arrived (at least) in the longitude of the modern Perovski, on the Lower Syr Daria. From this point, perhaps near Julek, the envoys made their way 'by certain mountains' *due south*[2] for seven days. These mountains were possibly the north-west outliers of the Ala-Tau, or Alexander Range, separating the Syr Daria basin from that of the steppe river Chu; and this note of the traveller's shows that he had now almost traversed the great featureless steppe of the Aralo-Caspian lowland, and was approaching the high country of Central Asia.[3]

On the seventh day of their new direction, the travellers sighted very lofty mountains to the south,[4] and entering a plain, irrigated like a garden, reached a Moslem town called Kinchat or Kenjek, in the Talas Valley (8th November).[5] Here the Mongol commander came out to meet their guide, who, as a young Tartar noble, must be received with befitting honour; the food and drink here offered was very welcome (especially the ale, a mark of respect to the messengers of the Khan); and the little band seems to have taken a short rest at this station. Meantime, from the 29th of September, the frost had begun, and ice was on the roads. Rubruquis vainly enquired the name of the 'province' he had now reached, but he could only learn the designation of the

---

[1] Which, unlike Carpini, he does not notice, even indirectly.

[2] For *due south* we should read probably *south-east*.

[3] Rubruquis, p. 278. It is at this point he mentions the wild ass of the desert (called *Kûlan* by the Mongols) and its extreme fleetness—a last remembrance of the vast uncultivated steppe which he was now leaving for a time.

[4] Probably these mountains were the modern Alexander range, running from about Aulie-Ata to the western end of Lake Issik-Kul.

[5] A steppe river in its lower course, lost in the plains on its way to join the Chu; it rises on south slopes of the Alexander range, and flows past Aulie-Ata.

town. Kinchat was very small, but lay in a well-irrigated neighbourhood, which produced the grape, and absorbed so much water from the River Talas, that it 'flowed not into any sea, but was sucked into the ground, forming many marshes.'[1]

On the next day (9th November) the mission came to another village nearer the mountains, which Rubruquis, from his knowledge of the classical geographers, supposed to be the Caucasus of Central Asia. This (according to common belief) ran right across the continent from west to east, and the envoy supposed that he had already crossed it in his passage of the southernmost spurs of the Ural range. Near the point they had now reached was the town of Talas, an important commercial centre for many centuries, where in recent times had been planted a colony of Teuton[2] slaves, belonging to Prince Buri, who perished in a family quarrel with Batu. These slaves had since been transported by Mangu Khan, with Batu's permission, to a certain town called Bolat, a month's journey eastward, where they were employed in the mining of gold[3] and the manufacture of arms. Later on, in his progress eastward, Rubruquis passed within 'perhaps three days' of Bolat, which may be located in the neighbourhood of Lake Sairam, slightly to the north of the modern Kulja. The mission

---

[1] Rubruquis, pp. 278-279.

[2] Visited by Hiouen-Thsang in the early years of the seventh century, and by many other Chinese as well as Moslem travellers. Zemarchus, the Byzantine, passed through Talas on his journey in 571, and here the Turkish Khan Dizabul entertained his Roman friends at a famous banquet, when defiance was hurled at the common enemy, Persia (Menander Protector, 228); *Dawn of Modern Geography*, i. 186-8. Its site is now uncertain, but it cannot have been very far from the valley of the Syr Daria, and Rubruquis believed it lay 'beyond' his route in the direction, of the mountains, six days' journey; conjecture at present hovers between the two extremes of Turkestan and Aulie-Ata.

[3] A tribute to the traditional skill of Germans in mining-work.

was, however, forbidden to visit this Teuton colony, either going or returning.[1]

Still proceeding eastwards and skirting the northern slopes of the Alexander range, the Friar now entered the proper domain of the Great Khan Mangu,[2] and after a few days, arrived at the Alps or High Land, where the Kara-Khitai used to live; here the party had to cross a great river (the Ili) in a ferry-boat. Thence along a valley, past a ruined camp, with mud walls,[3] in a district not wholly unfertile, through a 'goodly town' called Equius, inhabited by Persian-speaking Saracens,[4] and across the spurs projecting from the mountains in the great Southern range,[5] into a beautiful plain, with high mountains on the right, and on the left a sea or lake[6] twenty-five days' journey in circumference. The whole of this plain was fertilised by streams that flowed from the adjoining mountains[7] into the lake, whose southern shore was now traversed by Rubruquis, as the northern side was passed by him on the return journey. Both to the north and south were lofty heights, and in the plain had once been not a few towns, but most of these had been destroyed by the Mongols, to increase the area of pasture land, for there was excellent pasturage in that country. However, one large market-city remained, a place named Kailak,

---

[1] Rubruquis, pp. 279-280.

[2] Rubruquis notes, however, that in practical power Batu was superior, and his subjects, being more powerful, were less careful to show honour to Mangu's messengers than Mangu's people to Batu's envoys. Mangu, of course, was really the nominee of Batu for the titular headship of the Tartars.

[3] Near the modern Kopal.

[4] We may conjecture that this locality corresponded to the modern Lepsinsk, a little north of Kopal.

[5] The Ala-tau.

[6] The Ala-Kul. This lake in the Russian Province of the Seven Rivers or Semiretchi, is a great closed basin, north-east of Kopal and 190 kilometres east of Lake Balkhash; it is 75 kilometres from north to south; 36 kilometres from east to west. Its area is 1950 square kilometres, which is increasing of late years. It lies 237 metres above sea level. Its principal affluent is the Chamanti.

[7] Now the 'Zungarian Ala-tau.'

frequented by many traders; and here the company rested twelve days, till they were joined by a 'certain secretary' of Batu's, who, along with the travellers' guide, was to settle various matters at Mangu's Horde.[1]

This country had once possessed a language and letters of its own in which Nestorian books and rituals had been composed; but now it was all inhabited by Turcomans; and here for the first time Rubruquis saw the 'Idolatrous' Shamanists and Buddhists, of whom he gives a long account. Among the 'sects' or nations of this category the most educated, and (to a Christian) the nearest to the True Faith, were the Uigurs, living in the mountains east of the Ala-kul basin, and both in race and doctrine a mixture of Tartar, Nestorian, and Saracen, or, in other words, of Turco-Tartar, Syrian, and Persian elements. In Kailak the Idolaters had three temples; two of these the Friar entered, 'to see their foolishness';[2] he noticed that they worshipped towards the north, and that the Moslems abhorred this superstition so much that they would not even discuss the subject with him. Later on, at Karakorum, he saw a colossal image (of Buddha) 'as large as we paint Saint Christopher,' and heard of another, still greater, in Cathay, which could be seen at a distance of two days' journey, as he was told by a

[1] Rubruquis, pp. 280-281.
[2] In one of these he talked with a man who had a little cross in ink upon his hand, and who said he was a Christian. Heyd, *Commerce du Levant*, ii. 65, concludes that both Uigurs and Naimans had many Christians among them, but were not, as a whole, Christian; thus preferring the view of Carpini, 'an acuter observer than Rubruquis.' The latter's 'Krit and Merkit' are probably equivalent to the Keraits and Mekrits; cf. Carpini, 645.

The Keraits were Nestorian Christians at any rate in part; they lived to the south-east of Lake Baikal, on the rivers Orkhon and Tula. Abulfaraj gives a story of their conversion by the Nestorian Bishop of Merv in 1007. The Merkits or Mekrits seem to have been still closer to Baikal, in the valley of the lower Selenga; cf. d'Ohsson, i. 54; Rockhill's *Rubruquis*, Hak. Soc., p. 111.

Nestorian who had just come from the Far East. The Uigurs used bells like those of Christians, and their shaven celibate priests, dressed in saffron-coloured robes, and living in communities, had many points of likeness to the Catholic sacerdotal and coenobitic orders.[1] It was probably on this account, he adds, that Eastern Christians did not use bells, except indeed the Russians or Rutheni, and the Greeks in Khazaria or the Crimea. Their beads recalled to him the Christian rosaries; their repetitions of the Sacred Formula, *Om mani padme hum*, reminded him of Latin Prayers for which men 'expected as many rewards from God as they remembered God in saying them.' The Lama hats or mitres, barbarian as they were, and the shaven faces under them, almost made the stranger for a moment think that he had come among the Franks once more. Their very 'palls' or stoles were like those worn by deacons of the Western Church in Lent.[2]

The writing of the Uigurs, which, like Chinese, ran from top to bottom of a page, the lines following one another from left to right,[3] had been adopted by the Mongols, and it was in this script that Mangu Khan wrote to King Louis. Their temples were full of pictures and sacred sentences of magic virtue, hung up on separate slips; like the ancient Pagans, they burnt their dead, placing the ashes at the tops of pyramids. Those Mongols who had conformed to the Idolatry of the Uigurs nevertheless kept up much of their old wizardry or Shamanism, especially venerating their ancestors under the form of felt images, guarded by soothsayers, and placed in special carts when the encampment

---

[1] Their religious silence in choir puzzled Rubruquis not a little, and he tried in vain to make them speak.

[2] Rubruquis, pp. 282-286.

[3] The Tanguts, on the other hand, says Rubruquis, wrote from right to left, and from below to above, like Arabic. The Tibetan mode of writing was like the Latin, and the characters were similar (?).

was on the move, or arranged in circles within a house or tent[1] when the Horde had 'set down its dwellings.'[2]

On the other hand, those Uigurs, who lived much among Christians or Mussulmans, had mostly become worshippers of one only God. To Rubruquis this people was of especial interest, as inhabiting the cities which first obeyed Ghenghiz Khan and occupying the mountain land to the south of the pasture country of Prester John. Their writing was known to almost all the Nestorians; their language was the foundation of all the Koman and Turkish dialects;[3] and thus their civilisation had left its mark on the brutal conquerors who had almost exterminated them. Beyond the Uigurs to the eastward, but in the same mountain region, lived the Tanguts (of Kansu), once victorious over Ghenghiz himself, and famous for their yaks—very strong cattle with hairy tails like horses, which drew the big tent-dwellings of the Mongols. These strange beasts were lower on their legs than other oxen, but much stronger; even their bellies and backs were covered with hair. Their horns were slender, long, curved, and so sharp that the points were always cut off. Their cows, like bulls, furiously attacked red objects, and no one could milk them without a song. Beyond the Tanguts again were the misshapen and bestial folk of Tibet, who once made a practice of eating their dead parents, and still used handsome drinking-cups fashioned from the skulls of their ancestors, so that they might have them in mind even in the midst of their merry-making. Gold abounded in this land, but no one hoarded it, fearing the vengeance of God on such covetousness.[4]

Continuing his account of Asiatic peoples, Rubruquis

---

[1] Here no stranger might enter. 'I tried,' says Rubruquis, 'to force my way into one hut, but was most rudely treated,' p. 288.

[2] Rubruquis, pp. 286-288.

[3] Rubruquis, p. 289.

[4] Rubruquis, pp. 288-289.

next tells us about the 'Longa' and 'Solanga,' of Manchuria and Korea, whose envoys he saw at the Mongol Court, dragging with them huge carts harnessed to six oxen apiece. These people, the 'Solanges' of Carpini, the 'Sulangka' of Moslem geographers, apparently inhabited much of the basin of the Amur Proper, as well as those of its great tributaries, the Zeya, Bureya, Sungari, and Ussuri. They were small and swarthy, like Spaniards, and wore tunics and hats[1] somewhat in the fashion of dalmatics and bishop's mitres, evidently resembling the headdress which has survived in Korea. At Court their envoy always carried an ivory tablet, and in speaking with the Great Khan or other magnates, he appeared to read off from this what he had to say.[2]

Besides these quaint people, the Friar heard of another race called Muc, who dwelt in towns, possessed their herds in common, and were very suspicious of strangers, not permitting them to go about the country; the very scent of an Outlander, they declared, would madden the cattle. Here perhaps we have an allusion to the non-Chinese aboriginals of Kansu, of Szechuen, and of Yunnan, who now as then lived in fortified settlements on mountain-tops, herded their flocks together in the valleys near at hand, and practised a rigorous

---

[1] These hats were slightly lower in front than behind, not pointed but square at the top, and made of stiff black buckram, polished so that it shone in the sun like a mirror or burnished helmet. At the temples were long strips of the same stuff, fastened to the mitre, and standing out in the wind like horns. Sometimes these horns were folded across the mitre over the temples, like hoops over the head, and made handsome ornaments. Later in his work Rubruquis refers to the *islands* of the people called *Caule* and *Mause*, who are evidently the Kaoli, or reigning dynasty of Korea; in 1231 Korea had submitted to Ghenghiz, and the king fled to 'islands near the continent,' perhaps those which Rubruquis mentions as joined to the mainland by ice in winter. In 1241 Korea, after a revolt, submitted again, and sent a mission to the Mongol Court; this mission may have been seen by the Master William of Paris, who told Rubruquis of it.

[2] Rubruquis, pp. 290-291.

restraint, not unconnected with money profits, on any foreigners who came among them.[1]

Finally, in the same part of the world, was Great Cathay,[2] anciently called the Land of the Seres;[3] thence came those finest of silk stuffs styled Seric; and there also was a town named Seres. One of the cities of this country was said to have walls of silver and towers of gold;[4] many of its provinces were still independent of the Mongols. The Cathayans were small in stature, with the tiniest of eyes, a highly nasal speech, and great skill in all artisan crafts, fostered by an unbending hereditary tradition.[5] Their land was on the shore of the ocean; it possessed vines, but rice was used in preference to grapes for the making of wine. Here paper money was in use, and the writing of the people was performed with a brush like a painter's. In one figure they could combine the several letters of a whole word. Their doctors were well acquainted with the virtues of herbs, but used no diuretics; this Rubruquis had seen for himself.[6] Many of this folk lived at Karakorum, and those that were vassals of the Mongol Khan paid an immense annual tribute, no less than fifteen thousand marks a day, without counting supplies of food and silk.

All these nations, given to idolatry, from Uigur-land to Cathay, dwelt on the northern slopes of the Trans-Asian Caucasus,[7] as far as the Eastern Ocean, but on the south of that Scythia which was inhabited by the shepherd Mongols, lords and masters of all the races now described.

Scattered among them, and living after their manner,

---

[1] Rubruquis, p. 291.
[2] Or Northern China, pp. 291-292.
[3] An identification which Rubruquis, rightly enough, is the first to make explicit, p. 291.
[4] Singanfu (?).
[5] 'According to custom, all sons followed the same trade as their fathers,' p. 292.
[6] It is, however, false.
[7] In the classical exaggeration of this term, cf. Rubruquis, p. 292.

but in reality of utterly different race and habits, were to be found Nestorians and Saracens, as far as Cathay. In the last-named the Nestorians possessed colonies in fifteen cities, a bishop in one town,[1] and great influence everywhere, but as to the Christianity and character of these Nestorians the less said the better.[2] They were unutterably depraved, often Pagans or Saracens in manners and customs, sometimes polygamists, usually simoniacs, worse than the Mongols themselves. Words failed to describe their imposture, drunkenness, and ignorance; one of their priests, well known to Rubruquis, affected to expel devils with simple purgatives, being himself no cleric, but a cloth-weaver, innocent of the rudiments of letters. Others practised sorcery with swords, ashes, stones, or twigs; they even used chalices plundered from Christian lands in their heathenish divinations; it was true they had induced the Imperial family[3] to show honour to the sign of the Cross, and to observe the fast of Jonah (or the three days before Septuagesima), but this was no atonement for their evil lives and dogmatic shortcomings. For they ate meat on Fridays, kept this day as a festival, like the Mohammedans, had a married priesthood,

---

[1] Segin (Rubruquis, p. 292), apparently Singanfu or Chang'an, the capital of China under the Han dynasty and the site of the famous Nestorian inscription.

[2] Cf. Rubruquis, pp. 293, 323-324, 320. Among other things, Rubruquis accuses them of giving the benediction of the Christian religion without any proof of the acceptance of the Faith. Thus, among other things, he mentions the arrival of a Nestorian monk (who was a *prudens homo* into the bargain) at Karakorum, his introduction at Court by the Grand Secretary Bulgai, and the favour of the Grand Khan, who sent his children to be blessed by him. But Chabot points out how a Nestorian envoy from Arghun, Rabban Sauma, visiting Rome in 1288, was allowed by Pope Nicholas IV. to celebrate Mass, and was given Communion on Palm Sunday by the Pontiff himself. To the Nestorian Patriarch, Mar Jabalaba III., valuable papal presents were sent at this same time, *e.g.* a tiara, ring, and vestments.

[3] Mangu's eldest son had for tutor a Nestorian priest named David, who, whatever he taught his pupil, at any rate learnt from the Tartars the art of hard drinking.

and made use of sacred books in the Syriac tongue, a language which they had now forgotten, and only read and chanted like parrots. So with a parting outburst of Roman hatred against the Oriental Christianity whose depression was so long a cause of secret satisfaction to the Latin Hierarchy, the Friar returns to his Itinerary.[1]

On the 30th of November the mission left Kailak, passed through a Nestorian village,[2] and after three days came to the head of Lake Ala-Kul, tempestuous as the ocean.[3] The water was brackish; and from the shores the travellers saw the famous 'Island peak'[4] in the bosom of the lake. From lofty mountains in the south-east a valley opened towards the Ala-Kul; and among these mountains another lake was visible,[5] from which flowed a river, and along whose basin blew a wind of terrible strength. This valley they crossed, not without fear lest the blast should carry them into the water below; then, following a northerly direction towards some heights covered with deep snow, they hastened their pace so as to put two days' journey into one and arrive sooner at the Great Khan's Horde. From the 6th of December, along narrow paths and over a poor pasture-land, they hurried on through cold so extreme that they turned their sheepskins with the wool inside:[6] in another week's time they arrived (13th December) at a gorge celebrated for its devils; and here good words were needed to put the unseen enemy to flight. So the Friars chanted the Creed of Nicaea, and for their Tartar guides they wrote the shorter *Credo* and the *Pater Noster*, not as 'charms to be carried

---

[1] Rubruquis, p. 293.
[2] 'Singing joyfully, *Salve regina*.'
[3] Perhaps at that time it formed one and the same water with the now separate Sassyk-Kul.
[4] Or 'Ala tyube,' mentioned by various Chinese geographers.
[5] The Ebi Nor.
[6] Rockhill reads *outside*, the fashion worn in more moderate weather.

on the head'[1] but as 'saving words to be borne within the heart.'[2]

After passing these dangerous mountains, and the low land once inhabited by the Naimans, in a *northerly*[3] direction, the travellers crossed a final range and descended into the flat country of Mongolia itself, a plain vast as the sea (26th December). On the next day they arrived at the Court or Horde of the Great Khan, where they were wretchedly housed in a tiny hut, with scarce room enough for their baggage and their beds; their guide meanwhile occupied a large dwelling, and regaled himself with rice wine, which, to the Friar's discontented eye, seemed as good as the best wine of Auxerre, 'saving only the perfume.' Rubruquis was so troubled by the haughtiness of the Mongols, their contemptuous treatment of himself and his friends, and their obstinate belief in the submission of the King of France, that he felt he would gladly give his life to the preaching of a Crusade against these arrogant barbarians.[4]

The first reception of the envoys took place on the 28th of December; the bare feet[5] of the Friars caused great astonishment; and the crowd of onlookers gazed at the strangers as at so many monsters, asking if they had no use for their members, that they exposed them to the frost so recklessly. A Hungarian captive who was present recognised the habit of the Order, and explained matters as well as he could;

---

[1] After the usual Tartar manner; cf. Chang-chun and Hiouen-Thsang and their stories of aggressive goblins (who pulled the hair of passengers and were maddened like bulls at the sight of anything red) in the Thian Shan and Altai.

[2] Rubruquis, pp. 293-295.

[3] Probably north-east. These mountains were perhaps those now called the Tarbagatai. The Naiman country was along the upper basin of the Irtish and the shores of Lake Zaisan, from about Semipalatinsk to the present Russo-Chinese frontier.

[4] Rubruquis, pp. 296-300.

[5] Rubruquis' imprudence was not altogether without its reward; next morning he found the tips of his toes were frozen, p. 303.

then came Bulgai, the Grand Secretary,[1] who examined the visitors with the help of this Hungarian, and dismissed them to their lodgings.[2]

Returning from Court, Rubruquis visited an Armenian church, where he met a hermit, lank and swarthy, from the land of Jerusalem, who had come to promise Mangu the Empire of all Christendom if he would only embrace the faith. Envoys of 'Vastacius,' the Greek Emperor of Nicaea, were also at the Horde, as well as a Greek 'knight' and many other ambassadors from Western States. All of these bore witness to the genuine character of the new mission; on the 3rd of January 1254, the Friars were admitted to the presence of Mangu; and at the door of the Imperial tent the Latin visitors sang the hymn of the Nativity—

> 'A solis ortus cardine
> Et usque terrae limitem
> Christum canamus principem
> Natum Maria virgine.'

The felt hanging before the entrance was lifted up for the Khan to hear the better; but, none the less, the visitors, when they had done their song of praise, were searched for concealed weapons before they were permitted to enter the Court-room. This was all hung with cloth of gold, and had a great fire of wood and refuse, briars and saxaul roots burning in a grate in the centre of the dwelling. Mangu was seated on a couch, and appeared to be 'a little man, of moderate height,' aged about five-and-forty years, dressed in a skin spotted and glossy like a seal.[3] Among various

---

[1] A Nestorian who, through his acquaintance with a Latin prisoner, the goldsmith, Master William of Paris, had made some progress towards orthodoxy, and had even learnt to fast on Fridays. Rubruquis, p. 301.

[2] Rubruquis, pp. 300-301.

[3] This 'sea-cow' skin probably came from Lake Baïkal.

drinks Rubruquis chose a *cervisia*, or ale of rice, clear and flavoured like white wine; but while he sipped a little 'out of respect,' his interpreter was drinking deeply, and it was not long before the Friar had to dispense with his services 'such as they were.'[1] His unfortunate proximity to the Court butlers may have been the cause of this misfortune, but as he was always tipsy on critical occasions, and at the best had a feeble grasp of Western tongues, it did not much matter. Mangu himself was very incoherent, and the royal interpreter was only too proud to be like his master. So for a time the visitors could understand little or nothing, until at last the Khan roused their attention with a series of plain questions about the kingdom of France—its sheep and cattle and horses, and the possibilities of conquering the whole. Rubruquis had scarce control sufficient to conceal his indignation, but he returned an evasive answer, and gained the Khan's permission to stay two months at the Royal Horde. More than this, he was offered permission to visit Karakorum.[2]

At the Court Rubruquis met with quite a little colony of Europeans. Besides the embassies which had come to bespeak the favour of Mangu, there were various captives, like the woman Paquette[3] of Metz, who had been made prisoner in Hungary, and was married to a Russian tent-builder;[4] or the goldsmith, William Buchier of Paris,[5] who

[1] Rubruquis, pp. 301-306. The noble who had guided them from Batu's camp, on his arrival at Mangu's horde ceased to take further interest in the strangers, and devoted himself to his liquor.

[2] Rubruquis, pp. 306-309.

[3] Among other Europeans at the Tartar Court Rubruquis mentions in another connection, nearer the end of his work, (1) a Teuton girl slave, who had been thrown into a trance, lasting three days, by the Mongol diviners; (2) a Hungarian who by stealth had once been present at a devil-evocation of these same diviners.

[4] *Houses* (domos) is the word, but the ordinary house in these parts was, of course, the Central Asian tent or *yurt*. Elsewhere Rubruquis tells us that the houses in Karakorum of more solid construction were made by the Chinese.

[5] Master William had been taken

worked for the Tartars in Karakorum itself. These captives had suffered unheard-of misery on the journey, but they were now in better case, and made a good deal of money at times. Thus Mangu had just given William the Goldsmith 3000 marks and fifty workmen for a certain contract, and to finish this he was bound to work his utmost. He could not therefore spare Friar William the services of his son as an interpreter, though he would have been a good substitute for the tipsy knave whom Rubruquis had been employing.[1]

At this Tartar Court the Friar also met a very clever charlatan, a certain clerk from Acre, who called himself Raymond, but whose real name was Theodoulos.[2] This man had started from Cyprus with Friar Andrew of Longumeau, and journeyed with him as far as Persia; thence he had proceeded to the Court of Mangu with stories of one Otho, a bishop, who had received from Heaven letters in golden characters, announcing the future triumph of the Mongols over the whole world. With such flatteries Theodoulos persuaded the Great Khan that he would be a useful agent of Tartar ambition; he was entrusted with the Imperial passport of the Golden Tablet,[3] and sent back to the West

prisoner at Belgrade, in *Hungary*, where there was also a Norman Bishop from Belleville near Rouen; a nephew of the Bishop was also captured, whom Rubruquis saw at Karakorum. William was first in the service of Mangu's mother; then in that of the Great Khan's Brother, Arabuccha; finally in that of Mangu himself. The Grand Khan had given him 1000 marks for his works at Karakorum. Arabuccha had shown leanings towards Christianity; among other things he had restrained the railings of certain Saracens, because 'we know Messias is God.' At that moment, in confirmation of his words, a terrible wind arose, and devils were seen to rush along with it.

[1] Rubruquis, pp. 309-310.

[2] He suggests also, though he does not explicitly assert, the fraudulent character of another envoy, a Damascene Christian, who professed to bring offers of tribute and alliance from the Sultan 'of Mont Real and Crae.'

[3] This was put in the keeping of Theodoulos' Mongol companion.

with presents for the King of the Franks. In his company was a Mongol spy, charged to observe and report upon the roads, towns, men, and arms of Christian countries. As far as Nicaea Theodoulos made his way unchallenged, but he was unable to deceive his own people. The Greeks stopped him, examined him, and threw him into prison; the Golden Tablet was returned to Mangu; and at Erzerum, 'on the borders of Turkey,' Friar William passed the bearers of the dishonoured token on their way to Karakorum.[1]

While Rubruquis was at the Court another notable impostor, the Armenian monk and 'prophet' before-mentioned, whose name was Sergius, boasted that he would baptise Mangu on the Feast of the Epiphany next ensuing (6th January, 1254). But as a fact, the Christian priests, the Saracen Mollahs, and the Buddhist Bonzes were all equally confident of their ascendency over the Khan, were all permitted to enrol him among their devotees, and were all alike victims of delusion. For he believed in none of their creeds, and while they followed his Court like flies after honey, he only used them as a politic ruler, ready to welcome any means of quieting subjects of other nationalities and faiths. On the 12th of January, a week after the baptism, Friar William was present at a Nestorian service[2] which the Great Khan attended with some of his household; at the Imperial request the Latin strangers also chanted a

---

The Tablet was 'like a plate of gold, half a cubit long and a palm broad; he who carried it could command what he pleased without delay.' Mangu's presents to the King of France were a strong bow which two men could hardly bend, and two 'sounding' arrows, with silver heads, full of holes, that whistled in the air when shot (Rubruquis, pp. 311, 312);—examples of a famous Turco-Tartar weapon. In haughty and obscure terms Mangu offered the French King peace and the Western World, while the Tartars finished the conquest of the Saracens.

[1] Rubruquis, pp. 310-313.

[2] On the other hand, on the 8th of February, there was a Pagan or Shamanist divination at Court, by the blade-bones of sheep.

VISIT TO KARAKORUM

psalm;[1] and the whole ended with a sort of orgie—the Khan's ladies drinking, the Nestorian clerics singing and 'howling,' the Frank visitors partaking of the food which came so rarely to their mouths, and looking on at the rest with wearisome disgust.[2]

Some time after his arrival, Rubruquis found that the Court was about to visit Karakorum, and obtained permission to accompany its progress. On this journey 'northward' he verified the truthfulness of Baldwin of Hainault, who 'had been there,' and had told him, in Constantinople, how all the rivers of this tract flowed from east to west, and how the route was perpetually on the ascent. Still further, towards the sun-rising, lay the famous Cathay, at a distance of twenty days,[3] and on one side of the road was the home-land of Ghenghiz and his tribe, to the east and south-east of Lake Baïkal, ten days from Karakorum. But throughout all this region no city, town, or village was to be met with, but only the tents of the Nomade Kirghiz, and the hunters and trappers[4] who chased birds and beasts over the frozen snow and ice on shoes of polished bone. Yet nearer to the Pole lived in abject poverty some wretched tribes, who bordered on the west upon the Bashkirs of Great Hungary; and north of these again was a land of eternal cold, unknown to men.[5]

In spite of all enquiries, the Friar could discover no certain trace, or reliable news, of the monstrous races whom classic authors had placed in these regions, except only the

---

[1] The 'prose,' *Veni, Sancte Spiritus* In reward for their compliance Mangu's queen or chief lady offered them valuable presents, which they declined for themselves, but allowed their dragoman to accept. One of these, a piece of gold and silk brocade, he sold for 80 bezants in Cyprus; Rubruquis, p. 316.

[2] Rubruquis, pp. 313, 315, etc.

[3] According to the testimony of certain priests, from the same Cathay.

[4] Whom Rubruquis, p. 327, calls *Oengai*, perhaps the modern *Tunguses*.

[5] Rubruquis, pp. 326, 327, 334, etc.

Chin-chin or Kangaroo-men, who lived, and leapt, in the far Eastern countries beyond Cathay.[1]

On the 29th of March, while Mangu was still on his slow progress towards Karakorum, through a mountainous country swept by snowstorms and bitter winds, he received news of the completion of his new palace in that settlement, and at once hastened his pace. Early on the 5th of April (Palm Sunday) he and his train were in sight of the town, and at three in the afternoon of the same day the formal entry took place. Rubruquis and his friends raised their cross and banner, and passed through the Saracen quarter and the market to the church, where the Nestorians met them in procession. After service the Franks dined with their countryman, William of Paris, and his wife, a 'daughter of Lorraine,' but born in Hungary; one Basil,[2] the son of an Englishman, was also present. Master William kept the day as a festival, for his work at the palace was the admiration of the whole Court, and it was to see this that Mangu had come. Within the high boundary wall that enclosed the Imperial dwelling (like a monks' priory in a Christian country), were many buildings, like barns or storehouses, wherein the food supplies and treasures of the Khan reposed; and in the entrance the French goldsmith had made a silver tree, with four lions of silver at the roots of the same, and pipes or conduits from the foot of the trunk to the top of the branches, round which gilded serpents were twined, and above which, at the summit of all, stood an angel with a trumpet. In a vault beneath the tree a man could lie concealed, and by means of the pipes that led up to the angel he could blow the trumpet; this was a signal for the lions and serpents to pour out wine and kumiss, honey-mead and

[1] Rubruquis, pp. 327-328.
[2] Also 'born in Hungary.'

rice-ale, each drink being duly received in a separate bowl of silver.[1]

In its structure the palace within the enclosure was not unlike a church; it had a nave and two aisles separated by columns and three gates on the south side. Before the central gate stood the tree, and within, towards the north, but looking out to the sacred quarter and the silver tree, through the 'portal of honour,' stood the throne of the Great Khan, 'like a divinity,' on a daïs with two steps. The length of the palace was from north to south, not from east to west; on the right hand, towards the sun-setting, was the place of the male, on the left hand, towards the sun-rising, of the female, courtiers.[2]

But although Karakorum was the capital, if any capital there were, of the greatest Empire in the world, it was not as large as St. Denys, near Paris, a mere *bourgade* among French towns; and this boasted palace was but a tenth of the size of the great monastery at St. Denys. Besides the Saracen quarter to the westward, already mentioned, in which were the markets, and near which stood the palace, there was also an artisan quarter to the east, chiefly inhabited by men of Cathay. Beyond the town itself were the palaces of the Court Secretaries; other buildings of note were the twelve idol temples, the two 'mahummeries' or mosques, and the Christian church; the whole was surrounded by a mud wall pierced by four gates. Each of these ports had a market;— for millet and other grain at the east; for sheep and goats at the west; for oxen and waggons at the south; and for horses at the north.[3]

Rubruquis' ill-compacted and gossiping narrative gives many stories of his disputes with, and verbal triumphs over,

[1] Rubruquis, pp. 334-336, 337.   [2] Rubruquis, p. 336.
[3] Rubruquis, pp. 345-346.

the Nestorian and Armenian Theologians, and especially the monk, to whose hospitality he had been assigned, and whose Manichaeism[1] was alone enough to alienate his guest. These native Christians, according to Friar William, were as greedy as the lowest Mongols; not satisfied with getting drunk every day at Court, they thrust themselves upon the unlucky Western visitors whenever a present of food or wine had come from the Khan's table; and thus the writer learnt from experience 'what a martyrdom was the charity of the poor.' When in disfavour with Mangu, these braggart 'Christians,' showing themselves in their true light, thought nothing of an offer to do a vital injury to the Christians of Europe, and to go on missions as Tartar spies among the nations of the West, as far as to St. James of Compostella. Their 'miracles of healing' were only done by the aid of violent purges or emetics, which sometimes cured and sometimes killed, and to obtain a generous testament from the dying, they were known at times to kick and thrash them into obedience. Even theft did not come amiss to some of them; a crucifix which William of Paris had made for Bulgai, the Grand Secretary, was stolen by these missionaries.[2] Yet in

---

[1] 'Did not the devil,' he exclaimed once in argument, 'bring earth on the first day from the four parts of the world, and did he not knead it into clay and make the human body thereof, and did not God breathe a soul into the same?'

[2] As a matter of fact, the Nestorians, whatever their deficiencies, were the only Christians who made any impression on the Mongols, and procured a comparatively favourable treatment for the Christian subjects of the Khan. Thus Hulagu, whose wife was a Nestorian, usually spared the lives of those who belonged to this faith, while his slaughter of Moslems was almost as terrible as the massacres of Ghenghiz. King Hayton the Elder, of Little Armenia, has been wrongly counted as a Nestorian, but he seems to have been in close relations with the Nestorians and with their help influenced the Tartars to show greater favour towards subject Christians: still more powerful was the influence of a Nestorian doctor, Simeon, who, as early as 1241, appears as a prominent person at the Mongol Court. Cf. Heyd, *Commerce du Levant*, ii. 67.

spite of their heresies and their wickedness, it was some comfort to find that they accepted Rome as the Head of all the churches; that, if the way were open, they would gladly take their spiritual rulers from the Pope; and that it was only need which made them obey the Patriarch or Catholic at Bagdad, and receive from him their priests, their relics, their sacred oil, and their altar-tables.[1]

Like John de Plano, though perhaps not so poignantly, Friar William and his friends suffered from hunger, cold, the want of a good lodging, and every misery of barbarous surroundings, during the chief part of their stay at Mangu's Court. Occasionally the Khan sent them food and presents of clothing; and at the audiences he granted them they might have taken some handsome and costly gifts of silk, flax-cloth, and brocade, if their character had permitted it. But, as a whole, their privations were great; a poor thin sheep had to serve them for a week; on a day of plenty they were forced to share with a swarm of self-invited guests; and the season of Lent—when even fish was tabooed, and when the winds and snowstorms of Mongolia were most piercing—was only less terrible than the journey from the Volga in midwinter.[2]

It was, therefore, not unwelcome to find, as the month of May drew on, that the Mongol Court was preparing to dismiss the French mission, and to send by their hands Mangu's answer to King Louis. First of all, the Grand Secretary[3] examined the visitors as to their objects and the objects of their Sovereign: demanded whether any other embassies had been sent to the Khan by the King of France; and enquired whether they would escort a Mongol legation

---

[1] Rubruquis, pp. 314-315, 324-326, 331-332, 334, 342-345, 349, 354.
[2] Rubruquis, pp. 314, 315.
[3] Bulgai.

back to Europe. In answer, they repeated what they had said to Batu, and had already stated on their arrival at the Imperial Horde; referred to the visit of David and Friar Andrew; and professed their helplessness to undertake so perilous and political a charge as was suggested. Whatever their real feelings, they thought it well to declare that their wish was still to remain at Mangu's Horde, and to pray for him, his household, and his people. The formal decision was postponed for a few days; and meanwhile, the Khan was entertained by a series of theological disputes,[1] in which Catholics and Nestorians, Armenians and Manichaeans, Moslems and Idolaters, carried on a bloodless warfare, sometimes one side 'writing down' the words of another as 'false' or 'impossible,' sometimes bursting into laughter at the outrageous statements of their adversaries, the whole concluding in the usual manner with deep potations and loud singing.[2]

At last, on the 31st of May, Friar William was summoned to a decisive audience. Before he entered the presence, his new interpreter, the son of Master William, warned him that all was settled for his departure; and so he found it. Mangu was gracious, although Rubruquis had been accused to him of insulting his Majesty with the name of idolater; he readily accepted the Frank's disclaimer; and stretching out to him the sceptre-staff on which he leaned, bade his guest fear nothing. (If I had been fearful, murmured the Friar to himself, I should not have come here.) Then, helped by a not infrequent consultation of the flagon at

---

[1] Among other things, Transmigration of Being was discussed, a little after the fashion of Plato's *Meno*. Master William of Paris, at another time, told Rubruquis of a boy from Cathay, apparently about twelve years old, who was capable of all forms of reasoning, knew how to read and write, and claimed three separate incarnations; evidently this was a 'living Buddha' or 'Bodhisattva.'

[2] Rubruquis, pp. 355-359.

his side, the Mongol Emperor began to confide in his visitors as he had never done before. The Mongols believed in one God, and they perceived that as God gave the hand divers fingers, so He bestowed on man divers ways of life and of belief. For to the Christians He entrusted 'Scriptures that were not kept,' and to others He granted other things. No one, they thought, should wrangle with another; no one should pervert justice for money. At this glance of sarcasm Rubruquis protested his own pure and peaceful conduct, and the Khan, forestalling an exposition of the Catholic Faith which his visitors were eager to give him, announced his decision as to their return. They must go back to Europe; if they feared to conduct his ambassadors, would they venture to be the bearers of his letters or his messages? On their consent, he again enquired about the route they wished to follow, and, finding that they hoped for 'travelling charges' and a guard as far as Armenia the Little,[1] he reminded them of the all-powerful Viceroy in the West, through whose camp and with whose consent they must journey. 'There are two eyes in one head, but the two have one and the same vision; from Batu you have come, by way of Batu shall you return.'[2]

Rubruquis, though not without joy at the prospect of revisiting his King and country, was anxious to retain what influence he had gained, in truth or fancy, over the Khan and the Tartars; and before leaving the Court he offered himself as a priest to the 'poor slaves' at Bolat, near Lake Sairam. This colony of exiled Teutons he had not been allowed to visit as an envoy; perhaps he could gain access as a chaplain; in any case, had he the Imperial permission to return to the

---

[1] *I.e.* Cilicia.
[2] Rubruquis, pp. 359-361. At their first audience Mangu declared that, as the sun cast its rays over all the earth, even so his own power *and that of Batu* reached everywhere.

Mongol Empire? In vague and general terms Mangu assented; then, with a parting cup, and an exhortation (which must have sounded like irony) to comfort themselves with food and guard their health, he dismissed the Friars, and they saw him no more. To the last they hoped to hear from his lips some hint of his inclination to the True Religion, but it would have needed signs and wonders like those of Moses to extort such a confession and to humble the pride of such a monarch.[1]

There was not, indeed, much humility about the letters which Rubruquis soon after received as the result of his mission. Like Kuyuk's earlier epistle to the Pope, Mangu's words breathed a haughty and fanatical sense of overmastering power by the grace of God:—'This, by the virtue of the Eternal, throughout the great world of the Mongols, is the message of Mangu Khan to the Lord of the French. Wherever ears can hear, wherever horses can travel, there let it be heard and known: those who do not believe, but resist Our Commandments, shall not be able to see with their eyes, or hold with their hands, or walk with their feet. The Commandments of the Eternal are what we impart to you. If you will obey us, send your ambassadors, that we may know whether you wish for peace or war. But if you say, Our Country is far, our mountains are strong, and our sea is wide, then you shall find what we can do. For the Eternal makes plain what is difficult, and brings near that which is afar off.' Such was the substance, with many repetitions, of the reply which the One lord of

---

[1] Rubruquis, pp. 361-362. Many of the chief women at Court seem to have been Christians of a sort, e.g. the mothers of Mangu, Hulagu, and Kublai. These facts supported the idea of a wider Christianity among Mongols than really existed, and fostered hopes of alliance between Christendom and the Tartars, as well as of the conversion of the Great Khan.

Earth,[1] inspired by the One God of Heaven, deigned to make to a King whom he already regarded as a vassal anxious for his favour. As to the previous missions to and from the French Court in the Levant, they were nothing; David was an impostor, and Ogul Gaimish had no authority. For 'in affairs of war and peace and the welfare of a great realm, what could this woman, viler than a bitch, know or do?'[2]

At a concluding festival in Karakorum—on the 7th of June and the next three days—a week after the final audience, Friar William saw ever fresh tokens of the Mongol power:—the envoy of the Caliph of Bagdad carried in a litter between two mules; the ambassadors of an Indian Sultan who brought presents of hunting leopards and greyhounds; and the legates of the Sultan of Turkey (Rum or Iconium). These Indians followed the same route as Rubruquis, on his homeward way, for almost three weeks, going steadily west, and to indicate the direction of their country they pointed in the same quarter, a good example of the ignorance and carelessness which is responsible for so many of the local misconceptions of this Itinerary. Meanwhile, Brother Bartholomew of Cremona had decided to stay behind; for his part, soul and body could not endure the hardships of the journey; and Mangu was willing to keep and maintain him at the Court. William of Paris, who, for different reasons, could not leave the Horde, was entrusted with a fixed sum[3] to spend on the Friar's needs; an equal sum was given from the Khan's treasury for the expenses of the returning Franks; and about the 6th of July, 1254, Rubruquis set out for Batu and the Volga. With tears he parted from Bartholomew and from his good friend the

---

[1] According to the Mongol view, Ghenghiz Khan, *Son of Heaven*, was the one and only temporal lord from whom all authority was derived.
[2] Rubruquis, pp. 369-370-371.
[3] Five 'iascot.'

goldsmith (once, like himself, a subject of the King of France), who now sent to his old sovereign 'endless salutations' and a present of mysterious value—a girdle with a precious stone, such as the Tartars wore against lightning. After refusing all other gifts,[1] Friar William accepted Mangu's parting offer of 'three plain gowns,' lest it should seem that the royal bounty was treated with scorn; he had already permitted his interpreter, the 'Man of God,' to profit somewhat by the liberality of the Khan[2] and his ladies; and the *nasic* and buckram, the furs and silks and cotton cloths, which were carried back to Cyprus, sold for a good sum.[3]

In seventy days the mission accomplished the distance between Karakorum and the camp of Batu, apparently travelling (at least on part of the road) by a more northern route than on the outward way. Almost the whole journey was over desert country; for fifteen days their path lay by a stream winding through mountains; but, except for one tiny hamlet, and a number of graves, they never saw town, village, or building during the whole of the time; nor did they take any rest, save for the time allotted to eating and sleeping, and one day, when a forced halt occurred through the want of horses. For the most part, they came through the same 'nations,' but not through the same districts, as they had traversed before. Three weeks out from Karakorum they heard news of another wayfarer, whose visit had been long expected, and who was evidently considered by the Tartars to be far more important than the ambassadors of King Louis. This was Hayton, King of Little Armenia (the mediaeval

---

[1] Except some *pompion* fur-gowns in the depth of winter.

[2] Mangu was at first, Rubruquis thought, offended by their going to Sartach instead of proceeding immediately to the Imperial Court; but the Khan evidently liked the envoys more, and gave them better treatment, than Kuyuk did Carpini.

[3] Rubruquis, pp. 368-369, 372-374.

Christian kingdom in Cilicia and the Taurus); he and his house had thrown themselves cordially into the Mongol alliance; and a common enmity to the Saracens kept up a good understanding and a constant communication between these distant and ill-matched friends, between giant and dwarf. Much as he desired to meet Hayton, Rubruquis just missed the royal visitor, who reached the Imperial Horde in the middle[1] of September, when Friar William must have been nearly half-way on his journey to the Volga.[2]

By a strange coincidence the writer reappeared in the Horde of Batu on the same day of the same month as he had quitted it a year before.[3] Here he found Gosset and the others in safety, but miserable, poor, and hopeless; the Tartars were already preparing to make slaves and cattle-drivers of them; and but for the arrival and help of King Hayton, this would long ago have been their fate.[4]

Batu did not exercise the right which Mangu had granted him, of revising, abridging, or altering the Imperial reply to the French King; nor did he object to the southern route by Persia and Armenia; but a month passed before a suitable guide was forthcoming. At last an Uigur was put in charge of the party, and they started for the South, descending the banks of the Etil towards Sarai; when they crossed the great river near its mouth, they had to pass three main branches (each of which was nearly twice as large as the Nile at Damietta), and four lesser arms, composing a delta of seven channels. In the middle of this waste of waters, an

---

[1] The thirteenth. Hayton left Mangu, 1st November. On his way he had visited both Sartach and Batu; and on his return he turned out of his road to meet Sartach again, as the latter was then on his journey to Court and in the neighbourhood of Barchin, on the Syr Daria.

[2] Rubruquis, pp. 375-376.

[3] 16th September 1253 — 16th September 1254.

[4] Rubruquis, p. 377.

unwalled town, Summer-Keur by name, had resisted the Tartars for eight years;[1] Alans, Saracens, and even Germans, were among its people; and it seems not to have been far from Batu's new capital or camp at Sarai. With a good fortune that must have surprised the least despondent of his company, Rubruquis succeeded in recovering most[2] of the books and vestments that had been 'kept in charge' for him so long; the rest were left 'to please Sartach,' who really cared for none of these things; and on the 1st of November the legation set out for the passage of the Caucasus.[3] By the 15th of December they had reached the northern slope of the mighty mountain range; but during the first fortnight of November the way lay through a region absolutely desert,[4] and once they almost died of thirst. Both the Alan and the Lesghian tribes of the hill-country held out stubbornly against the Mongols, and sometimes even raided their cattle in the plains; to pass the Iron Gates or Defile of Derbent an escort of twenty men was furnished,—greatly to the delight of Rubruquis, who had often longed to see the Tartars under arms. This escort, however, was poorly appointed; of the twenty warriors only two had cuirasses, the rest were equipped with nothing better than bows and arrows, Persian caps, plates of iron, and jackets of leather, awkward and ill-suited for active work in combat.[5]

The Iron Gates, made by Alexander of Macedon, between

---

[1] How it had defended itself when all these waters were frozen in winter Rubruquis does not explain.

[2] He gives a considerable list, however, of articles he did not recover, e.g. a gold-fringed altar-cloth, a censer, the Psalter of the French queen, a Bible in verse, and the Arabic MS., worth thirty bezants, already noticed.

[3] 'Mountains of the Alans,' as Rubruquis calls them.

[4] Between the place where Rubruquis met Batu's Horde and Sarai, fifteen days' journey, the country was equally without inhabitants, except for a party out hawking, headed by one of Batu's sons.

[5] Rubruquis, pp. 377-381.

the Caucasus Mountains and the Caspian Sea, were marked and guarded by a town (the city of Derbent), more than a mile long, but so closely pressed between the hills and the shore,[1] that its width was never more than a stone's throw; it was strongly walled, with parapets and towers, but without moats; the only possible high-road lay through the very middle of the city. Two days' journey beyond Derbent, the travellers passed a Jewish colony at Samaron,[2] and here the route left the Caspian shore and struck inland, through a wild, hilly country, abounding in remains of Alexander's ramparts. Within certain other barriers in this region Hebrew exiles were said to be imprisoned, but of these the writer could learn nothing with any certainty. Still pushing on to the 'South,'[3] through Samag or Shamaka, once the capital of Shirvan, the travellers now descended into the great plain of the Kur and lower Aras, in which the famous cities of Tiflis, the 'head of Georgia,' and *Ganges*, the old 'metropolis' of the Khorazmian conquerors,[4] were to be found.

Friar William, for his part, made no excursions to celebrated sites, but pushed on steadily to the south-west 'towards Africa'; leaving Persia on the left or south and Georgia on the right or north; and following the valley of the Araxes, from the bridge[5] that spanned it, at its junction with the Kur, to the chief encampment of the Mongol army in the land of Hyrcania. This horde lay on the Upper Aras, with special authority over the conquered Georgians, Turks,

---

[1] Rubruquis is the only Western Christian traveller of the pre-Polo period who passes through the Iron Gates of Derbent.

[2] Edrisi's *Semmur;* cf. Benjamin of Tudela on the Hebrews transplanted by Shalmaneser Asher, ii. 158-162; also Carpini, 748, on the 'Jewish' Brutachi.

[3] 'South-west' properly.

[4] Of 1225 and following years; Rubruquis, pp. 381-383. Ganges, Gansh or Genje, is the present Elizabetpol.

[5] A pontoon, held together by great iron chains. Here Rubruquis quotes Virgil, *Æneid*, viii. 728:—
'pontem dedignatur Araxes.'

and Persians, but subordinate to the Tartar Prince[1] governing at Tabriz; their chief, Bachu Noian, entertained his visitors, not with Kumiss, as the Friar had hoped, but with wine, and sent them on to Naxua or Nakhitchivan, at the foot of Ararat. Here Rubruquis kept the Festivals of Christmas and the Epiphany in an almost deserted city; where two churches alone stood out of eighty; where first the Saracens, and then the Tartars, had worked their will on the luckless Armenians; but where the first resting-place of Noah was still proudly shown. Close by was the Sacred Mountain of the Ark, unscalable by mortal foot; for though in itself it did not appear too high for an ascent, a divine jealousy forbade the human race to explore the 'Mother of the World.'[2]

In this ruined town of Naxua, one Friar Bernard, a Dominican of Catalonia, was then staying, whose adventures had been in singular contrast to the success of Rubruquis. First he had been to Georgia with a Prior of the Holy Sepulchre, who had landed possessions in that country; then he had journeyed in the company of a Hungarian Friar to the camp of Argun at Tabriz; but permission 'to go to Sartach' was refused;' and he was kept in the Persian city, miserably inactive and solitary, his colleague having been sent back to Tiflis and the West, and his only companion being a German lay brother, whose language he could not understand.[3]

Friar William was detained by heavy snow at Naxua till the 13th of January; thence he travelled through the country of Sahensa, Prince of Ani, a most unwilling vassal of the Mongol, who professed allegiance to the Roman

---

[1] The famous Argun, before this time; now superseded by one of Mangu's brothers. Kars was at this time the chief station of the Mongol forces in this region.

[2] Rubruquis, pp. 384-387.

[3] Rubruquis, p. 387.

Church, and longed for the aid of the King of France. His capital was a strong fortress, one of the most ancient cities of Armenia, famous for its 'thousand and one' (or countless) churches, but now controlled and polluted by the presence of a Mongol bailiff; and here Friar William encountered five Dominicans, four of whom had come from France direct, while the fifth had joined them in Syria. They carried letters from the Pope to Sartach, Buri, and Mangu; but their position was simply that of missionaries; and their only request of the Mongol lords was permission to stay and preach. Rubruquis' advice was sound, but hardly inspiring: with the Papal letters they would doubtless be able to enter the Tartar realms, but they must take a good supply of patience and plenty of sound reasons for their coming; mere preachers would receive but scant courtesy among the Nomades; and, without an interpreter, they would find their lot a hard one. After this warning they started for Tiflis, where some other Friars of their Order were then stationed, and the writer heard no more of their mission.[1]

On the 15th of February, 1255, fifteen days' journey from Nakhitchivan, Rubruquis entered 'Turkey' or the Sultanate of Iconium, crossing the head waters of the Araxes, and passing through Marsengen, between Kars and Erzerum.[2] Near here he struck the upper course of the Western Euphrates, and down this stream he journeyed for a week to Camath or Kemakh, where the river turned 'southward'[3] towards Aleppo, and where the traveller passed into the ancient Cappadocia: thence over high snow-clad mountains

---

[1] Rubruquis, pp. 388-390.

[2] Though Rubruquis mentions Aarserum(-on) or Erzerum, 'a goodly city,' he does not say that he visited it, any more than Arsengan or Erzinghian; both, however, are probable stations of his route.

[3] From Kemakh the Euphrates continues a mainly westward course for some distance.

and through Sivas or Sebaste,[1]—the Cappadocian Caesarea,—and Iconium, to Curta or Corycus, on the coast of Cilicia or Armenia the Little, the practical end of the journey. In this last stage the writer saw the site of the Mongol victory over the Turkish Sultan, where terrible earthquakes had changed the face of the country, and the earth had opened its mouth to drink in the blood of Saracens. In Erzinghian alone these convulsions had destroyed ten thousand people of name and standing, to say nothing of the nameless poor. All through this land was evidence of the fulfilment of Isaiah's words, 'Every valley shall be exalted, and every mountain shall be brought low.'[2]

At Iconium Rubruquis was happily freed from his dishonest and troublesome Uigur guide, who, as he could get no pay from his convoy, used his commission and the awful name of Batu to extort money in every town. For here, at the capital of the Seljuk Sultanate, were several Christian merchants, and especially a Genoese, Nicholas of Santo Siro,[3] formerly settled at Acre, who, with his partner, the Venetian Boniface of Molendino, had completely monopolised the alum trade of Turkey, and carried on so flourishing and healthy a business, that what in former days fetched fifteen bezants was now sold for fifty. It was with the aid of this Nicholas that Friar William arrived at last on the Mediterranean (5th May, 1255).[4]

After a short stay at 'Curta,' Rubruquis heard that messengers had come from King Hayton to the Regent of the Kingdom; and these messengers reported that the King was now returning from the Court of Mangu, that the Great Khan had reduced the tribute, and that in

[1] In Lesser Armenia, says Rubruquis, strangely; Caesarea he reached on 4th April; Iconium on 19th April.
[2] Rubruquis, pp. 390-391.
[3] Cf. what Carpini, 772, says of the Italian Merchants at Kiev, p. 317 of this vol.
[4] Rubruquis, pp. 391-392.

future no more Mongol embassies, so oppressive and extortionate,[1] were to enter the friendly land of Little Armenia. The Friar now likewise hastened to the Court at Sis, told what he knew of Hayton's movements in Tartary, and obtained leave to pass over to Cyprus. Here apparently he trusted in his good fortune to find an easy journey to his Sovereign. But the Provincial of his Order would not permit him to quit the Levant; he dragged the illustrious traveller to Antioch and Tripoli, finally ordering him to stay at Acre, and hindering the delivery of his Record to Louis. Rubruquis therefore concludes by asking for deliverance from this petty tyranny, by a few more words about the recent history and political divisions of Asia Minor, and by some suggestions for a fresh Crusade along the overland route through Hungary.[2]

Carpini and Rubruquis are the foremost Western travellers of the Middle Ages, before the time of Marco Polo, and of late years something like a controversy has developed itself over their respective merits; it may be regretted that preference for the one should so constantly insist on depreciation of the other; yet this controversy has had the advantage of bringing out more clearly the many details in which both these Friars contributed to the knowledge of Christian Europe, the range and acuteness of their observation, and the epoch-making character of their journeys. And first of all, as to the later writer. Rubruquis' narrative has been deservedly praised of recent years; it is certainly one of the best records of travel in the Middle Ages; and the journey it chronicles is one of the most considerable ever performed

---

[1] The Mongol embassy to Iconium in two years cost for maintenance, exclusive of bread and wine, about £300,000 (Vincent of Beauvais xxx., ch. 28).

[2] Rubruquis, pp. 392-394.

by a Latin Christian before the age of the great discoveries. It is perfectly true that Friar William was an acute and intelligent observer, and that he collected a mass of valuable details on the geography, race-history, and manners, as well as on the religions and languages, of Asia. It is also evident that he was conscientious and thorough in preparing for this distant and arduous journey, and that his knowledge of Arabic[1] was of great service. This alone would show that he had something of the gift of tongues; but a more surprising proof of his accomplishments may be found in his acquaintance with the Mongol speech, of which he learnt sufficient to check and put to shame his poor and tipsy dragoman. But it would be unfair to treat him as the sole worthy predecessor of Marco Polo; to minimise the credit belonging to Carpini in order to give more to his successor[2]; and to put down to the French envoy the discovery of things which others had already noticed. Thus in geography it has been indiscriminately claimed for him that he was the first to indicate the true sources and course of the Don and Volga, the inland character of the Caspian, the identity of Cathay with the country of the classical Seres, the true position of the Balkhash Lake, and the character of that Central Asian depression, of which it occupies the eastern extremity; he is also credited with the first description of Karakorum, the first mention of Korea, and the first notice of some of the North Asiatic aborigines.[3] In natural history it may be more readily granted that he is the first Western writer to mention the wild ass, or *kûlan*, of the Central Asiatic deserts, and the great sheep of the Pamirs, now famous as the *argali*

---

[1] Like Carpini's acquaintance with Slavonic.

[2] Cf. Heyd's opinion, *Commerce du Levant*, ii., 65, on Carpini ('Jean da Piano de Carpine'), and his superiority to Rubruquis as an observer.

[3] Such as the *Tungus* or *Oengai*.

or *ovis poli*; as, in anthropology and philology, he deserves the highest credit for his account of the Crimean Goths, his identification of the Komans with the Kipchak and Kanglae,[1] or Kangitae, his careful distinction between the Tartars and the Mongols proper, and his recognition of the affinity of various dialects — those of Bashkirs[2] and Hungarians on one side, of Russians, Poles, Bohemians, and Vandals on another, and of Turks, Uigurs, and Komans on a third. Excellent, also, is his suggestion of the Volga origin of the Danube Bulgarians, and his account[3] of the Chinese, Tangut, Uigur, and Tibetan writings; and not less excellent is his description of the Christian communities within the Mongol Empire, their rituals and tenets. We owe, moreover, to him the earliest Western account of the Lama system of Northern Buddhism in all its details, and of the peculiar dress and ritual of the Uigurs.

But in this list there are several points in which the primary credit is certainly not due to Rubruquis, and several others where such credit may be easily overstated. As to the languages and dialects of the Mongols and their subjects, Rubruquis, no doubt, shows an advance on his predecessor, and it is beyond dispute that he mentions more of the animals of these wild regions; but, on the whole, there is something to be said for the old editors who treated Carpini as the primary figure in the group of missionary and diplomatic travellers who preceded Marco Polo. For Carpini's narrative is a masterpiece of clear and accurate description, without needless digression; and if its narrative of the author's actual journey is briefer and slighter than that of Rubruquis, its account of the Tartar manners, customs, and history is not only beyond comparison superior, but

---

[1] Carpini, however, clearly identifies Komans, Kipchak, and Kangitae.
[2] Or 'Pascatir.'
[3] With certain deductions.

is the best Latin treatment of the subject in the Middle Ages. Even as to the journey or Itinerary,—Carpini is so regular, orderly, and logical in his treatment of the matter in hand, Rubruquis is so bewildering from his verbosity and his incessant and long-sustained digressions—that the Frenchman's greater mass of detailed observation does not leave him so clear a supremacy after all. Carpini, in fact, is only put at a disadvantage because his instructions forced him to write the study of a people, rather than the account of a journey; and it is only natural that modern geographers should have preferred the rival narrative as more copiously and dominantly geographical. But it cannot be forgotten that in some places where Rubruquis has received credit for an 'earth-knowledge' beyond that of his time (as in the recognition of the inland character of the Caspian), very inaccurate notions (as of the nations on the Caspian coasts) likewise appear. His ideas of the geography of Eastern Europe are in some respects like those of the early mediaeval time, whereas Carpini, mixing more with the natives of the country, has learnt his lesson almost from a modern point of view. Sometimes, also, Rubruquis has been credited with scientific triumphs of a rather shadowy kind. Thus he says that the Don rises in a vast lake, and as it is true that its source is in a little lake, much has been made of the writer's insight. We must take this with what immediately follows, and then we might be staggered at the doctrine that this vast lake, under the name of the *Maeotid Marshes*, stretched away to the Northern Ocean.[1] Again, Friar William makes the salt water basin of the Sea of Azov a creation of the Don and a part of the same; he puts the Mountains of the Assassins or Mulehet on the east of the Caspian; and the compass-directions which he gives us

[1] Cf. the American 'Verrazzano Sea' Myth.

for his route are almost always astray — as in the section between the Crimean Sudak and Batu's camp upon the Volga, or in the reference to the position of India *west* of Karakorum. Rubruquis' account of his journey from the Caucasus to Cilicia may be compared with any part of Carpini's Itinerary by those who still doubt the superior lucidity and narrative-power of the older Friar;[1] and Brother William's references to the ruling family of Trebizond, to the journey of the Armenian King Hayton, to the wall of Alexander and the Iron Gates in the Caucasus, may be set off against John de Plano's confused language about the Volga and the Ural falling into the Sea of Greece.[2] Rubruquis gives an excellent description of the Mongol Court and of Karakorum; he was, moreover, fortunate in meeting with good information about China and Eastern Asia; but even here he is not entirely beyond criticism.[3] Thus, while in one place he puts Cathay on the Eastern Ocean, with only sea beyond it, in another place he makes other lands border Cathay on the east.

Above all, those who would blot out the older traveller in favour of the younger, forget that some respect is due to priority. Carpini went to the East eight years before William of Rubruck, and it is practically certain that the French ecclesiastic had read (or heard) the narrative of his Italian brother *Polycarp*,[4] if not in the longer form, at any rate in an epitome.

[1] This part of Rubruquis' route is almost impossible to reconstruct, and its confusions, digressions, and repetitions are endless; but nowhere does Carpini leave us in any doubt of his road.

[2] Carpini does not expressly say this; he enumerates Dniepr, Don, Volga, and Ural; says they were full of fish; and adds that 'these rivers' fell into the Sea of Greece. He may be thinking only of Don and Dniepr, but in any case, his language would mislead.

[3] He is not alone in his mention of Cathay, cf. Carpini, p. 653 (an admirable description); or of Turkey, cf. Carpini, p. 680.

[4] Cf. p. 268 of Rubruquis' text; p. 339 of this vol.

But if there are some grounds for refusing to oust the first great Latin traveller in Central Asia from the place he has long held; and if we may still ask for John de Plano the special honour of a pioneer and a place in mediaeval land-travel quite abreast of Friar William; yet, on the other hand, the latter must never again be treated with the neglect that has often been his lot. For the envoy of Louis IX., when all deductions have been made, was a man of extraordinary merit and extraordinary success. We may contrast the missions carried out by him and by Carpini with the abortive attempts of Brother Bernard the Catalan, and the other Friars whom Rubruquis encountered in the course of his journey home, or with those of Ascelin, Lawrence, and the rest of the official legates, who by their failures had already thrown some light on the difficulties of a Christian embassy to Mongolia.

It is unfortunate that we know nothing of his life, his appearance, or his character, except for the casual remarks of his own Itinerary on the heaviness of his body and the poverty of his dress. But from Roger Bacon we have one word more—a reference to a meeting of the English philosopher with the French traveller,[1] whose more important geographical passages are well summarised in the *Opus Majus*.

From the time of Bacon to that of Hakluyt no one seems to have taken any interest in Rubruquis; only with Bergeron's translation of 1634 did his own country begin to appreciate and study him; hitherto John de Plano had obscured him only too effectually. Thus neither Vincent of Beauvais, nor Wadding, the annalist of the Franciscans, nor Gerard Mercator in Hakluyt's day, had any clear ideas about the French traveller. It was not altogether unnatural that Rubruquis should be neglected by the purely ecclesi-

---

[1] In France, a few years after his return.

astical writers, for his journey was not of a purely ecclesiastical origin or character; he was not sent by the Pope, by his Order, or by the Church, but by the King of France; and, like the Friar Andrew of Longumeau, he has often been treated as a purely secular figure. But the indifference shown him by 'profane' historians and geographers has been far more inexplicable. Thus John Pits calls him an Englishman, Wadding makes him a native of Brabant, and others have credited him with two works upon the East— one an Itinerary, and the other a History of the Tartars, both being evidently sub-divisions or sub-titles of the report we have. Only in the middle of the present century has the Franciscan Order done justice in its records to one of the most eminent of its sons;[1] a little earlier (in 1839) the first adequate edition appeared of the text of Rubruck, as well as of Carpini. Yet at the beginning of the seventeenth century Purchas had referred to the journey of Friar William as a 'jewel of antiquity'; and in our own day both Peschel and Yule have drawn attention to the comparative freedom of both the Latin envoys from legendary twaddle, to their vividness of narrative and humanity of outlook, and to their admirable common sense. Both travellers had strong prejudices; both hated, and had cause to hate, the Mongols who bullied and starved them; both are naturally critical of the Nestorian rivals of the Church of Rome; but here Rubruquis passes beyond the limits of the candid friend or faithful dealer, and loads his pages with unconvincing abuse of men more successful than himself in the conversion and improvement of a hard-drinking, loose-living, and supremely arrogant barbarian race.

One other traveller of this time remains, one who is

[1] In Father Civezza's *Storia universale delle Missioni francescane*, 1857-1861.

closely connected with William de Rubruquis, through the evidence of the Friar's own narrative;—Hayton, King of Little Armenia. This mediaeval principality had been founded in Cilicia by exiles from Great Armenia[1] in the North, who had set up an anti-Byzantine state in the Taurus, and who, pressed between Constantinople and the Saracens, were glad to ally themselves both with Franks and Tartars—with the former, in faith as well as in arms; with the latter more vaguely, but not less skilfully. Leo II., in 1198, had taken the title of King, and received the holy rites of unction and coronation from a Latin prelate, Conrad of Wittelsbach, Archbishop of Mainz; he married successively two Latin princesses, and conquered part of Isauria; but his capital remained in the highland fortress of Sis, although the plain, from Tarsus to Alexandretta, now professed a like allegiance; and not less significantly it was at Sis that the new Patriarchate, created by the Southern Armenia as a rival to Echmiadzin, found its home. Hayton the traveller was son-in-law and successor of this Leo; he himself belonged to the family of Rupen, but he obtained the throne by his marriage with Leo's only child Isabel, and his father Constantine was but one of his subjects. Before his accession in 1224 he had been constable and 'bailiff' of the realm, and during a reign of five-and-forty years he had to keep every soldier in constant readiness, and to exercise every quality of a statesmanlike diplomacy, for the preservation of his little kingdom. All around him, in this time, went on a struggle of fell and mighty opposites, a conflict of powers so great that his own could hardly weigh in the scale; but early in his reign he perceived one safe rule of

[1] When the Byzantines killed King Gagik, one of his relatives, Rupen by name, escaped to the Taurus and established himself there, carrying on incessant war against Constantinople. The Zeitun Armenians are a relic of the Kingdom of Rupen, Leo II., and Hayton.

policy; and that rule, faithfully observed, brought his light barque through a tempest fatal to many heavier, grander, and more ancient vessels.

For King Hayton,[1] 'the pious friend of Christ,' was quick to discern the overwhelming power of the Mongols, and in spite of his Latin faith and friendship, he did not hesitate to make terms, and even something of an alliance, with the terrible Infidel. This good understanding had already lasted some years, when (on the accession of Kuyuk in 1248) Hayton commissioned his brother Sempad, the Constable of the kingdom, on a mission to his new overlord, to secure the continuance of friendly relations. Sempad was absent four years on this embassy, and during this time he despatched to his King a long letter, dated from Saurequant or Samarcand, and full of details which helped to support the current delusion as to the Christianity of large parts of Central and Further Asia. He describes the gathering of the Mongol lords and princes, for the election of Kuyuk, from India and from Kashgar, from *Chata* and from *Tanchat*, the home of the Three Kings of old. Here all the people were Christians, and 'believed in the Three Kings'; Christian churches[2] stood before the gate of the Khan's palace, and Christianity had become the religion of this great Sovereign[3] and his people. All over the East the true believers were to be found, though scattered, possessing fine, ancient, and well-built churches, many of which had been despoiled by the Turks. It was Kuyuk's grandfather, the great Ghenghiz himself, who first gave Christians liberty of worship, and protected them from Saracen insults; now Moslems were despised and put to nought; but the

---

[1] In the MSS. usually 'Hethum.'

[2] 'In their churches,' wrote Sempad, 'I have seen pictures of Christ and the Three Kings.'

[3] The Khan.

Christian faith made less progress than it would have done if its preachers in these parts of Asia had not been men of indifferent character, who deserved a signal punishment. Sempad concludes with a story of an Indian prince, a Christian in faith, who, with the help of the Tartars, had vanquished the Saracens; in this tale was perhaps contained both a suggestion to Hayton, who was pursuing a similar policy, and a justification of his past conduct.

King Hayton himself undertook a similar journey about two years after Sempad's return. On the accession of Mangu he received an invitation, or command, from Batu, the King-maker[1] and General established in the North, with a countless multitude, upon the banks of the great river Etil; the faithful and favoured vassal of the Khan was summoned to visit both the Western Horde upon the Volga and the Imperial Court in Mongolia; nor did he venture to disobey. But in order to pass safely through the neighbouring Turkish realms in the upland of Asia Minor, where he was hated and feared as a prime enemy of Islam and a client of the Mongols, he found it needful to disguise himself; and so, almost unattended, without baggage and without servants, after twelve days of hazardous travel, he arrived at Kars and the first Tartar camp in the early spring of 1254. Here Bachu Noian[2] commanded, and under his protection the Armenian Prince stayed some time at the foot of Mount Aragadz or Alagoz, near the Metropolitan Church of Echmiadzin, waiting for the presents[3] which were to be sent after him, and which soon appeared under the care

---

[1] 'Basileopator.'

[2] The *Baiju* of some writers, the Baachu of Rubruquis, 384; and the General to whom the Pope sent Friar Anselm in 1246–47.

[3] It is evident that Hayton, hurrying in disguise through the land of his Turkish enemies, would avoid encumbering himself with presents, or drawing attention to his journey by a numerous suite.

of a distinguished escort. This escort was apparently guided by Basil the priest, who had brought Batu's summons from the steppes; and among the party were the Abbot James, formerly an ambassador to Nicaea, who had negotiated for an alliance with the Emperor John Palaeologus; also the Abbot Mekhitar of Skerra, who had paid an earlier visit to 'the East,' a bishop named Stephen, and two priests.

With this suite, and the indispensable offerings they brought for the Tartar lords, Hayton resumed his journey, and passed through Albania or the 'land of the Aghovans,' the gate of Derbent, the fortress of Chor (beyond the giant wall of the Caucasus) and the North Caspian plains, reaching at last the camps of Batu and Sartach,[1] where he was welcomed with marked distinction, and the pretence, if not the reality, of cordial friendship. On the 13th of May he set out again on the next and most trying stage of his journey, to the Court of Mangu, the 'immeasurably long' road beyond the Caspian Sea. The narrative in this section [2] only stops to mention the passage of the River Yaik or Ural; the arrival of Hayton at a place called *Hor*, midway between the Hordes of Batu and Mangu; the country of the Aumani or Naiman on the Upper Irtish;[3] and the land of the Kara-Khitai close to the Imperial headquarters. On the 14th of September the Great Khan received his ally in 'glittering splendour,' accepted his presents, and for near fifty days

---

[1] Sartach was a religious Christian, declares the writer of Hayton's Itinerary, Kirakos Gandaketsi. In deference to Hayton, indulgence was shown to those of Rubruquis' party who had been left behind, and their hard lot was bettered in various ways. But for the King, these unfortunates believed, they would have been enslaved while Rubruquis was still in Mongolia.

[2] It is supposed by Klaproth and Yule that Hayton, on his outward route, went *far to the north* of the road of Carpini and Rubruquis; but this does not seem necessary.

[3] *Erthij*, in Hayton's narrative, *i.e.* the Mongol *Ertsis*.

kept him as an honoured guest; it was not till the 1st of November that Hayton started on his return, by a route so distinct from almost all other mediaeval Itineraries as to deserve a more prominent place than has usually been given it. After a month's[1] travel the King reached a place called Ghnmsghur, now unknown, but evidently between Karakorum and Barkul;[2] next he passed through Berbalekh or Barkul and Bishbalig or Urumtsi; and in the neighbourhood of the latter he describes[3] a race of wild men scattered over a desert region, naked and without speech or language, whose heads were covered not with human but with horses' hair. In the same country were wild horses, of black and yellow colour, larger in size than the steeds tamed by man; wild camels, also, with two humps, and other marvels.

From the *Pentapolis* represented by the modern Urumtsi,[4] Hayton passed through a number of towns whose position is now conjectural—Yarlekh, Kuluk, Henkakh, and Jambalekh. The last-named, however, is an important point in nearly all the Chinese Itineraries of Central Asia belonging to this time; and it answers to their Chang-bala or Fung-jun-fu, on the River Zeng, somewhere between the Thian Shan and the Lake Bulkatsi Nor. The next stages, Khutapai or Khutaiyai, and Anki- or Yangi-balekh, were probably in the same river basin; and both must be looked for west of Jambalekh,[5] and perhaps in the plain of Borotala, near Lake Sairam. After this the narrative places the 'entrance

[1] Thirty days.
[2] One of the chief points in the road from Lake Sairam and the Balkhash basin along the northern side of the Thian Shan and over the Gobi Desert to China. Urumtsi, due west of Barkul, is almost midway between this point and Kulja.

[3] Almost in the same language as Rubruquis.
[4] Bishbalig, 'the Five cities,' the ancient centre of the Uigur nation.
[5] On the main route from the modern Khutukbai, north-west of Urumtsi.

of Turkestan,'[1] and the town of Phulat, Pulad, or Bolat—the reported station of the Teuton captives, whom Rubruquis so ardently desired to see, a leading place in the Central Asia of the thirteenth-century Celestials, and to be fixed with tolerable certainty in the immediate neighbourhood of Lake Sairam, the Sut-kol of Hayton. Thence the way led over the Borokhoro Mountains, a north-eastern branch of the Thian Shan, and (leaving behind the 'Sea of Milk' or 'White Lake of Tranquillity') it brought the traveller down into the basin of the Ili,[2] on the western side of the great dividing range, and so to the Town of Apples and Town of Serpents,[3] near the modern Korgos, to the west of the present Kulja. From this point Hayton crossed a range called Thoros (probably one of the outliers of the Russian 'Alexander' Mountains), and appeared before Hulagu— the future, but already designated, conqueror of Bagdad— at Talas,[4] midway between the Syr Daria and the Chu. To Hulagu,[5] as Mangu's brother, was entrusted the supreme command in the south-west of the Mongol Empire—the army of the East or the Levant, as Hayton calls it—and he was to take the leading part in the next great forward movement of Tartar conquest. In this movement, against the Moslem world, it is not unlikely that King Hayton was to aid; and his journey to the Imperial Horde may have been connected with the discussion of the anti-Saracen projects dear to both the allies.

[1] *Tsekopruk* and *Dinka-balekh* are the places first mentioned in Hayton's Record after the 'entry' into Turkestan. Neither can be fixed at all with certainty, or even probability.

[2] Hayton's *Ilan-su*.

[3] (1) Halu-balekh or Almalik; and (2) Ilan-balekh; (1) on the banks of the Alimatu, or 'Apple-bearer,' which joins the Korgos Gol, a northern tributary of the Ili; (2) the Chinese Ili-bala, still marked by ruins on the banks of the Ilan-bash-su, or 'river of the Serpent's Head,' an affluent of the Tsiu.

[4] Probably not far from the modern Aulie-Ata.

[5] 'Hulavu' in Hayton.

From Talas,[1] Hayton turned north-west, following in the main the course of the Syr Daria, but keeping on the further or north-east side of the Kara-tau range, to Barchin and Sengak or Signak, on the 44th parallel of north latitude and a little above the modern Julek, in the Valley of the Jaxartes. Here Mount Thoros [2] began, and here was the old home of the Seljuk Turks; here also Hayton left his road and went off to meet that 'religious Christian,' Prince Sartach, who was on his way to Mangu, to receive the investiture of the Kipchak, according to one story; to arrange for taking part in the attack on Bagdad, according to another. After this digression, Hayton returned to Signak and retraced his course up the valley of the Syr—but this time in the basin of the river itself and along the south-west side of the Kara-tau—passing through the 'large city' of Savran, Yasun or Hazret-i-Turkestan, Otrar the famous, and Jizzak on the south-west of the Jaxartes, where he again reached a land well known to all Westerns, the country of Samarcand and Bokhara. Still moving south-west, he crossed the Oxus [3] (apparently near Charjui), and by way of Merv and Sarakhs, entered Khorasan, and approached the frontier of the *Mulehet* or Assassins, whose destruction had been already decreed by Mangu Khan,[4] and whose end was

[1] Up to this point, from Urumtsi, he had been following a mainly westerly direction, with only a little south in it.

[2] The Kara-tau is evidently meant; it is not really separated from the Alexander Mountains with which it unites by the Chatkal and Chichkan ranges.

[3] The other places he names between Talas and Signak: *Khutughtchin, Barkant, Sulghan, Urusoghun, Kayi-kant, Khuzarkh* or *Kamots*, and *Khendakhoir*, are mostly now unknown, even by guess, but they were probably along the north-east side of the Kara-tau. Hayton's course in this section of his route must have been very similar to Carpini's. On Otrar, cf. John de Plano, pp. 305-6 of this volume.

[4] While Rubruquis was still at the Mongol Court, Mangu was warned of an Assassin emissary then on the way to remove him, and in consequence the extirpation of the sect was immediately ordered, and executed in 1258.

not long delayed. Skirting the southern coast of the Caspian, and traversing the boundary province of Mazanderan, the Armenian prince seems to have visited the great cities of Rai (or Rhagae, near Teheran), Khazovin (Kazvin, or Kasben), and Tavrez (or Tabriz); thence crossing the Aras, he reappeared at the Court of Bachu, which had apparently now been moved from Kars to the fortress of Sitiens, near Lake Gokcha.[1]

Thanks to the powerful aid of his Mongol allies, Hayton's journey was rapid; it was only eight months after quitting the Horde of Mangu that he reached the borders of his kingdom (A.D. 1255); and here he at once resumed the powers of government. He continued to exercise them until his abdication in 1269, three years before his death; his scribe Kirakos, the actual compiler of the Royal Itinerary, died in the same year as his master (A.D. 1272); and thus passed from the world an intrepid and sagacious prince, who by his timely friendship with the Power in being, may have helped to avert a terrible storm from Christendom, and whose great journey, if recorded by a competent hand, might have ranked with the narratives of the two chief Friar-travellers of the West.

The scribe concludes with a glance at a few of the good stories which King Hayton was not afraid to relate on his return, and among these none was more curious than the following, in which modern students have easily recognised the mystical tenets of Buddhism. The idolaters of Cathay worshipped a clay image called *Shakemonia*. This image re-

---

[1] A substitute for Bachu or Baiju, one Khoja Noyan, seems to have taken the command of the main Tartar army at Kars. Both going and returning Hayton stayed in the house of one 'Kurd,' a Christian prince of Armenia, in the village of Vardenis, near Mount Arai and the celebrated patriarchal monastery of Echmiadzin.

presented a deity who had ruled the world for three thousand[1] years, and would still rule it for a hundred times that period, or even more ; at the end of these ages he would be replaced by another god, Madri ; and these names evidently represent the deified sage Sakya Muni and the coming Buddha,

[1] 3040 exactly, thus putting Gautama's birth about B.C. 1785 ; the period yet to elapse was 35 tumans (each tuman being equal to 10,000 years). The name of Sakya Muni is not recorded by any other Christian traveller or writer of this time.

Brother Benedict, the Pole, whose short summary of Carpini's journey was edited by Avezac in vol. iv. of the Paris Geographical Society's *Recueil de Voyages et de Mémoires*, 1839, pp. 774-779, adds very little to the narrative of John de Plano. The principal points have already been noticed in connection with Friar John's account. We may add here—(1) that according to Benedict 'another' Franciscan accompanied Carpini from Lyons, on Easter Day, 16th April, 1245. This was Stephen of Bohemia, as is shown by Wadding, *Scriptores*, 221. (2) Benedict joined Carpini, 'to act as interpreter,' at Breslau in 'Poland.' (3) Between Kurancha (Curoniza) and Batu (Bati), the journey took more than five weeks 'from the Sunday *Invocavit* to the Thursday *Coenae Domini*,' i.e. from the first Sunday in Lent, 26th February, 1246, to Holy Thursday, 5th April. Here is a difference of one day from Carpini's reckoning. (4) The great river Ethil, called Volga by the Russians, is believed to be the Tanais. (5) The Friars' presents to Batu were forty beaver and eighty badger skins. (6) The Friars refused to worship a golden statue of the Emperor or Great Khan, which stood in a cart at Batu's Horde, but were compelled to bow the head. (7) The Mission left Batu for Kuyuk on Tuesday after Easter, 10th April; Carpini says 8th April. (8) The travellers wrapped their legs in bandages to bear the long rides better. (9) Komania was full of wormwood (absincium), 'as Ovid speaks of Pontus, once the name of this very region.' (10) In travelling through Komania, the Friars had on their right the Saxi, believed to be Goths and Christians (in the Crimea); then the Christian 'Gazars,' in whose country was the rich town of Ornarum, flooded by the Tartars (apparently in a different region from Carpini's 'Ornas'); then the Christian 'Circasses' and Georgians. (11) In Russia they had on the left the Pagan Morduans (Mordvins), Bylers, and Bascards, 'the ancient Ungari'; then the Dog-Headed folk, and the Parossits (Parocitae), who had mouths so small that they could only suck up liquid. (12) The river Ural (Jagac) divided the Komans from the Kangitae ; in the latter's country were huge salt marshes which the Friars took to be the Maeotid Fens. (13) Thence through Turkey (Turkestan), where they found the great city of Ianckint (Yengui-kend). (14) The Pagan Kara-Khitai were once masters of the Tartars. (15) The Mission stayed four months near the

Maitreya, or Maidari, who was still waiting his time. The whole nation of the idolaters of Cathay were ordained priests, shaved their heads, wore a yellow dress, and lived according to a rule of moderation; they did not, however, abstain from marriage, which was the legal status of all these *Tuin* between the ages of twenty and fifty. Beyond Cathay, moreover, as the traveller had been told, there were to be found men like great hairy dogs, and women endowed with reason, a startling contrast to their sisters of the civilised world. There also was a sandy island, rejoicing in the possession of a sacred bone, called the Fish's Tooth, and shaped like a tree; and many another marvel was there, of which the royal traveller could tell, but which his secretary did not think fit to put in writing. As in the case of the Jewish wayfarer, Petachia of Ratisbon, a bald summary is all that has survived of an Itinerary little inferior in value and interest to any other, even of this time of freshly awakened activity and boundless enterprise.[1]

Imperial *tent* called Syra Orda. (16) Kuyuk was elected by 5000 mighty men, and received 3000 ambassadors. (17) In formal audiences the Friars had to put baldakin (purple tissue) over their gowns, like all other envoys. (18) The Syra Orda and its appurtenances described—the Imperial daïs, the screen-work above the daïs, the separate flights of steps. (19) The mother of the Khan received the Friars with kindness and courtesy. (20) The warlike Georgians, named after St. George their patron, were respected by the Tartars for their courage and strength: some of these Georgians frequented the Friars' society. (21) On their return the Friars were accompanied fifteen days' journey westward by the envoys of the Sultan of Cairo ('Babylon'), who afterwards struck off southward. (22) The Friars, on their way home, crossed the Rhine at Cologne.

[1] See additional note on p. 642.

CHAPTER VI

COMMERCIAL TRAVEL

WITH the fall of the Roman Empire in the West the maritime trade[1] of the Mediterranean rapidly decayed, and almost down to the commencement of the Crusading Era the commercial life of European Christendom, the heir of Rome, was surprisingly feeble; but the central period of the Middle Ages is marked by a mercantile development so great as to create a new era in politics and society; and in this time the trade of the great inland sea begins to spread into regions beyond the Mediterranean basin, thus preparing the way for the discovery of yet more distant regions. For the present chapter it will be necessary, first, to summarise the history of this development on the shores of the *Mare Internum* itself; and, secondly, to trace the connections of European commerce—through Egypt and Syria, the Euxine coasts and Asia Minor—with Central, Southern, and Further Asia, where the Western traders laid the foundations, or at least built the scaffolding, of an ever-growing Western influence in the future.

---

[1] The best account of mediaeval trade in the Mediterranean and adjoining lands is certainly that of Heyd, especially in the French edition, *Histoire du Commerce du Levant au moyen age*, 1886, superseding Depping, *Histoire du Commerce entre le Levant et l'Europe;* but a very meritorious study of the same is to be found in Wappäus' unfinished work on *Heinrich der Seefahrer;*—in full, *Untersuchungen über die . . . Entdeckungen der Portugiesen unter Heinrich dem Seefahrer*, 1842; ii. Abschnitt: *Skizze des . . . Seehandels. . . .*

It is in Italy that the commercial (like the intellectual) activity of mediaeval Europe is habitually centred; and the beginnings of the new mercantile life are best seen in the rise of the Italian maritime cities, through which the Imperial land renewed its youth even in the most troubled period of the Dark Ages. The Byzantine reconquest in the sixth century of so large a part of the Western Mediterranean did little to restore the activity of commerce. For the policy of Constantinople was not calculated to encourage trade; it often preferred to encourage extortion, monopoly, and the crudest forms of protection; and in spite of the matchless situation of the city on the Bosphorus, the masters of that city, with rare exceptions, looked on calmly at the advances of less favoured and more industrious rivals, acquiesced in their own growing infirmities, and were at last content to lean upon Italian traders for the very necessities of mercantile supply. Thus foreign ports came to control the finances and even the politics of the empire; acquired possessions on every coast-line of the Byzantine waters; guarded their factories by forts, and their forts by fleets, in the Archipelago, the Dardanelles, and the Bosphorus; and at last made the successor of Constantine a pensioner and a dependent of the haughty merchant princes of the West.

In the earlier Middle Ages the trade between Constantinople and the West of Europe seems to have been strictly limited; in periods of military activity (such as the reigns of Justinian, of the early Isaurians, or of Nikephoros Phokas and his successors) munitions of war, stores of food and clothing, and troops of mercenaries or volunteers, passed to and fro; but in quieter times the imperative needs of the Eastern Empire were usually satisfied from the corn lands of the Black Sea and the varied supplies of Asia Minor, Greece, and the Ægaean Islands. In this state of commerce there was little of that 'modern' spirit so active in mediaeval Italy, creating new wants, stimulating production, and multiplying consumers; there was indeed more of such activity in the dominions of the Eastern and Western Caliphs than of the Byzantine Caesar during this time;[1] and the conquests of the Saracens established in Syria and Egypt a more liberal commercial rule, and gave the central markets of the Levant a freer access to countries which supplied, and were supplied by, them.

The new mercantile life of Christian Europe began in the north-east and south-west of Italy during the ninth century. It arose in various maritime states, where local freedom had been achieved, and where a victory had been gained over the anarchy as well as the petty tyranny of the time. Thus from the beginning it was both liberal and ambitious, and in some ways it possessed a more highly developed character than most of the trade of the ancient world. For it claimed greater privileges on behalf of the trader, as such; it attached higher, and indeed supreme, importance to the free circulation of trade; it was more daring and speculative in its operations; it was less content with merely supplying articles of primary necessity; it dealt with conveniences and luxuries in an ever larger degree; and above all, it devoted a more strenuous and inquisitive energy to the discovery and conquest of promising markets, however remote.

In the upheaval of the Crusades, the first signs of the coming movement are given by the attacks of Genoa, Pisa, and the Christian fleets of South Italy upon the Saracens of Sicily, Sardinia, Corsica, and North Africa. The main line of communication between Christendom and its new outposts in the Crusading States of the Levant passed over the sea, and was almost entirely maintained by the maritime re-

[1] The pre-Crusading period.

publics of the Mediterranean. Their fleets supplied what has been called the *bourgeois* element in the Crusading armies; they also secured that indispensable command of the sea by which alone most of the Syrian ports could be reduced. The commercial cities of the West, moreover, had not only a part in the foundation, but also a great share in the organisation, of the Latin kingdoms and principalities in the Orient. Alongside of the normal feudalism which prevailed in these states was the republican constitution and active mercantile life of Pisan, Genoese, Venetian, Massiliot, and other colonies, planted in the heart of the freshly conquered harbour towns—Acre, Beirut, Tripoli, or Jaffa. The heyday of the Crusading power on land was also the heyday of the maritime prosperity of many Western cities; and at the conclusion of the religious wars the main result of the struggle was clearly to be found in the expansion of Christian trade, and in the assimilation of no small part of Oriental and 'Moorish' civilisation. For the meeting of East and West in this tremendous conflict brought little permanent gain to Europe and the Catholic world, in the political sense; through the medium of commerce, on the other hand, it directed the energies of the Christian nations to their true future. The frontal attack of the Crusaders was unsuccessful, but the Crusading struggle imparted a new culture and material prosperity, an increased knowledge, an immensely extended wealth, a restless but obstinate ambition, whose results were seen in the Renaissance of the fifteenth and sixteenth century, in the great discoveries both of geography and natural science, and in the final triumph of European arms and enterprise throughout the world.

In all this may be recognised the working of a spirit called out and fostered by the Mediterranean trade-republics; but ultimate success is only realised through the agency of more powerful and unified states, more adequate representatives of the collective strength of Christian Europe. The trade republics of the Middle Ages are nearer to the commercial life of the modern world than the Old Empire; but the Old Empire must be reproduced in the great modern nations before the commercial expansion could become a political and racial victory. The work of Venice and Genoa, Pisa and Amalfi, necessarily pass in the lapse of time to France and England, Spain and Portugal. The narrowness of resource, the lack of high political aim, the factious littleness of the maritime cities, prevent their assuming a part suitable to their wealth and industry, as the field of action broadens at the close of the Middle Ages. The ablest of their citizens, putting forward schemes too great for the efforts of a town, are obliged to carry their suggestions elsewhere; thus Columbus, Verrazzano, and the Cabots give to Spain, to France, and to England, new dominions beyond the sea which patriotism would have offered to Genoa, to Florence, and to Venice.

Venice in the north-east, and Amalfi in the south-west of Italy, were the first leaders of a commercial movement which cannot well be traced higher than the ninth century.[1] Like Naples and Gaeta, Amalfi still belonged in name to the Eastern Empire, but as early as about 820 it had secured free burgher rights and the position of a practically independent city-state. Twenty years later its mercantile marine is noticed in contemporary chronicles; its successes in the defence of Rome and the incessant naval war against the Saracens (notably in 847, 876, and 890) made it at the close of the Carolingian period[2] the foremost champion of

---

[1] But the lists of doges, practically independent republican chiefs, nominally vassals of Byzantium, begin in both cities much earlier, about 700.

[2] About 850-900.

Christian trade and sea power. At the opening of the tenth century, the Amalfitans possessed factories in Palermo, Syracuse, Messina, and other cities of Sicily, in Durazzo on the Eastern Adriatic coast, and in Constantinople. In the next two hundred years they developed a respectable trade with Egypt, Syria, the Byzantine Empire, and the Moslem world; at Antioch, even before 1098, they obtained a concession[1] or 'quarter,' similar to those afterwards granted by the Crusading conquerors; at Jerusalem they obtained a similar and more memorable privilege. By treaty with the Fatimite Caliph of Cairo they erected in 1020 a church, and a little later a hospital, in the Holy City; within this precinct, they exercised an undivided control down to the conquest of 1099, when the *College* (or *Society*) *of Amalfitan merchants* became the military and monastic order of the *Hospital*.[2] During the Crusading period the prosperity and power of Amalfi declined, and the Republic was only successful in obtaining trivial concessions from the new Latin masters of Syria; as in Laodicea (or Latakiyeh) from Bohemund III., in Tripoli from Raymond III., in Acre from the King of Jerusalem. The last-named was their principal settlement in the Levant; but even this was a small matter; and the Amalfitan commerce with Cyprus and Egypt in the twelfth and thirteenth centuries was a depressing contrast to that earlier period, when the *Defender of the Faith*[3] was the bulwark of Italian Christendom against Saracen attack, the mistress of the chief Mediterranean navy,[4] and the principal Western market for the products of Barbary, Egypt, and the Orient,—when the maritime laws of the *Tabula Amalfitana* were current among traders

---

[1] Including a hospice.
[2] Otherwise the Knights of St. John of Jerusalem.

[3] Title conferred on Amalfi by Pope Leo IV., 847-855.
[4] At least west of the Adriatic.

on every coast of the Inland Sea,—and when the coinage of the Republic was the chief medium of exchange between Latin Europe and the Levant.[1]

Prima dedit nautis usum magnetis Amalphis, said the tradition which ascribed to Flavio Gioja[2] of this city, the 'invention' of the compass; and although it is perfectly clear that the polarity of the magnet-stone, and the quality of definite direction residing in magnetised iron, had been noticed before the time of Gioja in France, Italy, and England, it may well have been a citizen of Amalfi who first brought the magnet (fitted with its compass card) into popular use, first among the mariners of his own state, and afterwards among other seamen of Southern Europe.

Meantime in the first half century of the Crusading age, 'Melphi' had fallen, like Naples, Salerno, or Brindisi, under

---

[1] On the Maritime Laws of the Mediterranean in the Middle Ages, cf. the Code termed *Il Consolato del Mare*, which, though claimed by some as Amalfitan, by others as l'isan or Barcelonese, has been referred to the tenth, the eleventh, and the thirteenth centuries by various students: it may contain matter from each of these periods; cf. Boucher's French translation of *Il Consolato*, i.72; Muratori, *Dissertation* 45; and Henry IV.'s charter to the Pisans of 1081 ('consuetudines, quas habent de mari, sic . . . observabimus'). Capmany refers the Code in its final shape to the merchants and lawyers of Barcelona under James the Conqueror; cf. *Codigo de las costumbres maritimas de Barcelona*, 1791.

On the prosperity of Amalfi, cf. William of Apulia, book iii. of his *Poema de rebus Normannorum*, given in Muratori, *Rer. Ital. Scriptores*, v. 267, and *Dissertation* 30.

'Urbs haec dives opum, populoque referta videtur, Nulla magis locuples argento, vestibus, auro. Partibus innumeris ac plurimus urbe moratur Nauta, maris coelique vias aperire peritus. Huc et Alexandri diversa feruntur at urbe, Regis et Antiochi. Haec . . . freta plurima transit. Hic Arabes, Indi, Siculi noscuntur et Afri. Haec gens est totum prope nobilitata per orbem, Et mercanda ferens et amans mercata referre.'

This is curiously parallel to Donizo's description of Pisa, but apparently more exaggerated. The smallness of the port shows that Amalfi could never have had a very great trade.

[2] Otherwise Gisia, born at Positanum, a town in the 'Dukedom of Amalfi,' and supposed to have made a discovery of the *pyxis nautica* in about A.D. 1320, or later,

the land-power of the Normans, while at sea it had undergone a decisive overthrow. The aggressions of Guaymar of Salerno and of the Norman conqueror, Robert Guiscard, did not permanently destroy the liberties of the place, but they were symptoms of coming destruction on one side, like the overshadowing growth of Pisa and Genoa on another. Roger II., King of Sicily and Grand Duke of Apulia, the patron of Edrisi, the fine flower of Christian culture among the sovereigns of the time, was execrated by the republicans who had survived the usurpations of Guaymar and Guiscard; for in 1131, on the surrender of the city, its constitution was sacrificed, the town was incorporated without distinction in the other dominions of the 'despot,' and its marine was committed to a fatal struggle with the Pisan enemies of Roger. The defeats of 1135 and 1137, accompanied by the sack of the port[1] itself, were symbolic of a decay that had already begun; trade had for some time been passing into other hands; and alike at Amalfi, at Salerno, and at Trani, the Norman rule, a boon elsewhere, could not hinder the slow progress of decay.

Yet in spite of this general and rapid decline, the merchants of Amalfi retained their settlement at Constantinople down to the overthrow of the Byzantine state in 1204; subject to the payment of a certain tribute to Venice,[2] they carried on that valuable trade in silk from the market of the Bosphorus which they had been among the earliest Latins to pursue; and their monasteries in the Imperial City, in Jerusalem, and at Mount Athos, long remained to them in spite of the ill-will of the Orthodox. Their factories and colonies in Sicily, and especially at Messina, had arisen on the Christian recovery of that rich island by Norman arms; the prosperity of these factories withered in the fierce commercial

[1] 4th-5th August, 1135.   [2] From A.D. 1082.

struggle with the greater harbours of Tuscany; but, like the Amalfitan *fontecci* in the Nile Delta, they seem to have retained some vitality into the thirteenth century. The Norman rulers, who could not permit the political independence of Salerno and Amalfi, were anxious to promote the interests of these towns abroad; thus in 1137, Roger II. promises to obtain a reduction of the dues paid at Alexandria, at least to the level of the tax on Sicilian goods.

At the sack of 1135, the Pisans were (falsely) said to have carried off a unique copy of Justinian's Pandects; they certainly bore away and transferred to themselves the glory and prosperity of their rival; but for these calamities there was some compensation in the gain of an important relic, the body of St. Andrew, translated to Amalfi in 1208; and as late as 1258 Amalfitan traders apparently continued to make voyages to Alexandria and Acre.

While Amalfi in the South-West was passing out of sight, Venice [1] in the North-East was rising to a position more lofty, to a power more extensive and sustained, than any other of the mediaeval trade-republics. In the Ostrogothic Kingdom of Theodoric and his successors, as under the rule of the

[1] On this Mediterranean trade of Venice in earlier Middle Ages, cf. Marin, *Storia civile e politica del commercio de Veneziani*, especially i., 118, 120, etc., 126-127, 145, 176; ii. 47, 59, 63, 162-164, 210, 216, 220, etc., 230, 232, 237, 246, 263, etc.; 256, 293-295; iii. 25-28, 62, etc., 66, 53, 201, 109-128, 167, 282; iv., 8, 246, 208-218, 253-254, 298-302; vi., 337. Also see Dandolo, *Chronicon*, in Muratori, *Rer. Ital. Scriptores*, vol. xii.; and the description of William of Apulia in book iv. of his *Poema de rebus Normannorum*, given in Muratori, *Rer. Ital. Scriptores*, v. '[classem] populosa Venetia misit, Imperii prece, dives opum, divesque virorum, Qua sinus Adriacis interlitus, ultimus undis Subjacet Arcturo. Sunt hujus moenia gentis Circumspecta mari, nec ab aedibus alter et aedis Alterius transire potest, nisi lintre vehatur. Semper aquis habitant. Gens nulla valentior ista Æquoreis bellis, ratiumque per aequora ductu.' Also see Marino Sanuto, *Vite dei Duchi di Venezia*, in Muratori, *Rer. Ital. Scriptores*, xxii. (already reprinted in the new Muratori, whose reissue from original MSS. has lately begun).

Byzantine exarchs who displaced the Ostrogoths, Ravenna was the official capital of Italy; its political importance was for a time accompanied by the commercial greatness of a port from which the sea has now retired. But within a hundred years after the victory of Narses, the artificial trade of Ravenna was already declining, all the more rapidly as a genuine commercial state of the first order was rising to prominence in close neighbourhood. The commerce of Venice was originally concerned with the fish and sea-salt which the boatmen of the lagoons found ready to their hand; but the beginnings of a wider traffic were not long in appearing. From the time of Charlemagne the Queen of the Adriatic began to take a place in the politics as well as in the commerce of the Latin world. Its situation had advantages beyond any other harbour town of Italy. Separated from the mainland by the sea, and from the open sea by the low fringing walls of its lagoons, surrounded almost entirely by shallows pierced only by a few deep channels, Venice was usually considered by its own citizens, as by foreigners, to be beyond attack; the political troubles of the Continent made it a refuge from the time of Attila; and the absence of any serious maritime rival on the Adriatic left open a valuable and extensive field of operations for commerce, for colonisation, and even for conquest.

As early as the eighth century the infant republic had not only begun to take its permanent constitutional shape; it had also adopted, consciously or unconsciously, the main rules of its policy in external affairs. No mercantile state has ever pursued a more constant, a more deliberate, or a more prudent course of action; and as in the case of Russia, and the fictitious will of Peter the Great, men invented some Testament, or Digest of fundamental principles, which they supposed to have been left by the earliest chiefs of the Republic, as their most sacred legacy to their successors. It is not necessary to suppose any formal document; but it is necessary to realise that, very early in its career, the Venetian State had reduced to practice certain rules of policy, which all experience confirmed. Among these rules the one that chiefly concerns us is that which goes to the root of all, an absolute devotion to trade and maritime intercourse, at the expense of all other interests, save only the independence of the city itself. The principles of siding always with the stronger, especially in maritime struggles, of maintaining an exemption from excessive papal interference, and of remaining on friendly terms, both with the Western and Eastern Empires, both with the Franks and with the Byzantines,[1] were, in the nature of things, liable to modification, and in later times we find Venice acting in clear opposition to some of them; but however things might change, the Republic always endeavoured to render itself indispensable to every Christian State which possessed a powerful frontage on the Mediterranean sea-board. Thus under the Doge Orso Ipato (726-737), Venice assisted the exarch Eutychius against the Lombards, and helped the Byzantines to recover Ravenna. In recompense for this her citizens obtained from the Greek Emperor many valuable rights of trade in the *Pentapolis*. Increased trade enabled them to strengthen their marine, to drive Ravennese competition off the Adriatic, and to combat their most troublesome enemies, the pirates of Dalmatia.[2] Their ever-growing prosperity now enabled them to appear in the first rank among the foreign mer-

---

[1] But above all with the latter, whose side the Republic must espouse, if unhappily forced to take part in war between the two Empires.

[2] Venetian dominion begins in Dalmatia, about 997, greatly increases during the eleventh century, and in 1082-85 is confirmed to the Republic by Alexius Comnenus in reward for aid against the Normans. From 1052 the Doge takes the title of Duke of Dalmatia.

chants who resorted to Constantinople; they were wise enough to side with Charles the Great in his Italian campaigns; and the help given by them at Pavia and elsewhere was rewarded by extensive privileges in the new Germano-Roman Kingdom. In 864 they won decisive victories over the Dalmatian sea-robbers, and about 870 the Doge Ursus gained as complete a success near Taranto over the Saracen pirates who then threatened the mouth of the Adriatic. Both the Eastern and Western powers of Christendom were grateful for this deliverance, and among other rewards they gained from the Byzantines a new 'indulgence'—free trade in the Black Sea. Owing to Saracen conquests and Saracen piracy in the Mediterranean; owing, also, to the control of the Euxine still possessed by the great Eastern State of Christendom; owing, lastly, to the continued importance of the ancient trade road from Central and Further Asia, along the Northern and Southern walls of the Caucasus, to the ports of Colchis and Lazica;—this concession was of peculiar value, and it was utilised to the full.

Liudprand or Liutprand, the brilliant and querulous Bishop of Cremona, who visited the Byzantine Court as the ambassador of Otto I. in 950 and 968, found a large number of Venetian citizens in the Greek Army and of Venetian ships in the harbour of Constantinople; their merchants exported the silks and stuffs nominally confined to the Empire by the regulations of the Custom-house; and the transport of letters between North Italy, Germany, and the Christian Orient was entrusted to the vessels of the Republic, until in 960 an angry political correspondence between the West and the East caused the abrupt suspension of this service by the Doge Pietro Candiano (IV.).

Down to the middle of the twelfth century, when the fatal quarrel with Manuel Comnenus altered the whole direction of Venetian policy, the Byzantine alliance and the Byzantine commerce were assiduously cultivated; but even this field did not exhaust the energies of a rapidly-growing marine; and in pursuit of fresh markets, Venice, like Amalfi, was ready and eager to traffic with the enemies of Christendom. Even in the time of Charles the Great there is evidence of a clandestine intercourse of this sort with Syria, Egypt, and North Africa, soon after detected and forbidden by Leo V., the Armenian.[1] The prohibition was naturally evaded; and when (in 971-972) a fresh attempt was made by the warrior-prince John Tzimiskes,[2] in alliance with the Pope, to render it effective, only the trade in ship-timber, iron, and military material was called in question. But even this, however readily conceded in theory, was difficult to enforce in practice, unless all intercourse with the Saracens was wholly broken off; and this commerce, though deeply repugnant to the moral sense of Christendom, and a standing danger to the *entente cordiale* with Byzantium, seems never to have been suppressed. Within twenty years after the compact of 972 the Doge Pietro Orseolo[3] sent ambassadors to 'all' the Moslem princes, at the same time that he despatched a special embassy to Constantinople, and obtained a reduction of dues from the conqueror Basil II., in return for the free use of the Venetian fleet, now, and in all future time, by the Byzantine troops passing over into Italy (March, 992). The same Doge, by

[1] At least as far as Syria and Egypt were concerned.

[2] The power of Constantinople at this time was shown in John's threat of burning every ship found transgressing this rule, and in the immediate stoppage of three vessels which were sailing for Tripoli in Africa, and for El Mehdia, the port of Kairwan, regions with which the Byzantines had little or no immediate concern.

[3] Pietro II., 992-1009. Perhaps only the Moslem Courts near the Mediterranean coasts are meant, viz., Aleppo, Damascus, Kairwan, Palermo, and Cairo. Cordova is more doubtful, and Bagdad improbable.

his victories over the Croatian and Dalmatian pirates (in the year 1000), made Venice the undisputed mistress of the Adriatic; and by his treaties with Otto III., he renewed and extended the ancient commerce of his people in the dominions of the Holy Empire. Germany and the North Italian Upland were dependent on the Republic for the products of the East; and when in 1017 four Venetian ships laden with spices suffered shipwreck, the event is noticed by a German chronicler[1] as a serious misfortune.

From its strict observance of a wise neutrality the State reaped benefits in certain quarters and at certain times; from its almost unreserved allegiance to Byzantium in the time of Byzantine revival (960-1020) it prepared itself to profit yet more in the beginnings of Byzantine decline. Various possessions of the Eastern Empire, especially on the Adriatic coasts, were resigned 'on trust' to so staunch an ally; and from the new acquisitions were derived not only strength of position, but increase of material resources, a fresh supply of good seamen, and stores of excellent wood for shipbuilding. With the reign of Alexius Comnenus, Venice reached the height of its ambition as the friend, the indispensable and favoured ally—not yet the supplanter—of Orthodox Constantinople. The fleet of the Republic played an important part in checking the progress of Robert Guiscard, who had already stirred up the Dalmatian coast against Venetian rule, and allied himself with Ragusa; the services thus rendered were great; and great were the rewards. It was now that the Amalfitan colonies in the Greek Empire were put under a tribute to the Church of St. Mark; it was now that all the chief ports of the Byzantine coasts and the principal markets of the interior were opened to the traders of Venice:—Latakiyeh and Antioch in Syria; Mopsuestia,

[1] Thietmar, in Pertz SS. III., 860.

Adana,[1] and Tarsus in Cilicia; Attalia or Satalia in Pamphylia; Strobilos in Caria; Chios, Ephesus, and Phocaea in Western Asia Minor; Heraclea and Selymbria on the Sea of Marmora; Chrysopolis near the mouth of the Strymon; Demetrias on the Gulf of Volo; Abydos on the Hellespont; Adrianople and Athens; Thebes and Thessalonica; Negropont, Nauplia, and Corinth; Corfu, Avlon, Durazzo, and several other harbours on the western coast of Greece and Epirus.

Thus, in the years immediately preceding the first Crusade, Venice had attained to a dominant position in all the waters of the Byzantine Orient, and could look down upon every rival. Amalfi, which alone could equal its antiquity, was clearly falling out of the race, and was now stamped by the law of Constantinople as the inferior and tributary of Venice within the Eastern Empire. Pisa, Genoa, and the other commercial republics were but *parvenus* beside the Bride of the Adriatic; their liberties, their institutions, their maritime developments were younger; their achievements in the great field of Mediterranean action less considerable. It was not unnatural that the Venetians should presume on their success, and take less advantage of the great movement of the Religious Wars than others, less wealthy and prosperous, who saw in this convulsion an unique opportunity for advancement.

Yet it was, perhaps, surprising to the rulers and citizens of the maritime states to find how important the over-sea routes became in the course of the Crusading struggle. The landways so extensively used on the first expedition of 1096(-99), proved wearisome, uncertain, and dangerous, con-

---

[1] Properly Adatia. Antioch was in Christian hands from 968 to 1084; and was recovered by the Crusaders in June 1098. Here, in about 1070, the Venetians procured the release of a Serbian prince, Constantine Bodinus; here also they met with citizens of Bari, who afterwards stole the relics of St. Nicholas from Myra in Lycia.

suming the time and dissipating the energies of the bravest warriors. For regular intercourse between the Latin conquerors in the Levant and their friends in Western Europe— for a regular system of supplies from the base to the outposts— naval communication was soon approved as the best possible by all concerned—even though in that age it might truly be said that no landsman went to sea unless obliged to do so, for a voyage was being in prison with the additional chance of being drowned. But the experience of the armies of Godfrey of Bouillon, of Bohemund, and of Tancred, overthrew Latin confidence in the Byzantines; and with remarkable energy the younger maritime states of the Mediterranean seized the chance thus offered them. If the Holy Places could only be held with the aid of ships and sailors, harbours and transports, the republics of Italy, of Provence, and of Catalonia would at least make this route an effective one. By serving the cause of Christendom they served their own; they multiplied, many times over, their carrying trade; they largely increased their export and import commerce; above all, they acquired a privileged, a more than half political, position on the coasts of the Levant. As time went on, and they became more indispensable to the Crusading princes, they were able to dictate their terms more freely, until the main burden of the Holy War rested upon them as the chief holders of power.

The concessions won by the mercantile cities of the Western Mediterranean in the eastern portion of the Inland Sea were in close geographical agreement with their respective fields of operation, with the sphere of interest claimed by this state or that.[1] Venice, for example, played a small part in North Syria, where Genoa and Pisa dominated; but from Tyre southwards Pisa was of no account; and in all lands

Cf. the Charter of the Ancona concession at Acre, 1257.

effectively subject to the Byzantine power no serious competition disturbed the primacy of the Old Ally, until the friendship of centuries was ruined by a short-sighted and intoxicated greed in the time of Manuel Comnenus.

The concessions granted to the Western traders in Levantine ports, and especially those of the new kingdom of Jerusalem, were of various kinds; so far as they applied to land, they might consist of scattered properties or small estates lying out of the towns, and sometimes at a considerable distance up-country; within port-towns, such as Acre, Jaffa, or Tyre, the grant was sometimes of a piece of waste land free for building; otherwise it might include a certain number of houses, an entire street, or even a large portion of the city. The obligations attached to such holding were usually defensive; offensive military service was rarely mentioned; but the grantees were bound to guard their own property against attack, and could not claim the aid of the feudal levies, except as a matter of grace and Christian brotherhood.

The cunning traders and seamen of the commercial republics looked on the Crusades with very different eyes from the average Catholic lords and labourers of the inland districts; they were not without religious enthusiasm, but they cultivated it rather as a useful commodity than as an inevitable state of feeling; and they felt but little of the blind hatred against Islam, and the passionate veneration for the Gospel sites, which sincerely animated the great body of the warriors and pilgrims they conveyed to Palestine. Mercantile interest was never absent from the minds of those who governed, bought, or sold in Venice and Pisa, in Genoa and Amalfi; and this will partly explain why the greatest of Italian ports, and the most worldly-wise of commercial oligarchies, did not embark in the new ventures with as much zeal as others who had less to lose. For the Venetians were far more anxious to preserve and improve their markets and their general predominance in the Byzantine Empire than to endanger their whole position by developing a Syrian commerce, viewed by the Greeks with no small jealousy. Yet their exertions were considerable. The competition of their rivals was alone sufficient to compel a certain amount of activity; and the friendship of Constantinople did not compel them to submit to a total exclusion from the new Syrian commerce. In 1099-1100 they sent a fleet of 200 ships to Jaffa; yet in return for their vigorous assistance, they only obtained from Godfrey and Baldwin a number of rather trivial mercantile privileges. Eleven years passed without any renewal of Venetian action on a large scale; but in 1110 an armament of 100 vessels helped (like King Sigurd of Norway) in the conquest of Sidon; in 1117 the mariners of the Republic gained a useful victory over a Saracen fleet near Jaffa; in 1122 another armada of 200 ships sailed to Syria, besieged Corfu on the way, arrived in time to rescue the Prince of Jerusalem from an awkward position, and at last compelled[1] the surrender of Tyre in 1124. Again, in 1153, Venice took part in the capture of Ascalon; both these ports[2] were special objects of Venetian ambition; and by agreements of 1117 and 1118, renewed in 1130, they obtained the promise of vast concessions in the same, proprietary rights over one third part of each city, and a large share of the customs revenue. By a vote of the Christian commanders at Acre in 1117 the Venetian people were also declared free of all dues and taxes in several of the recently-conquered ports,[3] but this (like some other charters) was too large a promise for easy fulfilment.

[1] They also helped in the capture of Haifa.
[2] Tyre and Ascalon.
[3] Especially in North Syria.

By about 1130 Venice had its trading factories in all the harbours of Palestine, and outside the proper domain of the King of Jerusalem they pushed their interests in North Syria; a considerable fraction of their trade was carried on with Aleppo and the Hinterland of Phœnicia through Laodicea or Latakiyeh; and both at Byblus and at Beirut they secured large concessions (when the Moslem tide was already returning) in 1217 and 1220.

Levantine travel, of a more extended, ambitious, and worldly character, was promoted by the growth of Western trade in these regions. Hitherto, at least in such stagnant times as the eighth century, pilgrims had been almost the only Europeans who journeyed in Syria, but now commercial interests drew the more daring traders over the Lebanon and the Jordan, and even (as the thirteenth century drew on) to Mosul and to Bagdad. About the middle of the Crusading period regular naval expeditions[1] were organised for the conveyance of peaceful and other voyagers from the Western Mediterranean twice in the year. The summer 'tour' usually began about the Feast of St. John Baptist, but was sometimes delayed into August or September; the earlier or spring fleet was due to start about Easter, though often subject to delays which ran well into the middle of May. Venice was the head-centre of this service, and profited more than any other city from its continuance; at the commencement of the thirteenth century the Magnificent Seignory authorised the addition of a winter sailing.

The Venetian aid given to Alexius Comnenus on the eve of the first Crusade was renewed in 1109, when Bohemund, the Norman, now Prince of Antioch, repeated his father's attack upon Durazzo, in alliance with the fleets of Pisa and

---

[1] Cf. Adam of Bremen on the voyage of pilgrims round from the North Sea to Marseilles in the later eleventh century.

Genoa. This aid was effective, and gained for Venice even more extensive privileges than before; the concessions in Pera, in Constantinople itself, in Durazzo, and in other cities of *Romania*, were enlarged; and an absolute immunity from dues and taxes, and from all but their own jurisdiction throughout the Empire, was bestowed upon the saviours of the New Rome. Hence resulted a natural increase of Venetian trade and power; now that one city enjoyed all the sunshine of the Byzantine Court, that city became ever more prominent as a link between Eastern and Western Christendom, and between the Christian world as a whole and the heathen world of the Orient. Incidentally, Venice now made a fresh start in commerce with Germany and the ultramontane lands; as the European staple of Mediterranean trade, she concluded an advantageous treaty with the Emperor Lothair III., and, intoxicated with past success and present ambition, she became reckless of her older cautions, and careless of the most valuable and ancient of her alliances. The friendship of Byzantium had been a rule of policy with the earlier doges and senators; to preserve that friendship the State had many times practised a self-denial irritating to the pride and damaging to the fortunes of many citizens; but in 1155 the Seignory passed into the open alienation of an unfriendly neutrality, refused aid to Manuel Comnenus in his war with William of Sicily, and negotiated a trade convention with the Norman. A common hostility to Frederic Barbarossa still united Byzantines and Venetians on one important point; Manuel had lately granted the latter certain fresh privileges, especially for the commerce with Crete and Cyprus; and the breach now created[1] was not

[1] Unless we follow Heyd (*Commerce du Levant*, i. 216-219) in supposing that the arrest of Venetians in Constantinople, 12th March, 1171, was simply

of his seeking; but he had already shown a disposition to admit other Italian traders to a share of favour, and this was an unpardonable offence. On the outbreak of the quarrel in 1155, as again on its renewal in 1171, the Roman sovereign showed something of the vigour of a Tzimiskes towards his troublesome Western 'vassals'; and punished Venice with a savage and unexpected energy, worthy of the last strong ruler of Christian Constantinople. Discarding the ingrates of the Adriatic, he immediately made terms[1] with the Genoese (1155), assigned them a quarter in the Imperial City, and granted them free trade throughout his dominions, with the exception of some Black Sea ports.[2] Pisa, Ragusa, and Ancona were also invited to take a share of the influence and commerce which Venice had sacrificed. At Ancona Manuel established stores and arsenals; at Zara and Pola, Spalato and Ragusa, his agents put fresh life into the anti-Venetian party; everywhere his hand was against the proud city, and heavy upon it. In reprisal, Venice incited Servia to make war upon the Byzantines, and aided William of Sicily in his struggles with Manuel's navy, and Frederic Barbarossa in his siege of Ancona. The quarrel was appeased for a time, only to break out again with greater violence in 1172; many Venetians in Constantinople and elsewhere suffered arrest and imprisonment; an immense quantity of goods was confiscated; and not until after the death of Manuel in 1180 was a hollow friendship restored.

The renewed prohibition of trade and intercourse between Christians and Saracens by the Lateran Council of 1179 was an advantage to Constantinople, which now

to satisfy the greed of the Emperor; was without any decent pretext; and was not caused, as Kinnamos says, by Venetian outrages upon the Genoese, insults to the Sovereign, and refusals either to apologise or to compensate.

[1] A treaty renewed in 1168-1170.
[2] Rosia and Matracha; cf. pp. 421.

again stood forth as the central market for the products of the East; the Venetians had lost more than they had gained from their quarrel with Manuel; and the accommodation of 1187 shows how many of their factories and settlements were placed on Byzantine soil and dependent upon a good understanding with Byzantium. At Philadelphia, Abydos, Rodosto, Adrianople, Plovdiv or Philippopolis, Saloniki, Thebes, and Corinth; in several of the Ægaean islands, such as Negropont, Chios, and Lemnos; above all, in Pera and Stambul itself, their interests were important, and in many cases beyond the reach of immediate support from the sea. Again, the recent war had enabled the Genoese to appear as formidable rivals upon the Bosphorus, the Pisans in Saloniki; German, French, English, Ragusan, and Amalfitan merchants had all flocked to the Queen of Cities in the later years of the twelfth century, during the season of Venetian eclipse; Anconitan, Provençal, Spanish, and Danish traders followed; even Armenians, Russians, Hungarians, Persians, Egyptians, and Syrians, Iberians of the Caucasus, and Turks from the Seljuk Kingdom of Rûm or Iconium, came to increase the prosperity of a market which Benjamin of Tudela extols as only equalled by Bagdad.[1] Venice had need of all its energy, and stooped to employ more than the usual subtleties of mediaeval policy, to regain lost ground. In particular, she watched the prosperity of Ragusa (trivial as it was by comparison) with keen suspicion and constant intolerance; asserting her supremacy in 1171, and on four occasions during the early thirteenth century; and doing her utmost to hinder the Ragusan intercourse with Nicaea and Trebizond (after the fourth Crusade), with the Tsars of Bulgaria (especially from

[1] Cf. Benjamin of Tudela on Constantinople; and Asher's notes, vol. i of his *Benjamin*, pp. 46, 49-51.

1197), and with the Sultans of Egypt, in whose port of Alexandria Rabbi Benjamin saw vessels from *Rakuphia*.

The nominal reconciliation of Venetian and Byzantine polity after the death of Manuel was not deep or lasting, as the terrible anti-Latin riots of 1182 sufficiently proved; many 'Adriatic' merchants resident[1] in Constantinople were then massacred; and the new Emperor Alexius (III.) failed to compensate the survivors for the losses they had suffered. What was worse than all, he continued to show favour to the Pisans and other dangerous competitors of Venice; generally speaking, the state of affairs was not satisfactory to those who had so long held a commercial monopoly of the 'Greek' dominions. Day by day, complained their enemies, the men of St. Mark pierced ever deeper into the 'Hemi-Pont' and Asia Minor; the commercial agreements of 1187 and 1189 showed a great practical advance on the 'privilege' of Alexius in 1082; but this was no compensation, in the mind of ambitious Venetians, for the increased success of their rivals. They did not hesitate, therefore, at a desperate stroke to recover their ascendency; and already in 1202 they were making ready for the great crime consummated in 1204 — the capture and sack of Constantinople and the overthrow of the dynasty. The decline of the Byzantine navy after the death of Manuel left the 'regions of Marmora' in a fatal state of exposure; in the last quarter of the twelfth century Pisan, Genoese,

---

[1] The Latin quarters, ἐμβολαι, or *embola*, all lay upon the Golden Horn; and the Venetian adjoined a part of the quay which perhaps had once been (before Theodosius II., 408-450) a Jewish leasehold in Pera. Here probably Rabbi Benjamin of Tudela lodged, if we may still suppose a Hebrew interest in this section of the city, as is implied by the Venetian concession of 1080, '*ab Hebraica ad Viglam.*' Heyd, however, denies the survival, *Commerce du Levant*, i. 249-250, and maintains that throughout the earlier Middle Ages the Jewish quarter of Constantinople was in the port of Galata (and of Galata only) adjoining the Bosphorus.

and other pirates desolated the coasts and islands of the Ægaean, and even ventured to within sight of the Bosphorus; and when Philip Augustus of France returned from Palestine in 1191 he found the Archipelago almost without inhabitants except for a few colonies of corsairs.

Venice, then, in 1204, was firmly planted at Byzantium; it was even proposed to transfer to the Bosphorus the seat of the Republic's Government; the Doge became Despot of *Romania;* the victorious city entitled itself Mistress of Two-Fifths of the Roman Empire. Pera was made an immediate possession, under a *bailo* or *podesta;* all the coast towns which could be secured shared the same fate, together with a large part of the Morea (valuable for its silk manufactures) Corfu, Crete or Candia, many small islands in the Archipelago, and a portion of Negropont. Thus a chain of Venetian possessions now stretched from the Adriatic to the Black Sea; a position of complete ascendency was secured in all the Greek waters, and the Republic was exalted to a rank among the world's chief powers. By the same conquests the Venetians acquired a command of the trade in Oriental and Arctic products, and especially in silks, spices, aromatics, and costly furs and stuffs; the commerce of the Empire was of all the greater importance, because by the Black Sea route came so large a proportion of the corn supply of the Mediterranean world. The new masters of the Pontus pressed their advantages in every way. They concluded treaties with the King of Little Armenia, and the Emperor of Trebizond; they opened an intercourse with the new Mongol power then overflowing Georgia and the east and north coasts of the Inhospitable Sea; and for the further development of their trade and influence they founded Tana,[1] at or near the mouth of the Don,

[1] At the very close of the Latin domination in Constantinople, probably not before 1260.

where in early times there had been a Graeco-Scythian market. Nor was this enough for Venetian energy and unscrupulousness. Gauging more accurately, as time went on, the extreme stubbornness of the Saracen resistance and the inherent weakness of the Latin states of Syria, unsupported by any large racial emigration from the West, the republicans who had assumed with so much unction the leadership of the fourth Crusade, soon gave up all pretence of devotion to the Holy War. On the contrary, they renewed their ancient and direct trade with the chief Saracen markets, and even entered into close relations with the principal Moslem sovereigns. Thus in 1219 the Venetian *podesta* in Constantinople concluded a treaty of commerce and friendship with Saladin of Iconium; in 1229 a similar agreement was made with the Sultan of Aleppo, an old and valued customer; and with the rulers of Egypt there was constant diplomatic and mercantile intercourse. The Egyptian connection is referred to by Saewulf, and illustrated by the treaties of 1158 and 1175, the latter signed by the great Saladin and the Doge Sebastian Ziani; it was cemented by the unnatural events of 1202-1204. The true centre of resistance to the Eastern Empire and the Crusaders was in Egypt; and the Egyptian Sultans felt a lively gratitude to the Christian city which had undermined one of the main bulwarks of Christendom.[1] Especial favour was therefore shown to Venetian traders; in 1215, out of 3000 Frank merchants in Alexandria, a great proportion were 'men of St. Mark'; and the renewed prohibition to

---

[1] To divert Byzantine assistance from the Crusaders, Saladin had cultivated a suspicious friendship with Constantinople; thus he sent presents to Isaac Angelus, 1185-95, and in 1189 the Saracens living in Constantinople are expecting to receive a relic (or *idolum*) which is captured on the way by the Venetians; about the same time two Genoese ships intercepted Venetian barks conveying an Embassy from Isaac to the Court of Egypt.

export wood and iron from Italy (issued by P. Ziani in 1226) was merely taken as a concession in words to Catholic prejudice, the fact of such export remaining, with undiminished profit to the good Catholics engaged in it. While the Latin dominion continued at Byzantium (1204-1266), Venice, as the controlling power in the Black Sea, controlled that traffic in slaves which was to Egypt what the corn trade was to other lands. In spite of prohibitions from the Church and from their own Government, and in spite of the scandal attaching to this commerce, Venetian merchants continued to furnish Circassian boys and girls, as well as European wood and iron, for the Cairo market; and by the contracts of 1207-8, 1225, 1229,[1] and 1238, they put the seal, as men said, on their disgrace.

In Syria, meanwhile, the establishment of her power at Byzantium, her frank abandonment of the Crusading cause, and her intrigues with the Saracens, almost of necessity compelled Venice to retire, gracefully or ungracefully, from the front rank; and after 1204 she only seems to have given much attention to her colony at Acre—where Genoese, Pisans, Florentines, Massiliots, and even English,[2] were also represented in the gloomy years which registered the slow ebbing of the Catholic tide and the approaching end of Acre itself as a Crusading citadel.

In the second part of this chapter we shall have to notice some of the evidence of the ubiquitous activity of Venetian citizens beyond the immediate basin of the Mediterranean,—in Bulgaria and at Iconium, at Kiev and Damascus, in the upland of the Lesser Armenia, at Trebizond and in the Crimea, at Tana and in the regions of the

[1] The treaty of 1229 is open to some doubt.
[2] Also Luccans, Siennese, Anconitans, and citizens of Piacenza and Montpellier.

Caucasus. Here it is enough to remember how the Republic, under the Latin Empire, profited by its position as chief intermediary [1] between the East and West, to conclude new arrangements with the Western Powers, helpful to its overland trade. In Italy, the cities of Bologna, Treviso, Padua, Mantua, and Ravenna; in Germany, the Emperors Otto IV. and Frederic II., entered into treaty [2] and gave special assurances of favour to the Venetian carrying trade; in return, German merchants were now permitted to acquire their own magazine and quarter in Venice, under the superintendence of three of the Grand Councillors.

From the first, Genoa [3] seemed marked by destiny to contrast with Venice. The latter, by its position and its history, was drawn towards the Levant, the Eastern Mediterranean, and the Empire of Constantinople; the former was as strongly linked with the western part of the Inland Sea, with France, and especially with the German or Holy Roman State. But the Ligurian Republic owed its greatness and its freedom to the Crusades, which, though Catholic and International in their character, were geographically something of an intrusion into the special field of Venetian influence, and offered a splendid prize to the swift and vigilant interloper. Genoa had been sacked by Moslem pirates in 935, had fought by the side of Pisa for the

---

[1] Together with Egypt. Venetian activity in pioneer exploration eastwards was remarkable; cf. Nicolo Conti in India, before 1444; the painter Francesco Brancaleone in Abyssinia (c. 1450), even before the Portuguese Covilham; cf. Bruce, *Abyssinia*, ii. 74, 92, etc.
The Venetian Consulate said to have existed in Siam at the end of the fourteenth century ('consul noster Siami'), is doubted by Heyd, *Commerce du Levant*, ii. 153, etc.

[2] 1209 and 1220.

[3] On Genoa's trade in the Mediterranean in the earlier Middle Ages, cf. Caffari, *Annales Genuenses*, in Muratori, *Rer. Ital. Scriptores*, vi.; Uberti Folieta, *Genuensium Historia*, in Graevius, *Thesaurus antiquit. Italiae*, i.

salvation of Sardinia,[1] and had joined in the great expedition of 1087 to the coast of Tunis; now it was ready to throw itself into the Crusading tide with a sincerity and thoroughness which Venice could not imitate. For Genoa had no double part to play and no Byzantine friendships to conciliate, until the change of Venetian policy threw the Greeks into alliance with a new friend.

The origin of the State as an independent body may be found in a *campagna* or political association, founded at the very close of the eleventh century, controlled by consuls freely elected, and supported by the bishops of the city against secular lords, such as the Oberti. In the pre-Crusading period Pisa, Amalfi, and Venice all enjoyed a more prominent position in Levantine waters; the gorgeous East was not yet held in fee by Genoa; and down to 1100 the future ally of the Palaeologi was content with Tyrrhenian commerce. The Norman Conquest of Sicily, and the heavy blows inflicted by the Tuscan republics on 'Barbary corsairs' during the second and third quarters of the eleventh century, may be said to have cleared the road to the East; the danger of flank attack was reduced to a negligable quantity; and from Norman Sicily the Genoese trader began to find his way to the Syrian and Egyptian ports.

But it is with the year 1097 that Genoa first appears in force upon these distant coasts; now, and again in 1099, she despatched warships and transports in great force to Palestine; she took a foremost part in the conveyance of troops by the maritime routes; her citizens fought bravely at the siege of Jerusalem; and, in 1100, a large Genoese flotilla, arriving at Laodicea, shared in the war on the Phoenician littoral. The capture of Arsûf, Caesarea, and Tortosa in 1101, of

[1] 1015, 1016, etc.

Gibelet in 1104, of Tripoli in 1109, and of Beyrout in 1110, rewarded their exertions, and furnished them with ample rewards. Gibelet was made over to their care; vast concessions in Tripoli were added; and these, with the advantages [1] they had already secured at Antioch in 1098 (a like reward for like assistance), constituted a kind of Genoese principality in Northern 'Aramaea.' At the fall of Antioch the Ligurian allies of Prince Bohemund shared with the son of Robert Guiscard the responsibility of future holding and defence (certain Provençal seamen from St. Gilles also pledging themselves to aid); and scarcely any recompense was too great for such alliance. Among other spoils of this region the Italian mariners captured an immense quantity of pepper, large enough to supply each sailor with two pounds of pungent treasure; the watchful foresight of their leaders took care that all privileges and properties bestowed upon the Genoese should be recorded in letters of gold at the Church of the Holy Sepulchre. The privilege granted by King Baldwin in 1105, and the new Antioch charter of 1106, also contributed to the concessions enumerated on this tablet; by the former, quarters in Jaffa and Jerusalem were acquired, with exclusive rights of administration and jurisdiction. Until 1112 Genoese power and influence steadily increased on the Levantine coasts, but then came something of a break; for attention was diverted from East to West by the growing urgency of the quarrel with Pisa, the battle for Corsica, and the renewed struggle with the fleets of the Spanish and North African Saracens. The dispute over their trading rights at Constantinople also helped to call off their energies from Palestine. The Republic, in fact, was driven to concentrate; its vaulting ambition threatened to overleap itself; and forty years were needed to put things on a safe

[1] Also a *quarter* in the city, 1106.

footing in home waters. In 1154 the tide returned; from this era Genoa speedily regained, by the ceaseless activity and skilful diplomacy of its consuls, aided by papal favour, the ground that had been lost in the Frankish States of the Orient; and what was even more important, a new chapter was now opened in its relations with Byzantium. Manuel Comnenus, disgusted with Venice, threw open to the most bitter rivals of the Old Ally a field which had long been barred[1] against them; but, in spite of large concessions in Constantinople, and a grant of free trade to all the ports of the Empire (except Rosia and Matracha[2] on the North Euxine coast), the Genoese were not content. The two harbours interdicted to their commerce (apparently at the Strait of Kertch and near the mouth of the Kuban) were valuable for the fur and grain trade of the North; the jealous prohibition of the Byzantine Government made them seem doubly worth entrance; and Manuel's price was high. In return for his favour, as displayed in the treaties of 1155 and 1169, he expected the aid of the Republic against the Hohenstaufen emperors of the West; but friendship with the Holy Empire (or German kingdom) was as much a rule of state with Genoa as the Byzantine alliance had been a principle with Venice, and both here

---

[1] The Genoese, however, had a factory in Constantinople at the beginning of Manuel's reign; this was destroyed in a furious riot by Italian rivals in 1162.

[2] Rosia and Matracha (τὰ Μάταρχα of Constantine Porphyrogennetos, the 'Matrica' of Rubruquis) practically commanded the North Euxine littoral. It was through Matracha that the Hungarian missionaries from the Theiss passed in 1230 on their way to their Pagan brethren in Great Hungary, near the Middle Volga. The region of Matracha was separated by a river from Ziccia, or Zichia, and the Zicci (mentioned by Carpini, in connection with Iberians, Georgians, and other nations of the Caucasus); on the west, according to Rubruquis, it was bounded by the Strait of Kertch. Zichia was reckoned among the most northerly territories of the Byzantine Empire.

and at Pisa the Comnenian policy was a failure; in fact, it was only at Ancona that Manuel's ambitions were completely realised. Both Venice and Constantinople wearied of a conflict as injurious to the one as to the other; and though the accord re-established at the close of Manuel's reign veiled a deep resentment and a revengeful ambition, it had at least the effect of restoring in great measure the ascendency of the Venetians, and throwing back the intruders from Genoa into their earlier obscurity.

The latter, therefore, had to wait their time in Greek waters, until the revolution of 1204 had created a final severance between parties, whose position, as reversed in 1261, formed the answer to every prayer, and the realisation of every wish, of the enemies of Venice.

Meantime the Ligurians vigorously maintained their interests in Palestine, and obtained a 'quarter' in Acre, and new charters and privileges in the dominions of Bohemund III. of Antioch (1169 and 1189). Good account was also made of an early friendship with the House of Ibelin, Lords of Beyrout, and afterwards Regents of Cyprus. At both these points Genoa soon outstripped all rivals; and no long time after the beginning of the thirteenth century they had the especial gratification of ruining the Pisan trade with the Isle of 'Khittim.'

In Egypt Genoa had a certain commerce as early as the tenth century,[1] and concluded a treaty of commerce with the Fatimite Caliphs; but it is only in the middle of the twelfth century that we have much evidence of this ancient intercourse, now evidently grown to large proportions. In documents covering the years 1155 to 1164

---

[1] If we may so understand the fragment of an Arab (Fatimite) diploma preserved at Genoa, promising protection to the citizens of a Western City. In 1063 Ingulf of Croyland returns from Jaffa in a Genoese ship.

Alexandria is mentioned far more often[1] than any other Mediterranean port as the object of Genoese voyages, and as furnishing the seamen of that city with most of their Oriental wares—cloves, Brazil-wood, alum, and pepper. In 1131 a Genoese ship, returning from Alexandria, was wrecked. Under the next year we still possess (if genuinely historical and not merely an exercise of rhetoric) the correspondence of two Genoese merchants upon details of the Egyptian trade; in 1177 a treaty with Saladin puts this commerce upon a new footing.

As with Egypt, so with Tunis. The twelfth century saw the old hostility rapidly fading away before the new disposition towards a mercantile intercourse profitable for both sides; and in 1250 Genoa concluded (almost at the same time as Venice) an agreement with that Moslem power which had so long harassed Christian Italy. Here the citizens of the former were nearer to their own special field of action, the Tyrrhenian Sea; for whatever they did, attempted, or neglected in more distant parts, they clearly grasped the fact that all else was nothing unless they maintained their rank in the Western Mediterranean. From about 1134 they had effectually weakened Pisa, and thrust their most dangerous rival into a lower place; then they took up their old quarrel with Saracen sea-power, conveniently labelled 'piracy.' In 1145 they attacked the Balearics, overran most of the islands, and destroyed many nests of corsairs. Soon after, they turned against the Moslems of Andalusia, stormed Almeria, and carried off (among other spoil) the marbles which still form the main front of the Cathedral of Genoa.

In the South of France they were very active. From 1132 (if not earlier) they possessed a factory in Narbonne;

[1] Sixty-six times.

in Montpellier they had already planted themselves before the year 1121; their Provençal policy was directed towards two objects. For one thing, they tried to hinder any direct intercourse between the ports of this region and the Levant; for another, they did their utmost to ruin the home trade of these cities, to take from them their carrying traffic, and to prevent their developing that internal prosperity which would enable them to struggle for new markets. Thus they allied themselves with the Counts of Toulouse in every attack upon the republican liberties of Marseilles and Montpellier, especially in 1143 and 1174; they struggled, with persistency worthy of a better cause, to shut the Provençal merchants out of Sicily; they hoped to make the Western Mediterranean a Genoese lake. The audacity of their pretensions may be gathered from their demand (of 1109) upon Bertram of Toulouse, that none but Genoese ships might enter the port of St. Gilles; and from the terms they sought to exact in 1143 and 1155, terms which forbade the vessels of Marseilles and Montpellier from calling at any port to the south and east of Genoa.

Meantime they pushed their interests in Christian and Moslem Spain. In 1127, a treaty with Count Raymond Berenger III. ensured them free entrance into all ports from Nizza to the Ebro, and protection in the same, on payment of a moderate harbour due. This impost was removed[1] by Raymond Berenger IV., together with 'landing customs' and 'anchor money,' after their zealous and effective aid in the campaigns of 1147 and 1148; and special rights of property and settlement were granted them in Tortosa, whose capture was mainly due to them. In return, they promised reciprocal favours to Count Raymond's subjects

[1] In harbours west of the Rhone.

in all the territories of Genoa; and here the Barcelona traders may have gained some real advantage.

The Genoese completed their work of making an open trade route from Italy to the Atlantic by their compacts with (resulting from naval victories over) the Moorish princes of Valencia (1149), of Majorca (1181, 1188), of Tunis (1250, etc.), and of the other Saracen principalities in North Africa and Southern Spain. Having cleared a path to the Straits of Gibraltar, the Genoese were the first Christian mariners to attempt serious discovery in the Western Ocean, and to seek for the long sea road to India, round the coast of the Dark Continent. Incidentally, as they worked their way towards the Pillars of Hercules, they created a dominant Genoese interest on the adjacent coasts,[1] and practically excluded Pisan competition; their seamen acquired an immense influence in the naval development of Aragon, Castile, Portugal, and Sicily; even in Provence, to which Genoa had been a true step-mother, Genoese merchants were exceedingly powerful in financial and commercial matters, nominally encouraging, often skilfully depressing, the trade of these vigorous rivals.

But as this period draws to a close,[2] the growth of great centralised inland kingdoms both in France and Spain was already foreshadowing a new period;—when the most wealthy and unscrupulous of mercantile republics would find itself overmatched by superior resources and equal craft; and when, under the patronage of the new continental states, navies of even greater power would arise, and discoveries

---

[1] The Genoese also opened a trade with Flanders very early; in 1224, a note of receipt in the Antwerp archives recorded the presence of a Genoese captain at Sluys.

The regular service between Venice and Flanders did not begin till 1317.

[2] About the middle of the thirteenth century.

ruinous to Italian trade would be made in distant seas.

Meanwhile the Genoese, secure in the West, attempted a master-stroke in the East, by the overthrow of the Venetianised Latin rule in Constantinople, the restoration of the Greeks, and their own establishment as the indispensable ally and commercial favourite of the Byzantines. As the Christian dominion in Syria, one of the most prosperous fields of past enterprise, was manifestly falling to pieces; as, moreover, it seemed daily more probable that the Western intruders on the Bosphorus would be displaced without any aid from others; as, finally, the barriers of common creed and race had been weakened by the action of Venice herself, and by the traffic of centuries both with Saracens and with heathen,—it was not in itself surprising that Genoa in the crisis of 1261 took its station by the side of Michael Palaeologus, and threw the whole weight of its resources on his 'scale of fortune.' But the success of the new policy was amazingly complete, and its consequences were very marked. It raised Genoa to a position of full equality with the mistress of the Adriatic, gave her for two centuries the command of the Black Sea trade, and more than compensated for the losses which inevitably followed on the ruin of the Christian kingdoms in Syria. Shortly before the Byzantine restoration events in Acre had made the Genoese more and more determined to transfer their attention from the Syrian coast to the Bosphorus and the Greek waters. For in 1258 a union of rivals, headed by the Venetians and Pisàns, drove the unpopular Ligurians from Ptolemais to Tyre (which they now made their head station in these parts), to Cyprus, and to Cilicia or Little Armenia, where they had secured their position by the commercial agreements of 1201 and 1215.

The origin of the Pisan Republic[1] was similar to that of the Genoese. In the course of the eleventh century the Marquis of Tuscany lost control over the city, and in 1085 a republican constitution was established by the mediation of Bishop Daibert; a charter of pacification was issued the same year, and consuls took the place of the feudal lords. But throughout the last two generations of feudal supremacy, Pisa, though still dependent, was gaining a famous name at sea. It was of some importance even in the tenth century; in 1017 it allied itself with Genoa, and achieved the conquest of the great island of Sardinia; soon after began the Pisan attacks upon the Barbary coasts. In 1035 the men of the Arno stormed Bona, the ancient Hippo Regius; in 1063 they forced the port of Palermo; in 1083 and 1087 they made two brilliant raids upon El Mehdia, the fortress-harbour of Tunis. With the help of these deeds of arms they extorted many trading privileges from the Moslem sovereigns of the Western Mediterranean; from Alexius Comnenus they won the grant of a most favoured nation clause in Byzantine waters; in return for these concessions the port of Pisa was freely opened to foreign commerce; and thus an amount of traffic was drawn to the city, so great as to promise it at one time the first place among the Emporia of the Mediterranean. Western Orthodoxy[2] was shocked by the 'marine monsters' from the ends of the earth who thronged the streets of the city; Pagans, Turks, Libyans, Parthians, and Chaldaeans defiled the town and

[1] On Pisa and its Mediterranean career in the earlier Middle Ages, cf. *Chronica varia Pisana*, in Muratori, *Rer. Ital. Scriptores*, vi. For the Syrian treaties and charters, cf. Muratori, *Antiquit. Ital.* ii., and Flam. dal Borgo, *Raccolta di Diplom.*

[2] Cf. Donizo, *Life of Countess Matilda*, i., 20, in Muratori, *Rer. Ital. Scriptores*, v. and Dissertation 31. 'Qui pergit Pisas, videt illic monstra marina Haec urbs Paganis, Turchis, Libycis quoque Parthis Sordida: Chaldaei sua lustrant moenia tetri.'

blackened its walls; here, most of all, was to be seen the triumph of commercialism over all the barriers of Latin exclusiveness, over race, religion, and language alike. The conquest of Majorca in 1114, following upon the occupation of Sardinia, seemed to establish, firmly enough, the supremacy of Pisa in the Tyrrhenian Sea, and its citizens turned with extraordinary energy to the Levant. They had already done yeoman service before Antioch in 1098; in 1099 and 1100 a fleet of one hundred and twenty sail, under the command of the city's own prelate, Daibert, afterwards patriarch of Jerusalem, co-operated in the sieges of Tyre and Acre. For such aid a rich reward was not to be refused; in Jaffa, Godfrey of Bouillon bestowed on Pisan merchants an extensive 'quarter' as a fief of the Patriarchate; Tancred, the Lord of Antioch, granted similar concessions, with exemption from export and import dues, both in his capital and in Laodicea. Later favours of 1154, of 1156 (from King Baldwin III.), and of 1157 (from Amalric of Ascalon), promoted Pisan trade, and protected the persons and property of Pisan merchants, especially in the case of shipwrecks, in the 'recognition' of wills and testaments, and in other matters of civil and criminal litigation.

In no other concessions do we possess greater detail and fuller explanation than in the Pisan, which stipulate, among other things,[1] for the building of churches and private houses, of stores, custom-houses, and courts of justice, of factories and magazines, of baths, mills, and ovens; for a fixed proportion of the customs revenues; and for the rights of using their own weights and measures, and of electing overseers or consuls of their own nation to safeguard their interests, both legal and commercial. No other

[1] Taking one with another.

of the Italian Republics, moreover, amassed a richer store of treaties, compacts, and agreements with the Crusading princes; nor could it well be otherwise, for Pisa devoted herself more zealously to the Syrian trade than any of her rivals. As her power waned, her energies, by a fatal error of policy, were even more concentrated on this field than before. Her fleet took the lead in attacking Alexandria and the Moslems of Egypt, the most formidable of all; King Amalric responded by concessions in Tyre (1165) and by the promise of great rewards on the conquest of the Nile valley (1168-69); fresh privileges were secured in Antioch and Tripoli from Count Bohemund III. (1170, 1172, and 1187); in the last days of the Latin Kingdom, Baldwin IV. and Conrad of Montferrat were equally generous at Acre, Jaffa, and Tyre (1182 and 1187-88). These three cities were the headquarters of Pisan trade in the Eastern Mediterranean; Jaffa contained the most numerous of their colonies, the leading *fundacum* or *fontecco;* and at Tyre Pisan merchants dominated throughout most of the twelfth century. Here their *Societas Humiliorum* was a trade guild, or commercial chamber, of special interest, and the men of the Republic helped to save the purple city from the Saracens in 1187. Even after Syria had really slipped out of Frankish hands Pisan diplomacy continued to exhort charters from the powerless kings, princes, and counts, as in 1189, 1191, 1216 and 1229; when Guido enlarged their property and privileges in Tripoli; when Conrad of Tyre and the Prince of Antioch confirmed their older rights in towns where only the shadow of former benefactions lingered; and when Frederic II. conferred useless favours, in Syria, on the city whose power could no longer maintain itself in the Tuscan seas.

At home the great enemy of Pisa was, of course, Genoa, the faithful ally of the eleventh century, the destroyer of the thirteenth; but in the Levant Pisans were more constantly at war with Venetians. Their conflicts in Byzantine waters were incessant; the Porto Pisano, on the Lycian coast, east of Myra, was a nest of Tyrrhenian 'pirates'; and the efforts of Pisan statesmen were long directed towards the securing of a direct route to Constantinople, partly by sea and partly overland, by way of Ancona, Zara, and Saloniki. This project, fortified by treaties with Ancona, Zara, and Pola (1188, 1195, etc.), was a direct challenge, like the earlier schemes of Manuel Comnenus, to Venetian ascendency in the Adriatic, and it was resisted with every weapon at the command of the Seignory; on their side the Pisans did not hesitate to assist every rebellion against Venetian authority (as at Pola in 1195); and their colony, grouped round the Church of St. Nicholas in Byzantium, and secured by the agreements of 1111 and 1192, was a focus of every anti-Venetian element in the Eastern Empire.

With Egypt, Pisan relations were marked by a brilliant and fortunate duplicity. On one side they shared in the Crusading enterprises against the Delta region, with such energy as to win recognition in the Syrian principalities and to gain a foremost place in the battle of concessions; on the other side they used every art to promote their trade with the Fatimite dominions.[1] In 1154 their envoy at Cairo, in his audience of the Caliph, promised that the city would give no more aid to the Crusaders; and in 1160

---

[1] In 1150 the famous Nicholas, Abbot of Thing-Eyrar in Iceland, a pilgrim to Rome and Jerusalem, saw some Egyptian ships in the harbour of Pisa. Relations probably began as early as about 1100. The Pisans claimed, also, to have made voyages to India by way of Egypt from 1175; this claim is doubted by Heyd, *Commerce du Levant*, ii., 154.

the Moslem ruler of Egypt was complacent enough to order the restoration of the Pisan *fontecco* in Alexandria; yet not many months had passed before Pisa inspired and led the great attack of September, 1163. In 1156 the King of Jerusalem had arranged with his allies that their Egyptian trade should not be used to cover contraband articles; iron, wood, and pitch found in Pisan ships thither-bound might be freely confiscated. As the Frankish power declined, and the greatness of Saladin began to overshadow the Levant, Pisa made great efforts to patch up a new treaty; between 1176 and 1180 they sent three embassies to Egypt; and their trouble was not without result. A new privilege covered many a broken covenant in the past.

With Spain and Sicily, the South of France, and the Barbary States, the city maintained, and in some places extended, its commerce during the twelfth century; its traders rivalled the Genoese in their activity at the fairs of St. Gilles and Fréjus; and a treaty was concluded with Montpellier in 1177,[1] which must have strengthened the Provençal cities in their struggle against Genoa. This struggle had been, since 1119,[2] an ever present danger of Pisa herself, and each succeeding generation deepened the enmity between the Tuscan ports. Till the very close of the thirteenth century the battle was undecided, and the Pisan power appeared fully equal to her foes, the Pisan Empire more extensive and more splendid, the Pisan city far more magnificent.[3] But gradually the prospect darkened.

---

[1] 6th February.

[2] The date of the first war between Pisa and Genoa.

[3] Pisa 'was the first Italian city that took pride in architectural magnificence.' Her Cathedral was built in the eleventh century; the baptistery, the leaning tower, and the arcades surrounding the Campo Santo, were mostly of the twelfth, some parts being of the thirteenth, century.

First of all, the Latin Empire in Constantinople, followed by the Greek restoration, substituted a Venetian, displaced in turn by a Genoese, monopoly, for the earlier and freer conditions under which Pisan trade had flourished among the Byzantines. The utter overthrow of the Crusading kingdoms in the Levant crippled the Pisans' commerce in another field, and their expulsion of the Genoese from Acre was of little use, for all Christians alike were driven out in the course of the next generation. Finally, the maritime campaigns in the Western Mediterranean tended more and more to failure, until, in 1284,[1] the fleet was ruined and the city too exhausted to equip another. The splendid possessions of Elba and Sardinia were soon after torn away, and the Metropolis itself fell under the heavy hand of fate.

The chief Provençal ports, Marseilles,[2] Montpellier, and Narbonne, were more important in the Crusading period than is often supposed. The dazzling fame of Venice, Genoa, and Pisa has blinded the eyes of the modern world, and some of the most interesting of the mediaeval maritime states have been treated with ill-deserved neglect.

During a large part of the twelfth century the Massiliots despatched their 'envoys' to Syria twice a year, probably accompanied by a trading fleet and some ships of war; they also took an active part in furnishing provisions, transports, and loans of money to those embarking for the Holy War. Marseilles became a very favourite point of departure for Western and Northern travellers to Syria; pilgrims from the

---

[1] In the Battle off the Island of Meloria.

[2] On the Mediterranean commerce of Marseilles, Montpellier, etc., in the earlier Middle Ages, cf. Heyd, *Commerce du Levant*, i. 20, 23, 125, 131, 146-148, 181, 185, 186, 188, 189, 312, etc., 319, etc., 327, etc., 338, 364, 420.

North Sea coasts and the military monks of the Temple and the Hospital specially favoured this route; and the Provençal pilots were among the first to make an open course out of sight of land across the Mediterranean. This shortening of landsmen's agonies to a voyage of fifteen days, between Provence and Acre, begins in the middle of the twelfth century, but is only attempted in summer weather, and by the lighter class of vessels; heavier ships continued to coast along Italy, to pass through the Straits of Messina, and to touch at Crete, Rhodes, and Cyprus. The devotion of Marseilles in the sacred cause had its reward. In 1117 the citizens received a quarter in Jerusalem, and in 1136 a general exemption from dues in the Kingdom of Jerusalem; in 1152 Baldwin III. conferred on them the right of possessing a street in each of the chief Syrian ports; and in 1163, they obtained a valuable mortgage in Acre from Bishop Rudolf of Bethlehem. In 1187 and 1190 they rendered effective naval aid at Tyre and Acre, and in return, gained an increase of their privileges, with the rights of possessing their own house of justice, of building or repairing ships in Syrian waters, and of strengthening their Palestine fleet at discretion. Thus it came about that the Massiliots possessed, in 1190, a sufficient navy for the transport of the army of Cœur de Lion; at this time their commerce with Egypt was flourishing; the attempts of Genoa to retard their progress had been repelled; and the suzerainty of the Counts of Provence had become nominal. The tariffs and statutes of Marseilles in the thirteenth century show a considerable traffic with the Moslem world, and especially with North Africa; with Little Armenia, Sicily, and South Italy; and indirectly with India.[1] The factories at Ceuta,

---

[1] In the dowry of a Marseilles girl of 1224, mace, ginger, cardamoms, and galangale are mentioned.

Bugia, Alexandria, and Acre were the leading outposts of the city's commerce; among all its local rivals only Montpellier remained in serious competition; St. Gilles (which traded mainly at Tyre), Narbonne (whose intercourse was chiefly with Acre), and Aigues Mortes (famous as the first sea-port of the French kings) no longer disputed the supremacy of Marseilles.

Yet, in the age of Frederic II. and Louis IX., the growing power of the inland monarchy checked the development even of the greatest Provençal cities as independent republics. The Moslem recovery in the Levant was not wholly beneficial to their trade and influence; and both Marseilles and Montpellier were made to feel, as the thirteenth century drew on, that their future was being more closely bound up with the new kingdoms of France and Spain. In 1204 Montpellier fell into the power of Aragon; but under the sane and vigorous rule of James I. 'the Conqueror,' its material welfare, far from suffering, was so increased, that the city was reckoned among the leading markets of Christendom. Alexandria, Beyrout, Tripoli, and Tyre were the favourite centres of its trade in the Levant; its commercial colony in the Delta, founded before 1250, was especially famous; and its Egyptian trade, like that of Barcelona, was actively furthered by King James. In 1267 two citizens of Montpellier were despatched by him on a mission partly diplomatic, partly commercial. The result of this is not clear; but it is evident that after the death of her great prince the town declined as a commercial capital, and that its carrying trade passed into the hands of Venetians and other Italians.

Narbonne is of peculiar interest in the earlier Crusading period as the medium of a trade between two distant countries, England and Egypt, whose connections were weakened in the later Middle Ages by the French conquest of this region. This commercial route passed through Bordeaux and Toulouse to the Mediterranean, and brought the tin and copper of Britain to exchange against the products of the Levant.[1]

The *Catalans*[2] were the first among Spanish peoples who developed on lines parallel to those of the Italian commercial republics. As early as the ninth century, the proceeds of land and water customs make an item in the revenues of the Catalan princes; the latter encouraged maritime industry among their subjects in every way; and by the twelfth century Barcelona had become a free port for all Mediterranean traders, and was entering upon a very prosperous future. The temporary conquest of Majorca in 1114 by the fleets of Pisa and Count Raymond Berenger III., the commercial treaty of 1127 with Genoa, the overthrow of Almeria in 1147, all helped forward the marine and commerce of Catalonia; Benjamin of Tudela, in 1159, witnesses to the resort of merchants from all parts of the world to Barcelona; many Italian merchants, by settling in the Catalan ports, still further increased their importance; and trade between the north-east and the south of the Peninsula attained considerable dimensions even before the re-conquest of Valencia and Murcia, of Cordova and Seville, by James I. of Aragon and St. Ferdinand of Castile.

The Catalans also, like the Amalfitans, devoted special attention, with excellent results, to the fixing and codifica-

---

[1] Ibn Said.
[2] On the Catalan trade in the earlier Middle Ages, cf. Capmany, *Memorias historicas sobre la marina, commercio, y artes de Barcelona*, and especially the *Collectio diplomatica* in tom. ii.; Navarrete, . . *Los Españoles . . . en las cruzadas, disertacion hist.*

tion of maritime and commercial law, to the system of consular representation for the development of trade, and to the elaboration of a scientific cartography, based on coast-surveys. The latter achievement, which gave us the Portolani—the first accurate maps of any part of the world—was perhaps due, in its inception, rather to Genoese or Pisans than to Catalans; but the seamen of Barcelona and the Balearics certainly had a share, and an important share, in this work. The conquest of Majorca by James I. of Aragon in 1227 freed Catalan trade from an old danger; and now a commerce which was already respectable made rapid progress. From 1230 Barcelona merchants began to resort to foreign parts in large numbers; from about 1260 their Levantine traffic became really important. Yet before the close of the twelfth century we hear of a joint-colony of Barcelonese and Provençals at Tyre, and the Aragonese dominions already imported spices from Syria and the Nile.

Almost wholly undisturbed by Church prohibitions, the Catalans maintained a vigorous trade with the Mohammedan states, both of Southern Spain and of Northern Africa, from the opening of the twelfth century. Even with Egypt the spice-commerce had so increased that in 1250 James I. concluded a treaty with the Sultan of Cairo, through the medium of two Barcelonese merchants engaged in this traffic. The same king, thirteen years earlier, in 1237, had guaranteed the shipbuilders of Barcelona state-protection for their intercourse with Egypt—although this wood for shipbuilding was among the articles which ecclesiastical censures had long forbidden Christians to export under any name to the unbelievers. No foreign vessel, James ordained, might be freighted in Barcelona for Egypt, Syria, or Barbary, while there remained in port any Catalan ship bound for any of these countries.

In the rise of Barcelona we see the drawbacks and the advantages of a state, not under independent or republican institutions—not aided, like Pisa, Genoa, or Venice, by its participation in the Crusading movement—but under the protection of powerful Continental princes, and uniting a measure of internal liberty with external obedience to a land-power able to protect the commerce of its shores by a far greater force than a maritime republic, pure and simple, could ever display.

In Northern and Western Spain the Biscayans and Galicians were, from an early age, hardy and daring fishermen, and chased whales in the Atlantic. Later on, some of their mariners were among the first to reach the American cod-banks; but, in the period that now concerns us, we have very little to record, beyond a fishing charter of St. Ferdinand in 1237, and the fact that these seamen aided in the capture of Seville in 1247, blockading the Guadalquivir with a flotilla of their boats.

But in Seville itself, Ferdinand III., from 1249 onwards, made serious attempts to create a Castilian Admiralty and a genuine commercial interest. Many Pisan, Catalan, and Genoese traders were induced by special privileges to settle there, and every effort was put forth to maintain and increase the old commerce of the city, which had been so active under the Moors. Regulations were framed, an admiral of Castile was stationed in the port, privileges of various kinds were granted to enterprising merchants; but the incomplete subjection of Andalûs, and the long wars which still had to run their course in Southern Spain, hindered the development of Seville. For the town was too near the scenes of conflict and danger, and suffered from its position, not on the sea itself, but on a river-channel, at a considerable distance from the Main, and always liable to be blocked by an enemy. It was not, moreover, till the following century that the growth of discovery and enterprise developed trade beyond the Straits of Gibraltar; and so opened to Seville a field for which nature had fitted it, lying as it did upon an oceanic stream. Before this its prosperity was too dependent on Catalan and Genoese middlemen to have an independent character of any value.

In that part of Western Spain, otherwise known as the county and kingdom of Portugal, commercial development, oversea enterprise, the spirit of adventure, the search for distant treasure, does not show much activity before the close of the Crusading Age. Yet even in the thirteenth century a certain trade with Flanders and the north and west of France had already begun, and the more distant Hanseatic[1] markets may have communicated with Lisbon and Oporto (as with Venice and Genoa) through the factories of London and Bruges.

The Crusading States of the Levant formed an advanced base for Western commerce and trade, as well as for Western politics and conquest. The former was, indeed, not inseparable from the latter; and after the 'Latin' Kingdoms had already fallen into decay, 'Latin' and 'Frankish' enterprise in trade and travel continued very active, reaching further into the Asiatic world than ever before; but with the complete extinction of the principalities of Jerusalem, Antioch, Edessa, and Tripoli, the old difficulty of a hostile and almost impenetrable Islam once more confronted the European nations along the coast line of Syria. The effect of this reaction, an effect also of the discoveries that had been made both in the Atlantic and in Asia, was seen in the fourteenth century diversion of Christian advance from land to sea.[2]

---

[1] Cf. Sartorius, Lappenberg, *Geschichte der Hansa*, i. 271; ii. 80-90.

[2] And from Western Asia to Western Africa.

The Kingdom of Jerusalem, in the narrower sense, was not in command of many of the trade centres of the Levant, and was not even close enough to the first-class Emporia to influence them directly; but, on the other hand, it was master of all the land routes that connected Egypt and Africa with Western Asia. 'Petra in the desert,' the Crusaders' Kerak, answering to the ancient Kir Moab; Montreal or Shaubek, 'le Crac de Montreal,' corresponding to the Edomite Petra; and Ailah or Elath, at the head of the Elanitic gulf of the Red Sea, controlled the ordinary pilgrim and caravan routes from Damascus to Mecca, the leading commercial roads for the sea-borne trade of Western Asia with the South and East. The Mediterranean coast route between Cairo and Damascus, passing through Ascalon, was also at the mercy of the Crusaders of Jerusalem; while their brethren at Edessa, Antioch, and Tripoli, possessing the whole coast of Phoenicia and the gulf of Issus, interrupted all the main lines of communication between the central body of Asiatic Islam and the Mediterranean Sea. Bagdad could only reach Cairo, the sacred region of the Arabian Haj, or the lands of the Western Moslems (in Spain, North Africa, or the islands), by an immense circuit, or by risking a passage over Crusading territory. It was significant that the crisis of 1187–88 arose from the attack of the Castellan of Kerak on a Saracen caravan, and the determination of Saladin, as ruler of Egypt, to break through the cordon from Elath to the Arnon, from the Red to the Dead Sea.

While the European dominion lasted, the flax of Egypt, the musk and rhubarb of Tibet, the pepper, cloves, aloes, and camphor of the Indies, the ivory of Africa and Hindustan, the incense and dates of Arabia, paid transit dues for their passage through the Crusaders' country,[1] and the chief ports of this country did a considerable trade in these precious wares. Acre, Tyre, Beyrout, Laodicea, and Jaffa were the principal harbours in Latin hands on the Levant coasts; then came Caesarea, Ascalon, and Haifa under Mount Carmel; Tiberias, as an inland lake port, was also of some importance; in all of these the Christian merchants seem to have been served with Oriental products by Saracen dealers. The fruits, cottons, silk, pottery, and glass of Syria were all worth export, and could be obtained without the aid of middlemen in the country mastered by the Crusaders. Thus, John of Salisbury[2] has a tale of a rich citizen of Canossa who provided his banquets with delicacies from Palestine and Tripoli. The more daring Western traders even went up country to seek their fortune in the great Mohammedan markets. It is curious that the Princes of Antioch,—though possessing such an excellent commercial situation, lying close to Aleppo and Hamath (just as the Counts of Tripoli lay close to Damascus and Hems), and commanding some of the most frequented trade-routes,—did not exert themselves more actively to promote commercial interests; their charters to trading cities, corporations, or individuals are few and far between; and one famous exception, the invitation to the Genoese issued by Bohemund III. in 1169, seems to imply in various expressions a past condition of improvident neglect. The fall of Acre into Saladin's hands, though it did not destroy Western commerce on this shore, greatly impaired the free and favourable meeting of Eastern and Western trade; and, after its recovery, the port of St. John

---

[1] As appears from the *Assizes of Jerusalem*.

[2] *De Nugis curialium*, viii. 7. Cf.

Heyd, *Commerce du Levant*, i. 178-179.

was more highly valued and more carefully guarded than before, especially by the regular organisation of European colonies, both mercantile and military in character.

The Moslem Hinterland to the Crusading Syria possessed four chief markets—Aleppo, Damascus, Hems or Emesa, and Hamath,[1] behind which lay the still greater mart of Bagdad and the lesser Emporia of Mosul and Bassora or Basra, covering the line of the Tigris. Aleppo was a head centre of the trade-route from the Abbasside Caliph's 'metropolis' to Antioch and Laodicea on the Western Sea. This route (which crossed the Euphrates at Rakka) Edrisi calls the grand avenue of the trade of Irak, Persia, and Khorasan; and the silk market of Aleppo proved its connection with the still more distant countries of the Far East. Even at the close of the thirteenth century many Venetian traders were resident here for the sake of the commerce in *Seric* goods, as well as in alum: the pigments and pepper found at Antioch by the Crusaders, on the capture of that city, also bore witness to an Indian commerce with the Mediterranean by this path; and the elder Sanuto is probably right when he says (at the beginning of the fourteenth century) that in old time most Oriental goods passed along this way to the Roman Sea.[2] Antioch, though not itself on the open coast, communicated directly with the latter by its harbour of Sueidieh[3] or St. Simeon; while Laodicea,[4] not yet superseded by Alexandretta, was among the best of Levantine havens; the smaller port of Gibel to the south, where the Genoese had planted themselves in force,

---

[1] To which may be added Bostra.
[2] Cf. Ibn Butlan (d. 1052) in Yakut; Wüstenfeld, *Zeitschrift deutsch. morgenländisch. Gesellschaft*, iv. 50; Kremer, *Wien. Sitzungs-Berichte*, ph.-h. kl., April, 1850, 239, 243;

Sanuto, *Secreta Fidelium crucis*, p. 22.

[3] *St. Simeon* of the Crusaders; *Sevodi* of the Armenians; *Sudi* or *Suetion* of the Greeks.
[4] Latakiyeh.

gave another outlet of minor importance for this commerce. The trade of Latakiyeh with Alexandria on the one side, and with Aleppo on the other, reaching beyond to the Red Sea and the Euphrates, brought many Christian merchants to its market, especially for spice, in which trade the Venetians took the largest share.

From Aleppo a route, which was often followed by a large amount of trade, ran to Damascus through Hems and Hamath, from North to Central Syria; at Emesa it was crossed by another path of traffic, from east to west, descending to Tortosa and Tripoli. These cities served as the harbours for all this region, the upper part of the older Coele-Syria; Tripoli, in particular, was very prominent, and the resort of many Armenian, Jewish, and Nestorian merchants; while the little port of Gibelet[1] or Byblus, lower down the coast, was notable as a point eagerly competed for by Western traders.

Damascus, then as always, occupied the position of a first-class market; no revolutions, sieges, sacks, or other disasters, could permanently affect the prosperity of the oasis, so plainly marked out by Nature as among her favourite spots of earth. Here was the starting-point, in Moslem times, of the Syrian Haj, of the pilgrim and trade-route to Mecca and Western and Southern Arabia; here, also, was the meeting-place of caravan roads from Mesopotamia, from Asia Minor, from Persia and Central Asia, and from the Mediterranean coasts. It had direct intercourse with the ports of the Persian Gulf, with Bagdad, and the other markets of the Tigris and Euphrates; its relations with Egypt, though necessarily more indirect, were still maintained during the Crusading period. For, however risky the passage, or however great the deviation, Damascus goods enjoyed so high a name in all the lands of Islam and among all the

[1] Cf. Edrisi, i. 356-359; ii. 130,137.

VI.]   TRADE OF ALEPPO, DAMASCUS, BAGDAD, ETC.   443

merchants of Christendom, that the most savage wars did not wholly prevent their circulation.

To the south-east of Damascus, Bostra in the Hauran, Ituraea, or Argob (the wild, hilly country beyond the Middle Jordan), had been an inland market of no small consequence, but it had now decayed, and the ancient and celebrated Fair of Meidan or Muzerib was not frequented as in old time;[1] even in the thirteenth century, however, European traders occasionally appeared at it (to say nothing of Saracens, from parts as distant as Irak); during the brief time when the region formed part of the Kingdom of Jerusalem, it would naturally be accessible to Christian merchants. In any case, it was well known under the names of *Sutar* or *Swide*[2] to the Crusading warriors, who made frequent raids into this country; and through it passed the great trade-route between Damascus, Arabia, and the Red Sea.

But all these towns, even Damascus, must be ranked below Bagdad in the commercial as well as the political sense. Until the fall of the Caliphate in 1258 the *Abode of Peace* retained that primacy of wealth, population, traffic, and honour which Benjamin of Tudela witnessed, and which after its fall passed in a measure from Southern to Northern Persia, from Bagdad to Tabriz. Among all the cities of Western Asia it was supreme; Cairo, Constantinople, Alexandria, and Cordova were probably its nearest rivals outside China. Upon it converged most of the great commercial routes of Asia; those which ran up and down the Tigris,—to Bassora and the Persian Gulf, and so to

---

[1] Cf. Edrisi, i. 352-353. The commencement of its decay may be put in the later twelfth century. Through here ('Mezeribe,' as he calls it), Varthema passed on his way to Mecca, April, 1503; see pp. 16-18 of the Hakluyt Society's Edition of Varthema.

[2] Arabic *Sâwet*.

Southern and Eastern Asia, on one side,—to Mosul, Amida,[1] Tabriz, and Armenia, on the other; those also which passed through Irak (by canal[2] or caravan) to the Euphrates, Rakka, and the Western Asiatic coast lands; those, finally, which crossed Persia, Afghanistan, and Central Asia to India and China. In Khotan and Ferghanah the merchants of Bagdad obtained the musk and rhubarb of Tibet; in Kabul and Ghazni they met the dealers of Hindustan; their still more distant journeys to the Celestial Kingdom were usually by sea to the port of Khan-fu;[3] if they chose the land-way they skirted the Takla-Makan desert either to the north or south, either by Khotan or by Kulja.[4]

West of Rakka one of the principal lines of commerce ran north-west through the Latin county of Edessa;[5] but in the fifty years of Crusading occupation there is no evidence of Western traffic in this region, and all the trade appears to remain in the hands of indigenous Armenians; perhaps this outpost of the Crusaders was in too dangerous a plight, and too often harassed by enemies, for the development of much peaceful activity.

There is but little positive evidence for the plausible belief that Venetian and other Christian merchants, before

---

[1] Diarbekr.

[2] *E.g.* the Nahr-Issa (Tigris-Euphrates Canal). Cf. Edrisi, i. 492, etc.; ii. 136, 144, 153, 214, etc.; Richthofen, *China*, i. 502, etc., 'an indispensable commentary.' Also M. Polo, ed. Pauthier, p. 45, etc. Polo's remarks about Bagdad trade are notable; they show that even the events of 1258 did not kill it; in the fourteenth century wares from China, Yemen, India, and the Zanzibar coast are said to appear at the Bagdad market once more.

[3] Answering to the modern Hang-cheu-fu.

[4] Cf. Heyd, *Commerce*, ii. 222, 223, etc. The routes south of the Thian Shan Mountains went from Khotan along the valley of the Tarim and so to Turfan, Khamil, and the end of the Great Wall. One path skirted Lob Nor; another avoided this lake and passed by Aksu and Karashar. The latter was followed by Hiouen Thsang, and Shah Rokh's envoys in 1420, as well as by Haji Mohammed, the Persian rhubarb merchant of the sixteenth century.

[5] 1098-1144.

the time of Marco Polo, made their way into the Euphrates region, and even beyond, to Mosul and Bagdad; yet Aleppo and Damascus they certainly visited; and a claim was made in the fourteenth century for Provençal enterprise at Rakka in the twelfth. Bearing in mind how individual Venetians, in the days of the Latin Empire at Constantinople, penetrated into the very heart of Asia Minor and into Northern Persia (where Pietro Vioni was residing in 1264), we can hardly reject[1] as impossible the suggestion of a less daring, extensive, and hazardous enterprise on their part in Northern Syria and Mesopotamia, under the most powerful of the Kings and Counts of Jerusalem, Antioch, Tripoli, or Edessa. At any rate, it would seem that the conquests of Ghenghiz Khan were first described to Christendom by the 'people of Count Raymond' on their return to Tripoli from various trading journeys in the upland, and especially by some merchants whose trade lay in spices and precious stones. It is also clear that about 1200 the Pisans were doing an active business at Aleppo, and it is probable that, by the douane-station of the 'Pons Ferri,'[2] over the Orontes, they had already advanced still further inland.

These Syrian principalities were the chief examples of the Crusading polity in the Levant, and (as we have seen) they afforded excellent outworks and advanced posts for Latin movement still further into Asia; but they were not the only fortresses gained by the inrush of West upon East, from which new movements of war and commerce took their start. Syria was a meeting-place for most kinds of Oriental goods, but Constantinople was no less a world-emporium, and the

[1] Heyd, however, refuses to accept it, though he gives, *Commerce*, ii. 110, the case of Pietro Vioni's (Viglioni's) will, made at Tabriz, 10th December, 1264.

[2] This would not necessarily take them beyond Aleppo, but it points to a further advance. Anyhow, they crossed the 'Pons Ferri' and paid dues at this point.

revolution of 1204 gave a practical monopoly herein to the Venetians. From this point of vantage they started afresh, and pushed on to exploit the wilder regions east and north of the Bosphorus. Thus John de Plano Carpini, returning in June, 1247, from the Tartar Court, met with several of his countrymen trading at Kiev, and among them a Venetian named Manuel, and two other merchants, both probably of the same city, Jacobus Venerius or Reverius [1] 'of Acre,' and Nicolaus Pisanus.[2] Had Friar John touched at the famous ports of Rosia and Matracha on the north-east littoral of the Euxine, he would have found there also a number of the men of St. Mark, as in other parts of the Taman peninsula, at Sudak or Soldaia in the Crimea, and at Trebizond and Sinope. The ancient and important trade-route which crossed the Caucasian isthmus to the south of the great range led down,[3] by the valley of the Rion or Phasis, to the Black Sea; but it threw off a branch which traversed the Armenian plateau and united at Sivas or Sebaste in Cappadocia with several other avenues of commerce, from East, North, South and West, from Mesopotamia and Persia, from Syria and Little Armenia, and from Sinope, Nicaea, and Iconium. After 1214 Sinope was dependent on the Seljuk Sultans of Rûm, who attached a high value to their window on the open sea; through Sudak it had intercourse with the north coast of the Pontus and the Kipchak lands; and the rulers of Iconium encouraged this commerce to the utmost, even undertaking expeditions (as

[1] *Venerius*, Heyd suggests, *Commerce du Levant*, i. 319, refers to Giacomo, a member of the great house of Venier which had a branch at Acre; the *Pisanus* here named was probably a member of the Venetian house of Pisani.

[2] Besides one Michael of Genoa, from the great rival port. All these Italian traders at Kiev had come direct from Constantinople. Cf. the text of Carpini in the Paris Geographical Society's *Recueil*, iv. p. 772.

[3] After ascending from the Caspian shore to the watershed near Tiflis, by the valley of the Kur.

in 1227[1] to the Crimea) to chastise outrages upon their subjects.

It must remain uncertain whether the Turkish, Seljuk, or 'Rumi' ports in the Black Sea and the Mediterranean were open to Western trade during the time of Latin domination at Constantinople; but there is abundant evidence for the visits of Italian merchants to the kingdom of Iconium, for their successful enterprises in the same, and for the commercial understanding between the Sultans of Rûm and the Republics of Venice and Pisa. In 1219, Venetian traders are recorded at Lampsacus, and a few years later in 'Turkey' or *Turkmenia*[2] itself: the commercial rights of the Western visitors are guaranteed by treaties of 1210, 1215, and 1220. The Provençals seem to have also gained access to Turkey by means of a harbour on the south coast of Asia Minor; this town, lying 'opposite' Cyprus, was known in the West by the names of *Lo Proensal*, *Prodensalium*, or *Port of the Provençals;* and some account must also be taken of Satalia, a haven which remained in the hands of the Greeks till 1204, was then seized by the Aldobrandini, and finally passed under Seljuk rule in 1207.

For pushing their commerce in the upland of Asia Minor the Latins found some assistance (and perhaps equal hindrance) from the Greeks and Armenians who still lived there, not yet overwhelmed and extinguished, though already surrounded and oppressed, by the Turkish majority. These Eastern Christians were keen traders, and took especial interest in the export of alum; but when Rubruquis passed through Iconium in 1255 this trade was mainly controlled by the newcomers from the West; in particular, a 'long firm' of strange constitution, under a Genoese and a Venetian, Nicolas de Santo Siro of Acre, and

[1] Under Alaëddin Kaïkobad. [2] The Seljuk dominions.

Boniface of Molendino,[1] had monopolised all the alum in 'Turkey,' binding the Sultan to sell only to them, and raising the price from fifteen bezants to fifty.

Politically, the Venetians were the most deadly enemies of the Greek empires of Nicaea and Trebizond, the relics of the great dominion they had shattered in 1204; but even with these states the indefatigable merchants of the Adriatic contrived to do some business as early as 1219. Here, however, they were naturally at a disadvantage, and the traders of Genoa and Lucca appear at Nice, from about 1230, with no indistinct foreshadowing of the future alliance, which was to expel the Latins from Constantinople. This alliance took shape about 1239; the Bulgarians joined hands with their Orthodox brethren against the heretical Franks; and the Genoese loss of Rhodes in 1250, after a bare two years of possession, added fuel to the flame of their hatred against the intruding masters of the Archipelago and the Bosphorus.

The kingdom of Trebizond,[2] like those of Nicaea and the Lesser Armenia, was a 'gate' of trade for East and West alike—of exit for Oriental, of entrance for European, products and dealers. Long before the revolution of 1204 Trebizond, as a frontier city of the Byzantines, had been a meeting-place of traffic from many quarters;[3] and as a Euxine port its commerce with the north and east coasts of the Pontus was ancient and respectable. The Mongol destruction of Bagdad gave a new opportunity to its enterprise,

---

[1] Cf. Heyd, *Commerce du Levant*, i. 302, and note, where 'Molendino' is explained as referring to the Venetian family of Molino, and the Boniface in question is identified conjecturally with a condottiere in the service of the Sultan of Iconium, 1242-1243; cf. Sanuto, *Secreta Fidelium Crucis*, p. 235, etc.

[2] Formed as an independent state, after the overthrow of Constantinople in 1204, by a branch of the Comnenian family.

[3] As noticed by Masudi, Al Istakhri, and Ibn Haukal.

VI.] CHRISTIAN TRADERS IN THE BLACK SEA 449

as the desolated trade-routes of the South yielded for a time to their Northern rivals; and the restoration of Greek rule at Constantinople put it once more in unimpeded communication with the West.[1] In 1266 two Massiliot traders arrived at Trebizond with letters to the Emperor from Charles of Anjou, Count of Provence; with the aid of the Comnenian Government these envoys hoped to reach the Grand Khan of the Mongols, and deliver to him the letters of their Lord; but we do not know that they ever penetrated into Tartary. If so, they probably followed the North Persian trade-route, which came into fashion on the advent of the Mongols, from Trebizond to Erzinghian, Erzerum, Lake Van, and Tabriz,[2] and thence along the south coast of the Caspian to Bokhara and Samarcand.

As to the other commercial tracks of Asia Minor, the most famous and popular led from south-east to north-west, and from north-east to south-west,—from Antioch through Little Armenia[3] to Iconium, Nicaea, and Constantinople, or from the Caucasus and the Armenian Plateau through Sivas and Kaisarieh[4] to Konieh[5] and the sea ports of Lycia or Cilicia; along the latter Friar William of Rubruck journeyed on his return from the Mongol Courts on the Orkhon and the Volga.

But the conquering Venetians of the thirteenth century (and other Latins in their train) did not only push their traffic into the centre of Anatolia from their new base at Constantinople; they also advanced, as we have seen, to the North Euxine coast, to the mouths of the Dniepr, Don,

---

[1] The merchants of Trebizond, however, only undertook 'petty' commerce, leaving higher things to Western visitors.

[2] This was afterwards traversed by Clavijo, on his way to the Court of Timûr, pp. 64, 69, 78, 89, etc., of Hakluyt Society's edition.

[3] Cilicia and the Taurus.
[4] Sebaste and Caesarea.
[5] Iconium.

Kuban, and Volga, eastward to the shores of the Caspian, northward to the Russian and Finnish lands above the steppes of the Kipchak or Komans. As early as the twelfth century, the Russian *Song of Igor* boasts that the glory of the hero Sviatoslav had reached Venetians as well as Greeks; in the thirteenth, Carpini and Rubruquis found the Western traders active at Kiev and throughout the Crimea. It would seem that the vague suzerainty of the Byzantine Emperors over the North Euxine littoral,[1] marked by fortresses in the more vigorous times of Justinian, had been reduced to vanishing point by the Komans, under whose rule Latin trade soon gained a share in the export fur-trade hitherto absorbed by Saracen dealers. The Koman rule was replaced by that of the Mongols in the second quarter of the thirteenth century (1223-1239); but this change of masters did not destroy Western commerce; and the chief port of Komania—Sudak or Soldaia, near the still more famous Kaffa,[2]—retained its shipping, its old administration, and its Greek bishop, paying a moderate tribute, and sheltering a busy population of many nationalities, among which Goths or Germans are noticed by Rubruquis. Under Manuel Comnenus, as in earlier times, Rosia and Matracha, Kherson and Tanais,[3] had con-

---

[1] None of these regions are mentioned by the Frankish and Venetian spoilers of the Eastern Empire in the 1204 treaty of partition; some are found later to be nominal possessions of the Trebizond emperors. At Alusta and Ursuf Justinian's Crimean fortresses still stand in good repair.

[2] Or Theodosia. Soldaia is otherwise called Sugdaia, Sodaia, Soldachia, Soltadia, Scholtadia, and Sudagh. From its numerous Goths the whole region was sometimes called Gothia; cf. Rubruquis, p. 219, vol. iv. of the Paris *Recueil*.

[3] Sevastopol and Azov. This last was a very ancient town, cf. Strabo, xi. 2, 423; Tana (not on the *same* site) was probably a foundation of the later thirteenth century; there is no clear trace of it earlier, as in Rubruquis; but the name occurs on some of the earliest portolani (*e.g.* that of Pietro Vesconte), of. c. 1309-1310. It was eighteen miles from the mouth of the Don, and on the east side of the river, according to Bembo.

trolled the trade of this coast; later, as in the fourteenth century, Tana and Kaffa were supreme; and Ibn Batuta (c. A.D. 1334), while praising the port of Sudak as 'one of the finest' known, admits that the city, though 'formerly very large,' had been mostly destroyed in wars between the Greeks and Turks.[1] It was still apparently at the height of its prosperity when Nicolo and Maffio Polo,[2] on their way to sell their jewels at the Court of the Khan Berké, landed on the Crimean shores (1260). As their brother, Marco Polo the Elder, then trading at Constantinople, possessed a house of business in Sudak, and as no other port in the neighbourhood could then pretend to the same importance, it was probably here that the travellers disembarked; the Polo property in Soldaia was, doubtless, typical of a numerous class of Italian holdings in the Euxine; and the careful preservation and maintenance of the same was also characteristic. When old Marco retired from Constantinople to Venice he left his Crimean house or houses in the charge of his son Nicolas and his daughter Marocca; by his will of 1280 he bequeathed them his distant estate; in the same manner the Genoese retained and transmitted family holdings in Kaffa and Matracha.

The latter city, in the Taman peninsula, near the Strait of Kertch and on a branch of the Kuban,[3] was the political and commercial centre of a large district covered with villages. It had a port of fair depth, accessible to large vessels, and as early as the tenth century it was a chief

---

[1] According to Abulfeda, Christianity had been displaced by Islam at Sudak, since Ibn Said (1250-1274); in other words, between the time of Rubruquis and his own.

[2] Their way lay through Sarai (to Bulgar and back, up and down the Volga) along the west coast of the Caspian to Bokhara, and so to the Great Wall of China, which they were the first Europeans to cross. Their way back brought them to Lajazzo. We leave this journey to another volume.

[3] The *Sakir*, according to Edrisi, ii. 400.

town of the Khazars, well-known to the Imperial author[1] of the tract on the Administration of the Byzantine Realm. In 966 it was captured (according to a somewhat dubious tradition) by Russian raiders, led by Prince Sviatoslav, and became part of the Slavonic principality of Tmutorchan, which disappears in the first quarter of the twelfth century. Edrisi,[2] however, knows nothing of this Russian occupation, but tells us that the people of Matracha were at constant feud with those of Rusia or Rosia, probably the celebrated town of that name on the Sea of Azov. The ruling family or families in *Matrica* during the thirteenth century seem to have been Greek in speech and Christian in religion; many Greek laity and clergy were settled there about 1230; while merchants[3] from Constantinople (among whom Venetians must have taken the leading place) traded in the harbour, and thence crossed the Azov Sea in light boats to the mouth of the Don, where good purchases of fish might be had (c. A.D. 1250).

In the days of Byzantine power, as late as Manuel Comnenus, the Greek commerce with these regions of Rosia

[1] Constantine VII. (Porphyrogennetos) calls it Τὰ Μάταρχα or Ταμάταρχα. In the fifteenth century 'Matrica' was still of some importance, and was held for a time by the Genoese, under the suzerainty of the neighbouring Circassians.

[2] Edrisi, ii. 400. The port of *Rosia*, probably the same here and in Manuel's treaty with the Genoese in 1169, is mysterious; Edrisi's Map marks a *Rosia* at the mouth of the river of the same name, flowing from Mount Kokaïa. The Strait of Kertch, both in Edrisi and Rubruquis, is considered as the mouth of the Don. *Rosia* probably stood either (1) near the site of Kertch, Bosphorus, or Vosporo; or (2) on the other side of the strait; or (3) at the true mouth of the Don, near the present Azov. Possibly it is another name for the ancient *Tanais*. Between *Matracha*, on the Taman Bay, and *Sudak*, Edrisi places two stations—(1) *Butra* or *Buter;* (2) another, near Matracha, adjoining the town of Rosia, and much in the position of the modern Kertch; perhaps he misunderstood a map he was copying; cf. the *Casal* (or *Cassar*) *degli Rossi*, of the fifteenth century portolani, at the true mouth of the Don; and see Reinaud's *Aboulfeda*, ii. 320; Frähn's *Ibn Foslan*, pp. 31, 32.

[3] Rubruquis, p. 215.

VI.] CHIEF PORTS OF THE BLACK SEA 453

and Matracha, was highly valued by the princes and people of Constantinople. Here was one reason for the jealous exclusion of foreign interlopers (as in Manuel's prohibition to the Genoese); another explanation of the same lay in the very real danger that the Western merchant might sell arms to the natives of the interior, and thus equip a powerful force against the unwarlike Greeks.

Kaffa, 'the fort'[1] of Constantine VII., the Theodosia of the older Greeks, though its fame was of later time, could boast a high antiquity. Originally a colony of the Milesians, it seems to have been re-built by one of the Graeco-Scythian kings of Bosphorus; at the opening of the fourth century, in the time of Constantine the Great, it marked the frontier between the dominions of these princes and the 'Roman' territory of Kherson or Sevastopol. As a natural harbour it was only surpassed by one other upon this coast,[2] and it was directly planted upon the commercial route from the Bosphorus to the Volga and the Don. The Genoese colony may have been founded before the events of 1204;[3] but it probably resulted from the Greek restoration of 1261, and the consequent destruction of Venetian ascendency in the Pontus. Like Sudak and the rest of the Crimea, it was treated by the Mongols as an immediate possession of the House of Batu; thus in 1265 the fugitive Sultan of Rûm[4]

---

[1] ὁ καφᾶς, Constantine Porphyrogennetos, *De Administrando Imperio*, 552, 555, Bonn edition; not in Edrisi.

[2] The long neglected inlet of Sevastopol itself.

[3] A collar is still shown in the Imperial Treasury at Moscow, which is said to have been won as *spolia opima* by the Grand Prince or Duke Vladimir Monomach in single combat with the Genoese Governor of Kaffa (c. A.D. 1100). The Imperial Ambassador to Moscow in 1517 and 1526, Sigismund von Herberstein, is the first writer to give this story, in his *Rerum Muscovitarum Commentarii* (i. 16 in edition of 1841 by Starczewski); the Polish Chronicler and geographer Strykowski, also of the sixteenth century, corroborates; but it must remain at present very doubtful; cf. Heyd, *Commerce*, ii. 161.

[4] Or Iconium.

was given an asylum in *Armenia maritima* (as it was often called from its Armenian settlers) together with the town of *Sudagh* as a fief. The first clear view we have of the Genoese settlement is only in 1316, when its statutes are drawn up; from the terms of the concessions given and withheld by Manuel Comnenus in 1155 and 1169, it does not seem likely that the Ligurians were then established either at Kaffa or at any other point on the Northern Euxine littoral. At the most, they were occasional visitors.

On the same coast, or in the immediate neighbourhood, the towns of Solgat, Porto Pisano, Sebastopolis, and Kherson figured humbly in thirteenth-century trade. The first of these, also called *Krim*, was in later time the residence of the Emirs who governed the Crimea in the name of the Mongol Khans of Sarai; it lay a few miles west of Kaffa, and was greatly frequented, especially for its furs, silks, and spices, by the merchants of that port. *Porto Pisano*, defined by Pegolotti, in the fourteenth century, as the first harbour touched at after quitting Tana, may be on the site of the present Siniavka; in any case it was near the true mouth of the Don;[1] it is mentioned by a portolano of 1318; and its name, which sufficiently indicates its origin, is to be found even on maps of the seventeenth century. *Sebastopolis*, near the modern Sukhum Kalé in the Colchis region, on the extreme east of the Pontus, was an important Byzantine fortress under Justinian and later emperors; after the collapse of Imperial power it became the residence of a Georgian prince; its name occurs in some of the earliest portolani; and it seems to have been visited by Western traders from about 1250.[2] The inquisitive traders of the

---

[1] Pegolotti, *Practica della mercatura*, p. 39; Heyd, *Commerce du Levant*, ii. 166-167.

[2] A small Latin community here is noticed in 1330.

West also made occasional visits to the best of all the North Euxine harbours, the ancient *Kherson,* the modern *Sevastopol,* whose various names have caused so much confusion, whose site was famous for the martyrdom of St. Clement of Rome, but whose merits as a port were never realised in the Middle Ages.[1] They were left for Catherine II. to discover, or at least to proclaim and to use; the Genoese might have made it a greater Kaffa, if they had been able to extend their influence along the western shore of the Crimea; but they never seemed to have gained a firm position beyond Balaclava, and their dominion was confined to the southern coast rim between the mountains and the sea.

The mouth of the Don, whatever the port might be, which at one time or another commanded the estuary, was the starting-point of a trade-route, not unimportant during the earlier Middle Ages, but brought into far greater prominence by the Mongol victories and settlements. Hence, through Sarai (or later, through Astrakhan), and round the north coast of the Caspian, this commercial road brought the dealer to the Aral Sea, the Syr and Amu Daria, and the rich land of Khiva, Khwarezm, or Urgenj,[2] and so along the course of the Jaxartes to Otrar, Ferghanah, and Almalig.[3] From this point the way to China lay through the pass of the Iron Gates in the Thian Shan and brought the traveller to Lake Sairam, Urumtsi, Khamil, and the western extremity

---

[1] Jordanes (Jornandes) ascribes its foundation to the Greeks of Heraclea Pontica; its livelihood was trade with the Scythians; and 'here the greedy trader brought the rich produce of Asia.' The Huns (Hunuguri) supplied this market with furs in Jordanes' time; but it never could have aspired to rival Tana, Kaffa, or Matracha. The neighbouring Balaclava did some trade in the fourteenth century. Inkerman is apparently mentioned in an inscription of 1247 as 'Kalamita.'

[2] Also called Jorjan and Jorjaniah by Arabs and Persians; distinct from the other Jorjan to the south-east of the Caspian. Otrar was probably near Talas, the modern Hazret-i-Turkestan.

[3] Or Kulja.

of the Great Wall.[1] Finally the merchant,[2] passing on into the Pacific plain of Asia, would reach his goal either at Kinsai or Pekin, where his purchases would be mostly of silks, brocades, gold-threaded stuffs, spices, musk, rhubarb, and galangale; the export tea-trade appears to belong to a later age, and was clearly unknown in the time of Marco Polo.[3]

A South-Caucasus and trans-Caspian route to the Far East was also in use between the Don estuary and the Lower Syr and Amu Daria by way of Sukhum-Kalé or Sebastopolis in Colchis, Derbent at the Caspian Gates, Trestrego near the modern Krasnovodsk, and Khiva or Khwarezm.

Through the new Christian state of Little Armenia[4] on the Gulf of Issus, Western trade penetrated into the Upland of Asia by another route. The kingdom of Rupen, at its greatest extent, under King Leo II. (1187-1219), possessed the south coast of Asia Minor from near Satalia to Alexandretta and even beyond, and in this shore-line were several good harbours, notably Palli,[5] Lajazzo, Corycus or Curco, Tarsus, Adana, and Mamistra or Mopsuestia. Several

---

[1] At Sucheu in Kansu.

[2] Most traders followed the path along the north, in preference to the south, of the Thian Shan.

[3] For he never mentions it; Pauthier (*M. Polo*, p. 384, note; 386), thinks he means to allude to the tea-tree in Assam (Gaindu), but this is very indistinct. Cf. Heyd, *Commerce*, ii. 231, etc., 252-253; Yule, *M. Polo*, ii. 37, etc.

[4] Founded near the time of the first Crusade by a migration from Armenia proper; made important by Leo II.; its traditional policy was one of alliance with Latins against Moslems and Byzantines. Leo westernised his realm, and introduced a feudal constitution; the Court was modelled on Western examples; the crown was accepted from Henry VI. as a fief of the Western Empire; the Church was united with Rome; branches of the Western Orders of Chivalry were instituted. The Mongol conquests brought them right up to the borders of Little Armenia. The new routes and new markets opened by the Mongol conquests have not been sufficiently noticed.

[5] Pals or Plas. Also called *Portus Palorum* (or *de Pallibus*), the modern Porto di Plas. From Lajazzo, a trade-route to Tabriz in North Persia is described by Pegolotti in the fourteenth century: *Pratica della mercatura*, ch. vi.; see pp. 299-301 of Yule's *Cathay and the way thither*.

trade-routes crossed the country ; from Aleppo or Antioch to Iconium and the Bosphorus ; from the Upper Euphrates (by Marach) to the centre and West of Anatolia ; from the harbour of Lajazzo to Northern Persia and Tabriz ; and from Egypt, Arabia, and Syria to the North. The southern and eastern roads met at Alexandretta, passed through the coast towns of Mamistra and Adana, and crossed the Taurus by the pass of Gulek Boghaz.

The products of Little or Southern Armenia were varied and highly prized, especially its cottons, camelots, and iron ;[1] the most favoured among its Latin visitors were the Genoese, whose embassy in 1201 was well received, and gained an increase of trading privileges by the treaty of 1215 ; a diploma of Leo II., issued in March of this year, fixed the dues payable by the Genoese at the passage of the river Jihan, and provided against any extortion. In earlier days, after the Greek emperors of the Byzantine revival (960-1025) recovered this province, the Venetians had been foremost ; and their rights, as guaranteed in 1082, were confirmed by treaties of 1201 and 1245.

Syrians, Persians, and other Moslems did business in Little Armenia ; thus the Bagdad firm of Yussuf had a branch for its cotton trade in Lajazzo ; and in 1268 a galley full of rich wares belonging partly to Armenians, partly to men of Acre, Tyre, and Antioch, and partly to various subjects of the Mongol Khan of Persia,[2] was seized in the harbour of Corycus.

Both from Lesser Armenia and from the northern and southern coasts of the Black Sea, from Lajazzo as well as from Trebizond and Sudak, new or extended trade routes were opened by the levelling conquests of the Mongols ; the way from the Taurus to Tabriz and the Caucasus ran

[1] From mines in Mount Taurus.   [2] Abaka, successor of Hulagu.

through Sivas, Erzinghian, and Erzerum; and the rulers of Sis and Tarsus were in a specially favourable position to promote the traffic on this road. As Christian sovereigns of the Roman faith, and as favoured 'allies' of the Great Khan, they formed a bond between the Latin and Tartar worlds; the Emperors of Trebizond occupied a similar but less happy position; for they were adherents of the 'Greek schism,' on one hand, and counted rather as vassals than as friends of the Mongols, on the other. It is not surprising, therefore, that by the time of Marco Polo the Genoese seem to have pushed on to the Caspian [1]—like the Russian pirates and pedlars of the tenth century—and to have begun a maritime trade, especially concerned with silken stuffs, upon its waters; in the light of this fact, we shall better understand the remarkable knowledge of the inland sea shown in the early portolani.

Egypt, as we have seen, was of capital importance, both in the politics and in the trade of the Crusading Era; for the Fatimites were the chief leaders of the Moslem resistance to Western attack; and among the three leading markets of the world the Delta of the Nile must certainly be counted, only inferior (if inferior at all) to Southern Mesopotamia, or the Bagdad region, and the Bosphorus, or neighbourhood of Constantinople. Both from the Mediterranean and from the Red Sea traffic flowed into the 'market of two worlds,'[2] as William of Tyre surnames Alexandria; by the former Egypt communicated with the most distant countries of Europe—Flanders, England, the

---

[1] Cf. Frähn, *Ibn Foslan*, pp. 1-23, 70, 147, 168, 226, 247, 258, 266; ibid, *De Chasaris*, pp. 591, 601, etc.; Masudi, ii. pp. 9, 11; Ibn Dasta in Rössler, *Romänische Studien*, p. 362. The Bulgarians of the Volga, however, usually carried most of the Russian goods down to Itil (answering to the modern Astrakhan); cf. Masudi, ii. 15; Heyd, *Commerce*, i., 60-62; Saveliev, *Handel der Bolgaren* in Erman's *Archiv*, vi. 96-98.

[2] Properly, *forum publicum utrique orbi*.

Hanse Towns or the Scandinavian lands; by the latter it received the products and dealers of Yemen and Oman in South-Western and South-Eastern Arabia, of India, and even of China. But this last was carefully guarded against Christian intruders, and all European merchants who wished to do business in Egypt or with the regions of South-Western Asia had to take what opportunities were offered them at the market of Alexandria. The Frank conquest of Syria interrupted the old pilgrim and trade route across the Sinai peninsula, and compelled Egyptian commerce and travel to fall back on the old road of the Ptolemies from the Middle Nile to the ports on the west coast of the Red Sea, the ancient Myos Hormos and Berenice, the mediaeval Aïdab. To this shore came the import traffic from Aden and the Indian Ocean; a short passage of seventeen to twenty days with camel-caravans brought one to the Nile, at Syene [1] or at Kus; and from either of these harbours it was an easy river-journey down to Cairo and the Delta.

The importance of the Nile in this connection was in part the explanation of the popular belief that the river of Egypt, supposed to be identical with the Gihon of Scripture, flowed from Paradise, and brought the spices, aloes, and other products of Eden down to the lands of fallen men; [2] in the religious and book-learned world of Christendom this view long survived the spread of juster notions in commercial societies; and even the latter may have regarded it with some satisfaction as attaching a greater

[1] Assuan.
[2] Cf. Edrisi, i. 28, 35-39, 49, 51, 127, 132, etc., 152, 313, 331, etc. M. Polo's description of this commercial route from Aden to Alexandria is somewhat inaccurate, especially in his reckoning of time required by various stages. Kus and Assuan, not Korosko, served as the Nile rendezvous for the Red Sea trade. James de Vitry speaks of vessels coming direct from the Indies to Damietta, whereas three kinds of craft were usually employed on the journey— Ocean vessels, Red Sea vessels, and Nile boats.

value, and so a higher price, to the articles imported from Alexandria. At the great port of the Delta an amazing quantity of spices was constantly being collected and distributed; Benjamin of Tudela, William of Tyre, James de Vitry, and Burkhard the ambassador of Frederic Barbarossa,[1] are but the leading witnesses, among Latin writers and travellers of the twelfth and thirteenth centuries, to the greatness of this 'market of all nations.'[2] Rosetta and Damietta, though at a long interval, followed Alexandria in wealth and prosperity; the admirable waterway of the Nile, the close proximity of the Red Sea,[3] the central and commanding position of Egypt both for war and commerce, gave the ports of the Delta an unquestionable superiority over the Syrian harbours.

But great as were the profits which the Egyptian trade brought to Europe, this commerce had another side, wherein European imports appeared no less indispensable to Egypt. That wealthy country was wholly dependent on outside supply for its wood and iron; and in spite of every patriotic and religious prejudice, and prohibition, Christian states continued to furnish the Fatimite dominions with these vital articles from the tenth to the twelfth century. After the time of Saladin this commerce rather increased than diminished with the growing sense of commercial interest; the efforts of the Popes (even of Innocent III.), of the Venetian Doges, of the Genoese Consuls, of the Aragonese kings, were in vain directed against it,[4] with a zeal often feigned and never

---

[1] 1175 A.D.

[2] Cf. Benjamin of Tudela on Alexandria; see pp. 261-3 of this vol.

[3] Connected with the Mediterranean Sea by a freshwater canal at least as old as Necho, and constantly renewed and as constantly abandoned, down to the middle of the eighth century; cf. *Dawn of Modern Geography*, i., p. 162.

[4] Cf. *e.g.* the prohibitions by (1) The Doge P. Ziani, 1226; (2) the Consuls of Genoa, 1151; (3) James I. of Aragon, both for Montpellier and Barcelona (1231-1274). The Pisan secret treaty with Egypt in 1173 is a good example

EGYPT AS A TRADE CENTRE

effective; and while official proclamations were being issued against a 'blasphemous trade,' it was secretly extended by fresh treaties. The Italian and other Mediterranean smugglers might perhaps claim that, if they retired, their place would be taken by the Flemings, Saxons, Danes, and other Northerners[1] who now made their appearance in the markets of the Delta. In spite of the rage and terror which Saladin's conquests produced; in spite of the growing sense that if the Holy War were ever to succeed, Egypt must be conquered; and in spite of the 'Egyptian' character of all the later Crusades,[2] commercial ambitions steadily gained upon political enmities. Thus Wilbrand of Oldenburg[3] replaces Godfrey of Bouillon; and the indifferentist, Frederic II., nominal leader of a Crusade, maintains so close a friendship[4] with the Sultan of Egypt that German merchants (it is said) were able to travel in the company of Egyptians to the Indies.[5] Even if this story may be in part accepted, it is probable that the journey in question must be understood as no more than a voyage to Aden,[6]—

---

of a numerous type; cf. Werlauff, *Symbol. Geog. med. aev.*; Hanotaux, 404. Even little states like Ancona, Bari, Brindisi, Trani, and Barletta, and other small ports of the Norman Kingdom and Central Italy traded with Alexandria, Rosetta, or Damietta in the last century of the Crusading period, 1170-1260.

[1] Cf. Benjamin of Tudela on Alexandria, see pp. 261-3 of this vol. He adds, according to one emendation of the (here doubtful) text, the *Russians*. Asher corrects this to *men of Roussillon*, and this seems more probable; but Heyd defends the old reading, and quotes Constantine Porphyrogennetos and Ibn Khordadbeh, as implying that the Russians navigated as far as Syria in the tenth century—an exceedingly doubtful theory. Normans and Burgundians are also named by Benjamin of Tudela among Alexandria merchants; it is curious that as late as 1174 (as in 1153-1154) Norman fleets attacked Egypt.

[2] From 1202. The fourth Crusade was officially announced as an attack upon the Nile Delta.

[3] 1211.

[4] In 1225-1240.

[5] So Matthew Paris, *Chr. Maj.*, v. 217, ed. Luard.

[6] Cf. Ibn Alathir, in *Rec. des hist. des Crois., hist. orient.*, i. 597, and Edrisi, i. 51; the latter specifies, among the articles of the Aden market, musk, pepper, camphor, galangale, cocoanuts, aloes, ebony, ivory, and porcelain.

the 'market of India and Zanzibar, of Abyssinia and Oman and Persia,'—where, from remote time, Indian and Chinese traders had often come to meet those of Arabia and Egypt, and where Indian produce was usually transferred to the lighter barks of the Red Sea.[1] A stricter criticism must be applied to the Pisan claim of an 'Indian voyage' by way of Egypt in 1175, and not even the later Consulate of Venice in Siam[2] can put this tale beyond question.[3]

[1] Makrizi, in Quatremère, *Mémoires sur l'Egypte*, ii. 162, etc.

[2] At the end of the fourteenth century.

[3] Heyd, *Commerce*, ii. 153-156. As to the Indian trade, cf. the strange narrative in the *Conosçimiento de todos los reynos y tierras y señorios que son por el mundo*, etc. (c. A.D. 1330) of a journey of Sorleone Vivaldo to Magadoxo in East Africa, on the way to India; however fabulous the story may be in many parts, the mention of Magadoxo in this connection is important, referring apparently to the close of the thirteenth century, and suggesting a well-known trade-route, in use at the close of this period of the Central Middle Ages (c. 1260). In the early fourteenth century (c. 1315), another member of the Vivaldi family, one Benedetto, seems to have gone to India by the overland route from Lajazzo to Tabriz and Ormuz.

The growing persistence, daring, and success of the Christian trader of the Crusading period corresponds to an ever-increasing weakness and decay of Moslem commerce, which, down to the close of the eleventh century, unquestionably controlled the purse-strings of the world. How great the mercantile ascendency of the Arabs must have been in the pre-Crusading time may be partly guessed from the amount, the diffusion, and the chronology of the Mohammedan coins which in our own day have been found buried in European hoards. These range from A.D 698 to 1010; they have been discovered in the neighbourhood of Kazan and Lake Ladoga, and in the valley of the Petchora, in the Crimea and the vicinity of Mainz and of Frankfort-on-Main, in Iceland and England, and in almost every part of the Baltic coasts, from the mouth of the Oder to the Gulf of Finland, from the ancient 'Jumna' to 'Aldeigia-borg.' In Esthonia, Livonia, and Courland, more than 13,000 pieces have been unearthed, and the vast majority of the finds have been within the borders of the modern Russian state, especially on the Upper and Middle Volga. As to this, we need only recall the early mediaeval importance of Bolghar and the early Islamism of the Old Bulgarians of this region, who, according to Ibn Dasta, were paid in dirhems, and bought whatever they wanted with marten skins, constantly descending the Volga (a river-journey of two months) to trade at Itil or on the shores of the Caspian, and even making a regular business of caravan-trade with Khwarezm or Khiva. Through the medium, chiefly, of Arab

The development of piracy in the Persian Gulf during the thirteenth century naturally increased the importance of Aden and the Erythaean route; the junks of Fo-kien, under the Sung dynasty, stopped at Muscat and Aden, in Oman and Yemen, and thus Egyptian commerce in Southern and Further Asia was developed. An embassy from Ceylon to Cairo in 1283 seems to have been, not the commencement, but the continuation, of a direct commercial intercourse dating from the middle of the century.

Thus the Red Sea avenue to India, and beyond, enjoyed from time to time during the Middle Ages a decisive and almost modern preponderance over the maritime route from the Persian Gulf; yet the latter was of great importance during most of this period (950-1250), and it was merchants, Byzantine jewellery and money found its way to Bolghar, and Ibn Foslan observed the use of the same, not only among the Bulgarians, but also among the less civilised and commercial Russians. The conversion of the Bulgars (or at least their de-Islamising by the Russian princes and others) in the eleventh century broke up a trade-route which had depended on the close connection of Bolghar with the Mohammedan world, and especially with the Central Asiatic and Persian lands from which come the vast majority of the Arab coins referred to. Rich Moslems seem to have had for many generations a peculiar weakness for the furs of the North, and, to a less degree, for the amber of the Baltic; commerce as well as proselytism brought Arab wanderers into mid-Russia; and if religion was the dominant interest with Ibn Foslan, mercantile conceptions seem to underlie the remarks of Masudi on the Ludaaneh, that 'largest tribe of Russians,' whose trade, he declares, extended not only to Khazaria but also to the Bosphorus, the New and the Old Rome, and even Spain. Much of this commerce, if we regard it as authentic, must have passed through Moslem channels; but these were still the days of the cosmopolitan activity of Islam and the Arabs, whose simultaneous presence in Spain and the Narbonnese, in South Italy and in Sicily, in Korea and Japan, as well as on the Middle Volga, testified in the tenth century to a long-lost expansive and civilising force in the faith and system of Mohammed. During the 'Macedonian' revival of the Byzantine State a great increase of trade was recorded between Constantinople and the Moslem world, especially of Spain; it may also be noticed how close and active was the commercial intercourse between Spain and Egypt in the tenth century. Both were anti-Abbasside states, one Ommeyad, the other Fatimite; and Chasdai, the Hebrew minister of Abderrahman III., not

intimately connected with the overland ways already noticed to Tabriz. From the latter traders could descend through Persia by way of Yezd and Kerman to Bassora and the Gulf; the voyage thence to China lay past Ormuz and the Malabar coast.

only worked for his own race and religion, but for his sovereign and the Western Caliphate; it was clearly in the interest of Cordova to cultivate friendship, and, if possible, alliance with Cairo on one side and Khazaria on the other. In this policy Constantinople formed an almost indispensable link; here was one of many reasons for the traditional amity of the Eastern Caesars and the princes of 'Magreb.'

Cf. Paul Saveliev, *Numismatique mahométane*, and map showing diffusion of Moslem coins; ibid, . . . *Handel der Bulgaren* . . . in Erman's *Archiv.*, vi.; Frähn, *Bulletin de l'Academie de St. Pétersbourg*, t. ix. Nos. 19, 20, 21; ibid, *Ibn Foslan*, pp. 1-23, 70-71, 147, 168, 174, 247, 258, 266; Masudi, ii., 9, 11, 15, 18; Rafn, *Antiquités russes*, i. 295, 317, 426, 432; ii. 119; Heyd, *Commerce du Levant*, i., 47-49, 50, 53, 58-59, 61-63, 67, 70-71, 74-76. Heyd identifies the 'Ludaaneh,' not with any Russians proper, but with the Slavic Lutchanes or Luzanians of Lutsk in Volhynia, or of Veliki Luki near Old Novgorod; other writers have usually translated the term either by 'Lithuanians' or 'people of Ladoga.'

CHAPTER VII

GEOGRAPHICAL THEORY AND DESCRIPTION

IN the former volume of this history [1] an attempt was made to illustrate the chief types of European study or speculation in geography during the first six hundred years of Christian supremacy. The Statisticians, the Cosmographers, and the Fabulists were examined in the writings of Dicuil and the Ravennese, of Cosmas and of Solinus, of Capella, Macrobius, Basil, Æthicus, and others. In the Central Middle-Age period that now concerns us, Patristic geography passes into Scholastic; and while practical exploration steadily advances with the travels of pirates, warriors, merchants and diplomatists, theoretical 'earth knowledge' shows likewise an unmistakable progress. Of this we shall take three examples, one Byzantine, another German, the third international; and in the manual of Constantine Porphyrogennetos *On the Administration of the Empire*, in Adam of Bremen's *History of the Church of Hamburg*, and in the chief Maps of the tenth, eleventh, twelfth, and earlier thirteenth centuries, we shall hardly fail to recognise a very different spirit from that which overshadows the Dark Ages proper. Constantine VII. indeed belongs to a time which, for Western Europe, is often considered to be among

[1] *Dawn of Modern Geography*, i., pp. 243-391.

the gloomiest of epochs; but he represents Oriental Christendom in the full tide of a revival (under the 'Macedonian' Emperors), which in many ways led up to, aided, or announced the mediaeval Renaissance of the West. With Adam of Bremen, the signs of that Renaissance are unmistakable; we have reached the clear light and true shining of a new life, soon to be expressed in action by the Crusades, and in thought by the Schoolmen and the Universities. Finally, in some of the Maps, even of the tenth and eleventh centuries, and more conspicuously in the later 'chorography' of Matthew Paris, we have, for the first time since the ruin of classical science, a draughtsmanship which is not merely fanciful, and which attempts to co-ordinate and express observation of material fact. At the same time, we must not forget that at the very close of the Crusading time, distinguished authors may still be found expressing themselves in the spirit of Solinus rather than of Ptolemy or Strabo. In the middle of the thirteenth century, the scientific spirit and careful enquiry of Roger Bacon and Albertus Magnus may be contrasted with the bald and superstitious traditonalism of most 'philosophers'; geographical mythology and pseudo-science continue to flourish till a much later date;[1] and even to the end of the Middle Ages, the Romance of Nature paralyses the enquiry of a majority of students. From the ordinary maps of this time, moreover, and in spite of certain marked developments, one may see how little the increase of intellectual activity is associated with truer views of the earth surface among untravelled scholars; books and charts are multiplied, but they often seem only to repeat the vain babblings of the wonder-seekers; we have yet to wait for the time when achievement and thought, action and reflec-

[1] And may be illustrated by such works as the Borgian Map of c. A.D. 1450.

tion, discovery and enlightenment, are normally and adequately connected. Only by the labours of a few was the mist dispelled from the face of the earth.

§ 1.—*Byzantine Geographers—Constantine VII. (Porphyrogennetos)*

Among the Byzantine geographers of this central mediaeval period, the Imperial Encyclopaedist, Constantine VII., better known by his surname of *Porphyrogennetos*[1] (911-959), holds a primacy as unquestionable as that of Procopius in the age of Justinian. The Dark-Age time, whose beginnings are illustrated by the one writer, and whose close is adorned by the other, can scarcely boast of any names more illustrious in its literary and mental history than these Byzantine authors, 'so accurate for their own times, so loose or fabulous for preceding ages.'

Constantine VII., as the clear judgment of Gibbon[2] acknowledged, is more to us as a geographer than as a lawyer, moralist, economist, tactician, or master of ceremonies. For though his Treatise *On the Public Administration of the Empire*, with its detailed account of the subjects, vassals, and neighbours of the Byzantine state, is not free from the defects of the *Tactics, Ceremonies, Themes, Geoponics,* or *Basilics,* of the same royal author, yet it is certainly distinguished by 'peculiar merits.' For it abounds in details of contemporary (if not always first-hand) information upon the Barbarian world of Europe to the north and north-east of Constantinople; and as this Barbarian world of the tenth century was in great measure

---

[1] *I.e.* Born in the *Porphyry Chamber,* reserved for the use of the pregnant empresses. This room was hung with purple (Greek *porphyry*), the imperial colour.

[2] *Decline and Fall,* ch. liii.

a lasting (and not as so often before, an ephemeral) group of nations, the photographic sketch of the *De Administrando Imperio* likewise remains as a record of abiding value.

For the history and geography of several of these nations, it is in point of time the primary authority, a parallel for South-Eastern Europe to the Mosaic Record for the Hebrews, or to the *Germania* of Tacitus for the regions East of Rhine and North of Danube. Its chief objects are to survey the neighbours and vassals of the Empire; to describe their countries, to point out the various tribes which might be considered friendly or hostile, useful or dangerous, to Byzantine interests; to draw out the rules of Imperial policy in dealing with these barbarians; to explain how they might be weakened by divisions fomented, by subsidies and promises liberally despatched, from Constantinople.

In the tenth century, the early eleventh, and the middle of the twelfth, the Byzantine State was perhaps the most important, the most civilised, and even the most powerful of Christian Kingdoms. In the course of its age-long and obstinate resistance to the forces both of Moslem and Barbarian invaders, it thrice renewed the energies which befitted a *Roman* nation; and among the sovereigns of the Isaurian, Macedonian, and Comnenian dynasties, there were some not wholly unworthy of the throne of Justinian. It is the second of these three revivals that now concerns us, when Basil I. saved the State from Michael the Drunkard; when the 'Macedonian' or Slavo-Armenian House won back province after province of the shrunken Empire; and when an imperial scholar, in defining the external policy of Constantinople, revealed for the first time some of the new races and future masters of Eastern Europe. This sketch, though political and diplomatic in its essence, is incidentally a geographical monument of the first importance; and, unless it is studied with the prejudices so long traditional in Western Europe, it will appear to give certain obvious lessons on the place of Byzantium in the world of the Dark Ages. For it will show us how the Eastern Empire, autocratic and centralised in its Government, influenced on one side by Levantine luxury and on another by Scythian savagery, fighting with diplomacy more often than with arms, and with spiritual weapons more effectively than with temporal, was yet able to extend its influence over an area[1] scarcely less wide than the sphere of its Western rival, in the days of the Saxon Emperors.

Out of the hordes of Slavic and Turco-Tartar tribes on the North-East, Byzantine civilisation helped to create the Christian States of Bulgars, Magyars, Croats, Serbs, and Russians. To the South-East, after the losses of the seventh century, the Byzantine resistance maintained the line of the Taurus against the forces of Islam; saved Armenia,[2] that all-important vantage-ground of Western Asia, from an exclusive Moslem domination; and by the side of the decaying Caliphate of Bagdad (ruined as a political force from about 950) exhibited afresh something of the ancient life and strength of Rome. Yet on the Asiatic side the Byzantine recovery left no permanent effects; it delayed the triumph of the earlier and more civilised Islam in the nearer East,—it did not avert such a triumph under the leadership of barbarous hordes; and it repelled the Arab only to fall a prey to the Turk,—Seljuk and Ottoman,—for ever incapable of founding an Eastern Cordova, a new Bagdad, or a Northern Cairo. Thus it was only in the precious moments of delay, in the time

[1] From Kiev to Abyssinia, and from the Caspian to Sardinia.

[2] The mountain system of Asia has two great centres or knots, one in the Pamir, the other in the Armenian plateau, and the possession of either is of course a question of the first strategic value.

it saved for the development of the European nations, that the Eastern Rome can be said to have rescued Christendom from the Saracens. But on the side of Europe the Byzantine tradition, inspired by the subtle, attractive, proselytising statecraft of ages, was not a mere incident of History. Constantinople drew the Slav world into the Christian Society. The new races of Sarmatia, some of them destined to a long career and an abiding place in the European family, gained from Byzantium an alphabet, a religion, the beginnings of a literature, and a great body of new conceptions in government, law, and social organisation.[1] And in the most virile and expansive of the Slav peoples the Byzantine tradition, though strongly marked elsewhere, reaches its most definite expression. Bulgars, Serbs, and Croats owed much to the Bosphorus, and the influence of Constantinople was a great part of their national life; but the Russians owed still more, and it was in the Principality of Kiev that Byzantinism was most exactly reproduced.

In Geography, as we have seen, it is the Byzantine Constantine VII. who gives us the first glimpse of the

[1] It is hardly necessary to do more than refer here to the Byzantine function of handing on ancient Greek learning, both to the Arabs and Latins directly, and in a still larger measure to the latter through the former. Byzantium was truly the 'keeper of ancient treasure'; the greater part of their science was learnt by the Arabs from the Greek books they found in the conquered Byzantine towns of Syria, Egypt, etc.; and from this Graeco-Arabic learning came (largely by way of Spain) the first or mediaeval renaissance of letters in the eleventh century (cf. Gerbert's life at Cordova as a student). The second renaissance of the fourteenth and fifteenth centuries, as inspired by Byzantine scholars flying before the Turks, is a commonplace of history; it is more interesting to speculate, with Nordenskjöld, how far the Portolani compass-maps, or coast charts, of the Western Mediterranean were originally derived from Byzantine sources, either before or soon after the first Crusade. The transference of this germ of science cannot well in any case have occurred before the eleventh-century awakening of Western intelligence, but remembering the close connection of Venice and Amalfi with Constantinople, it is difficult to ignore the suggestion altogether.

habitat and movements of the nations which were destined to prevail in Eastern Europe. The scanty rays of light which the Scandinavian Sagas, following the track of Rurik, begin to throw a little later from the North-West towards Novgorod, are preceded by the far clearer beams of a knowledge which radiates from the South-West and the Imperial Chancery of Stambul to Kiev, the Dniepr, and the Volga.

Western Europe betrayed and ruined the Eastern Empire in the fourth 'Crusade' of 1202-1204, and thus having made a fatal breach in the chief Levantine bastion of the Christian fortress, abandoned to the Turks the kingdom and the city which for so many centuries had kept out the unbeliever and the barbarian, and which had not only been the protector, but in a great measure the educator, of the new and still halting civilisation of the German and Latin peoples. Having accomplished the first two acts of this brilliant drama, the public opinion of the progressive nations consummated its work in the third,—burying under contempt and obloquy the history and the achievements of a people which, at least in art,[1] must be ever memorable, and which in the science of diplomacy, the faculty of racial absorption, and the field of law, was not unworthy of the great name that it inherited. In some ways, the fate of Christian Byzantium has been like that of Poland, but as its services were greater, and its place in civilisation far more eminent, so its fate has been more terrible and more despised. The Slav

[1] For a long time Byzantine art is a dominating influence in many regions west of the Adriatic, and from Perigueux to Ravenna, from Venice to Trebizond, and from Palermo to Moscow, the influence of its architecture may be traced. St. Sophia at Constantinople must always rank as one of the greatest human buildings, and, until the thirteenth century, Christian architecture had nothing to put in the same class with it. In mosaic work the Byzantines were the teachers of all other artists.

peoples, to whom the Lower Empire gave letters, and law, and religion, and the elements of civilisation, alone preserved some feeling of gratitude and reverence for their Mother, the Second Rome.

Constantine VII., fourth of the Macedonian line, was himself an expression of the union of races within the Empire, of the possibilities within the grasp of the most distant provincials. His family, partly Slav, partly Armenian in origin, had risen (through the talents of Basil I.) to the headship of the State; and the intruding House had the longest as well as the most glorious reign among Byzantine dynasties. Under their rule, the principle of hereditary succession was firmly established, the monarchy of Justinian was in great part re-established, the legislation of the sixth century was amplified, and a remarkable series of state-manuals was drawn up, which described in the most minute detail [2] the provinces of the Empire, the vassals who adjoined it, and the hostile and dangerous nations who pressed upon its borders or had established themselves within its area. From a geographical stand-point, Constantine is, of course, most important to us in the last of these sub-divisions, as the describer of Bulgarians, Magyars, Russians, Petchinegs, Khazars, and Arabs; but in his treatment of the Italian, Dalmatian, Armenian, Crimean, and Caucasian allies or dependents of Byzantium, the imperial author is hardly less valuable. It was fortunate for his work that the circumstances of history gave it a permanence it could not have had in earlier time. In Europe the Empire was racially complete; the last foreign invasions and immigrations (except for Turks and Mongols),[3]

---

[1] 868-886 A.D.

[2] To say nothing of the Treatises on Court Ceremonial, Military Tactics, Agriculture, etc.

[3] Both insignificant in Europe from an ethnographical point of view.

had been accomplished; and the age of change was over. From the tenth century, the population of Byzantine lands, on this side of the Straits, grows constantly more homogeneous. Beyond the limits of the Empire, once more, the same broad fact is discernible, through the whole of South-Eastern Europe; here also the wanderings of the nations, the ever-recurring waves of invasion which pass over the great plain from East, North, West, and even South, during seven centuries,[1] are drawing to an end; and the lands north and east of Danube are beginning to settle down into a condition ethnographically, if not politically, similar to the present. We have now to deal with the settled and permanent masters of the soil in 'Sarmatia,' 'Dacia,' and great part of 'Russia.'

The author begins (in chaps. 1-8) with the nation whose friendship and alliance he considers of most importance — the Petchinegs or Patzinaks of the modern Rumania and South-Western Russia. Having laid down the principle that, with this race on the side of Rome, she need not fear what other men could do to her, Constantine proceeds to develop this thesis by showing that Russians,[2] Hungarians,[3] and Bulgars all feared these invincible savages, and were all checked and held in by them, lying as they did on the front of the first, on the flank of the second, and on the rear of the third. The Petchinegs were also useful for the commerce of Cherson, the great Crimean dependency of the Empire, and could transport Greek wares, or Roman Embassies, into the more distant lands of the Khazars and the Zicci on the Don, Volga, and Kuban.

---

[1] From the third to the tenth; from the time of Decius to the age of the Macedonian and Saxon Emperors.

[2] οἱ Ῥῶς. *De Adm. Imp.*, ch. ii.
[3] τὸ τῶν Τούρκων γένος. *De Adm. Imp.*, ch. iii.

Cherson is emphasized as the best starting-point of all Byzantine trade and diplomacy on the North Euxine shores to the East of the Dniepr; but another route is indicated for envoys from the Bosphorus to the Western parts of 'Patzinakia,' by the rivers Danube, Dniepr, and Dniestr.

The Russians (separated from the Black and Azov Seas by the Petchinegs, from the Caspian and Lower Volga by the Khazars) are next described; and the path of their raids, from Old Novgorod[1] and Tchernigov, down the Dniepr to the Black Sea, is detailed; a list of the cataracts which break the lower course of the Borysthenes is also given (chap. 9).

Next came the Khazars (then masters of the whole South-East of modern Russia-in-Europe), and evidently held in great suspicion by Constantine. Their neighbours, the Uzes and Alans, it is noted with satisfaction, can both do them considerable damage and cut off their supplies. The Fortress of Cherson (by the site of the present Sevastopol) and that of Sarkel (at the mouth of the Don) were well protected from Khazar assailants, as long as a good understanding prevailed between 'Romans' and Alans; this understanding was therefore vital. *Great, Black*, or *Old*, Bulgaria (on the Kama and Middle Volga) could also attack the Khazars (chaps. 10-12).

The first part of the *Administration* concludes (chap. 13) with an account of the nations that bordered on the Turks or Magyars,[2] and the writer then passes to speak of the insatiable greed and curiosity of these Northern barbarians, and the necessity of never granting some of their favourite demands. Such were requests for the secret of Greek Fire,[3]

---

[1] *Nemogarda* in chap. ix. of the *Administration*, ἀπὸ τοῦ Νεμογαρδάς.
[2] The placing of the 'Patzinaks'
to the *North* is curious.
[3] Πῦρ ὑγρόν.

VII.]   THE 'ADMINISTRATION OF THE EMPIRE'   475

for marriage alliances with princesses of the Imperial House, and for crowns and robes of pre-eminent dignity;[1] the authority of Constantine the Great and Holy was to be quoted in support of that firm refusal which was always to be given.[2]

After the North, the South. In the next chapters (14-25) we read of the Arabs; of the descent, career, and creed of Mohammed (here ludicrously misconceived); of the race of the Fatimites; and of the date of the 'Saracen Exodus' (the Hegira, Hijra, or Flight of the Prophet to Medina, is apparently intended). A reference is made to the *Chronicle* of Theophanes for an account of the death of Mohammed, the succession of Abubekr, and certain details of the doctrine of Islam, especially as to Paradise and its sensual joys. The reigns of the first Caliphs; Omar's capture of Jerusalem and the 'blasphemous' temple therein erected by the victor; the Saracen capture of Rhodes, and removal of the Colossus; their early attacks on Constantinople; their invasion of Spain; the vast extent of their conquests; and the geography of the Iberian peninsula;[3] are illustrated from Theophanes and other writers, mostly of slender value.

From the South and the extreme West, we now move round to Italy and Illyria; and here the *Theme* of Lombardy and its governors, the melancholy story of decaying Roman influence in the Peninsula, the brighter story of the founding of Venice, and the geography, races, and history of Dalmatia and the neighbouring Serb and Croat lands, are dealt with in turn (chaps. 26-36).

---

[1] Called καμελαύκια.

[2] Exceptions, however, certainly occurred in the marriage-alliance rule (*e.g.* cf. Leo 'the Khazar') and, *à fortiori*, they must have occurred in the robe-and-crown prohibition. The secret of Greek Fire was probably guarded with care, but Saracens are found possessed of the discovery from time to time.

[3] Oddly defined at the beginning of ch. 24, as forming two provinces of *Italy* ('Ιταλίας ἐπαρχίαι), *i.e.* of the Roman dominion.

476 GEOGRAPHICAL THEORY AND DESCRIPTION [Ch.

Having described at great length all the Illyrian peoples, Constantine returns to the Petchinegs. In his opening chapters he spoke of the value of their alliance to Byzantium, and of their relations with surrounding nations; now he adds various details about their country, their former migration from the Volga and Ural basins,[1] and the eightfold division of their nation, four tribes dwelling to the West, and four to the East, of the Dniepr. A word is added about the six ruined cities to be found in Western Patzinakia, and the traces therein afforded of a vanished Christianity and a former Roman civilisation (chap. 37). In the same way, the *Administration* recurs to the Turks or Magyars (chap. 38), their racial affinities, former habitat, and recent movement into the plains of 'Great Moravia'—the present Hungary.[2] Something is also said about the geography of this region, the social organisation of its new masters, their eight tribes, and the traces of old Roman rule in their country (chaps. 39-41). An important topographical chapter follows (No. 42), setting forth the chief features of all the lands from the Danube to the Caucasus; and especially describing the positions of the Danubian Belgrad, of Cherson and Bosporos[3] in the Crimea, of Sarkel at the mouth of the Tanais, of Tamatarcha or Matrica on the Asiatic side of the Strait of Kertch, of Nikopsis near the site of Sukhum Kalé, and of the adjoining land of the Zicci, separated from Matarcha by the Uruk or Kuban. Beyond or above the Zicci lay Kasakia, or the Cossack country; 'above' Kasakia the Caucasus Mountains; 'above' the Caucasus the nation of the Alans. The Russian land on the upper Dniepr, the lesser rivers between Danube and Dniepr, especially the Bûg or Bogu, the Maeotid Marsh or

[1] Atel, Etil, or Itil, and Geêch, or Yaïk.
[2] The name of Kangar or Kanka (Κάγγαρ) affected by three of the Petchineg tribes, is also mentioned.
[3] Otherwise Panticapaeum, the modern Kertch.

Sea of Azov, the shallow gulf of *Nekropyla* or Karkinit (the *Nigropolo* of later Italian traders) on the West side of the Crimean isthmus, are carefully described; the distances are given from one point to another; and the imperial author moves on to the East and to Armenia. The races and countries of the North, he considers, have been sufficiently described; it remains to speak (in chaps. 43 and 44) of Daron or Taron (to the west of Lake Van), of Manzikert (to the North of the same), and of the fortress of Kars (in the heart of the Armenian plateau), then the residence of the titular chief of all Armenia, the 'King of kings,' who, in Byzantine phrase, appeared only as 'First among *Archons*.' The past relations of all these petty states with Constantinople are recounted; and special attention is given to the affairs of the Georgians or Iberians, an interesting, orthodox, but troublesome, nation of allies or vassals, who figure largely in the pages of Constantine (chaps. 45, 46). Their fortress of Adranutzion,[1] near the modern Olti, had long been coveted by the Greeks; its military strength was great; and its market was a centre for the trade (as its ramparts were the 'keys') of Trebizond, Iberia, Abkhasia, North Armenia, and the South Caucasian slopes. The region of Arzen, or Erzerum, is identified with the district of Adranutzion, and the story is added of that diplomatic struggle for its possession which ended in the victory of the Iberian princes.

Two chapters about Cyprus and its re-colonisation (Nos. 47 and 48) form a transition to the concluding part of the *Administration*, which is almost entirely occupied with internal affairs, such as recent changes in the organisation of certain *Themes* (chap. 50), details of Byzantine naval service (chap. 50), and an account of the invention of Greek Fire (chap. 48), ascribed to the reign of Constantine Pogonatus, and

[1] Otherwise Adranutzê (τὸ Ἀδρανούτζη).

the genius of one Kallinikos of Heliopolis. The concluding chapter, however, on the history of Cherson (No. 53), recurs to the field of external interests. It is much the longest in the whole work, and was perhaps a separate tract in its original form. Beginning with a record of the aid rendered by the men of Cherson to Diocletian and Constantine the Great (when they combated the 'rebel Scythians' on the banks of the Ister, and defeated the aggressive Bosporans near Kaffa)[1] the author tells us of the rewards showered on the faithful city, and concludes in a different spirit by suggesting various measures that are to be taken against its trade and livelihood in case of a revolt.

I.—In the foreign politics of the Byzantines, during most of the tenth (as of the ninth) century, the *Bulgarian* question was paramount. This Finnish people—whether or not to be connected with those Huns[2] who, after the death of Attila, had fallen back to the basin of the Don and the Azov sea-coast; whether or no they may be recognised in this region about the year 463; whether or no they received in the fifth or sixth century a reinforcement of their countrymen from the Older Bulgaria of the Middle Volga; whether or no they owned the supremacy of the Avars in the lifetime of Heraclius and Mohammed—at any rate appeared upon the Danubian frontiers of the Empire in the reign of Constantine IV., surnamed *Pogonatus*, or the Bearded. They crossed the border river in 679; they gradually overran, conquered, and settled the whole of the country north of the

---

[1] The story of Gykia, who saved the city from the meditated treason of her lord, related by Constantine, in ch. 53 of the *Administration*, as of a recent matter, is referred by some to a much earlier period—about 36-16 B.C.; cf. R. Garnett, *E. Hist. Rev.* 1897, vol. xii. p. 100, etc.; Finlay, *Greece*, ii. 354-357.

[2] Cf. Gibbon, chap. lv.; Kruse, *Uebersicht d. Gesch.*, tabs. 9-16.

Rhodope Mountains and east of the Albanian Highlands; and thus they became masters of a kingdom roughly corresponding in area to the modern Principality, as restored in 1878-1885. But as they seem to have retained their power over a vast region to the north of the latter, over all the Eastern part of our Hungary, over the Banat of Temesvar, over Transylvania, and over Moldavia and Wallachia, the Bulgars of the pre-Magyar time were perhaps the most powerful nation of Eastern Europe, and, except for their north-western neighbours, the Moravians, they might be said to be as much the leading race among Sarmatian peoples as the Franks had so long been among Germanic. In 811 they defeated and killed the Emperor Nikephoros I., whose skull was for generations a favourite drinking-cup of the Bulgar princes; and the 'Roman' frontier shrank within limits even narrower than (though often curiously similar to) the reduced Turkey [1] of our own days. Adrianople and Thessalonica were close to the borders of the diminished state, and almost within sight of the Bulgar outposts. The Pontiff of Old Rome, anxious to secure the rising power to Latin allegiance, and indifferent, if not hostile, to the rights of Constantinople, sent to Tsar Simeon (888-932) an *imperial* crown.

But the skilful intrigues of Leo VI. and the everlasting migrations of wandering warrior-tribes from the East, cut short this greater Bulgaria,[2] at the close of the ninth century, by the inroads of the Magyars and the Petchinegs;[3] all the lands beyond the Danube (comprising some of the purest,

[1] Especially to the borders of European Turkey, as defined by the San Stefano Treaty of 1878.

[2] Arnulf of Carinthia pursued the same policy, on behalf of German interests, towards Great Moravia (then the most civilised of Slav states) which he crushed altogether, with Magyar help; cf. Rambaud, *L'empire grec* (edition of 1870), 328, based on *Codex diplomaticus Moraviae* (edition by Boczek), pp. 52, 70-73.

[3] From about 889.

or *blackest*, of Bulgar settlements) were now lost; and the German alliance which an Embassy to Ludwig the Pious, or Louis the Debonnair, may have attempted to secure (in 822), was not forthcoming. This flirtation with the Western Empire, like the advances of Krum and Simeon[1] to the Saracen powers, probably aimed at the final overthrow and partition of the Greek Empire; in any case, this ambition was notorious enough to suggest to the threatened State the most desperate remedies—even an invitation to more distant and more savage hordes. For a moment this policy appeared little short of suicidal: the Bulgars, stripped of half their empire,[2] and at the same time consolidated into a firmer and better compacted nation to the South of the Danube, threw themselves with double fury against Stambul itself.

But Simeon failed in his great attack of 924; a truce of many years followed; and when the struggle was renewed in 966, Byzantine arms completed the triumph of Byzantine diplomacy. Nikephoros Phokas, John Tzimiskes, and Basil II., by successive stages, accomplished in the next half century the conquest of all that remained of independent Bulgaria, and restored the Danubian frontier of the Empire. Yet the conquest was perhaps as much a peaceful as a military one. Bulgarian merchants, even in the time of Simeon and Leo IV., traded at Constantinople and Thessalonica; restrictions on their rights formed an excuse for the invasion of 889; and the more the Barbarians saw of Roman life and manners, the more powerless they became to resist the manifold attractions, the subtle enervations, of

---

[1] *E.g.* in 812, Krum attacked Mesembria with siege engines directed by an Arab: in 925 Simeon signed a treaty with the Fatimite, Al Madhi, to partition the Byzantine Empire.

[2] From Trajan's Bridge westward, the North Danube lands fell to the Magyars; east of this, to the Petchinegs. Cf. Rambaud, *L'empire grec.*, pp. 320, 323, 324, 329, 331-337.

the civilised society they professed to despise. Like their Finnish brethren the Magyars, they brought with them from the Urals a marked capacity for a higher culture, for a settled and commercial existence as well as for a military and roving life. Masters of the ports on the West Euxine coast, such as Odessos, they had already developed by 892 a profitable salt-trade with the Moravians;[1] and for generations they played a notable part as intermediaries between German and Byzantine trade. They may perhaps be credited with maintaining 'Consuls' in the great Byzantine markets, during the later years of Constantine VII.; in any case, their representatives were admitted to the Court festivities with an honour shown to no other envoys; and Liutprand[2] vainly protested, in the name of the Western Emperor, against the precedence accorded to this most favoured nation.[3]

II.—Like the Bulgars, the 'black swarm' of another Finnish race left behind it an older settlement in the Eastern Volga basin; in their progress to South and West the Hungarians absorbed into their eightfold nationality many Russians, Komans, Petchinegs, and even *Ishmaelites* (or Moslem Bulgars,[4] Khazars and Alans), who formed in later time a compact colony of *Unitarians* in Pesth. The migration and final home of this composite, but rapidly unified, people was a close parallel to the movements and locations of the Avars,

[1] *Codex diplom. Morav.* 52, 70-73.
[2] Cf. Liudprand (Liutprand), *Legation*, ch. 19; Constantine Porphyrogennetos, *Ceremonies*, i. 82; ii. 429, 430.
[3] Cf. Rambaud, *L'empire grec*, 346-358; 358-363. This was at the banquet on the 'Feast of the Holy Apostles,' 968. The Bulgar envoy was shaved in Hungarian fashion girt with a brazen chain, and apparently a catechumen. Yet this man was placed above the bishop who represented the Holy Empire of the West.
[4] Otherwise Bulgars of the Volga, whose conversion to Islam occurred c. 920 A.D; cf. Masudi, ii. 58.

three centuries earlier; their inroads carried them far into Spain, Italy, and Flanders; yet for many years their princes owned the suzerainty and received the investiture of the Khazar Khan.[1] Their victory over Oleg outside Kiev[2] was a passing triumph, the forcing of a barrier on their westward march; but in destroying the Moravian state and cutting short the Bulgarian, in planting themselves upon the middle Danube, as the new and permanent masters of Western 'Carpathia,' they changed the face of Eastern Europe, thrust a wedge between the Slavs of the North and South, and so crippled the energies and blighted the prospects of that race for ages.

To the Byzantines, whom they saved from the Bulgar deluge, under whose 'patronage' they first crossed the Carpathians and the Danube (888-889), and whose Christianity they seemed to favour, they were never so repulsive as to the nations of the West. One Magyar bravo[3] was famed, or fabled, to have struck his battle-axe into the Golden Gate of Stambul; and the Venetian and Dalmatian vassals of the Empire suffered no little misery at their hands; but these things were not so terrible as the sack of Bremen[4] and St. Gall; the baptism of two Hungarian princes at Byzantium in 948 filled the Orthodox with hope; and before the mission of Piligrinus in 973, the Greek Church appeared to be winning as valuable a triumph among these 'Gog-Magogs'[5] in the West as the conversion

---

[1] Cf. Rambaud, *L'empire grec*, 346-351.

[2] Cf. Nestor, *Chronicle*, a. 6406; 'Anonymus,' ch. 8.

[3] Botund by name; cf. 'Anonymus,' ch. 42.

[4] Cf. Adam of Bremen, i. 54 (43, Lapp.); *Annals of Corbey*, A.D. 915; on the Magyars and their relations with their barbarous neighbours, especially the Moravians, and with Constantinople, see *De Administrando Imperio*, chs. 3, 4, 13, 38, 39-42.

[5] This name was very generally applied in these times, not merely to the Magyars, but to other enemies of Christendom, or even of Islam, *e.g.* Vikings; cf. the term *Magioges*, ap-

of Olga and her Russians effected in the East. But appearances were deceptive; and the monk Hierotheos, sent by Constantine VII., and consecrated by the Patriarch Theophylact, as Orthodox Bishop of the Turks, found his career impeded and his prospects threatened at the Hungarian Court by a Latin emissary from the West, a monk of the Black Forest. In the result, it was only possible for the Byzantine to secure a minority of the converts; the victory of Otto the Great at the battle of Augsburg practically decided the question in favour of German and Latin influences; and the marriage of King Geysa with the Polish Adelheid helped to increase the predominance of Papal Christianity. The education of St. Stephen, a determining factor in the future of Hungary, was decisively Latin,[1] and only a fraction, though a large one, continued to adhere to the Eastern Church. Yet the religious buildings of the Orthodox Magyars were numerous and splendid; their pilgrim hospice at Constantinople rivalled those at Ravenna and Rome; and the Apostolic Crown[2] of Hungary itself preserves the memory of Byzantine influence, and apparently the name of Constantine VII.

Not less dramatic than this later encounter of Greek and Latin missionaries at the Hungarian Court was the earlier meeting of envoys from Constantinople and Cordova in the

plied in 843 to the Northmen who attacked Lisbon (cf. Dozy, *Recherches*, ii. 252-340; Rambaud, *L'empire grec*; Condé, *Arabs in Spain*, Part ii. chs. 45, 49 (i. 289-291, 299).

[1] Rambaud, *L'empire grec*, 362, 363.

[2] It is formed of two Crowns, one given by the Pope, the other by an Emperor Constantine who cannot well have been other than Constantine VII. This last is expressly recorded upon the Crown itself, the lower part of which seems to be of Greek, while the arches are of Roman workmanship; so Count Mailath, *Gesch. d. Magyaren* (1828), *Anhang*, 7, from personal examination. The coronation-robe of Hungary is another piece of mixed workmanship, partly Byzantine, partly Arab. It was presented in 1031 to the Church of St. Mary at Stuhlwissenburg by Gisela, consort of St. Stephen.

presence of Otto I., after the battle of Augsburg.[1] But this encounter was wholly amicable, for Greeks and Saracens, from the two extremes of Europe, joined in congratulations upon the overthrow of the Magyars. The arrows of the Hungarians had troubled the frontiers of Galicia[2] and Thrace alike; yet in the Balkan Peninsula they had appeared rather as freebooters than as butchers; and from the time of Nikephoros Phokas, Hungarian mercenaries began to serve under the imperial standards.[3] The commerce of the Danube basin and of the Northern Adriatic, which had been ruined by the Pagan Magyars, revived under the first Christian Kings of Hungary; the corn of the Hungarian plain became a valuable export to Constantinople; and furs and iron[4] were exchanged for silks, gems, wines, and chased armour.[5] Merchants and pilgrims from the West were again free to cross the ancient Dacia,[6] as they had not been for seven centuries, and the re-opening of this new avenue gave an immediate and considerable impulse to the Crusades.[7]

III.—Constantine VII. is, of course, supremely valuable

---

[1] With this we may connect the journey of Liutfrid, a rich merchant of Mainz, to the Byzantine Court, as Ambassador of Otto I., in 949; his route was through Venice; cf. Liudprand, *Antapodosis*, in Pertz, M. G. SS., iii. 338; and Heyd, *Commerce du Levant*, i. 80.

[2] Cf. Masudi, ii. 58, 64.

[3] *E.g.* in 968, Nikephoros Phokas takes forty Magyars with him to Syria; on the other hand, Tzimiskes, in the battle of Silistria, fights against Magyars allied with Russians, Petchinegs and Bulgars; cf. Liudprand, *Legation;* Leo the Deacon, vi. 12, who calls them Οὖννοι; Rambaud, *L'empire grec*, 360, etc.

[4] But the export of horses, slaves, and salt was strictly forbidden (Rambaud, 361).

[5] Some of this, apparently of Byzantine manufacture, was used as early as the Battle of Augsburg.

[6] Even the Great Moravia and Greater Bulgaria, which were shattered by the Magyar invasion, did not afford the same security to trade and travel.

[7] As early as 1092 Mohammedan merchants are also found in Hungary; cf. *Ladislai regis decretum*, i. cap. 9, noticed in Heyd, *Commerce du Levant*, i. 83. Earlier than this an Hungarian colony is settled in Constantinople, for whom St. Stephen builds a church.

for his treatment of the Russians. Practically, he introduces this people to European literature and history in the middle of the tenth century, just as Ibn Foslan introduced them to Moslem and Asiatic notice a generation earlier. More than a hundred years were yet to elapse before any clearly articulate voice[1] comes from within the Russian people itself, in the shape of 'Nestor's' *Chronicle.*

It is plain, from the *Administration*, that the Byzantine chancery was fairly well informed upon the geography, ethnology, and politics of the new Slavo-Scandinavian tribes from Kiev to Novgorod, and from the Carpathians to the Upper Volga.[2] The Imperial author knows the names and positions of the chief Russian towns, the distances between various parts of the country. He is acquainted with the story of the Varangian immigration of Rurik and his men; most of the Slavic, Finnish, and other tribes[3] who owned a vassalage to the Grand Prince of Kiev are to be found in his pages; and no less than twenty-two princelings of the Rurik dynasty occur in the *Ceremonies* as joining with Igor to sign the treaty of 944. He is even capable of giving us

[1] In c. A.D. 1110-1114. 'Nestor' is only a traditional name for the authors of this chronicle. Except possibly for some folk-songs, and the pilgrim narrative of Daniel of Kiev, there is no earlier monument of Russian literature.

[2] The north-eastern boundary of the Russian tribes at this time may be roughly traced from the source of the Volga to Tver, thence south, or a trifle south-west, to about Kaluga, Orel and Kursk, thence south-west to the Dniepr a little below Kiev (Moscow then lying outside these limits.) Along the northern part of this line the Rus principalities were bordered by Finnish tribes, along the southern part by the Khazars, the Pechinegs controlling all south of a line drawn from the middle of the Carpathians to the Dniepr near Kiev, across the upper waters of the Sereth, Pruth, Dniestr and Bûg.

[3] *E.g.* Serbs, Slovenes, Drevlians. The same number of Varangian *sub-reguli* (22) is given by Nestor (ann. 6453) in reference to this treaty. At the reception of Olga (956 or 957) Constantine Porphyrogennetos himself (*Ceremonies*, ii. 15) made presents to the deputies which these vassals sent in the train of the Princess Regent; cf. also Rambaud, *L'empire grec*, 366.

a good sketch of the social system which had super-imposed the military rule of a few Viking aristocrats upon a mass of anarchic and quarrelsome tribes of lower race. He traces clearly and well the Itinerary from Novgorod to Constantinople, by way of Smolensk, Tchernigov, Kiev, and the cataracts of the Dniepr. Of these famous but not very formidable waterfalls, he enumerates seven under both Slav and Varangian [1] names; and in three cases his terms have survived.

The importance of Kiev as the meeting-place of inland —especially riverine—traffic, and the value of the Crimean coast, of the Isles of St. Gregory,[2] of the Danube delta, of the Bay of Varna and of the White River (of Akkerman) in the Euxine, as half-way stations for Russian and other pirates, are not forgotten. Like the Anglo-Saxon Kings, the early princes of the House of Rurik collected tribute and fed their Court by means of constant progresses; it was on one of these that Igor was killed by the Drevlians, an episode which became famous in early Russian history and the theme of many songs. This important and typical custom of princely touring, a mixture of freebooting and tax-gathering, is not omitted by Constantine.[3]

Our author (when he wrote the *Administration*) could hardly have foreseen the complete and rapid change which

---

[1] *Russian*, as he calls them (*Administration*, ch. 9), understanding by this word the Scandinavian aristocracy of the Russian lands. In the same way Constantine Porphyrogennetos (*Administration*, ch. 9) gives for Kiev the Slav name of *Kioba* or *Kioaba* (-*ava*) and the Scandinavian term of *Sambatas*. Cf. Rambaud, *L'empire grec*, 364-366.

[2] Constantine Porphyrogennetos (*Administration*, ch. 9) mentions especially St. Etherios, which also occurs in Nestor. In all the Byzantine treaties with the Russian princes, stipulations were inserted against the military use of these points of vantage. Cf. Rambaud, *L'empire grec*, 370.

[3] *De Administrando Imperio*, ch. 9. Cf. also Leo the Deacon, vi. 11, who tells how Tzimiskes in Bulgaria, 970 A.D., recalls to Sviatoslav how his father Igor had met his death 'in an unjust war against the *Germans*.'

Byzantine Christianity and civilisation would effect in these cruel and detested foes of the Empire. It is true that he records the journey of the Princess Olga in 956 (or 957), but he does not allude to the baptism which (in the ordinary tradition) made it memorable; nor could any one anticipate that the conversion of the Regent of Kiev would so speedily carry with it that of the whole nation. Up to this time there was nothing to indicate the peculiar connection of the future between Constantinople and the Russian people; their savage and daring raids upon the Bosphorus — of Askold and Dir under Michael III., of Oleg under Leo VI., and of Igor under Romanus Lecapenus—had left only a memory of horror. Their attacks on the Black Sea ports, and especially on Cherson or Sevastopol, dated from very early times,[1] were a constant danger, and kept alive the Byzantine alliance with the Petchinegs and Khazars, masters in name of all the Northern littoral of the Euxine. The still earlier tradition of friendship and alliance, instanced by the Russian Embassy of 839 (to contract an alliance with Theophilus and perhaps even with the House of Charles the Great)[2] had been disregarded from the time that Varangian rule was firmly planted in Kiev; and of all the heathen enemies of Eastern Christendom, no single tribe (except the

[1] *E.g.* 842, 852, 854. Cf. Rambaud, *L'empire grec*, 371.

[2] *E.g.* According to Nestor, *Chronicle*, ann. 6366, these envoys of the *Khagan* of the *Ros*, making a long detour to avoid the Petchinegs, joined the company of some Byzantine Ambassadors on their way to Louis the Debonnair, and so came to Ingelheim, where they were arrested as Vikings, relations of the Northern Pirates who were even then ravaging the German coast. See also *Annals of St. Bertin*, 839 A.D.; Liudprand,

Antapodosis, v. 6. This is the earliest certain reference to the Russian name and people, and that on its Scandinavian side only. Theophanes in a doubtful, and perhaps corrupt, passage, reckons them among the Byzantine allies in 774, but this is very doubtful; obviously mythical is the story of Kij, founder of Kiev, one of three Slav brothers established on the Dniepr, his visit to Stambul, and his honours from its *Tsar*, long before Rurik.

'friendly' Petchinegs) seemed more irreclaimable. Yet time would show that the political and religious traditions of Byzantium were to find in the Russian one of their aptest pupils;[1] that in the most barbarous of the Slavic folks the Orthodox Church would acquire its firmest bulwark and most powerful missionary; that in these savage foes the civilisation of Europe would gain a new rampart against the incursions of Asiatic hordes.

The Scandinavian restlessness, the Viking capacity for rapid, effective movement over vast distances, was shown not only in the Rurikides' repeated dashes upon the Bosphorus, by way of the Dniepr and the Euxine, but also in their descents of the Volga, their appearance upon the shores of the Caspian (as in 913), and their raids as far as the mighty mountain-wall of the Caucasus, and even beyond it, for in 944 their warriors encountered a Moslem foe at Berdaa[2] near the modern Elisavetpol. And the same qualities, with different applications, are to be seen in the journeys of Russian envoys and princes to Germany in 959, 973, 1040, 1042, and 1075;[3] as in the commercial wanderings which

[1] Politically, Venice is an even better pupil.

[2] Cf. Masudi, ii. 18, on the raids of 913 and 944, and the alleged early Russian trade to Spain (?), Rome (?), and Syria (?), as well as Khazaria. Their Syrian intercourse is also noticed in a somewhat doubtful passage of Benjamin of Tudela; in Ibn Khordadbeh, ed. Barbier de Meynard, *Journal Asiatique, serie* vi. t. v. p. 514; in Constantine Porphyrogennetos himself, *De Administrando Imperio*, p. 180, ed. Bonn; as well as in Nestor, who adds *Black* Bulgaria to the countries visited by Russian merchants. Cf. the treaty of 971 between John Tzimiskes and Sviatoslav. See Heyd, *Commerce du Levant*, i. 70, and pp. 262, 458, 461 of this volume; also Rambaud, *L'empire grec*, 374-379. The geography and trade of the Russians are treated in chs. **2**, **9**, 13, 37, 42, of the *De Administrando Imperio*, as well as in ch. 4, where the 'Turkish' Magyars are also treated.

[3] This journey of 1075 was performed by the grand prince Isiaslav or 'Isaeslav,' who came from Kiev to Mainz to ask the aid of the Emperor Henry IV., and brought with him a number of valuable presents for that end; cf. Heyd, *Commerce du Levant*, i. 79.

planted the Russian trading colony of St. Mamas[1] at Stambul, and brought their adventurers to the fairs of Cherson, of Trebizond, of Petra in Colchis, and of the Khazarian Itil, near the site of Astrakhan. For of these enterprises, as of similar mercantile ventures to both the Bulgar lands (on the Danube and the Volga alike), and to the marts at the estuary of the Don, there can be no doubt; even if we discredit the traditions of their tenth-century visits to Syria, Spain, and Italy.

IV.—The Petchinegs, or Patzinaks, were then regarded as the most powerful nation or group of tribes to the North of the Pontus; they ruled the whole country between the Tanais and the Danube; northward they controlled the Steppe country up to nearly 50°, dominating the modern Podolia, Bessarabia, and Ukraine; and they are treated by Constantine, in the first eight chapters of the *Administration*, with marked attention.[2] Their alliance, as we have seen, he regards as of vital moment for the Empire; together with the more suspected but less barbarous Khazars, they commanded every opening into the Euxine from the North; and no Varangian raiders could descend the Don or Dniepr in the teeth of their opposition; unluckily, they often preferred to wait till the plunderers returned from the Bosphorus, or other parts of 'Rumania,' laden with booty. They were too fond of gain, moreover, to refuse all traffic in horses, oxen, and sheep with the northern bandits in whose

---

[1] In 1043 a serious riot occurred between the citizens of Constantinople and the Russian merchants trading in the city; cf. Cedrenus, ii. 551; Heyd, *Commerce du Levant*, i. 71. The death of a Russian in this tumult was the cause of, or excuse for, the war that followed. A similar Russian commercial colony seems to have existed at Itil in the tenth century; cf. Masudi, ii. 9, 11.

[2] Constantine Porphyrogennetos, *Administration*, chs. 1-8.

land these animals *could not breed*;[1] otherwise the Russian marauders might almost be starved into submission.

Two chief routes led into the country of the Petchinegs; of these the Eastern path started from Cherson, and was the most suitable for reaching the tribes between the Don and Dniepr; the Western began at the mouth of the latter, or at the estuary of the Dniestr.[2] These worthy and useful allies did not, however, present many attractions, save to the diplomatist or the soldier; six ruined cities recalled a tradition of ancient Greek or Roman settlements; but the new masters of the soil were Nomade shepherds, and hated town life, as they hated Christianity, social order, and the morals of settled peoples.[3] Whether cannibals[4] or no, their alliance was to Constantine all-important for the safety of Cherson, of the Bosphorus, and of the entire Euxine; in his judgment, they were also invaluable as guides and middlemen for Byzantine trade in Khazaria, Russia, or the Caucasus;[5] and no trouble or sacrifice (except the secret of Greek Fire) was too much to conciliate their friendship.

Byzantine diplomacy had already used them with telling effect against the Greater Bulgaria of the ninth century, but as a defensive instrument against the Russians they were less serviceable. It suited their interest to overrun the Wallachian and Moldavian lowlands; it did not suit them equally well to hinder the Vikings of the North from forays against the Empire; it was even to their advantage to join in any well-organised attack upon their rich ally. Thus in

---

[1] This odd statement is in *De Administrando Imperio*, ch. 2.

[2] Note the Slav names of these rivers, Δάναπρι, Δάναστρι, in Constantine Porphyrogennetos (*Administration*, ch. 8), as in Carpini.

[3] *De Administrando Imperio*, c. 37. The six towns are called Aspron, or White Stone City, Tunkatai, Kraknakatai, Salmakatai, Sakakatai, Giaiukatai.

[4] So Matthew of Edessa, c. 85 (Dulaurier).

[5] Especially *Zichia*, or the land of the *Zicci*, *De Administrando Imperio*, chs. 6, 42.

934 they were to be found beside the Hungarian ravagers of Thrace; in 944 they followed Igor in his second raid against the *heaven-protected* city of Constantine; in 970-971 they joined Sviatoslav in that great onslaught of the Barbarian upon the Civilised World, which was repelled by John Tzimiskes. Eleventh-century history records in 1026, in 1051, and at other times, fresh incursions of these highly-salaried 'sheep-skins' upon the lands of their friend and pay-master; and the persistent forgiveness shown by the Imperial policy in their case can only be explained by the overpowering dread of the Bulgars at Constantinople. This danger was unquestionably lightened by the advance of the Petchinegs to the Lower, as of the Magyars to the Upper, Danube; and for Byzantine friendship it was then enough to be the traditional and necessary foe of the subjects of Krum and Simeon.

V.—The Petchinegs were the spoilt children of the Byzantine Chancery; the Khazars, on the other hand, though a far more interesting and redeemable nationality, were treated with coldness and suspicion. Yet their alliance, if sedulously fostered, might have been more useful than that of the undependable Nomades to the North of the Danube. The Judaism of their kings, the Mohammedanism of a large part of the people, and the natural and necessary menace (to Cherson) implied by their domination in the Crimea, probably contributed to this reserved attitude on the part of the Christian Caesars in the tenth century. Yet Heraclius, Justinian II., and Leo the Isaurian[1] had not disdained to treat of alliances with the Royal House of this bravest and most cultured nation of *Scythia*;[2] and Leo IV.,[3] by his sur-

[1] On behalf of his son Constantine V.
[2] Justinian II. married a sister of the Khazar Khan; Constantine V. a daughter of another *Khagan* and *Khatun*.
[3] Son and successor of Constantine V.

name of 'The Khazar,' confessed his origin, and bore witness to the merits of both Roman and Barbarian stocks; for if the one was worthy to give an Emperor to Constantinople, the other proved its power of government over so many different nations by a marriage-policy of such liberality and wisdom. Even the dress of the Khazars came into fashion at Byzantium soon after Leo IV.[1] But in the time of Constantine VII., suspicion had taken the place of the older friendship, and in 1016 the Byzantines found themselves at war with a people who had usually respected, and sometimes protected, the Roman possessions of the Chersonese;[2] who in 833 joined with the Greeks in the building of Sarkel, at the estuary of the Don;[3] whose soldiers formed a regular contingent of the imperial army; and who fought under the banners of Rome in Italy, in Thrace, in Crete, and in Armenia.[4] The great conqueror of the Macedonian House, not satisfied with his overthrow of the Bulgarians and his subjugation of Armenia, crushed[5] the Khazar Empire, in alliance with Vladimir of Kiev and the savage tribes beyond the Volga, almost at the same time that the dominion of the Petchinegs began also to decline. The realm which under King Joseph (c. 948 A.D.) had reached from the Caspian to the Volga near Tver, from the delta almost to the

[1] Constantine Porphyrogennetos, *Ceremonies*, i. 1.

[2] And left their name to the Crimea as *Gazaria*, or *Khazaria*. Cf. Rubruquis, p. 214.

[3] For the better protection of the Azov sea-board. Sarkel seems to have stood on the south side of the Tanais estuary, a little below the later town of Azov.

[4] Cf. the forty-seven Khazars who took part in the South-Italian expedition in the regency of Romanus Lecapenus; Constantine Porphyrogennetos, *Ceremonies*, ii. 44. The Bulgarian Tsar Simeon, after his victory in 889, furious to find Khazars fighting against him (*Kazar* was traditionally the brother of *Bulgar*), cut off all their noses. The Khazar contingent, as a picturesque element in the Byzantine army, was constantly on show, especially at the reception of foreign envoys; cf. Constantine Porphyrogennetos, *Ceremonies*, ii. 15.

[5] Cf. Nestor, *Chronicle*, ch. 12; Cedrenus, ii. 464.

source of the greatest European river;[1] which had established a kind of order and civilisation from Eupatoria to Derbent; which stamped its name for many ages on the Crimea, the Euxine, and the Caspian;[2] and which claimed tribute from nine tribes on the Volga, fifteen nations of the Caucasus, and thirteen peoples of the Black Sea littoral,[3] was now reduced to an uncertain and miserable existence in the Steppes of the Kuban and the Terek. Even in the time of Constantine VII. the Khazar Empire had been reduced by almost half its area, and the Imperial diplomatist was probably the less inclined to cultivate the friendship of these Barbarians, as he had witnessed the steady shrinkage of their power. In the eighth century their western frontier had been marked by the Carpathians and the Dniestr; northward their rule had extended over all the lands hereafter known as Russian. A little later, by a singular error of policy, they dislodged the Petchinegs from the Steppe lands of the Lower Volga and the neighbourhood of Itil; but the only result of this was to plant these same enemies more firmly to the west of the Don.

Four chief cities seem to have flourished in the Khazaria of the tenth century. First came Itil at the mouth of the Volga, the political[4] and commercial capital of the country,

---

[1] Though interrupted in the middle course by Old or Great Bulgaria, which controlled the river from about Nijni Novgorod to below Samara.

[2] Commonly called 'Sea of the Khazars,' especially in the Arab Geographers, e.g. Masudi, ii. 19, edition of Barbier de Meynard and Pavet de Courteille; Ibn Khordadbeh, pp. 471-472; Edrisi, ii. 332, ed. Jaubert; cf. also Nestor, under 6472-3 A.M. = 964-5 A.D.; 6367 = 899; 6370 = 862; 6292-3 = 884-5, on the limits of the Khazar power. The Black Sea was also called at times after the Khazars. See Rambaud, L'empire grec, 394.

[3] So King Joseph boasts to Chasdai, Vizier of Abderrahman III. of Cordova. Even the Magyars, during most of the tenth century, accounted themselves vassals of the Khazars.

[4] Joseph's letter to Chasdai speaks of three royal residences, but Itil seems to have been the chief.

in a position where (as shown by the present Astrakhan) a great city seems marked out by natural conditions.[1] Here Ibn Foslan (in about 920 A.D.) found thirty mosques; a number of synagogues, baths, and bazaars; and a royal palace of brick, in the midst of a great collection of felt tents and clay huts. Here, too, Arab merchants met with the distant Bulgarians of the North [2] and Russians of the North-west; boats descended from the Kama, the Oka, and the Volga, or put in from various parts of the Caspian; all found security and a ready market under a Jewish ruler. The other towns,—Taman or Tamatarcha, on the ruins of the old Greek colony of Phanagoria; Bosporos or Panticapaeum, on the other or European side of the Strait of Kertch; and Sarkel [3] or the *White House*, at the mouth of the Don, answering to the Hellenic Tanais,—were all on the site of Greek Colonies or outposts; and besides these we have in Arab authors the names of some lesser settlements in uncertain positions.[4]

VI.—Byzantine relations with the Arabs of East and West

[1] To the south-west of Astrakhan the ruins of *Madchar* are conjectured to preserve a trace of a *Magyar* settlement, planted here and abandoned in the course of the Hungarian movement from the Ob and the Urals to the Danube.

[2] From this point men journeyed up the Volga to Bolghar, near Kazan, not only for trade, but sometimes for the sake of seeing the marvel of a 'Midnight Sun.' This was a special object of Ibn Batuta (who also purposed a sledge journey to the Arctic Ocean) in the fourteenth century.

[3] Built by the Khazars in alliance with Petronas and a Greek force sent by the Emperor Theophilus in 833; here the exceptional policy of taking measures *against* the Petchinegs was followed.

[4] One called 'Khazar,' another Semender, another Bilandiar or Balandiar: Semender is, perhaps, Edrisi's *Semmur* on the Caspian, some fifty miles from Derbent. It must be remembered that while the Khazar kings were Jewish, while many of the people were Christians (by the time of Constantine Porphyrogennetos), and while their relations with Byzantium were important, still the mass of the nation seem to have been Mohammedan, and their intercourse with Bagdad on one side, and with Khwarezm or Khiva, Samarcand, etc., on another, was even more prominent than with the Eastern Empire. Cf. Masudi, *Meadows of Gold*, ch. 17.

CONSTANTINE ON THE ARABS

were normally governed by a policy of friendship towards the more distant, of hostility towards the nearer and more dangerous. An active commercial, intellectual, and diplomatic intercourse subsisted between Constantinople and Cordova;[1] the *Ceremonies* record how a Spanish (Moslem)[2] embassy arrived in the Bosphorus in 946; Liudprand mentions another on his first visit to the Imperial City in 950, just as he notes his meeting in Venice with a Byzantine envoy, Salamon the eunuch, returning from Spain (948 or 949). A common fear of the Hungarians may explain some of these missions; for Salamon, on this same journey to the West, had also visited the Court of Otto, the great foe and future queller of the Magyars. But a more obvious reason may perhaps be found in the Greek schemes of crushing the Mohammedan pirates of Crete, for which the friendship or neutrality of the Spanish Moors, elder brothers who had stayed at home, was obviously desirable. At all events, this purpose is clearly discernible in the legation of 949[3] (when Stephen the *Ostiarius*, with a squadron of three ships, sailed

---

[1] Another evidence of the widespread interest of Spanish Moslems may be found in the travels of Abubekr Mohammed, of Tortosa (hence surnamed 'Tortushi') born in 1059, who (sometime before 1083) visited Mainz, where he saw and described some silver coins of Samarcand, struck in 913-915 by the Samanid prince Nassr II., son of Ahmed. Cf. Quatremère in *Journal Asiatique*, série v., t. xvii., pp. 147, etc.; Frähn, *Mémoires de l'Académie de St. Pétersbourg*, série vi., *sciences politiques*, t. ii., pp. 87, etc. (1834); Heyd, *Commerce du Levant*, i. 50, 79-80. After 1083 Tortushi settled in Egypt for the rest of his life. Kazwini, who tells us of Tortushi's visit to Mainz, speaks of the spices of India and the Far East—pepper, ginger, cloves, galangale, etc.—being obtainable in that city; cf. Wüstenfeld's edition of Kazwini, pp. 350, etc., 1843.

[2] There is no clear evidence of intercourse at this time between Byzantium and the Christian kingdoms of Spain, but some of these embassies may of course have come from the latter.

[3] For in this very year, 949, occurred the great attempt of Constantine VII. to recover Crete, which failed; Nik. Phokas succeeded in 960, while Abderrahman III. was still reigning (912-961).

to 'Iberia,') and perhaps in that from which Rabbi Chasdai gleaned his information of the Jewish dynasty reigning in Khazaria (c. 958 A.D.).[1]

It was probably by agreement with Abderrahman III that the Byzantine fleet joined with Hugh of Provence to extirpate another nest of Saracen pirates at Fraxinetum about 941 ; just as in 916 the league of Pope John X., King Berengar of Italy, and the Empress Zoe, had cut out a third cancer from the body of Christendom, by destroying the brigands of the Garigliano in the neighbourhood of Rome itself.

With the Fatimites of Africa, on the other hand, there was incessant trouble;—a wearisome but indecisive struggle in South Italy was ended for a time by the mission of John Pilatus in 951;[2] but there remained the haunting terror of a joint attack from Bulgarian soldiers and 'Barbary' sailors upon the very heart of the empire. The faithful Calabrians in 925 captured the ambassadors of Simeon and Al Madhi,—the former returning from a successful mission to the Infidel,—sent them to Constantinople, and thus enabled Lecapenus to avert the danger by skilful diplomacy and the prompt payment of a liberal subsidy.[3]

In the Eastern Mediterranean the struggle between Byzantines and Saracens (of the Abbasside or Bagdad obedience), continued obstinately throughout the tenth century. Geographically, there is a certain interest in the steady advance of the 'Roman' armies into the Armenian and

[1] Cf. Rambaud, *L'empire grec*, 406-407; Liudprand, *Antapodosis*, vi. 2 ; Constantine Porphyrogennetos, *Ceremonies*, ii. 15, 45, 48. The *Administration*, ch. 25, shows a good acquaintance with the three main divisions of the tenth-century Moslem (' Syrian ') world under the Abbasside, Fatimite, and Ommeyad Caliphs or 'Αμερμουμνεῖς.

[2] The war, however, broke out again in 956, when a storm destroyed the Arab fleet off Palermo. Cedrenus, ii., 359-360.

[3] Cf. Cedrenus, ii. 356-357.

Caucasian regions from about 925. When the tide had once turned, it ebbed fast; for two centuries it had borne up the *Unitarians*, after a first hundred years of rushing and resistless victory; now it receded from Erzerum [1] and the Lake of Van, from Iberia or Georgia, from the Upper Euphrates, from the whole Armenian plateau. In Crete (as in Cyprus), Islam still held its own; the Greek attack of 949 in vain marshalled against it Russians, Armenians, Petchinegs, Khazars, and other allies or mercenaries from the most distant regions; but the re-conquest of both islands was only retarded for a decade.

VII.—The Vassals of the Empire, as surveyed by Constantine VII., form a ring of defensive outworks, a series of buffer-states, protecting the *Roman* world—or that fragment of it which still remained effectively subject to Constantinople—from the full brunt of barbarian attack. On the other hand, to a more ambitious policy, they appeared as distant dependencies needing constant support, as advance-posts from which Imperial authority might be re-established in lost provinces, as protectorates which a spirited policy might transform into possessions.

Five chief groups of vassal states,—the Italian, Dalmatian, Crimean, Armenian, and Caucasian,—extended the influence and the nominal supremacy of the Byzantines from Sardinia to Georgia and the basin of the Kur. When the *Ceremonies* and *Administration* were compiled, the great western island had practically fallen away from the Christian cause and the standard of the Caesars; yet even in 968 Nikephoros Phokas included among his servants, as he de-

---

[1] Arzen or Theodosiopolis, captured in 928, and 949, by the Greeks, and held by them till after 1071. On these campaigns, cf. Rambaud, *L'empire grec*, pp. 422-427.

498   GEOGRAPHICAL THEORY AND DESCRIPTION   [CH.

clared to Liudprand, the *archon* of Sardinia, as well as the rulers of Amalfi and Gaeta, of Venice and Naples, of Salerno and Capua.[1] For the most part, however, these vassals (even the Croats and Serbs of the 'Sirmian isthmus' and the towns of the pirate-haunted Dalmatian coast) obviously belong to the better-known European and Christian world, and cannot even be considered as lying on the outskirts of it. The Byzantine relations with them do not therefore possess any definite geographical interest, except as connected with the commercial movements of the Mediterranean towards outer regions, which have been already noticed. The customary journeys of Venetian Doges and sons of Doges to Constantinople, and their receptions and honours in the Imperial town (receptions and honours which were viewed as a confirmation of their present authority or future prospects); the joint efforts of Venice and Byzantium to put down the Croatian and Dalmatian piracy which so often interrupted the diplomatic and mercantile connections between the Suzerain of the Bosphorus and her ally of the Lagoons; the Saracen raids, which compelled the disobedient Ragusa (as in 871), to implore the aid of its mistress; the dim and curious history of the *Strategoi* of Dalmatia,[2] their revolts [3]

---

[1] *Archons*, of the first pair; *dukes* or *doges*, of the second; *princes*, of the third. Cf. Liudprand, *Legation*, c. 27; Constantine Porphyrogennetos, *Ceremonies*, ii., 48; Rambaud, *L'empire grec*, 439-441, 441-450.

[2] On this Adriatic and especially Illyrian history, cf. Rambaud, *L'empire grec*, pp. 450-484; especially 444, 457, 460, 472-474, 480, 481-482. Even the Graeco-Istrian town of Justinianopolis is accused before the Suzerain of piracy in 933 (by Venice).

[3] Such as that headed by Paul of Zara in 806; in the Byzantine

'Notitia Dignitatum' the Strategos of Dalmatia ranked with the Strategos of Cherson; cf. Constantine Porphyrogennetos, *Ceremonies*, ii. 50, 52. Maius, under Basil II. (986), is *Prior* of Zara as well as *Proconsul* of Dalmatia; and he seems to have been *Strategos* when about 998 the protection of the Empire's Dalmatian subjects and the policing of the Adriatic was committed to Venice, as when, after the completion of his Bulgarian triumphs, Basil II. resumed his sovereignty; Lucius, ii. 4-9.

and their submissions; the tribute paid by various little groups of South-Western Slavs[1] either to the Eastern Emperor direct, or to Venice as representing the Empire;[2]— all testify to the wide-reaching but vague character of Byzantine pretensions; they do not show the influence of Constantinople penetrating into that outer world with which exploration was concerned.

Something, however, of a more definitely civilising and geographical character accompanies the Imperial connections with the Crimea, the Caucasus, and Armenia. The ancient Greek colonies on the North and East coasts of the Pontus still continued (in some instances) to act as advance-posts of civilisation in the tenth century after, as in the fifth century before, Christ. But the glory of Bosporos or Panticapaeum (the modern Kertch), of Olbia (at the mouth of the Bûg), even of Odessos (near the site of the present Varna), had been eclipsed by Cherson[3] which in the tenth century rivalled Trebizond as the leading port of the Euxine, and continued firm in its Byzantine traditions till the eleventh century. Here the Emperors Maurice and Basil I., Leo VI. and Constantine VII., Romanus II. and Nikephoros Phokas, Alexius and Manuel Comnenus, coined money.[4] Here Themistos governed in the time of Diocletian, Diogenes in that of Constantine I.; to this place were exiled Pope Martin,[5] various enemies of Leo IV. and of Theophilus, and (most notable of all) Justinian II., afterwards the city's mortal enemy, whose

---

[1] As, for example, those of Krka and Osero.

[2] See *Monumenta spectantia historiam Slavorum meridionalium*, 1868.

[3] The Russian, though not the Byzantine 'Sevastopol.'

[4] Cf. De Saulcy, *Essai de classification*, plates iv. xviii.-xxi. xxvii.-xxviii.; on Cherson, cf. Constantine Porphyrogennetos, *Administration*, ch. 53; Rambaud, *L'empire grec*, pp. 484-494, especially 484-485, 486-489,

[5] In 654.

onslaught for the moment threw it into alliance with, and subjection to, the Khazars. On Cherson, in reward for its many services, the great Justinian conferred exemption from all imposts, and Constantine VII. expressly reaffirms these privileges [1]; we have already noticed his suggestion of this port of call as the best route for diplomatic and other missions from Constantinople, on their way to Khazars, Petchinegs, or Russians. Here the Greek language was maintained: here the Barbarian invader was kept at bay; behind its massive walls of squared stones the Ionic columns of the ancient Cherson perpetuated the memory of a past which the ephemeral Nomade could not but dimly reverence. Goths, Avars, Turks, Uzes, Huns, Bulgars, Petchinegs, Khazars, Komans, 'Scythians' of every kind and name, successively wandered across the plain outside the ramparts of this *eternal* city, gazed upon its towers, threatened its defenders, and disappeared in distance and oblivion; but sixteen centuries witnessed the Greek Colony, beside the finest of Euxine harbours, still maintaining its Hellenic tongue, blood, and allegiance, its Hellenistic Christianity, its racial distinction, and its commercial prosperity:—

> 'A Homer's language murmuring in its streets,
> And in its haven many a mast from Tyre.'

As the Middle Ages drew on, the importance of Cherson declines; under the Comneni it is of small account; but in the time of Constantine VII. it appears (like Zara in Dalmatia), as the seat of an Imperial Strategos or Proconsul. Bosporos had long deserted the cause of Greece and Rome, dropping its classic name,[2] affording a ready welcome and good market to Pontic Pirates, and enriching itself with the plunder

---

[1] Justinian, *Novellae*, 113, ch. 2; Constantine Porphyrogennetos, *De Administrando Imperio*, ch. 53,

'Ιστορία περὶ τοῦ κάστρου Χερσῶνος.
[2] Panticapaeum.

of the Roman coasts; it had been recovered for a short time by Justin I.; but now, in the middle of the tenth century, it was definitely a Khazar town. Tanais, at the estuary of the Don, no longer survived; its place was taken by Sarkel or by Rosia; Phanagoria, opposite Bosporos on the Taman side of the Strait of Kertch, perished as a Hellenic town in the sixth century, and had only of late revived as Matrica or Tamatarcha, so famous under the Comnenian dynasty; Odessos, Istriopolis, and the other outposts of the old civilisation to the west of the Crimea had all fallen into the hands of Bulgars or Petchinegs. Cherson alone remained from this shipwreck of Pontic Hellenism, thanks in part to the energetic measures of Justin I., of the great Justinian, and of Theophilus. The latter sovereign, besides building Sarkel (in alliance with the Khazars) as a barrier against Petchineg incursions, replaced the old Republican and semi-independent government in 'St. Clement's town,' by the Byzantine provincial administration, and inaugurated the rule of *Strategoi*, in addition to the *Proteuontes* and *Archontes* of the past (c. A.D. 833).[1] More than ever the city now became an advance-post of Constantinople in the Scythian world, a watch-tower against all movements of the Northern barbarians. But the Imperial government was never very tyrannical at Cherson; the status of vassal-city was never entirely exchanged for that of subject-town. Yet Constantine VII. directs the seizure of all the *karabia* of the port in case of revolt, the stoppage of supplies, and the imprisonment of captured Chersonese; their prosperity depended, he believes, upon their position as carrying-traders between the Empire and the Steppes; and his view was too well founded for any permanent breach to subsist. From Cherson the Byzantines had the first news of a Russian descent of the

[1] Cf. *De Administrando Imperio*, ch. 42; *Ceremonies*, ii. 50

Dniepr,[1] or of any other movement among the nations of the North Pontic Regions. All diplomatic intercourse with 'Scythia' was managed through the medium of Cherson, and by Chersonese advice; it was a native of this city[2] who conducted the delicate negotiation by which Nikephoros Phokas drew on the Russians to attack Bulgaria.

The tenth-century Byzantines therefore cannot be accused (as Strabo accuses the Augustan Romans) of neglecting their Black Sea interests, even in the far outlying Crimea. All the three Russian treaties of the 'Macedonian' Dynasty—with Oleg, with Igor, and with Sviatoslav—expressly provided for the safety of Cherson; the Barbarian chiefs swore by the Gods[3] of the Slavs that *Korsûn* should be left in peace; the treaties with the Khazars contained similar stipulations; and the distrust of these 'royal Scythians,' shown in the pages of Constantine, seems to have been based upon a conviction that they were, after all, the most dangerous of possible aggressors in *Taurica*. Alans of the Caucasus, Russians of the Dniepr, Bulgars of the Kama and Middle Volga, were all salaried by Constantinople against the day when 'Khazaria' should attempt to sweep the Romans into the sea. At Cherson the Khazar tongue was spoken as much as Greek, and at certain times (as when the fury of Justinian II. threatened the city) a Khazar governor[4] with a strong garrison held its walls and forts. But most of the Barbarians around were interested in the fate, and anxious for the prosperity, of the city. The Zicci of the Kuban, the

---

[1] Cf. Nestor, ch. 27.

[2] Kalokyr, son of a Proteuon of Cherson; cf. Leo the Deacon, iv., 9; Cedrenus, ii. 372. Chersonites probably had exceptional qualities as navigators and sea-warriors from their experience of the stormy Euxine and of naval dangers; cf. Liudprand's account of Michael, commanding for Nikophoros Phokas at Corfu, in 958 (*Legation*, end).

[3] *E.g.* Perûn and Volos, cf. Nestor, A. M. 6415-6420, 6453.

[4] *E.g.* one named Tudun.

Highlanders of the Caucasus, and even the brutish Petchinegs of the Steppes, were to be found in its markets, and exchanged furs, hides, and honey, for the silks, linens, muslins, purples, pepper, spices, and brocades of Byzantium and the South.[1] After the days of Constantine VII., both Sviatoslav and Vladimir I. of Kiev seem to have gained possession of Cherson, but neither injured it; for the citizens (according to Leo the Deacon) had themselves invited the Russian Princes, partly from discontent at the Imperial rule, partly from desire for autonomy. It was a Christian priest, Anastasius by name, afterwards Bishop of Kiev, who opened the gates to Vladimir; on his conversion, the conqueror built a Church in the city, and professed to hold the place in right of his Byzantine wife.

This Russian leasehold, however, like the Khazar occupation, proved but temporary; and in spite of such occasional lapses, Cherson retained the aspect of a Greek settlement, and a certain real connection with the Empire, down to the close of the twelfth century.[2] To the religious world it was venerable as the traditional scene of the martyrdom of St. Clement; here Cyril, the Apostle of the Slavs, stayed to learn the language of the Khazars before entering their country; and here he discovered the bones of a Saint[3] who was to become one of the Patrons of the Slavonic world. Like Bosporos, Cherson was the seat of a Bishop or Archbishop;[4] the Chersonites were eager proselytisers; and their occasional acceptance of Barbarian rule may perhaps

---

[1] Constantine Porphyrogennetos, *De Administrando Imperio*, chs. 6, 53.

[2] In 1190 Eupeterios, the 'illustrious General and Duke of Cherson,' figures in the Byzantine epigraphy. Böckh, *Inscript. Christ.*, No. 8740, also Nos. 8742, 8757.

[3] Clement.

[4] Cf. Constantine Porphyrogennetos, *Ceremonies*, ii. 54, which adds Nikopsis (about half-way between the present Novo-Rossiisk and Sukhum-Kalé) as a third See in the 'eparchy' of Zichia.

be explained by this missionary enthusiasm, which maintained an active ecclesiastical intercourse with Kiev, and perhaps with Itil. But from the beginning of the thirteenth century, the colony sinks into complete obscurity; first Sudak and then Kaffa[1] drain away its trade and wealth and people; and in the sixteenth century its walls and towers, and public buildings, with their marble and serpentine, stood like the edifices of a magic city in the Arabian Nights, without a living soul to use them.

In Armenia and the Caucasus, as in the Crimea, Constantine VII. throws real light into the darkness. Since the rise of the Caliphate the civilised world had been almost confined to Arab notices of these regions for the maintenance or extension of the knowledge which classical learning had bequeathed; and until the Byzantine revival, under the Macedonian or Armenian dynasty, scarcely anything was done for the description, the exploration, or the civilisation of these regions, from the Greek or Roman side.[2] But from the regency of Romanus Lecapenus (himself of Armenian race, and probably well acquainted with the Armenian language), an active and profitable intercourse was re-opened between Byzantium and the Ibero-Armenian lands. Thus King Aschod, suzerain of all the Armenian principalities, comes to the Bosphorus to beg for the aid of Constantine and Zoe against the Arabs; in 923 the Marshal or *Curopalata* of Armenia, also appears at Court; about the same time arrives

---

[1] Theodosia.

[2] The travels and descriptions of Sallam the 'interpreter' (841-846 A.D.), reported by Ibn Khordadbeh and Edrisi; of Ibn Foslan, in 921; of Masudi, in 943-946; of Al Isstakhri, about 951; and of Ibn Haukal, at various times between 942 and 972; are the chief Moslem works dealing with these countries before the Crusades. Sallam is thought by some to have reached even beyond the Bashkir country; Ibn Foslan probably did not go beyond Bulgar and the lower Kama; the last three at most touched the lower Volga region (Itil, etc.).

an Iberian functionary of the same title, one Aderneseh; and nearly all the members of the great House of Daron make similar journeys.[1] On the other hand, Byzantine agents—diplomatists, interpreters, or dragomans—were constantly maintained in Armenia;[2] and Byzantine Generals, in the victorious advance of Constantine's later years, possessed themselves of many of the chief strongholds of the country, such as Erzerum.[3] The same time is also full of treaties and agreements, the results of Byzantine diplomacy. Thus in 920 the Patriarch John VI. writes to Nicolas, the *Holiness* of Constantinople, to beg the aid of Constantine VII., his former pupil;[4] the Princes of Armenia and Iberia, not belonging to the favoured House of Daron, express a pained surprise at the exclusive favour shown to the latter; the nature of the Byzantine tenure at Erzerum is defined by treaty;[5] and the threatened occupation of Adranutzion (whose capture by the *Patrician* Constantine is immediately announced at Byzantium, and recorded in the Imperial archives) is warded off by a remonstrance from the Georgian chieftains.[6]

[1] On the Armenian journeys to, and relations with, Byzantium in the time of Constantine Porphyrogennetos, see *De Administrando Imperio*, chs. 46, 45, 44, 43.

[2] *E.g.* Sinutes, Constantine Krinites, Constantios, Theodoros, etc. John Kurkuas or Gurgen, the captor of Erzerum, and himself of Armenian race, was the leading figure among the *Roman* military leaders on the eastern frontier at this time.

[3] Otherwise Karin or Theodosiopolis.

[4] Cf. the *History*, chs. 100, 101, 107, 108, of John, the Catholic, or Patriarch, of Armenia in question. The letter of Nicolas is short and business-like; it urges Armenians to forget their divisions and unite against the enemy of their faith, joining with the Kings of Iberia, Abasgia, and Albania; the two former of these have already been exhorted to the same effect from Byzantium; the *Emperor crowned of God* will send imposing forces. John writes at great length to thank the Emperor for his support. All Armenia desires the alliance of the Emperor; if any do not, they are not of Christ's fold.

[5] Especially with the Iberian Princes, who were somewhat jealous of any other power at *Karin* (Erzerum).

[6] The Iberian chancery must have been a wonder for the time, repeatedly producing for the information of the Byzantines documents which the

After the Arab disaster at Palermo had relieved Byzantine anxieties in the West, the new Eastern policy was vigorously seconded by a fresh mission; the susceptibilities of Bagdad were ignored; and an imperial Embassy brought presents of splendid lamps to the tomb of St. Thomas in the valley of the Kur.[1] Times had changed since the Armenian Princes had been content with the protection and suzerainty of the Caliph; under Basil I. they had begun to seek once more the Roman investiture as well as the Arabian; during the tenth century this divided allegiance passed into one reserved for the Basileus alone.[2]

Under Constantine VII., the Armenian and Iberian *archons*, the lesser chiefs of Colchis or Abkhasia on the Euxine, of Albania on the Caspian, of Vaspuraçan to the east of Lake Van, of Gogovid or Gokovid[3] in the region of Ararat, are often bound to the empire rather in a religious than in a political obedience. The latter is still capable of various interpretations, and may still at times be divided between Bagdad and Stambul; but as the 'spiritual sons'[4]

latter have not preserved;—*e.g.* a treaty signed by Romanus Lecapenus and by Constantine VII.; cf. *De Administrando Imperio*, chs. 46, 45, 43.

[1] Rambaud, *L'empire grec*, 432; Contin. Constantine VII., ch. 32.

[2] Basil II. completed the victory, and raised European influence in the Armenian plateau and the Caucasian isthmus to its zenith, as far as the earlier Middle Ages are concerned.

[3] Also of Moex or Mogh, to the South of Van and Mount Niphates; of Vaitzor or Sisagan in the mountains of Siunia; of the Serbots or *Black Children;* and of Khatsen or Artsakh adjoining Sisagan and Lake Gegharma. As to the Serbots, John the

Catholic asserts the existence in the Caucasus of Chekhs, Serbs, Croats, Bulgars, as well as innumerable other races. On the nine Armenian and five Iberian vassals of the Empire, cf. Constantine Porphyrogennetos, *Ceremonies*, ii. 48; *De Administrando Imperio*, chs. 43, 44, 45; Rambaud, *L'empire grec*, 507-515.

[4] Cf. Constantine Porphyrogennetos, *Ceremonies*, ii. 48, and a letter of John Tzimiskes to Aschod III., *archon of archons*, or supreme head of the Armenian chiefs, as head of the Bagratid or Pagratid House, whose residence was at Kars from 928 to 961, at Ani, from 961 to 1041. Cf. Matthew of Edessa, ch. 16.

VII.] THE GEORGIAN VASSALS OF BYZANTIUM 507

of *Constantine Augustus*, as officers in the army of that 'Soldier of God crowned by Jesus Christ,' the position of these princes is clearly defined; they are members of the Church whose earthly head was the Caesar-Pontiff of Byzantium.

Both politically and ecclesiastically, Georgia or Iberia stood apart; it owned no subjection to Armenia, and, unlike the latter, its orthodoxy was unimpeachable. Its rule extended south-west into the Chorok valley, and included the harbour of Batûm, but (then as now) the real seat[1] of the Georgian state and nation lay in the basin of the Kur and the plain of Tiflis.

Among all these vassal princes, and their more distinguished subjects, the Byzantine Court distributed titles, pensions, honours, and promises; some even possessed their town-houses[2] or property in Constantinople; the scholarly distinctions of Doctor in Medicine,[3] or in Philosophy, were employed to reward the loyalty of others; but sharp reproaches were always ready for those vassals who had been backward in good works; and Constantine VII. deals very faithfully with a Daronite kinglet and an Iberian Marshal, who were suspected of preferring an Arab to a Greek master at Erzerum.

As the tenth century advanced, and the power of the Caliphate and of Levantine Islam as a whole was more sensibly weakened, the Byzantines took up a stronger

---

[1] Rambaud, *L'empire grec*, 512-514, notes at least five Iberian or Armenian principalities not noticed by Constantine Porphyrogennetos; among these Gaban and Harar appear the most important; Varujnani, Andsevatsi and Aghdsnikh are more obscure. Mtzkhet, the ancient capital of the Iberian people in the fourth and fifth centuries, had long ceased to be more than the St. Denis of Georgia.

[2] One called the *House of the Barbarian* was owned by Gregory the (Taronite or) Daronite; *De Administrando Imperio*, ch. 43.

[3] Ἀρχιατρός and ὕπατος φιλοσόφων. The former title was given to the famous physician, Leo, by John Tzimiskes.

and more imperious position, and occupied many of the strategic points which the *Administration* enumerates [1] to the West and North of the Lake of Van. The wild Highlands of the Ararat plateau became well known to Greek soldiers and statesmen, as they had been to few since Xenophon; and it was here that in 1071 Romanus IV. faced the advancing Turks on the fatal day of Manzikert. Constantine VII. had witnessed, if he had not consented to, the evacuation of the Georgian Adranutzion by his generals, and even to the seizure of Erzerum by the Iberians; in 943, the attempt of the Orthodox or Byzantine party to consecrate a church in the royal town of Kars, according to their rite, was defeated by heretic Armenians. But Nikephoros Phokas and John Tzimiskes compelled a more respectful attitude; the latter exacted a loyal and effective service from his Armenian allies in his Eastern wars; and Basil II., everywhere pursuing his policy of conquest, occupied the Iberian province of Taikh,[2] and the Armenian province of Vaspuraçan; seized many of the fortresses so ineffectually coveted by his grandfather; and placed Daron, Ani, and most of the allied territories under Byzantine governors. For a short time the chroniclers of the Empire could describe the new provinces with a facility and thoroughness of opportunity given by political domination.

The alliance in faith and arms was accompanied by a certain commercial intercourse; Adranutzion is praised by Constantine VII., not only as a military position, but as a market well placed at the meeting of routes from Iberia, Abkhasia, the Trebizond coast, and all parts of the Armenian

---

[1] Kars, Ani, Erzerum, Manzikert, Adranutzion or Ordchenhagh, Percri, Khelath, Ardshish, Mastat, and Abnic, near Erzerum, were the chief of these. Cf. *De Administrando Imperio*, 44, 45, 46. All these places were attacked, and most of them were captured, by Basil II.

[2] Considered to represent all Iberia.

highland.[1] The manufactured products of skilled Greek workers (vases, instruments of music, purple and gold stuffs) were exchanged here and elsewhere for the raw materials which alone were to be had in this poor barren land.

The ceremonial influence of Byzantium was still more clearly felt; the petty courts of Ani, Kars, Mush, Van, and Tiflis imitated the ritual and ornaments, the dignities and procedure, of Constantinople. Georgia had its *curopalatae* and *protospathairoi*; the former of these were in constant correspondence with the Imperial Court, as directors of Iberian policy; while one of the latter (then an official named Zurbaneles) is recorded to have paid a visit to the Bosphorus[2] in the later years of Constantine VII.; he seems to have been charged with a regular diplomatic mission (c. 952 A.D.)

In the remote highland of the Caucasus itself, at the Western foot of which Greek settlements had been planted at such an early age, and where the Byzantines maintained a slender foothold till the close of the twelfth century,[3] Constantine mentions three chief vassals of the Empire, the Alanian, Abasgian,[4] and Albanian princes. The two latter are constantly found in alliance with Constantinople against the Arabs; the Albanians distinguished themselves by their enthusiasm for Orthodoxy and the Greek rites against Armenian pravity; an Archbishop of Alania

---

[1] Cf. *De Administrando Imperio*, c. 46; Rambaud, *L'empire grec*, 521.

[2] Like the Iberian *curopalata*, Adranaseh, or Aderneseh, in about 923; (cf. *De Administrando Imperio*, c. 45, apparently composed in 952); perhaps this was one of many visits, mostly unrecorded, on the part of such officials. Adranaseh had received the dignity of 'Curopalata' from the Emperor Leo VI.

[3] Greek was still spoken in Lazica in the sixth century, thanks to Greek merchants; cf. Rambaud, *L'empire grec*, 485.

[4] Or 'Abkhasian.' Abkhasia, like Armenia, had its Patriarch or *Catholic*, and Constantine Porphyrogennetos, *Ceremonies*, ii. 48, gives the formula of address for this ecclesiastic.

is recorded by Leo VI.,[1] at the opening of the tenth century.[2] Leo the Isaurian in 710, Thomas the Usurper in 802, both appealed to the loyalty of these 'children and dear allies'; under Constantine VII. they received a subsidy for helping to guard Cherson and threatening the rear of the Khazars.

In this country of Caucasia,—where every branch of the human family had its representative; where every wandering nation and every exiled race left a fragment of itself; and where the varieties of language were 'known only unto God,'—it would be difficult indeed to follow out the lesser tribes of the Byzantine obedience.[3] The Zicci and Circassians of the West or Euxine coast (two famous races who stand on a very different footing from the Mokan, whom Constantine so strangely places by the Maeotid Marsh, or Sea of Azov), were well known to the Byzantines; but they do not occur in the list of Caucasian vassals given us in the *Ceremonies*. Yet the Circassians were regularly visited by the merchants of Trebizond, and as regularly frequented the annual fairs of the Black Sea ports, where 'Romans,' Moslems, and Armenians met together;[4] while the Zicci possessed a distinct value in Byzantine policy as a check upon the rear and flank of the Khazars. Christianity had

---

[1] In his *Hypotyposis*.

[2] But in 932, according to Masudi, II., 43, its misguided people drove out their prelates, and broke off communion with the Church of Chrysostom.

[3] The Krebatades, the Kedonians (perhaps a tribe,—the Jidans of Masudi,—who lived near the Khazar town of Semender), the Tzanarians between Tiflis and the Pass of Derbent, the Sarban (or Serbians?), and so forth. Add the Archons and peoples of *Chrysa*, *Vretza* (in Kakhetia), *Mokan* and *Azia* (or *Ossetia*, in the neighbourhood of the Caspian Gates or Pass of Dariel). On the Caucasian vassals, cf. Constantine Porphyrogennetos, *Ceremonies*, ii., 48; *De Administrando Imperio*, 10, 11; Rambaud, *L'empire grec*, 524-527. Some of these, the Sarban, for instance, are placed by Constantine in the North-Central part, others towards the East, of the Caucasian chain.

[4] Ibn Haukal, in d'Ohsson, *Les peuples du Caucase*, p. 26.

declined in the Caucasus since the days of Justinian: Islam had made many converts and boasted of more;[1] some tribes still practised their ancient nature-worship, or the cult of Fire; others observed a selection of Jewish, Moslem, and Christian beliefs and rites; but, nevertheless, the religious tie was a strong one, even in the tenth century, between these regions and the capital of Eastern Christendom.

The Arabian vassals of the Empire, some of whom are placed by the *Ceremonies* in Egypt, Persia, and Khorasan, while others are evidenced by the diplomacy of the Macedonian House in Syria and Mesopotamia, are not of any great geographical interest. Basil I., in his campaigns against the Manichaeans of Tephrice, received the submission[2] of the Arab town of Taurus or Tarax, and the principality of Locana; both these were enrolled as allies of the Empire. A little later, an Arab emir near Adapa, under Romanus I., the Emir of Melitene, submitted to Byzantium. John Tzimiskes, the conqueror of Jerusalem, Damascus, and Edessa, had a great number of Arab servants; his letter to Aschod III. of Edessa (A.D. 971) is famous; but in all this we only hear of well-known localities, far within the borders of Moslem civilisation.

VIII.—The Subjects of the Byzantine empire, in distinction from the Neighbours and Vassals of the same, offer comparatively little of geographical interest; nor is the survey of Constantine VII., in his treatise[3] on the *Themes*, or military

---

[1] *E.g.* the Jidan (or Kedonian ?) Kings, living near the Gate or Pass of Derbent; also the rulers of the Kaitaks; cf. Masudi, ii., 39, 40, 43, 45-46. The Circassians were Fire-worshippers, according to Masudi. Ibn Haukal's story of the *Moslem* King of Gurj or Georgia, reigning over Christian subjects, sounds most improbable, and is without confirmation.

[2] Rambaud, *L'empire grec*, 528-529.

[3] Cf. Rambaud, *L'empire grec*, 175-253.

subdivisions of the European and Asiatic Provinces, as useful as his treatment of the neighbouring and vassal nations. It is noteworthy, however, both from his pages and other sources, that the New Rome, at the end of the first millennium of Christian history, still showed something of that cosmopolitan character which had marked the rule of the Pagan Caesars. Latins and Teutons from the West, from the Italian and Dalmatian Themes, and from the countries beyond the Byzantine limits; Arabs from East, West, and South alike; Armenians and Iberians; Turks and Finns; Slavs and Caucasians—all found a place in the service, in the armies and fleets, even in the hierarchy, of the state. 'Like the College of Cardinals,' the official nobility of Constantinople was 'recruited from the notabilities of the world.'[1] Once baptised, the barbarian was admitted in theory, often in practice, to the dignities of State and Church alike.[2] Among the troops led by Belisarius and Narses had been Goths, Heruli, Vandals, Lombards, Persians, Moors, and Huns; Thomas the Rebel and Usurper, in his rising against Michael II., is followed by Indians, Persians, 'Assyrians,' Zicci, and countless nationalities of the Caucasus; the legions of Constantine VII. and his successors, Nikephoros Tzimiskes, and Basil II., include Varangians or Russo-Scandinavians, Magyars and Khazars, as well as Frankish, Slavonic, Bulgarian, Venetian, Amalfitan,[3] and 'Patzinakian'[3]

[1] Rambaud, *L'empire grec*, 531-540.
[2] In spite of Rambaud's caution, *L'empire grec*, 276, we may probably reckon among the Byzantine Mercenaries Sabaeans, Shamanists, Odinists, and Nature Worshippers, even a few Jews and Moslems.
[3] Cf. Liudprand, *Legation*, p. 357 (Partz), on the Italian troops; Nikeph. Greg., vii. 3, on the

Catalans. Rambaud, 276-277, doubts the Petchinegs, as being such obstinate Pagans. The Empire preferred to enrol Christian soldiers among its Barbarian troops. But from the language of Constantine VII., one can hardly doubt that his most valued allies fought in the Byzantine armies as well as in the Byzantine cause.

auxiliaries. Under the Comneni, Spanish Catalans fight for the Eastern Rome in Asia Minor. The great names in Byzantine military history are more often 'barbarian' than not: from the Armenian Narses and Basil I.,[1] John Gurgen and Nikephoros Phokas, to the Spanish Guzman[2] under Constantine XI. (1066), the Slav generals of Basil I. and Leo VI., and the Frankish, Scandinavian, or Russian captains who served Constantine Monomach and Michael Stratiotikos in Georgia and Armenia, in Mesopotamia, and in Asia Minor.[2]

Above all did the Empire profit from its absorption of two races, Slavs and Armenians. To the former belonged Justin I. and his dynasty, as well as a part-interest in the Macedonian House; to the latter a preponderating share in the blood of the Basils, of the family of Nikephoros Phokas, of John Tzimiskes, of the Regent Romanus Lecapenus. Even Michael III., whom Basil I. displaced in 867 (in order to found a mainly Armenian dynasty, lasting for two centuries), was himself a 'half-blood' of the same nationality.

The poor, half-educated, barbarian peasant appears again and again in Byzantine history upon the very throne of the Caesars; beginning as an immigrant settler, and soon finding an employment suitable for his native energy, he drifts to the great city, to the palace-guard or ministry, to the immediate surroundings of the Emperor; he becomes the trusted, indispensable adviser and protector of a weary debauchee, or a superstitious invalid; thus finally he reaches the summit. Short of this, he or his descendants fill the highest offices of government—in State and Church; the patriarchate is not withheld from the scholar and the devotee; the supple and

---

[1] Basil I. must also be considered as partly Slav. Basil II. is of course descended from this Armeno-Macedonian stock.

[2] Cf. Cedrenus, ii. 606, 616, 624, 630. On Guzman, cf. Matthew of Edessa, ch. 91.

statesmanlike ecclesiastic, whatever his origin or nation, can always find a career. A *Notitia Dignitatum* of the Byzantine realm, in the days of Constantine VII., would give us a mosaic of races.

Constantinople shared with the centres of Arab learning —such as Cordova and Bagdad—the honours of ancient Athens in the eyes of the earlier mediaeval world. Gerbert goes to Spain, but Liudprand eagerly accepts an embassy to New Rome: the legation is a secondary matter, beside the chance of seeing the greatest of Christian cities, and of learning the philosophy and literature of Greece. For this one might well sacrifice the half of one's fortune.[1] So in the eleventh century Adam of Paris, afterwards Bishop of Spalatro, visits Byzantium to perfect himself in the art of writing; so the physician Honainus, in the time of Basil I., visits 'Greece' to learn the medical language of civilisation.

§ 2.—*Latin Geographers—Adam of Bremen*[2]

Adam of Bremen is not merely the Annalist of the Churches of Bremen and Hamburg; he is the first and best historian of Northern Germany in the Middle Ages; and in geography he occupies a still more important place. The earliest writer who mentions Vinland, the one contemporary who has preserved a record of the Polar voyage of Harald Hardrada, our primary authority on the pre-Hanseatic trade of the North Sea and Baltic coasts, Adam shares with King Ælfred, or, more properly, with Ohthere and Wulfstan, the pioneer honours in one great field of geographical knowledge.

Each of the four books, or main divisions, of Adam's work (variously called *Ecclesiastical History, Deeds of the Bishops of*

[1] Liudprand, *Antapodosis*, vi. 3.   [2] Fl. 1070.

Hamburg, or *History of the Chiefs*[1] *of Bremen*) has a different content. The first deals with the conversion of these 'Saxon' regions, and the lives and Northern journeys of the first Bishops of Bremen. The second recounts the general history of Germany, and especially of the German Church; the wars of Germans and Slavs; and the attacks of the Northern invaders on various countries, for about a century before Adam's time.[2] The third book gives a life of Archbishop Adalbert of Bremen, Adam's great patron. The fourth is devoted to the geography of the Northern parts of the world, and has been edited as a separate tract *On the Position of Denmark, and of Other Regions beyond Denmark*.[3]

The word *Geography*, here used by Adam himself of his fourth and concluding section, is not very common in the writers of the Middle Ages; the term *Geometry* (applied to one part of the Quadrivium) was used in exactly the same sense,[4] but with the express inclusion of what is now called Ethnology or Anthropo-Geography; and among other mediaeval expressions for this study we have the common *De Natura rerum, De Natura locorum, De Mensura Orbis Terrae*, the equally common *Cosmographia*, and the rarer *Cosmimetria*.

Adam lived on the borders of the patristic and scholastic periods, and his geography may be said to look back upon the one and forward upon the other. Born, according to the generally accepted tradition, in Old Saxony on the North

---

[1] *Praesules* (here 'Bishops'). The various general titles, *Historia Hammaburgensis Ecclesiae, Gesta Hamb. eccl. pontificum*, etc., in some editions are made to cover bk. iv. *Descriptio insularum aquilonis*, in some not.

[2] C. A.D. 940-1040.

[3] Cf. Augustin Bernard, *De Adamo Bremensi Geographo*, Paris, 1895; K. Kretschmer, *Physische Erdkunde in Mittelalter*; Peschel's *Geschichte der Erdkunde*, especially pp. 80-82 and 90, has many excellent remarks on Adam, and suggestive extracts from him.

[4] As we see in Raban Maur, who applies *Geometria* to *Totus Orbis et diversae ejus facies*.

German coast, he became a Canon of Bremen in or about 1069; he was also master[1] of the Cathedral School and Chronicler of the Church; and he died, according to the Diptychs of the See of Bremen, on October the 12th, probably of the year 1076.

Two eminent men assisted him in the compilation of his Annals, both directly and indirectly, both by their own experience and by the varied advantages which their patronage and hospitality secured. For it was at the Courts of Archbishop Adalbert, his own diocesan, and of Svein Estrithson,[2] King of Denmark, that Adam picked up and pieced together the fragments of new knowledge which give so exceptional a value to his work. Though not himself a great traveller, Adam seems to have had a sufficiently good personal knowledge, not only of 'Saxony,' and Thuringia, but also of Denmark as far as the Sound. It is evident that he possessed the geographical instinct; almost every mention he makes of persons, places, or nations is accompanied by some definition of their habitat or position. The circumstances of the time were such as to invite some attention to earth-knowledge, and the Bremen chronicler was not deaf to the invitation. Everywhere old boundaries were being broken down before the slow and troubled but steady progress of Christianity eastwards along the North German plain, and in the woods, hills, and heaths of the Scandinavian countries.[3] At the commencement of Adam's own century not only Denmark, Sweden, and Norway, but the far distant Iceland and Greenland had accepted the new faith, although the Slavonic Wends of the South

---

[1] One Vilecinus appears as schoolmaster in his place in 1125.
[2] Or Astridson.
[3] The Church of Hamburg was founded, c. 831; destroyed by Northmen, c. 845; and united with the Church of Bremen, c. 847; from c. 860, under St. Ansgar, it claimed authority over all 'Northern Regions.'

Baltic coast still struggled obstinately, and in 1066 for the third time began a 'Pagan War.' In 1055 the first Bishop of Iceland, by name Isleif, was appointed as a suffragan of Bremen; to the same Province were also attached (in the course of the eleventh century) all the homelands of the Scandinavian peoples, to say nothing of Orkneys, Faröes, Hebrides,[1] and other scattered islands of the Northern Ocean. Under Adam's patron, the great Adalbert, Bremen was therefore at the height of its power and influence· Political and ecclesiastical matters were then almost inseparable; and Adalbert, practically a Patriarch of the North, wished to obtain the formal grant of the title, all the more as he shared with Hanno of Cologne the guardianship of the young Emperor Henry IV.[2] His ambition was not gratified, but he made Bremen an Arctic Rome and his court the greatest centre of Northern learning. Italians, Franks,[3] Scots, Greeks, and Jews were all to be found there; likewise physicians, astrologers, painters, and musicians, and among them such famous men as John Scotus Erigena,[4] Gualdo the Frenchman,[5] Trasmundus the painter-monk, the musician Guido of Italy, Aristo the Greek, Adamatus the physician of Salerno, and the much-travelled Bovo, who had thrice visited Jerusalem, knew 'Babylon of the Saracens' or Cairo of Egypt, and had wandered through various other parts of the world. Many

[1] It is curious that the Papal letters of 1027 make no reference to the conversion of Iceland and Greenland; this is first mentioned by Leo IX. in 1053. In 1103, 1110, and 1112, Denmark, Sweden, and Norway were successively separated from Bremen, in the ecclesiastical sense. On Adam's use of *Orkneys*, apparently including *Hebrides*, cf. *Hist. Hamb. Eccl.* iii. 11. etc. [ch. 127, Lappenberg], and at end of bk. iii. (additions); also his *Descriptio insularum* in bk. iv. These *Orkneys* in former times had had English and Scottish bishops, but now Adalbert consecrated one Turolf as Bishop of 'Blascona' (a now unknown site).
[2] The future penitent of Canossa.
[3] Lit. *Gauls*.
[4] 'John of Ireland.'
[5] 'Gaul.'

Slav and Scandinavian merchants and pilgrims also passed through Bremen, and were entertained by Adalbert, whose court-circles (and Adam among the rest) were thus able to hear the latest news of Rome, Compostella, and Jerusalem on one side; of Iceland, Greenland, Norway, and the Baltic coasts on another; even catching now and then a hint of the new-found lands in the Far North-West.[1]

In the midst of this brilliant society, Adalbert moved like a king. The great mediaeval Churchman was seen to full advantage in this man, of splendid presence, lofty ambition, high lineage, rare eloquence, and fervent devotion.[2] Of low amours he was free, an uncommon praise in that age from the mouth of enemies; and, like the High-Souled Man, in whom Aristotle sketched the ideal of ancient philosophy, he was full of gentleness and affability to his inferiors—to the poor marvellously humble (often washing the feet of thirty beggars)—but to his equals haughty and unbending, as became one who claimed relationship with the Imperial families both of the West and East.[3] His splendour and ostentation in household management, in buildings both ecclesiastical and civil, in the maintenance of body-guards and personal state, sound like descriptions of an earlier and German Wolsey, and his faults were not unlike the proud, liberal, brilliant, unscrupulous,

---

[1] On the position of the See of Bremen under Adalbert, as a centre of Northern Civilisation, cf. Adam, iii. 23, 32, 57, 70 [chs. 142, 150, 183, 201, Lappenberg], and Schol. 91.

[2] On Adalbert, cf. Adam, bk. iii. *passim*, and especially chs. 1, 2, 5, 7, 9, 10, 30, 31, 33, 35, 36, 38, 39, 45, 48, 57-59, 68, [chs. 118, 119, 120, 122, 124, 126, 127, 148, 149, 150, 151-156, 162, 165, 183-185, 196-199, Lappenberg]; also Lambert of Hersfeld, in Pertz' edition of 1874, pp. 56, 57, 60, 68, 69; under A.D. 1063, 1065, 1066. Often, says Lambert, in the service of the Mass, 'when he offered the Saving Victim to God, his emotion would overcome him, and he wept as he stood before the altar.

[3] Alike with Otto III. and Theophano, his wife.

much-hated minister of Henry VIII. of England. Alike in kindness and in anger both were regal; alike in merits and defects both were examples of a vigorous and gifted human nature, of an energy and capacity, a love and hate, a generosity and a harshness beyond that of common men. During the life of the Emperor Henry III., Adalbert had what he needed; a firm master, who honoured his talents, stimulated his abilities, and controlled his arrogance. The death of his judicious patron diverted the Archbishop more and more from his own proper work[1] to a purely political life; he first aspired to a share in the custody and education of the young Kaiser, then to the exclusive control of his sovereign; in these aims he succeeded; but in the hour of his success he threw aside all restraint, and vented his open scorn and hatred upon the stupid, greedy, boorish nobles, whom he had long ridiculed in the privacy of his inner circle.[2] The bishop's household might enjoy their laugh, but their lord was ill-advised in publishing his sarcasms, in braving too wide an enmity. What was worse, he began to neglect business for alchemy, spiritism, jugglery, or dice; he seized lay lands and annexed them to the estates of his Church with rather too liberal an eagerness; and after a few years of power the sceptre broke in his hands at the diet of Tribur, a few months before the Norman Conquest of England. What we may call his religious exploration of the North is of special interest. Scarcely did he become Archbishop of Bremen when he despatched letters to 'the most remote of his clergy and to

---

[1] As in his missions to the Orkneys and Iceland.

[2] Yet he was very proud of his family connection with one of these 'boors,' the Count Palatine of Saxony; but whereas Hanno of Cologne, as a man of obscure birth, practised a shameless nepotism in bestowing Church patronage, Adalbert despised such vulgar means of support, and made little attempt to create party.

other bishops of the Arctic regions';[1] in these letters he does not forget to allude to the islands in the depths of the sea; and he even expresses the hope of imitating his apostolic predecessors,[2] Ansgar and Rimbert, by a personal journey to the distant parts of his vast diocese. In some of his earliest letters, Adalbert warns the Bishop of Iceland that he will soon make a visit as Metropolitan both to that island and to Greenland; but from this purpose he was dissuaded by Svein of Denmark, who, as a practised traveller, put before him some of the difficulties and dangers of the way.[3] Yet his interest in the countries of Northern Europe was not thrown away; he inspired the geographical writings of his retainer, the master of the Cathedral school of Bremen; and we may fairly suppose that it was to his permission, or perhaps suggestion, that Adam owed his knowledge of the Archives of the See.[4]

Adam, in all probability, owed more to his bishop than to all besides;[5] but he owed not a little, as we have said, to another great personage of his time, the subtle, fortunate, and successful King Svein Estrithson, who, after a life of many adventures and much profligacy, a visit to England, and a residence of eleven years[6] in Sweden, securely established the independence of Denmark against powerful and dangerous neighbours. His learning was extraordinary for a layman at that time, and gained the especial praise of Hildebrand; no less remarkable was the king's well-stored memory, preserving as in a book the record of so many

---

[1] Adam, iii. 11 [ch. 172, Lappenberg].

[2] Adam, iii. 70 [ch. 201, Lappenberg].

[3] Adam, iii. 23, 70; iv. 35 [chs. 142, 201, 244, Lappenberg].

[4] Including the lives of the wandering missionaries Willibrord, Liudger, and Ansgar; and perhaps a diary and itinerary of the last-named apostle, no longer extant.

[5] But Lappenberg maintains that Svein, as the more practical man, gave Adam his best information.

[6] 1031-1042.

of his people's exploits.[1] Vessels sailing to the Baltic from the 'outer coasts,' rarely appear to have called at Bremen in Adam's day, but the city was a place of passage and stay for many travellers from the Lower Rhine basin to the Danish and Wendish markets of Sliaswig or Schleswig, Roschald or Röskilde, and Jumna. A large and constant stream of visitors was thus attracted to Adalbert's court and town, and although the broils which marked the later years of his episcopate caused a decrease in the number of merchants who resorted there, yet throughout the lifetime of Adam, the Fair Haven of the Weser was indisputably supreme among the harbours of the North German shore.

Meantime, during the whole reign of Svein Estrithson,[2] Denmark was steadily growing in commercial strength; its markets were becoming famous among Moslem traders; and first among the Scandinavian kingdoms it was forming intimate relations with the rest of Christendom. For Danish merchants now made their way from the Baltic to the Black Sea and the Russian upland, along the courses of the Neva, Dvina, and Dniepr; Scandinavian visitors were not infrequently found at Novgorod and Kiev, as at Byzantium and in the Mediterranean; and in the *Vaering* or *Varangian* body-guard of the Eastern Emperors, Russians, Norwegians, Danes, Swedes, and English exiles found a common employment and a common field of honour and glory.

Adam's sources of information were partly oral and partly written, and among the former we must reckon not only his great patrons, Adalbert and Svein, but also many lesser acquaintances, travellers whom he met at the Court of Denmark and in the household of the Archbishop of Bremen. Besides his patrons Adalbert and Svein, them-

[1] Adam, iii. 53; ii. 41 [chs. 171, 84 Lappenberg].
[2] A.D. 1047–1076.

selves repositories of information collected from many friends and visitors, we find the historian especially relying on an (unnamed) Danish bishop; a Nordalbingian noble; the companions of Bishop Adelward the Younger; a Christian traveller who had witnessed the Pagan sacrifices at Upsala; William the Englishman, Bishop of Zeeland, who had been Chancellor of Cnut the Great, and an intimate of Svein Estrithson; and Adelward, the Dean of Bremen,[1] formerly 'Bishop of the Goths in Sweden.'

But like all men of his time, Adam put the authority of classical writers far above this modern testimony, or the vulgar witness of ordinary sense. He assumes that all the regions of the North were known to the Romans, but less definitely, under different names; he displays no small anxiety to base his knowledge of *Scandia* upon the sure testimony of Virgil, Lucan, and Horace; and except for a few writers of the age of Charlemagne and other 'Dark' periods, his book-learning is mainly of the Julian and Augustan time. Sallust[2] especially is employed, his style imitated, his views and expressions introduced. Of the early Christian or late Pagan writers he uses Macrobius[3] and Capella, Solinus and Orosius, Paul the Deacon and Gregory of Tours,[4] Einhard and Bede. The *Annals of Fulda* and the *Deeds of the Saxons*[5] are responsible for some of Adam's history; his general theory in physical questions is largely derived from Bede's tract *On the Nature of Things*,[6] and still more from Capella's *Nuptials of Philology and Mercury*,[7] that 'fount of all science' (in the enthusiastic

---

[1] Adelward the Elder, as Adam calls him.

[2] As by Widukind and Lambert of Hersfeld.

[3] The *Commentary on the Dream of Scipio*.

[4] Perhaps in Fredegar's *Epitome*.

[5] Really Rudolf's *Translatio S. Alexandri*, wrongly ascribed to Einhard by Adam.

[6] Adam also uses Bede's *De temporum ratione*. Cf. p. 574 of this vol.

[7] Especially bk. vi., *On Geometry* (including *Geography*).

language of Gregory of Tours), already accessible to Germans in their own language by the middle of the eleventh century. The Solinus used by him was no doubt the fifth-century recension[1] with which the new title of *Polyhistor* was associated. In all these respects Adam resembles other mediaeval writers, and shows little superiority to his time, except in a more guarded use of fabulous material.

Adam was probably ignorant of Greek, although some Greek words occur in his work under Latin forms; at this time the knowledge of Greek was rare indeed among Western Christians, and the revival of the study did not begin till about half a century later, when William the Monk brought a fresh stock of Greek manuscripts from Constantinople to Paris. Yet Adam would seem to have known something of Ptolemy, if only the shadow of that great name which is also to be found in Cosmas,[2] Cassiodorus, the Ravennese Geographer, and Ælfred the Great; for until the fifteenth century it is the name rather than the works of the Alexandrian physicist which can be discerned in Christian literature.

In particular, the *geographical* matter of Adam's work may fairly be derived from these three sources of personal knowledge, contemporary information, and writings of earlier date, of the first century before Christ, or of the fifth, eighth, and ninth centuries after Christ, by preference. And first, as to his general theories, on the form of the earth and the ocean. There are only two passages in Adam's writings which can be said to bear upon abstract or mathematical

---

[1] This probably contained the Irish additions to which Mommsen has drawn attention.

[2] Cosmas Indicopleustes, iii. 177, 182, Montfaucon; Anon. Rav. iv. 4 [edition of Pinder and Parthey, p. 175, lines 2, 11]. Cosmas of course is well acquainted with the Ptolemaic *system*, and most of the *Christian Topography* is an attack upon its spherical and geocentric theories.

geography;[1] one relates to the opinions of King Svein and others on the length of the night in the *island* of Halagaland; the other to the views of Solinus and Bede upon the length of night and day in Thule, and the necessary inequality[2] in the seasons of light and darkness, arising from the roundness of the earth, which nullified the little inequalities of valleys, mountains, and seas.[3] Adam is quite free from the wild theories of the northern heights behind which the sun was hidden, although these venerable myths, derived from Indian 'philosophers,' had enjoyed the support of Cosmas, of the Ravennese geographer, and even of Dicuil.[4] In this also our annalist owes a measure of true guidance to the eighth century, and especially to the vigorous and manly work of Bede and Bishop Virgil,[5] who in the midst of almost universal scientific depression had done something to re-establish the experimental study of nature. From the time that the great name of Bede could be cited in its favour, the globular view of the earth had a better chance of mediaeval acceptance; for scientific theory was no longer monopolised by the dangerous speculators of the ancient world whose heathen doctrines the true Catholic must needs suspect; and thus the naturalism which had been confined to an insignificant minority,[6] was, in the eleventh century, already winning its way to a recognised position and ultimate victory. As to the face of the earth and the distribution of land and sea, Adam considers that *terra firma*

---

[1] Ad. iv. 35, 37 [chs. 244, 246, Lappenberg], and Schol. 152.

[2] Demonstrated by the Northumbrian philosopher.

[3] Bede, *De rerum natura*, c. 46; *De temporum ratione*, c. 32.

[4] In a measure.

[5] Of Salzburg.

[6] K. Kretschmer suggests that before Bede the globular view of earth was professed only by a very small band of 'faddists'; after Bede, he seems to think it immediately became the creed of the majority of learned men—rather too sudden a transition.

was entirely surrounded by the infinite and terrible ocean;[1] the northern part of this ocean was covered with ice and darkness, and was known as the frozen, glutinous, or darkling sea;[2] it was also conceived as 'thick' or 'stiff' with salt, and covered with black and aged ice, so dry as to burn in fire; the darkness of this region is illustrated by a story.[3] Harald,[4] the King of the Norsemen, and the most experienced of their chiefs, not content with a life of matchless adventure and good fortune, must needs examine the breadth of the Northern Ocean, must even try and penetrate to the Pole itself. Some Frisian mariners seem to have drawn him on by reports of their extensive wanderings on the sea, and Harald, who always ventured beyond the utmost of other men, reached the very limit of the earth,[5] and was only saved with the utmost difficulty from falling headlong into the depths of that 'profound chaos,' wherein the ebb of ocean was swallowed up and again poured forth as the flow or rising tide. Into the vortex of this appalling maelstrom 'at the darkling end of the failing world' the reckless Frisians (gripped by the ice and covered by the darkness) were suddenly drawn; but the ever-shifting current or *Euripus* of the Main was so mighty and so variable, that

[1] Ad. iv. 10, 34 [chs. 217, 243, Lappenberg]. This all-surrounding ocean is not only a classical, but a German and Scandinavian tradition prominent in the heathen mythology.

[2] Adam's (a) *Mare concretum*, also qualified by him as (b) *caligans*, and (c) *spissum a sale*; (a) is common enough, a regular traditional expression, *e.g.* in Capella and Dicuil; (b) is perhaps a Latin translation of the *Dumbshof* of the mythologies. The expression (c) is no doubt connected with the favourite doctrine of the scholastics that the saltness of sea was due to stagnation, a theory true enough as regards lakes. The word (d) *libersee*, glutinous, also applied to the northern waters, has special reference to the half-frozen ocean, such as exists off much of the Lapland coast. It is perhaps connected with the *libberig* of the Bremen dialect, used in the same sense as *klebrig*.

[3] Ad. iv. 38 [ch. 247, Lappenberg].

[4] Hardrada.

[5] 'According to report.'

while it destroyed it could also save, and while it swept the victim down to the depths with one motion, with another it carried him again into the upper air.

Like Macrobius, the Venerable Bede had clearly re-stated the results of Greek science and Phoenician observation upon the tides' 'obedience' to the moon, yet Adam never gives us this solution. He quotes various remarks of both these writers[1] upon the phenomenon, but carefully avoids the point.

Among the marvels of nature we find sufficiently plain references to icebergs and a somewhat confused and symbolic allusion to the volcanos of the Icelandic fells and the currents of the Icelandic shores. From these volcanos was apparently procured an article of merchandise, called 'Olla Vulcani,'[2] an important item of trade at the Wendish mart of Jumna near the mouth of the Oder.[3] Here it was known by the name of *Greek fire;* and as the Greeks, like all the 'Barbarians,'—Franks and Northmen, Slavs and Saxons,—were well acquainted with this 'greatest of the cities of Europe,' and frequently visited its market, the name is scarcely to be attributed to sheer ignorance. Adam shows far too great a knowledge of Iceland—its situation, seasons, and commerce with Bremen—for us to suppose that he was ignorant of its volcanos, and his allusion to the black and burning ice near the great northern island

---

[1] Among others, Lucan is also quoted.

[2] 'Vulcan's jar.'

[3] Some have thought 'Olla' is a misreading for 'Wollin' at the mouth of the Oder; another theory derives it from 'Hölle' (Hell) volcanos being constantly spoken of in the Middle Ages as mouths of Hell; cf. Willibald, *Hodoeporicon,* end (on Vulcan's Isle in the Liparis); *Dawn of Modern Geography,* i. 154. This conception, coming from the Greek-speaking Mediterranean, perhaps explains the use of the term 'Greek fire' in the text here. On Iceland, cf. Saxo Grammaticus, who calls it a place for the torture of tyrants; one theory makes the 'Greek fire' of the text an expression for the torment of Greek Church heretics, who were numerous at Jumna (or Jumne).

can hardly refer to anything but lava running down from the fiery mountains to the 'frozen, dark, and burning' ocean.[1]

Once more, just as this *Greek fire* of the North may be connected with Iceland, so we may link with the same island the famous and obscure passage about the 'triple Neptune,' which, with three straits, washed its shores,[2] one 'green' (or ice free), the second 'whitish' (or ice bound), the third 'black with fierce tempests.' It is true that the first recorded eruption of Hekla was a little after Adam's day;[3] but other volcanos of Iceland are known to have been active in far earlier times, even in the ninth century, shortly after the first settlement of the Northmen.

From these allusions to abstract geography and to certain phenomena of geographical interest, we must turn to Adam's treatment of the countries of Northern Europe, in other words, to his detailed topography. The unquestionable value of the chronicler's descriptions is somewhat impaired by his necessarily vague estimates of distance; land-travellers and mariners of that time had no sure means of determining positions; for, at the very earliest, they cannot be shown to have used the compass before the closing years of the next century.[4] In the case of some of the more important Northern cities reckonings of latitude may already have been made from the length of the longest day;[5] but at sea the course of ships could only be directed by empirical observations of

---

[1] Ad. iv. 35 [ch. 244, Lappenberg], and Schol. 149. An apparent misplacement of the text would seem to imply that this 'Olla Vulcani' was actually obtained at Jumna or on the shore of Pomerania, but there can be little doubt that the question is of an article of import trade.

[2] From the text (*fretis alluitur illa insula*) the reference is apparently to Jumna, but no doubt in reality to Iceland.

[3] In the year 1104.

[4] About A.D. 1180.

[5] Thus at Röskilde such a reckoning was made in 1274, and in other cities similar computations were made by order of the bishops, as in the *Book of Röskilde*. The same reckoning is used by the Sagas of Red Eric, etc., in estimating the position of Vinland.

the stars and of the flights of birds, such as the ravens sacred to Odin and useful to the navigator from the length and steadiness of their flight. Adam's usual computation is by the day's journey of twelve hours,[1] an obviously uncertain measure, but assumed (on the Northern Seas) to average about fifteen geographical miles. Still more difficult and uncertain was such a reckoning on land; for there all depended on the state of roads, which in Denmark were excellent, in Sweden poor, and in Norway and the Slav countries of Eastern Europe almost non-existent.

It must remain doubtful how far the Northern knowledge of the world extended in Adam's time, at least as regards some of the more distant and obscure regions; and it is clear that while he preserves a notice of the Vinland or American discoveries to the West, he has no conception of the White Sea and Biarmaland explorations of the Northmen to the East. The *Scythia* of the Saxon geographer includes the whole of Eastern and Northern Europe beyond German limits, and comprises not only Wendland and the modern Russia, but sometimes Denmark, Sweden, and Norway as well. Thus Birca in Sweden is a *Scythian* town, and a great resort for Danish, Norse, and other *Scythian* traders; while Harald Hardrada, in his Byzantine service, was constantly at war with *Scythians*.[2] Such a use of the term is hardly satisfactory from a geographical, historical, or racial point of view, and it is complicated by Adam's employment of the designation *Greece*, to cover all Christians of the Eastern Church, from Russia to Byzantium. As to the latter, our author makes no secret of his Roman prejudices. The Greek schismatics, he believed, were quite

---

[1] The same, according to Dr Storm, *Studies on the Vinland Voyages*, is the regular meaning of the expression in all the Sagas, the night never being included.

[2] Cf. Ad. i. 62, 64; iii. 12 [chs. 47, 49, 130, Lappenberg]. See also Schols. 63, 64.

as far removed from the Christian Faith as the Barbarians,[1] and he feared that they would some day be the leaders in the seduction of all Christendom, in that dark hour when Antichrist would triumph, when Gog and Magog would rage against the saints, and when the Lord would send fire upon all those who dwelt at ease on the islands.[2] This time of trial Adam sometimes considered to be very near; the end of the first millennium, and the tyranny of Harald Hardrada, were both suggestive of the accomplishment of the Reign of Evil, after a thousand years; and although in each case the fulness of time had not arrived, there were many signs that the last day was at hand. For one thing, the Slavs were now all going back to idolatry; and no long time before the historian wrote, the city of Hamburg had been destroyed[3] by its Pagan enemies.

As to the Scandinavian peoples, Adam's language is confused and uncritical; he is content to identify them all with the ancient Hyperboreans; and in the spirit of classical and mediaeval philology he mingles together Goths and Getae, Danes and Dacians, Swedes and Suevi. The last-named, according to the powerful testimony of Orosius and Solinus, once occupied the greater part of Germany, and their rough mountainous country stretched far into Scythia up to the Riphaean Hills.

The Bremen geographer is the first author of Latin Europe to give us the (Slavonic?) name of 'Baltic,'[4] but he has

---

[1] Under the name of Barbarians, Adam sometimes groups all who are not of the Latin Church, both Pagans and Greek Christians.

[2] Ad. i. 28 [ch. 22, Lappenberg]; cf. Ezekiel xxxix. 6.

[3] An earlier fulfilment of the prophecy of Ezekiel is seen by Adam i 28 [ch. 22, Lappenberg], in the capture of Rome by the Goths, A.D. 410. Ezekiel's words are specially connected with the supposed meaning of *islands* attached to the word *Magog*.

[4] Almost certainly from *Bieli* (white), though most Latins connected it with *Balteus*, 'a belt.' In Lithuanian the name is '*Baltas*.'

somewhat distorted notions of its shape, and consequently of the Scandinavian peninsula, island, or archipelago; for in treating of this vast region the writer commonly hovers between these three conceptions.[1] In one passage, however, he frankly surrenders himself to the influence of the personal testimony of 'some who knew the country well,' and asserts roundly that the Baltic was entirely encompassed by land. His informants declared that they had been overland all the way from Sweden to 'Greece,' and but for the barbarism of the peoples dwelling on this route, not so many would be found to brave the dangers of the sea passage. In other words, men preferred to cross the Baltic sooner than pass through the Finnish tribes of the Eastern shorelands.

It is strange that on the question of the Northern Euxine, where Adam had abundant information from Danish and other contemporaries who had sailed over a great part of it,[2] he yet attaches so much importance to the traditions of ancient writers usually destitute of all first-hand information or original value. At the same time, he does not neglect (as some would have done) the true guidance that offered itself; he only weakens, he does not obliterate it by his traditions; and herein[3] we have an excellent example of his method and a good illustration of his characteristic merits and defects. Martianus Capella declared that the Maeotid Marsh, washing the *Scythian shore* or

---

[1] As was the case with his more ignorant 'Authorities' of antiquity; several of these made Scandinavia a group of isles; Ptolemy drew it as one large island; with Adam begins the tendency towards the true or peninsular conception, iv. 15 [ch. 222, Lappenberg. 'Some well acquainted with the localities assert that men have gone overland, from Sueonia even to Greece']. On the other hand, Adam makes Halagland or Heiligland (Halogaland) a separate island from Scandinavia, and groups it with Vinland, iv. 10, 37, 38 [chs. 217, 246-247, Lappenberg].

[2] At least as far as the Gulf of Finland to the north-east.

[3] Viz. in the Baltic question.

desert of the Getae, was the limit of that fourth gulf of Europe, which commenced at the Hellespont; this Marsh, according to the same authority, was itself a bay of the Northern Ocean; and Adam is evidently disposed to think [1] that the Baltic (otherwise the *Eastern, Barbarian,* or *Scythian* Gulf) formed the connection between the Marsh in question and the Northern Seas, which by these channels might be said to encircle Scythia. Einhard, the biographer of Charlemagne, is also regarded as an important authority here; true, he had given no name to this long and narrow gulf, which stretched for an unknown distance from west to east,[2] while in breadth it never exceeded a hundred miles; but in the main his description[3] is quoted and adopted, without dispute in some places, and with slight modification in others. In the spirit of his teacher, Adam places Birca in Gothic Sweden, just opposite the town of Jumna and the mouth of the Oder, and declares that the breadth of the Baltic from South to North was very moderate, and that the length had never been properly computed. Yet the Bremen chronicler is aware that the Baltic was not entirely confined to a main channel running due east and west, but that it threw out arms or branches far and wide; he was also aware that its breadth increased as it penetrated into the Continent. Once more, the vague dogmatism of Einhard upon the 'unknown length' of the gulf (though confirmed by King Harald [4] of Norway and Ganuz Wolf the Dane) is corrected by the narrative of some Danish mariners, who themselves related how with a favouring breeze they had gone in a month from

---

[1] On the Baltic cf. Ad. iv. 10, 11, 16, and 20 [chs. 217, 218, 223, 229, Lappenberg], which last follows Capella and Einhard more doubtfully, and Schols. 115, 117, 121. The quotations are mostly from Einhard's *Life of Charles the Great,* ch. 12.

[2] 'Above the Maeotid Marsh.'

[3] Ignoring the northern bend of the inland sea, as a whole.

[4] Har.hada, who had in vain attempted to reckon the size (whether the area or the volume is not clear) of the Baltic.

their own country to Ostrogard in Russia.[1] Around the Baltic Adam groups—correctly enough—a number of peoples and races:[2] Danes, on the southern shore towards Saxony; Slavs, within and without the diocese[3] of Hamburg, from the Elbe to the Oder; Pomeranians, Poles, and Wends, beyond the Oder; Russians in the Far East; Northmen and Swedes on the North. The kingdom of Russia, or 'Ruzzia,' he regards as the most distant and extensive province of the Wends, terminating the Baltic Gulf on the East; to the north of Russia, and in the north-east corner of the Baltic, he seems to place the Land of Women (or of Amazons) and the home of the Turks, perhaps supposed to lie between Russia and Sweden.

Adam was well aware that the Baltic joined the Western, Britannic, or German Sea, itself a part of the all-encircling and infinite ocean, by which men could sail round from Scandinavia to the Byzantine lands; one section of this journey, the passage from Denmark to England, might be made in three days with a favourable wind.[4] To the south of the Britannic Sea was Saxony (itself bounded on the *south* by the Rhine, as 'Slavia' by the Elbe) and Frisia, sometimes reckoned as a part of Saxony; to the west were the Britons; to the east Danes and Northmen; to the north the Orkneys and Iceland, 'at the extremity of the world'; thence the encompassing ocean flowed westward by Ireland,

[1] On the Baltic investigations of Harold Hardrada and Ganuz Wolf, cf. Ad. iv. 11 [ch. 218, Lappenberg], and Schol. 116 on Ruzzia and Ostrogard.

[2] On the peoples of the Baltic littoral, cf. Ad. ii. 16; iv. 12, 13, 14 [chs. 62, 220-222, Lappenberg].

[3] *Parrochia*. Among smaller Slav tribes Adam mentions *Wilzi* and *Leuticii* immediately west of the Oder; with the *Turks* he couples the Slavonic Wizzi, Lami, and Scuti.

[4] On the Western Ocean or North Sea in general, cf. Ad. ii. 50; cf. also, iv. 10; iii. 70 [chs. 89, 217, 201, Lappenberg], the last especially refers to Adalbert's projected journey to Denmark, Sweden, Norway, the Orkneys and Iceland, from which Svein Estrithson dissuaded him.

eastward by Norway. The northern part of the sea was called the Ocean of Darkness, and the chief islands herein were Thule, Gronland,[1] Halagland,[2] and Vinland, all apparently supposed to lie due north of Norway, close to the edges of the abyss. But even in these distant lands (where Adam remarks on the 'wild vines and self-sown wheat' of Vinland), and in regions nearer home, he does not merely give us a list of names, but he clothes these dry bones with something of a body. For in most of the countries he touches on he has many a detail to tell us about the soil and its products, the animals and the people of various regions, their mode of life, their language, and their real or supposed relationship to other races; in all this he is constantly reminding us of his superiority to the common mediaeval writer.

So far, we have followed Adam in his treatment of Northern lands from two centres, or geographical points of vantage,—the Baltic and the North Sea; but his most elaborate descriptions are compiled from a third—his own town of Bremen and country of Saxony. A good deal of repetition, and even contradiction, is caused by this continual re-arrangement of his scheme, probably due to revision of his material at different periods; but repetitions and contradictions are notes of mediaeval literature. As to Saxony, he is content to repeat from Einhard that it contained no small part of the German lands,[3] and was bounded by Frisia on the west, by Franconia and Thuringia on the south, by Normania on the north, and by the country of the Obodritae on the east. It is equally curious that he says nothing of the Harz Mountains, unless he includes them

[1] Greenland.
[2] Halagaland or Halogaland.
[3] Ad. i. 1-5, from Einhard, *Life of Charles*, 15; cf. also, Ad. ii. 19, [chs. 1-4, 66, Lappenberg], the last mainly concerned with the Slavonic peoples to the East of Saxony as far as Kiev 'Chive.'

in the 'few hills' that broke the Saxon plain. The rivers of this region, and especially the Elbe, now maintained (according to Adam) a large ship traffic; but the province grew no vines; and here we have no doubt an allusion to other German districts whose wine was already famous. From his own observation, or theory, he adds the comparison of Saxony to a triangle, one angle stretching south to the Rhine, another north to the country of Hadelohe, the third east beyond the Elbe.[1] But the province is not always defined in the same manner; for both a wider and a narrower conception of *Saxonia* occur; in the latter sense, it is bounded eastward by the Elbe;[2] in the former it reaches beyond that stream and includes the Nordalbingians and the Slavonic Sorabi, whose land stretched far into Altmark. Officially, however, the eastern border of the Saxon land was defined with sufficient clearness by Charles the Great and later Emperors; and this border is minutely described by Adam, from its 'southern' point on the Elbe, near Lauenburg, to its 'northern' on the Baltic, near the Bay of Kiel. The details here given may have been drawn from an old map at Bremen or elsewhere, in the Bishop's library, or the Chapter archives; but if so, the map in question cannot have been very scientific. For Adam's sketch distorts Nordalbingia no less than Saxony; and as the latter has the Rhine for its southern boundary, so the former has the Elbe.

To the west, beyond the Wapling Marsh[3] and the

[1] Ad. i. 1 [ch. 1, Lappenberg]. From angle to angle was a journey of three days, except for the part beyond the Elbe, where the Sorabi lived, the Nordalbingians dwelling on this side of the river.

[2] As in Einhard. On the wider sense, cf. Schol. 19, which speaks of the Sorabi between the Elbe and Sala, and 'others' of this tribe as living even beyond the Ara in Altmark. Perhaps this was a result of knowledge acquired by Adam during his residence in Magdeburg.

[3] The modern river Wapel.

mouths of the river Wirraha, *Saxonia* was bordered or continued by Frisia, a maritime region inaccessible from its pathless marshes.[1] Orosius had said the same thing about Saxony itself, but Adam knew the country too well to endorse the statement for his own time; woods and marshes are indeed features in his picture of this region, cutting off one *parrochia* or one mark from another, but the province, as a whole, watered by three large rivers,[2] is no longer a 'pathless waste.' The chronicler touches with a light hand, and apparently with little interest, the geography of Western countries. His eye is fixed on the wild and little known regions to the east and north, and in this characteristic, rare enough in later times, and extremely remarkable in a German ecclesiastic of the eleventh century, lies no small part of his value to the modern reader.

To the East of Saxony, *Slavonia* or *Slavania* is defined as the most extensive province of Germany, inhabited by Vandals, Wends,[3] or Winuli, lying between Elbe and Oder, and shut in by strong barriers of woodland, marsh, or river. Reckoning with it the countries of the cognate Poles and Bohemians, who lived on the other side of the Oder, used the same language, and were characterised by the same habits, it was fully ten times greater than Saxony. Eastward it stretched infinitely far to Hungary,[4] Bulgaria,[4] and 'Greece' (or the Lands of the Eastern Church); its breadth was measured from the Elbe to the Baltic, from south to north; and on its western extremity it touched the central districts of the great diocese of Hamburg-and-Bremen. It

---

[1] Sch. 3, which gives to Frisia 47 'pagi,' and adds to the other boundaries mentioned in the text, the marsh of Emisgoe and the Ocean.

[2] The Elbe, Wisurges, and Emisa.

[3] The name *Winuli* seems to occur first in Adam, and is perhaps his variation of the common form *Winedi* or *Veneti*. *Winuli qui olim Wandali*, ii. 18 [ch. 64, Lappenberg].

[4] *Ungria* and *Beguaria*.

was divided in two parts by the river Peene or Panis,[1] and the Western Slavs on 'this' side of the stream [2] belonged to the jurisdiction of Hamburg. It was of these Western Slavs, near Magdeburg and Hamburg, that Adam knew most; he was less acquainted with those between the middle Elbe and Oder; he had but a vague idea of the trans-Oder nations. Yet the trade of Bremen with Jumna and other Baltic havens had spread a certain knowledge even of these more distant races, and Adam is perhaps not free from the common tendency of reserving his highest admiration for the remote, obscure, and strange. The 'noble city' of Jumna [3] was strongly hostile to the Faith; and Saxons were only allowed to live there if they dissembled their Christianity, and did not insist on practising the rites of the Church in public. But so notable a store of merchandise was there to be found, and so many rarities could only be obtained in its market, that peoples of every nation and climate flocked to it. From Hamburg and the mouth of the Elbe was seven days' journey by land; and on the other side men took ship from Jumna eastwards to the Land of the Prussians [4] and

[1] On the divisions of Slavania and the river Panis, cf. Ad. ii. 18; iii. 18 [chs. 64, 137, Lappenberg], the last on the progress of Christianity beyond the Elbe; Giesebrecht and Lappenberg, in their controversy about Adam's *Slavonia*, seem to ignore this division by the river Panis.

[2] As to whom Adam's detailed descriptions contrast forcibly with his meagre and obscure notices of the more distant Slavs. What he says of the latter is generally introduced with a *dicitur*. Of course the pagan Slavs were really outside his province, as historian of the Church of Hamburg.

[3] Called *Jumneta* by Helmondius, i. 2; it must have been on the site of, or near to, the modern *Swinemünde* or *Wöllin*, at the mouth of the Oder. Here the products of Byzantium, and even perhaps some of India, were to be found; cf. Heyd, *Commerce du Levant*, i. 77-78; Storch, *Gemälde des russischen Reichs*, iv. 45.

[4] On the Province (miscalled *isle*) of Semland (=Zemlya ?), and the other Baltic islands adjoining the Slav coasts, cf. Ad. iv. 18 [ch. 225, Lappenberg]. On the Vistula, cf. Ad. i. 2; ii. 78 [chs. 2, 117, Lappenberg]. Einhard, *Life of Charles*, 15. The name Wissula, Wisara, or *Visula* (=*Vistula*) is applied by Adam not only to the great Wendish river, but also to the Saxon stream Weser or

VII.] ADAM ON SLAVONIA, PRUSSIANS, AND RUSSIANS 537

the Russian port of Ostrogard, a voyage of fourteen days. Adam knows nothing except the name of the Vistula, but he has a good deal to say about the Prussians. The *island* of Semland or Samland, which they inhabited, was on the borders of the Poles and Russians; and as a people they were beyond dispute the most humane of mankind, ever ready to assist ship-wrecked mariners, setting little value on gold and silver, but much on their store of furs, and especially on their marten skins, which they bartered for articles of clothing.[1] Excellent in all other respects, they had a fierce hatred of Christianity, guarding their sacred groves and fountains with jealous care against the approach of visitors. They were blue-eyed, red-faced, and hirsute in appearance; their food was horse-flesh; their drink mares' milk or even blood; and their government was utterly democratic, for they endured no lordship among themselves. In all this Adam is perhaps (like Tacitus) contrasting the savage virtues of a primitive ideal with the civilised corruptions of a more 'refined' society.[2]

Ruzzia or Russia, the last and greatest province of the Wends, called *Ostrogard* by the barbarous or pagan Danes, and by some others *Hungard* or *Chungard* (as being the original home of the Huns) is described by our author as one might speak of a fertile garden,[3] abounding with every good thing.[4] The metropolis of the country was the proud,

---

Visurges [ii. 78; ch. 117, Lappenberg], while Einhard likewise appears to interchange these terms.

[1] Adam especially mentions the women's garments called *faldones* or *feldr*.

[2] In this connection we may notice the journeys of the famous merchant Vidgaut of 'Samland,' who travelled widely in the Russian plains, c. A.D. 1110-1120, as well as the 'Samland' ships, which, according to Adam himself, frequented the Swedish mark of Birca; cf. Rafn, *Antiquités russes*, ii. 134, etc.; Adam in Pertz, M.G. SS. vii. 305; and Heyd, *Commerce du Levant*, i. 75-76.

[3] The Scholiast, 116, expressly calls it so.

[4] Adam, iv. 11, 13, 19, and Schs. 63, 118, 116, on Ruzzia and Ostrogard [chs. 66, 218, 221, Lappen-

ambitious Kiev or 'Chive,' emulous of the sceptre of Constantinople, that brightest glory of Greece.

All these nations beyond the Oder,—Pomeranians, Poles, Prussians, Bohemians, and Russians,—are included in Adam's 'Slavonia' in the wider sense of that word; for (as in the case of Saxony) his use of the term varies greatly in extension.[1] On the other hand, it is in the narrower sense that he is speaking, when he tells us that Churches had been set up everywhere in Slav-land, and that the country was divided into eighteen districts.[2] There are few mediaeval subjects more confusing than the distribution of the Slavonic peoples and their proper nomenclature. For although their tribes appear to have been fixed in their habitations, they have all the evasiveness of Nomade races, through the incessant combinations and dissolutions of national groups.

Through the cloud of names and details given by Adam, in his detailed summary of the South Baltic Slavs, we may discern the definite existence of this race in Eastern Holstein,[3] in the islands of Fehmarn and Rügen, in Mecklenburg, in East and West Prussia, and in Pomerania. The people of Rügen, near Jumna and the mouth of the Panis, were unique[4] in their valour and their enjoyment of kingly government; they were powerful, piratical,[5] and fiercely

berg]. The extreme aptness of Adam's reference to Kiev or 'Chive' and its ambitions needs no emphasis.

[1] For the wider sense, cf. Ad. ii. 13, 19; iv. 13; for the narrower, ii. 24, 40, 46-47, 69; iii. 19 [chs. 60, 66, 221, 69, 138, 83, 86-87, 105, Lappenberg].

[2] Perhaps the sub-divisions of the diocese of Oldenburg.

[3] About the modern Oldenburg and the port of Heiligenhafen.

[4] Among the Slavs. Ad. iv. 18, and Sch. 117 [chs. 225, 226, Lappenberg]. Rügen was not far from Jumna.

[5] Like the men of Fehmarn or Fembre, cf. the *Imbra* or *Fembra* of Ad. iv. 16, 18 [chs. 223, 225, Lappenberg]. Fehmarn is called Imbra on a chart of the fourteenth century. Elsewhere, ii. 19 [ch. 66, Lappenberg], Adam seems to place the Runi, Rugi, or Rügen folk at the mouth of the Panis,—*in hostio Peanis fluvii*,—adjoining the town of Dimine. He is the first to locate

pagan; later writers celebrate their devotion to trade, and notice the resort of fishermen to their island in the November of each year. The town of *Retra*, four days' journey from Hamburg, and apparently situated between the Oder and the Panis, was the capital of the chief Slavonic tribe and the metropolis of all the Slavs; and here was a famous idol-temple, where a Christian bishop named John had paid with his life for his ill-timed preaching.[1]

Adam quits Slavonia with a short reference to the Oder, that greatest of Slavonic rivers, which rose in the depths of forests like the Elbe, and near its source was not so far removed from the latter, though in its lower course it diverged widely. The chronicler now passes to Scandinavia, the next great division of his topography, which he considers as 'almost wholly divided into islands,' but already grouped in three great realms, federations, or peoples,—of Norway, Sweden, and Denmark. Of these he takes the last first, probably as the nearest to his home, and as one of the lands with which he was most intimately acquainted. It was bounded to the south by the Eyder,[2] and was severed from Norway by the strait called *Otto's Sound*,[3] where the Restorer of the Empire

this tribe; Widukind (iii. 54), is the first to mention them. No tribe of Slavs was more celebrated in the twelfth-century chroniclers. On *Magnopolis, civitas inclita Obodritorum*, and the Mecklenburg Slavs, cf. Ad. iii. 20, 19, 50 [chs. 138, 139, 166, 167, Lappenberg]; *ibid*, for *Razzisburg* or *Razzispurg* and other cities of the Obodriti. The Lingones, beyond the Obodriti, mentioned by Adam, occur in the ninth-century *Annals of Fulda*. At the mouth of the Panis was the town of *Dimine*.

[1] Ad. iii. 50 [ch. 167, Lappenberg]. The temple was dedicated to *Redigast*, the *Riedegost* of Thietmar, vi. 17; it seems to have been on an island in a lake or marsh.

[2] As fixed by the Emperor Conrad and by Cnut the Great. The region between the Eyder [cf. iv. 1, ch. 208, Lappenberg], and the gulf and March of Slia (or the Sleswig March) divided Denmark from Saxony and Nordalbingia. This strip of German territory was now possessed by Denmark; Ad. ii. 3, 54 [chs. 50, 93, Lappenberg].

[3] *In eum angulum, qui Wendila dicitur*, Adam, iv. 1 [ch. 208, Lappenberg]. In iv. 16 [ch. 223, Lappenberg]. Wendila is an *island*.

made his way to the Northern Sea at the extremity of Jutland.[1] The wood of Isarnho in which the Eyder rose was a more effectual wall of partition between Germany and Denmark than the river; to the south-east the forest stretched away to the Slav town of Liubicen, or Old Lübeck.

Jutland is characterised by Adam (almost in the terms of a modern handbook) as sterile, the soil thickly impregnated with salt, the population thin and scattered; its only towns of any note lay upon inlets of the sea.[2] From one of its three chief ports called Sliaswig or Schleswig, ships sailed to Slavonia, Sweden, Semland, and 'Greece' (in which Ostrogard or Russia is obviously included); from another, Ripa, men journeyed to Frisia, England, and Saxony; from Aarhus was the route to Norway and the Danish Isles, the chief of which was 'Seland' or Zealand, in the depths of the Baltic, containing the royal seat of the Danish Kings at Röskilde[3] and the gold hoards of the Vikings.[4] Thence one might look across to Scania or the southernmost and peninsular province of Sweden, then reckoned as the 'last part' of Denmark.[5] Finally, among these outlying Dane-

---

[1] At Lymfiord, the modern Vendsyssel; cf. Saxo Grammaticus, bk. x., Sch. 95. Jutland is *Dania cis-marina* in Adam, ii. 3; iv. 1 [chs. 50, 208, Lappenberg]. *Dania trans-marina* includes the Danish Isles and Scania. In Adam, iv.5 [ch. 212, Lappenberg], a number of distance-reckonings are given between Seland or Zealand, Jutland, Fune, Scania, Aarhus, Lund-(ona), and the Strait of Norway (the Kattegat) on the north, as well as to the *Sinus Slavonicus* on the south.

[2] *E.g.* the three Bishoprics of Sliaswig (Schleswig), Ripa, and Aarhus, cf. Bede, *Hist. Ecc.* i. 15; Adam, i. 27, 59, 61; ii. 3, 34; and iv. 1 [chs. 21, 46-47, 50, 77, 208, Lappenberg]. Sliaswig was named from the arm of the Baltic called *Slia* on which it stood; its harbour was also named *Heidaba*. Very little is added to Adam's description of Jutland by Helmoldus, Saxo Grammaticus, and Arnold the Slav chronicler, except that the latter places the *Amazons* here.

[3] 'Roschild' or 'Roschald' in Adam.

[4] Of Copenhagen, or *Hafnia*, Adam says nothing. Saxo Grammaticus, c. A.D. 1180, is the first to mention it; cf. Ad. iv. 5, 6 [chs. 212, 213, Lappenberg].

[5] Ad. iv. 7 [ch. 214, Lappenberg].

lands, besides Bornholm[1] (*Holmus* or *Hulmo*) in the Baltic, famous as a good harbour for ships on their way to 'Greece' and the Barbarian lands of the East, comes Helgoland or Heligoland[2] in the North Sea, confused by Adam with one of the Frisian islets, and identified with the old cyclic story of a country ever shrinking before the storms of ocean.[3]

And now, passing beyond the Danish Isles, 'another world,' in our chronicler's imagination, was revealed in Sweden and Norway, 'extensive regions of the north hitherto almost unknown,' where King Svein, the son of Estrith, for twelve years had fought and journeyed. From his report of the land, if that might be trusted, it was certain no one could traverse Norway in less than one month, or Sweden[4] in less than two; but in this reckoning the Danish Scania is not included; the latter was commonly assumed to extend to the narrow, hilly, wooded neck of land called Smaland, beyond which 'Gothia' and 'Suedia' began.

At the very end of the world, stretching into the extremest region of the north, was the mountainous and infertile Norway, commencing at the 'prominent rocks' overhanging the *Baltic Straits*, and then bending backwards towards the Pole, and encircling the shore of ocean with its winding coasts, till at the Riphaean Hills, covered with eternal snow, it ceased together with the weary earth itself.

---

[1] Ad. iv. 8, 16 [chs. 215, 223, Lappenberg].

[2] Otherwise 'Föhr' or 'Farria,' Adam, iv. 3, and Sch. 104 [ch. 216, Lappenberg]. The older name of *Fosetisland* is also given by Adam, who places it near to *Frisia* or *Wisara*, three days' rowing-voyage from England, over against *Hadeloa* or *Hadeln*, to the south of the Elbe estuary. Its length, according to Adam, was now scarcely eight miles, its breadth four.

[3] A story also associated with Taprobane in Solinus, etc.

[4] This lesser Sweden included the whole region between $56\frac{1}{2}°$ and $62°$ N. Lat., and for this Svein's estimate is not excessive. Cf. on Scandinavia, Ad. ii. 22, 23; iv. 24, 25, 30, 31, 32 [chs. 68-69, 232-233, 239-241, Lappenberg]; and Schs. 122-132.

In his 'Sweden,' a land rich in cereals and honey, richest of all in cattle, Adam follows carefully the local divisions of 'Svithiod' and 'Gauthiod,'[1]—the former including the region about Lake Mälar, the cradle of the Swedish kingdom, and separated from the latter by the marshy depression in which Lakes Wener and Wettern lie. Among the towns of Sweden the most notable were Birca and Upsala,[2] of which the former, in the 'middle' of the country, looked straight across the Baltic towards Jumna; for, as we have seen, Adam's conception of the Northern Pontus requires a main direction east and west, though allowing for an arm or branch which tended towards the Pole. To this Birca, perhaps the Isle of Björko in Lake Mälar, resorted Danes, Northmen, Slavs, Sembi, and other *Scythian* peoples engaged in commerce, all attracted thither by its admirable position, almost equi-distant from every Baltic coast.[3] Upsala, or Ubsola, accounted a holy city (just as Birca was reckoned a leading market), and possessing a famous heathen temple, where Odin, Thor, and Freya were appeased with human sacrifice, is but slightly noticed in the Bremen

Much of this is from Solinus, chs. 15, 17, 30; Paul the Deacon, *History of the Lombards*, i. 5; and Martianus Capella, 663, 666, 683. As to its chief town Trondhjem (Nidaros), and the routes from Norway thither, Adam gives us nothing of special value.

[1] Except that 'Gauthiod' is pushed by Adam a little too far north, and Sudermania included in it; iv. 21, and Sch. 126; cf. also iii. 14 [chs. 132, 230, Lapp.]. The river *Goth-Elba*, the Albis or Elbe of Scandinavia, flowing 'through the midst of the Gothic peoples' into the ocean, on whose right were Goths and Danes, and on whose left Northmen only, is the modern Götha-Elf.

[2] Ad. iv. 20, 26, 28; i. 62; Sch. 121 [chs. 47, 229, 234, 236, Lapp.]. The Gothic capital *Scarane* (the modern Skaraborg) near Lake Wener, ii. 56, may be added. Adam divides *Gothia* into *Westra-Gothia* and *Ostro-Gothia*, the former lying nearest to Scania, the latter stretching to the neighbourhood of Birca.

[3] *E.g.* from Scania, five days, from 'Ruzzia' (Ostrogard), five. Its greatness seems to have ended about 1220, when it was sacked by Esthonian pirates; but even in the twelfth century Visby, in Gothland, had already quite eclipsed Birca; cf. Ad. Sch. 138. On Upsala, cf. Ad. iv. 26, 27, 28 [chs. 234-237, Lappenberg].

Annalist; yet brief as are his words, Adam seems not unmindful of the unique position of the city, the first true centre of Swedish nationality, one of the most ancient homes of Northern letters, and even more sacred (and so more prominent) in Pagan than in Christian times.

Between Norway and Sweden, but to the north of the latter, Adam places various Finnish tribes inhabiting the western parts of Halland and reaching down to the Gulf of Bothnia, the 'Scrito-Finns' of several mediaeval writers.[1] Still further towards the north-east lived pagan 'Northmen,' of monstrous life and visage, and given to magic and incantations, by which they could even draw great fish out of the sea. Their dwarfish stature, bearded women, huntsmen's habits, and rare and precious furs—black fox, black hare, white marten, and so forth—are suggestive (in Adam's description) of the Lapp branch of that great Finnish

---

[1] Cf. Ad. iv. 24, 25 [chs. 232-233, Lappenberg]; Saxo Grammaticus, *Historia Daniae*, bk. i., *Pref.*; Paul the Deacon, i. 5, who explains the term 'Scrito-Finns' by the German word, *schreiten (stride)*, probably referring to the movements of these people on snow-shoes; cf. the proverb *Die Sonne scheint, der Schnee fällt, der Finne schreitet*. Weinhold, *Alt-Nordisches Leben*, p. 307, and Peschel, *Erdkunde*, p. 88. Saxo places *Helsingland*, the capital of the Skrik-Finns, or Scrito-Finns, on the west coast of the Gulf of Bothnia. In iv. 25 [ch. 233, Lappenberg], Adam repeats a summary of the borders and neighbours of Sweden; to the south the Baltic, to the north the Scrito-Finns, to the east the Riphaean Hills, etc.

The Ravennese Geographer (c. A.D. 650) places the 'patria Scirdifennorum et Rerefennorum' in a cold hilly country on the Scythian Ocean (iv.12; edition by Pinder and Parthey, p. 201); even Procopius, in the middle of the sixth century, had heard the name of the 'Scrithiphins,' *Anecdota*, edition by Isambert, 1856, p. 602; King Ælfred (c. A.D. 890) has much more definite knowledge of the 'Scridafinnas,' and in the 'Geography of Europe,' which he adds to Orosius, evidently means to place them in our Lapland; while Olaus Magnus, five centuries after Adam of Bremen, describes 'Scricfinnia' as the land between Finmark and Biarmia (*e.g.* between Hammerfest and Archangel; cf. *Historia de gentibus septentrionalibus*, i. 4. With this agrees Sebastian Cabot's world-map of 1544, and Gerard Mercator's map of Russia in his 1595 Atlas. See also Hakluyt, *Voyages*, i. p. 283 (edition of 1598-1600).

stock which in the eleventh century covered all the north of Eastern Europe and Western Asia, from the Latitude of Novgorod. As to some of these Hyperborean peoples, ancient wisdom had recorded that many virtues long since driven out of the civilised world still lingered among these 'outcasts,' and this tradition the historian did not venture wholly to reject; unhappily, he knew the Danes too well to suppose that vice was absent from their nature. Of Swedes and Norwegians[1] he knew less, and so thought more. Their temperance and devotion he declared were alike marvellous; their life was patriarchal and 'most noble'; untouched by luxury, and living in a bracing, barren, and hilly country, they were the bravest of men in war and the most righteous in peace; the faults of piracy and passion were alone to be recorded.[2] Thus in many ways these Arctic races (like the Prussians) are idealised by the Bremen chronicler, as the Old Germans had been by the moralising criticism of Tacitus, eager to find a contrast to the wickedness of a luxurious society. Yet even here Adam does not surrender himself altogether to romance; and he carefully distinguishes between the more prosperous and settled families of the Scandinavian farming class, who

[1] *Northmanni*, Ad. iv. 30, 31 [chs. 239, 240, Lappenberg]. Masudi also refers to the black fox skins which the Arabs brought from the Northern peoples. Among other furs Adam mentions the skins of the white bear and *urus* (both of which lived *under* water), and of the *bubalus*, *elax*, and *bison*, iv. 31, end.

In iv. 25 [ch. 233, Lappenberg], Adam describes the strong, agile, dwarfish people of the Far North, evidently Finnish Lapps, but unnamed, about whom King Svein had told him; their home was in mountains, but they incessantly ravaged the lowlands.

[2] On the peoples, manners, etc., of Scandinavia and Prussia, cf. Ad. iv. 18, 21, 22-24, 25, 26, 27, 30, 31 [chs. 225-227, 230-235, 239-240, Lappenberg], and Schs. 118, 123, 125, 134, 135, 137. The virtues of the Prussians, as Adam thought, had been celebrated by Horace under the name of *Getae* (Ad. iv. 18). The one fault of the Swedes was *in mulierum copula modum nesciunt*, iv. 21 [ch. 230, Lappenberg].

lived by tillage, the poor herdsmen, who eked out their subsistence by piracy, and the barbarous hunters and trappers of the furthest North, whose lands stretched away eastward to those Riphaean Mountains where they chased their prey over the snow[1] with marvellous agility and strength. Near the regions of the last-named an ancient and reliable tradition (which had been lately verified beyond possibility of doubt by Adam's contemporaries) placed the Amazons, the Cyclopes, and other monstrous races, as well as the more human Turks and Alans; in all this there is an evident confusion of traditions derived, on one side, from the Caucasus, on the other, from the Ural Mountains and the Central Asian Plains.

Under the title of 'remote *islands*' in the Baltic, Courland and Esthonia[2] are also noticed as possessions of Sweden; both were inhabited by cruel pagans; and while the former was not far from the great Scandinavian market of Birca,

[1] 'On strange *vehicles*' (vehiculi) adds Saxo, *Preface*, and bk. ii., evidently alluding to the snow-shoes of the Lapps. On these subjects Adam may have learnt something from Bishop Adelward of Skaraborg, who told him about the *Terra Feminarum* of the Northern Amazons, etc.; cf. Ad. iii. 15; iv. 14, 19, 25; Sch. 119 [chs. 134, 222, 228, 233, Lappenberg]. Whether or no the proper derivation of 'Cwenland' is from the 'Quaines' or 'Cayani,' as some have thought, the Cwens were doubtless a branch of the Finnish stock; their country certainly lay upon the Gulf of Bothnia (called after them the Cwen Sea), and towards the northern angle of the same. They are usually distinguished by mediaeval geographers from the Lapps or Ter-Finns; but they were certainly of the same race. The legend of a tribe of Finnish Amazons, living separate from men, is based on a misunderstanding of the name of 'Cwen' ('Woman'). To perpetuate the race these female warriors were said by the wilder mythologists to associate at fixed intervals with the Cynocephali of the furthest North, monsters often hunted by the Russians, according to some veracious annalists; cf. Æthicus of Istria, edition by Wuttke, p. 15. J.R. Forster, *Entdeckungen in Norden*, p. 75, was the first to show how the Northern Geographers, knowing only that Cwen meant 'female' in the Germanic, Scandinavian, and Anglo-Saxon dialects, took the Finnish 'Cwens' to be a northern counterpart of the Amazons.

[2] 'Churland' and 'Æstland.'

the latter was near Cwenland, the Kingdom of Women or Region of the Amazons,[1] apparently located near the modern Helsingfors.

Lastly, as to the 'islands in the ocean,' Thule, Greenland, Halogaland, and Vinland, Adam is really possessed of better information than he could always command in his description of the Baltic coasts. For as yet (about 1070) these regions belonged to the metropolitan province of Bremen, and the household of Archbishop Adalbert was naturally concerned with their temporal and spiritual affairs.

Of England, Ireland, the Orkneys, and the other members of 'Albion's Archipelago,' the chronicler has little to say beyond the traditions of Strabo and Solinus. Like the former, he places Hibernia ('the land of Scots') to the *north* of Britain; but he adds a few more modern estimates of distance[2] among these 'numerous and not despicable' isles opposite Norway.

Thule,[3] the last of lands, now called Iceland,[4] from the ice which here congealed the ocean, was in the depths of the

[1] Cf. Ad. iv. 16, 17 [chs. 223, 224, Lappenberg]. Giesebrecht, *Nordlandskunde*, p. 183, assumes that Gothia and Sweden proper are conceived as islands by Adam, but the latter nowhere commits himself to a choice between an island and peninsula as far as Scandinavia is concerned. Wulfstan, in King Ælfred, on the other hand, commits himself to the peninsular view, speaking of the *isthmus* between the ocean and the Baltic, still more explicitly described in Saxo Grammaticus, *Preface*.

[2] *E.g.* from Trondhjem to the Orkneys the navigation of a day (*per diem*); from the Orkneys the same distance either to Scotland or England. Here Adam is very different from Solinus and Martianus Capella. Britain is hardly noticed by Adam except in reference to the wars of Cnut the Great.

[3] On' Britain, Ireland, etc., cf. Adam, ii. 50, 51, 53, 55, 59 [chs. 89, 90, 92, 94, 96, Lappenberg]; on the Orkneys or *Orchades* (called by the barbarians *Organes*, and dispersed through the Ocean like the Cyclades), Adam iv. 34 [ch. 243, Lappenberg]. On Iceland, see Adam, iv. 35; on Greenland, iv. 10, 35, 36; on Halogaland, iv. 37, and Sch. 152; on Vinland, iv. 38, and Sch. 37 [on Toki *dux Winlandensis*]; on the Frozen Ocean beyond Vinland, iv. 39, 40 [chs. 217, 244-249, Lappenberg].

[4] Island.

VII.] ICELAND, GREENLAND, VINLAND, ETC. 547

awful and infinite sea beyond Norway, and thus hidden from the view of other countries. Yet both Roman and Barbarian writers, and especially the excellent Pytheas, had written concerning the same not a little worthy of note. No crops grew here, and the supply of wood was scanty; like the Swedes, the inhabitants were pastoral; they dwelt in caves beneath the ground, and led a life of holy and blessed poverty, envied by none. Their bishop was their king, and they had all things in common.

Still deeper in the ocean was Greenland, lying over against the mountains of Sweden, or the Riphaean Hills. From Norway the distance was about the same as to Iceland, a sail of five or seven days. It is clear that Adam puts Greenland somewhat to the east of Iceland, for while the latter is 'north of' Britain the former is 'opposite to' Norway. The name of the country is derived from the *Caerulean Sea* that surrounded it; and from the same came the colour of its people.

Halagland or Halogaland, on the contrary, was nearer to Norway, and some said (but this view of the matter is not endorsed by the annalist) that it was the 'last portion' or province of Norway itself, adjoining the land of the Scrito-Finns, but inaccessible from its cold and the roughness of its mountains.

Lastly, Vinland,[1] so called from its vines of spontaneous growth, was also famous for its self-sown wheat. King Svein himself had told the author how many navigators had

---

[1] Which Adam is the first mediaeval writer to mention by name, thus carrying its literary history back to c. 1070. The later writers of the Middle Age (outside Scandinavia) who refer to Vinland are only two: (1) Ordericus Vitalis, who died c. 1145 (and whose ecclesiastical history reaches to 1142), in bk. x. of same, mentions 'Vinlandia' in connection with Greenland, Iceland (Island), and the Orkneys (Orcades); (2) Albert the Great, who died 1280 in his Treatise *De Natura Locorum* xxvii. p. 6.

discovered this region in the recesses of the Northern Ocean, and how the well-assured opinion of certain Danes, apart from all fables, attested the same. Adam does not, however, attempt to give the position of Vinland, nor to estimate its distance to other countries; but his ideas are here affected by a confusion similar to that which appears in some of the Sagas. For whereas the peculiarities he himself records in the products of the soil point to a more southerly region than Saxony[1] (where the vine could not grow), he declares that beyond Vinland all things were full of unbearable ice. It was here, in the black darkness of the frozen ocean, that Harald Hardrada and the adventurous nobles of Frisia had run such risks in the days of Alebrand, the predecessor of Adalbert.[2] This earliest of German expeditions to the Polar countries had started from the mouth of the Weser,[3] had passed the Orkneys, and so had come with joyful song[4] to Iceland. Thence they rowed on to the very axis of the North, where they were almost engulfed in the abyss. The strange island which they described as fortified with high rocks like a town, and peopled by a dog-like folk who lived and barked in caves beneath the ground, may have been the east coast of Greenland; in any case, the record shows us one of the first enterprises of geographical discovery which started, not from Ireland or Scandinavia, but from the North German shore, where the League of Hanseatic traders was already forming.

---

[1] The Sagas add the detail, that from the length of daylight as well as from the nature of the soil, Vinland was further south than Iceland or Greenland.

[2] This properly refers to the voyage of the Frisian nobles, which perhaps preceded that of Harald. Alebrand was Archbishop of Bremen and Hamburg, 1035-1045.

[3] Wirraha.

[4] 'Proceeded with *Celeusma*,' properly the call ($κέλευσμα$) of the rowing master of a Greek trireme which gave time to the rowers.

§ 3.—Maps [1]

THE first group of maps that concerns us here is that which starts from the design of the Spanish priest, *Beatus*, in the latter years of the eighth century. The author was famous in the general Church history of Spain as a leading opponent of the *Adoptionist* [2] heresy of Felix of Urgel and Elipandus of Toledo, and along with Etherius, Bishop of Osma, he maintained the eternal Godhead of Christ in opposition to the view of those who taught that the Son had been *adopted* and received into Divinity by the Father. For some time Beatus appears to have led a monastic life under Abbot Fidelis, of St. John of Pravia, near Oviedo; his death in 798 took place at the Benedictine House of Vallecava [3]

---

[1] For all these maps of the Central Mediaeval Period, the best work of reference is undoubtedly Konrad Miller's *Mappaemundi; die aeltesten Weltkarten;* Heft, i., *Die Weltkarte des Beatus*, with four map reproductions, and a scheme of the Beatus group, 1895; Heft, ii., *Atlas von 16 Lichtdrucktafeln*, 1895; Heft, iii., *Die kleineren Weltkarten*, with 78 illustrations of maps, and several schemes showing cartographical relationship, 1895. Miller has thrown great light on the connections, probably or certainly existing between many examples of mediaeval cartography, which formerly had been treated as quite distinct; he has been the first to draw attention to several interesting designs; and he has brought to bear upon this study an amount of conscientious labour, critical acumen, and synthetic ingenuity, never before devoted to mediaeval maps. His weakness is to insist overmuch on conscious intellectual relationship from small coincidences, *e.g.* in maps of widely different dates and origins. Santarem's *Essai sur Cosmographie et . . . Cartographie* (1849-1852), and Lelewel's *Géographie du moyen age*, have been in a great measure superseded by Miller's writings, though they may still be consulted with advantage, and were in their time invaluable. Among more recent enquiries or collections of permanent value may be mentioned Jomard, *Monuments de Géographie*, 1862; Cortembert, *Trois des plus anciens monuments de Géographie*, 1877; Bevan and Phillot's *Hereford Map*, 1877; Philippi's *Rekonstrüktion d. Weltkarte d. Agrippa*, 1880; and Walleser, *Die Welt-Tafel des Ravennatem*, 1894.

[2] The Adoptionist controversy begins in 782.

[3] Before his death, Beatus seems to have become Abbot of this house.

or Valcavado, near Saldanha, in the Asturias. Queen Adosinda, the wife of King Silo of Oviedo (774-783), was a patron and firm friend of Beatus, her father-confessor; by one tradition, he was also teacher of the famous Alcuin, though another story [1] describes him as a deaf mute, hardly capable of imparting instruction by ordinary methods. At any rate, he shares a place with the celebrated scholar of Charlemagne's Court in the pillory of Adoptionist polemic,[2] and especially in the rich vocabulary of Elipandus. His great work, the *Commentary on the Apocalypse*, appeared about the year 776, and among the many sumptuous illustrations of this volume was a map of the world which (there is no good reason to doubt) also came from the hand of Beatus or was drawn under his supervision. The map in question seems to have had a special object. It was probably executed to portray the spread of the Christian faith over the earth, in allusion to the texts which compare the world and the kingdom of heaven to a field sown with seed.[3] This idea was expressed most clearly in a series of pictures of the Twelve Apostles, each in the locality where tradition fixed his preaching and his diocese. A note of the Latin Commentary tells us plainly that these vignettes were an essential feature of the map from the beginning; that they illustrated the preaching or sowing of the Word 'in the field of this world. One Sunday he terrified his hearers so that all of them fasted till the ninth hour, expecting the consummation of the ages every moment; a worldling, named Hordonius, at last suggested that it was as well to meet death with a full meal. Cf. Letters of Elipandus in Florez, *España Sagrada*, v. 509, 537, 543-557, etc., edition of 1763; K. Miller, *Weltkarte des Beatus*, pp. 4, 5.

---

[1] Only in Eguren, *Mem. desc.*

[2] *Asinus silvestris, doctor bestialium, foetidissimus Beatus, carnis immunditia foetidus*, are among the vigorous allusions of the Archbishop of Toledo, who pictures his opponent returning *ad thorum scorti*, and compares him to a madman, drunk with wine, thinking himself a prophet, appointing an abbot for the animals of the district of Liebana, and foretelling the immediate end of the

[3] Cf. Matthew xiii. 1-9, 18-23, 24-32.

THE (BEATUS) MAP OF 'ST. SEVER' OF C. A.D. 1030.

[*To face p.* 550.

world'; and, by implication, that the apostolic portraits were placed in certain definite cities or regions, to which their work had special reference.[1] Further, this Distribution (or *Divisio Apostolorum*) was conceived as follows:—To Peter was given Rome; to Andrew, Greece or Achaia; to Thomas, India; to James, Spain; to John, Asia; to Matthew, Macedonia; to Philip, Gaul; to Bartholemew, Lycaonia; to Simon Zelotes, Egypt; to Matthias, Judaea; to James, the brother of the Lord, Jerusalem was assigned; while for Paul there was no such definite location, as his mission was to all the world. His portrait, however, appeared upon the map along with that of St. Peter, at Rome, as a co-founder of the Apostolic See. The Commentary of Beatus (mainly based upon St. Isidore of Seville) describes all this in writing; but the pictures themselves have only survived in one of the ten existing copies of the map, that of Osma (1203 A.D.), and of course the so-called portraits are all of one type, and that an intensely sacerdotal one.

The ten copies of the Beatus map are of widely different date from the tenth to the thirteenth century; some, like that of Turin, have been known for a considerable time; most have been brought to light within recent times; all of them till lately have been treated as distinct and isolated specimens of Dark-Age cartography. They are, however, certainly connected; they all occur in manuscripts of the same work, the *Apocalypse Commentary;* and there can be no doubt they are all derivable from the common original, the design of the

---

[1] *Et hii falcibus haec seminis grana per agrum hujus mundi metent. Quod subjecta formula picturarum demonstrat.* Apostolic 'titles' also occur in *e.g.* Lambert of St. Omer, the Ebstorf and Hereford maps, the Byzantine Oxford of 1110, Matthew Paris, and a plan at Velletri, but nowhere with the same completeness. St. James in Spain, St. Peter in Rome, St. Thomas in India, are almost the only cases. Osma leaves out (in its pictures, though not in its text) both Thaddeus and Judas the brother of James, making up the full number with Paul and Matthias.

priest Beatus, in or near the year 776. The first in order of date is that of 970, traditionally the work of a copyist named Obeco, and known in recent times as the 'Ashburnham' map, from its last possessor. But this, although the earliest example, is of small value; it omits nearly all the rivers, cities, and hamlets given on other copies, and turns the oval form,[1] which was almost certainly that of the original, into a right-angled one. Next comes the 'St. Sever' (now at Paris), a work executed at the above-named Aquitanian convent by the order of Gregory de Muntaner, Abbot from 1028 to 1072; this is the most valuable, the most carefully executed, and the richest in content of all the transcripts.[2] It is probably the nearest to the original type, and is therefore primary in any attempted reconstruction of that type. The copies which follow, known as the 'Madrid,' of 1047, and the 'Valladolid,' of 1035, seem to be, in great measure, derivatives of the work of 970; they are unimportant, very debased, and further removed from the original of 776 than any other, except their immediate ancestor. The 'Gerona' of about 1100, and the 'London' of 1109, like the 'Paris' of 1150, do not call for special remark;[3] but our next copy, the 'Turin' of the twelfth (?) century, which is more closely related to the Gerona map than any other, is perhaps the best-known example among maps of a strictly Dark-Age character. It is not so ancient or so important as once supposed; but it has been before the modern world for a century and a half, far longer than any

[1] 'Ashburnham' alone is quite square; 'Paris' of 1150 and 'Turin' alone quite circular.

[2] The name of the artist (or of one among a group of executants responsible for the map) may be conveyed in an entry on fol. 6 of the Paris MS., Bib. Nat., Lat. 8878 (containing 'St. Sever'), viz. *Stephanus Garcia Placidus*. The measurements of this codex are 370 millimetres by 290.

[3] Except that 'Gerona's' picture of Jonah seized by a cuttle-fish off the Spanish coast probably represents an original feature.

THE (BEATUS) 'TURIN' MAP OF C. A.D. 1150.

[*To face p.* 552.

other 'Beatus,' and it has become almost classic. Its remarkable peculiarities have naturally made it a favourite subject for reproduction. For though the celebrated wind-blowers, so prominent here, are also to be found in other mediaeval *mappe-mondes*,[1] their execution on the Turin example is far more vigorous and detailed,[2] and supplies us with the most striking artistic feature in Beatus Geography.

Last in point of date, though not of importance, are the 'Osma' copy of 1203 and the 'Paris' of 1250. The latter, indeed, presents a frightful confusion of lands and seas, placing Palestine in the interior of Africa and Southern Italy next to Jerusalem; but it contains, nevertheless, some valuable reminiscences of original matter not so prominent elsewhere. The map of Osma has much in common with St. Sever, especially in its general form, and, after the Aquitanian copy, it is certainly our best example of this school or group of designs. In some points it must indeed take the first place, as more directly representing the prototype. Thus in its pictures of the Twelve Apostles it is unique, and explains to us the very *fons et origo* of the Beatus scheme: like the 'Paris' of 1250, it represents the *Skiapodes*, or Shadow-footed men, of the Southern Continent, an important feature (in all probability) of the primitive work; while none of the Beatus

---

[1] As in the 'Paris' of c. 1250, the latest of the Beatus copies; cf. the angels in Henry of Mainz.

[2] Where each wind-spirit is seated on a sack or Æolus-bag, out of which he is squeezing a lively 'blast of air.' This has become a favourite illustration of mediaeval thought. It is certain, however, that in this shape the picture was not a feature of the work of 776; the simpler form of the wind-blowers in the thirteenth-century Paris has greater claims in this respect. The *idea* must be recognised as occurring in both the 'Osma' and 'Valcavado' families, which greatly strengthens its pretensions to originality.

copies gives so good a delineation of the Lighthouse-Towers at Alexandria and Brigantia.[1]

These ten copies of the map of 776 have been suggestively classed in two main families,[2] those of Osma and of Valcavado. To the former[3] belong the examples of 1030[4] (St. Sever), of 1203 (Osma), and of 1250[4] (Paris); to the latter, those of 970 (Ashburnham), 1035 (Valladolid), 1047 (Madrid), and 1109 (London), with the Gerona of the closing eleventh or early twelfth century, the Turin of about the same date, and the Paris of 1150.[4] The parting of these families probably took place in the ninth century, and each appears to have been immediately derived from certain lost intermediates of the tenth century, such as the two executed in whole or part[5] by Emeterius of Valcavado between 968 and 978.

There is no graduation on any of these copies, although certain lines 'showing through' from the ruling and writing on the other side of the page have sometimes been mistaken for horizontal and vertical indications. In all the designs, save one, the East is at the top; the Paris of 1250, which substitutes the South (probably through Arabic

---

[1] Here probably original Beatus matter is reproduced. Osma is especially valuable in its representation of the Levant or Eastern Mediterranean, of the Nile, of Mount Taurus, and of Taprobane. Among other maps of the Central mediaeval period the 'Cottoniana' offers the greatest likeness to Osma.

[2] By Professor Konrad Miller, *Weltkarte des Beatus*, especially pp. 24-7. Miller also subdivides the Osma family into stem (A), including 'St. Sever' and the 'Paris of c. 1250' or 'Paris II.'; and stem (B), including the 'Osma' of 1203; while the Valcavado family is parted into stem (C), comprising the 'Ashburnham' of 970, the 'Valladolid' of 1035, the 'Madrid' of 1047, the 'London' of 1109; and stem (D), including the 'Gerona,' 'Turin,' and twelfth-century 'Paris,' otherwise 'Paris III.'

[3] Much the more important ;—(1) as ampler, (2) better scientifically, (3) nearer to the original type; both the St. Sever and Osma copies belong to this branch.

[4] These dates are approximate.

[5] Some of these intermediate copies may yet be recovered in the recesses of Spanish convents or elsewhere.

THE 'LONDON BEATUS' (SO-CALLED 'SPANISH-ARABIC') MAP OF A.D. 1109.

[*To face p.* 554.

influence), makes this substitution with such inconsistency, in regard to other parts of the map, that it is clear the copyist is here departing from his original. Paradise is placed in the extreme East, not on an isle (as in many other designs), but on the mainland,[1] encircled by unscalable mountains, and accompanied by pictures either of Adam and Eve or of the Four Sacred Rivers.[2]

The division of the Continents is usually the same as that of the so-called T-O maps, Asia occupying the upper half, while in the lower part Europe has the left-hand quarter, Africa the right-hand. The western border of Asia is formed by a series of rivers and narrow seas, from the Don[3] to the Nile, while to the south of Africa, separated by a strip of ocean, is[4] the Southern, Australian, or Antipodean Continent of ancient theory.[1] According to this view, Africa did not reach to the Equator, which was covered by the Ocean zone above noticed; this zone was of unbearable heat; and beyond it again was a land of non-human monsters. Of the Antipodes, in the strict sense, implying the earth's rotundity, Beatus[1] gives no sign; but on the other hand, he utters no condemnation of a theory permitted, though not endorsed, by the authority of St. Augustine.

In the Beatus designs the ocean is usually ornamented with pictures of row-boats and fishes, which appear to follow regular courses, as if to indicate the direction of currents, or the periodical wanderings[5] of tunny or herring shoals. Among the various divisions of ocean it is noticeable that the *Red Sea* is extended by nearly all the copyists,

---

[1] Following Isidore of Seville.

[2] Pison, Gihon, Hiddekel, Euphrates.

[3] Tanais.

[4] On all but one of our ten copies, and that the latest and most corrupted (Paris of 1250). Even this gives the 'Skiapod' in a corner, and hereby implies the southern land.

[5] So, following Santarem, we may perhaps guess from St. Sever and the B. Mus. Map of 1109.

not only along the whole southern coast of Asia,[1] but along that of Africa as well. The latter, of course, in all the earlier mediaeval maps, has its length from East to West, its breadth from North to South.

The chief towns of the Habitable World were probably marked on the original of 776 by vignettes such as those of the Peutinger Table; these have survived on the Osma of 1203 and the Paris of 1250; and are developed with considerable sumptuousness and artistic beauty in the Paris of 1150. Among these city-pictures the most noticeable are those of Rome, Antioch, Jerusalem, and Constantinople, with the beacons or lighthouses of Brigantia and Alexandria.

There is not much of purely fabulous matter in the Beatus designs; the two most prominent of such indications are the Phoenix of Arabia and the 'Skiapodes' of the Australian Continent.[2] Jerusalem is never made the centre of the earth,[3] although the latest of our copies, the Paris of 1250, shows a tendency in this direction. It is in the 'Sallust' maps of the twelfth century and in the T-O design of 1110 that this point is first expressed with perfect clearness in cartography; but of course it appears in written descriptions or allusions of a far earlier date, and among these we may instance that of the pilgrim Arculf, in about 690.[4]

Beatus himself, no doubt, considered his map as primarily

---

[1] As so often in classical geography.

[2] Both these features were probably on the original Beatus of 776, like Mount Sinai, and Jonah and the fish (surviving in the Gerona copy). It is curious that no reference is made to the Ark, the races of Gog-Magog, the Tower of Babel, or the rivers of Spain, the native country of the designer.

[3] This conception does not acquire definiteness and fixity till the Crusading Period. One of its very earliest instances is the Byzantine-Oxford T-O map of 1110.

[4] Cf. *Dawn of Modern Geography*, i. pp. 133, 338-9.

illustrative[1] of the Old and New Testaments,[2] and of the spread of the Catholic Faith; but in addition to the Scriptures, he seems to have used two main authorities. One of these was St. Isidore of Seville; the other was a Roman province-map,[3] bearing some resemblance to the Peutinger Table. From the great Spanish theologian come almost verbatim most of the longer inscriptions, or *legenda*; but we must not forget that Isidore himself usually derived the matter of his geographical dissertations from the cosmographies of the later Roman period. On the other hand, no earlier source is known of the 'Apostolic' pictures of Beatus, as far as embodiment on a *mappe-monde* is concerned; and this detail may well be a refinement supplied by the priest of Valcavado himself; but of course the apostolic locations or dioceses are found in very early Christian tradition, and most of them are probably true in fact. As to the Roman province-map, it is perhaps from this that Beatus and his copyists derived most of their 'profane' geography.[4] The Caspian Sea, the Alexandrian Pharos, the Nile inscription, and the desert where the Children of Israel wandered forty years, as we have them in Beatus cartography,[5] are closely parallel to the representations of the Peutinger Table; and the relationship between these works is the key to all satisfactory study of the 'Spanish designs.' The connection is shown in many other details, as in the names of the peoples, cities, hills, and rivers of various countries, and in the Indian,

[1] From the manuscripts it is clear that the Map was merely one ornament of a splendid picture book, in which every Apocalyptic incident and monstrosity was duly figured.

[2] Yet the Biblical loans are very extensive, *e.g.* on Osma some thirty-seven names, exclusive of the Apostles.

[3] Apparently Beatus tries to mark two of the Metropolitan towns in each province. This is especially shown in the Paris of 1150.

[4] Though very few, if any, of their longer inscriptions or legends.

[5] Especially in St. Sever.

Syrian, and African legends; while in Gaul not only are the same provinces named and the same divisions made, but the more striking omissions of the Table occur also in St. Sever. Of the one hundred and thirty-three names of towns in the Beatus maps, more than ninety agree with 'Peutinger,'[1] and among these ninety parallels are to be found all except two of the important places marked by pictures;[2] on the other hand, the great vignettes at Rome, Constantinople, and Antioch (as we have them in the Table), if existing in the work or works used by Beatus, have undergone transformation and reduction to a much humbler level. Further comparisons may be drawn from the similar *arrangement* of towns in more than one region, and especially in Greece or Achaia. The *Francia* of Beatus, lying wholly to the east of the Rhine, is evidently copied from a map or description of a date far earlier than the eighth century; in certain instances, where St. Sever (for instance) gives names to places[3] which are depicted, but not titled, in Peutinger, the relationship of the Beatus group with an old imperial map is still more pointedly suggested.[4]

On the other hand, the Beatus designs have nothing similar to the Roman Itineraries, properly so called, or to the station-, distance-, and road-markings of the Table. Nor,

---

[1] Seven or eight later than Peutinger Table; seven Biblical; nine ancient names not in the Peutinger Table; fourteen old Gallic names represented by more modern forms on our copy of the Peutinger Table;— make up the rest. Beatus' use of the Peutinger Table (or similar old Roman map) was very eclectic anyhow. The names peculiar to the Paris of c. 1250 are mostly in the Peutinger Table or in Julius Honorius.

[2] Usually little houses.

[3] *E.g.* Catana, Cibale, Sardes, Cartusium (?).

[4] Besides the position of 'Francia,' we may note that the Bactrians in India, three rivers in the Balkan Peninsula, four in Italy, one in Africa, and the streams of the Caspian basin, are identical in Beatus and the Peutinger Table. Among mountains, the Atlas and Cyrenaean ranges, and Mount Emodus in India, may also be compared. The arrangement of names in various provinces, *e.g.* Achaia, is likewise closely similar. On the Beatus maps cf. Appendix, pp. 591-605.

of course, can the latter's six hundred references to pagan temples and worship be found in these plans of the tenth and subsequent centuries. But, in spite of all differences, we appear to have, in the various works derived from the Spanish priest of Liebana and Vallecava, a Dark-Age reflection of one or more cartographical works of the Old Empire, free from all additions of the Crusading period, and of inestimable value as a link between the ancient and the mediaeval worlds.

The 'Cotton' or 'Anglo-Saxon' Map [1] now in the British Museum, though of small size, is among the most interesting of all mediaeval world-pictures, and gives us a design far above the average. It portrays, with comparative fulness and accuracy, various places, regions, and natural features elsewhere omitted—or misunderstood—until a much later date; and its delineation of coast (and other) outlines forms a striking contrast with such works as the Ashburnham or Valcavado of 970. In fact, there is hardly any map of the Middle Age, before the appearance of the Portolani, which can be compared, in the general contour of the great land-masses, with the 'Cottoniana.' [2] In its presentation of the world as a whole, this design adopts a roughly square form, and in this it recalls some of the most debased examples of the Beatus group. But in the execution of this right-angled scheme there is all the difference between the narrow ignorance of an uncompromising symmetry, and a certain respectable, if not highly developed, knowledge and scientific insight.

[1] 'The last known monument of the Roman geographical school,' Lelewel strangely calls it, *Géographie*, pp. 10-13. It measures 21.2 by 17.6 centimetres.

[2] Though of course later maps of separate countries (*e.g.* Matthew Paris' 'England') are in some instances much superior.

560   GEOGRAPHICAL THEORY AND DESCRIPTION   [CH.

The 'Cottoniana' does not appear to belong to any one of the ascertainable families of mediaeval maps. It is far removed from all the members of the Beatus group; it is equally far removed from the school to which 'Henry of Mainz,' the 'Psalter,' the 'Hereford,' and the 'Ebstorf' plans appear to belong. Nor has it any relation with the various types of zone- (or climate-map) which we know under the names of 'Macrobius,' 'Sallust,' or T-O sketches. A certain likeness, especially in the British Isles, may, indeed, be descried between the 'Cottoniana' and the 'Matthew Paris' maps of the thirteenth century; but this is not the detailed and conscious resemblance of works really in touch with one another, as model and copy; it is rather the unstudied likeness of good work in the same subject at widely different periods. Matthew Paris, moreover, has all the advantages of his time. He lived in one of the most flourishing and civilised of mediaeval epochs, whereas the Cotton map comes upon us as a surprise from one of the gloomiest Dark Ages of the Latin world.

Some [1] have pushed back the date of this example to the ninth century and the time of King Ælfred; it is more probably of about a century later, the time of Archbishop Sigeric of Canterbury (992-994). The map itself occurs in a copy of Priscian's *Periegesis*, a fifth-century manual of geography, based upon an earlier treatise;[2] but it stands in no special or obvious relationship to the work it professedly illustrates. It is, indeed, more closely linked with Orosius; it has also certain obligations to Pomponius Mela, St. Isidore of Seville, and the topographical writings of St. Jerome; finally, it bears some indications of a much later time, the

[1] *E.g.* Lelewel and Wuttke; Santarem puts it as late as 1020.
[2] Of Dionysius, surnamed *Periegetes* after the *Periegesis* compiled by him perhaps about A.D. 100. On the 'Cottoniana,' see Appendix, pp. 608-12.

THE COTTON OR 'ANGLO-SAXON' MAP OF THE TENTH CENTURY (END).

age of the discoveries and migrations of the Northmen in the eighth, ninth, and tenth centuries. The correspondences of various entries and delineations in the 'Cottoniana' with certain names and descriptions in Adam of Bremen[1] afford at least a possibility that the former sometimes drew from the same originals as the great northern annalist, while some of the names in the British Isles, in Gaul, and in the Far East and North-East, support the tenth-century date, which most scholars are now inclined to accept. The possibility of so good[2] a work emanating from so dark a time, and so uncultured a society, has been disputed; but it is impossible to press this sentimental objection in the face of so much positive evidence to the contrary; and from other designs,[3] equally belonging to this Dark-Age period, we see that even then it was possible to draught a respectable sketch of a limited region. The map of St. Sever, a work of even greater elaboration, though of less scientific character, was produced before the dawn of the Crusading or Middle-Age Renaissance; while the monstrosities of the Hereford and Ebstorf maps disgraced the last and most splendid period of the same;—for the goodness of a map is by no means an absolute matter of date; even in the tenth century something belongs to the copyist himself and his immediate original.

The comparative excellence of the 'Cottoniana' is perhaps due to its being the production of an Irish scholar-monk living in the household of the learned and travelled Arch-

---

[1] Fl. 1070.

[2] One of the most startling excellencies of the Cotton is its insertion of the Slav name of the Dniepr, *Naper fluv.*

[3] *E.g.* even the Albi map of c. 730, in which some of the Mediterranean coasts are at any rate better than in most Arab maps. The Mosaic map of Madaba, c. 550, of course belongs to a highly civilised period and country, Syria under Justinian.

bishop Sigeric, with whose *Itinerary*[1] the present design has several curious resemblances. In the British Isles of the pre-Norman period, there is no school of learning, art, or science comparable to that which sprang from the Irish Church of Patrick, Colomba, and Aidan; and the insertion of the name of *Armagh*, so rarely found in mediaeval maps, strengthens the view that here we have the handiwork of a student who was trained in Irish schools, or derived his knowledge from men so trained.

The chief authority is the geographical chapter of Paulus Orosius,[2] but a certain amount of antique material also reappears, which is different from anything now to be found in the Orosian writings. The scribe or draughtsman tells us that he found, in the manuscript of Priscian's *Periegesis* used by him, a map which he supposed to have been drawn by Priscian himself;[3] this map, however, may well have been a copy of the lost Orosian *schema terrae;* and in any case it was obviously the immediate original of the 'Cottoniana.' Thus the latter design, unique among its contemporaries, and therefore mysterious in its superiority, is brought into a possible relation with an earlier school of no contemptible attainment; and, here as elsewhere, we may find a process of evolution, of inter-connection, and of historical development, bringing together the most surprising and distant parallels, and throwing light upon a field where, as in natural science, the mistaken ideas of separate and unconnected existence had long prevailed.

A Roman province-map may have been the source of the

---

[1] From Rome to the English Channel. See Appendix, pp. 612-4.

[2] Out of one hundred and forty-six legends, seventy-five occur in Orosius. These seventy-five contain the textual basis of the whole map, and all its names of countries, with very few exceptions.

[3] If the designer's story be accepted. On Orosius, cf. *Dawn of Modern Geography,* i. pp. 353-355.

VII.]   MAP OF HENRY OF MAINZ   563

divisions so clearly marked in Asia Minor, in Central and South-Eastern Europe, and in North Africa; while the Biblical loans of the Cotton Scheme may be traced not only in many names but in certain aspects of the general plan. Indeed, it is obvious that here the design was not merely indebted to the Scriptures for details such as almost all mediaeval maps exhibit, but was to a large extent devised for a special Biblical lesson. For just as the *mappe-monde* of Beatus had for its radical purpose[1] the delineation of the Twelve Apostles, their dioceses, and their distribution over the Habitable World as 'Sowers of the Word,' so the 'Cottoniana,' perhaps based on a lost design of Paulus Orosius, had for one of its main objects a picture of the settlement of the Twelve Tribes of Israel.[2]

The map of Henry of Mainz (c. A.D. 1110) apparently belongs to a family of cartographical works which may be compared with the more closely-knit members of the Beatus genealogy. In the present group are also included the tiny Psalter map (of about 1230), and the Hereford and Ebstorf examples from the later thirteenth century—huge wall pictures which represented, in size though not in execution, the possible eleventh-century original more closely than their elder but smaller brethren or cousins. Lastly, the so-called 'Jerome' maps, of about 1150, may be collaterally referred to the same family, through the medium of the Mainz design.

The last-named is to be found in the *Imago Mundi* of a certain Henry, probably the same person as a Canon Henry (Heinrich) who in 1111 appeared[3] before the Episcopal Court

[1] In part at least.
[2] The *colouring* of the 'Cottoniana' is grey for most seas; red for the Persian and Arabian Gulfs, the Nile Valley, and the Lakes of Africa; bright green for all mountains.
[3] As a witness.

of Mainz;[1] possibly he is the same as the Archbishop Henry, who ruled this church between 1142 and 1152.[2] In any case, the map accompanies a work which was written about 1110, and was dedicated to the famous and unfortunate Matilda, wife of the Emperor Henry V., daughter of Henry I. of England, and mother of Henry II. This work, the *Imago Mundi de dispositione orbis*, a compendious description of the world, containing also a short chronicle of universal history,[3] was copied and interpolated, but not originally composed, by Henry of Mainz; it was really the work of a contemporary, Honorius of Autun. The map, however, is apparently the addition of the scribe Henry, and is not derived from Honorius, although it is based on another and older design.

It is oval in form, of small size,[4] and contains two hundred and twenty-nine legends or inscriptions, together with a large number of unnamed cities, mountains, and rivers, whose titles can for the most part be ascertained with the aid of its younger relatives, the Ebstorf, Psalter, Hereford, and Jerome (?) plans. Although the present world-scheme is apparently intended to illustrate the *Imago* copied by Henry, the connection between the two is but slender; for (as in the case of the 'Cottoniana' and the text of Priscian it accompanies) the peculiarities of the chart are often not in the manuscript, nor are those of the manu-

---

[1] In 1121 this Canon Heinrich appears as Priest of St. Victor; in 1133 and onwards as a Priest attached to the Cathedral (of Mainz). He is mentioned in various charters.

[2] This prelate was deposed on the charge of opposing the election of Frederic Barbarossa as German King, and died in 1153.

[3] In seven books: of these (1) *De imagine mundi*; (2) *De temporibus matthesis;* (5) *De aetatibus mundi chronicon;* (6) *De luminaribus sive scriptoribus ecclesiasticis;* (7) *De haeresibus;* are the chief. Honorius died, according to some, in 1140; according to others, after 1152.

[4] About 12 inches by 10; 29½ by 20½ centimetres. See Appendix, pp. 614-7.

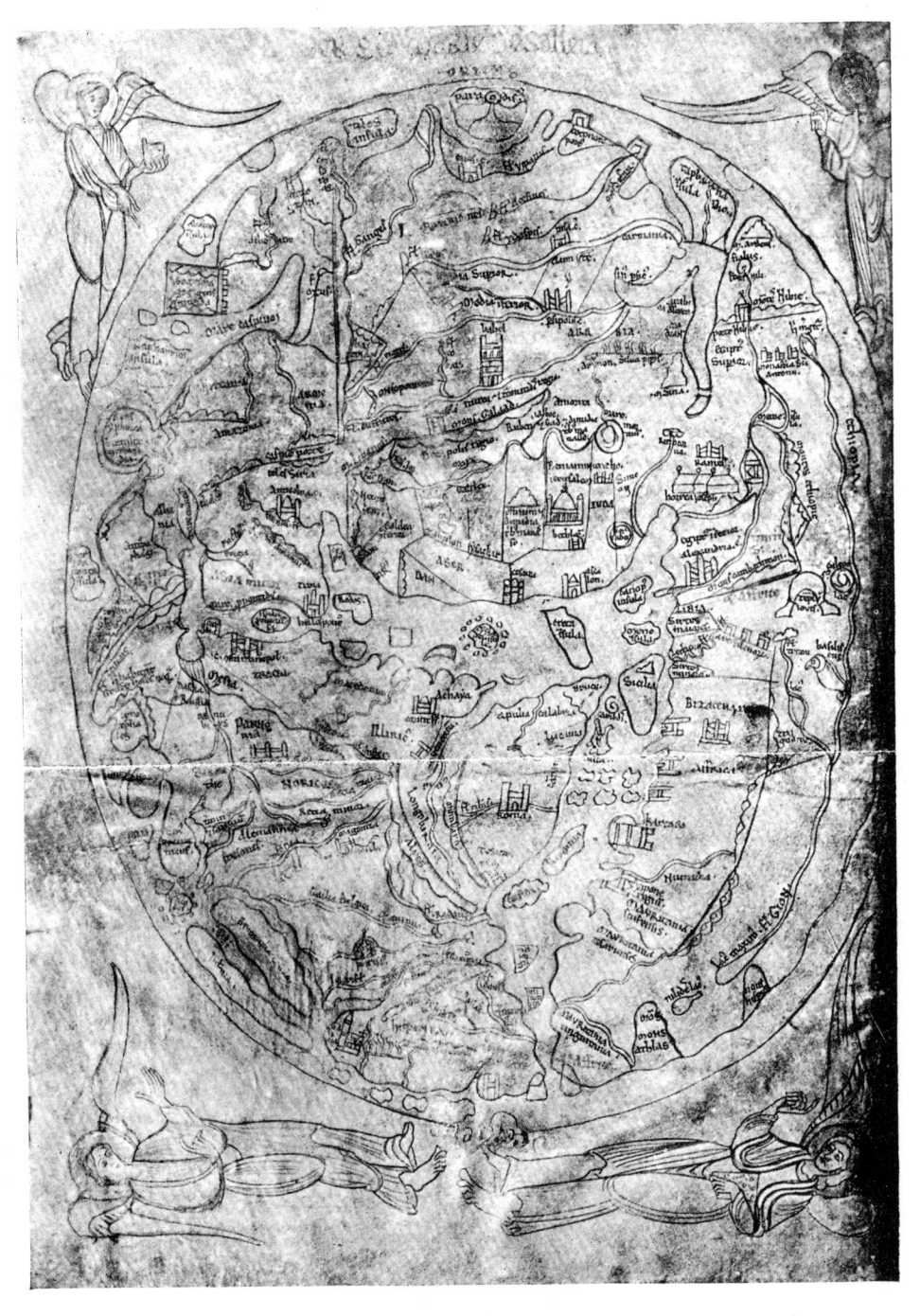

THE WORLD-MAP OF HENRY OF MAINZ, C. A.D. 1110.

[*To face p.* 564.

script usually represented in the map. On the other hand, the Mainz design is obviously related to the Hereford map, as an elder to a younger brother; and the similarities of detail in these two works may be traced in almost every part of the world and in nearly every important feature of the draughtsmanship. Thus both have the same widening of the Mediterranean at its Eastern extremity, the same projecting horns to represent the angles of the Levant, the same elongation of the Black and Azov Seas, the same approximation of the last to the Northern Ocean. Once more, both have practically the same Nile system and the same representation of African mountains, Asiatic rivers, and Oceanic islands; both give the boundary between Asia and Africa in much the same way; both omit to specify any definite boundary between Asia and Europe; both agree in their arrangement of the surrounding ocean, in their drawing of the chief parts of the continental coast-line, and in various typical details.[1] At the same time, the greater size of the 'Hereford' enables it to admit a far larger content; and this greater mass of material appears still more clearly on the 'Ebstorf,' while the little 'Psalter' is naturally more limited, though far more crowded, than the Mainz copy. All these works probably spring from a great wall map of the eleventh century; of this original, Henry's transcript is more accurate but less complete; the 'Hereford' fuller, but less true and scholarly.[2]

The relationship between 'Henry' and the so-called 'Jerome' maps is almost as close[3] as that between Henry

[1] Such as the British Isles, the Caspian and Baltic Seas, the neighbourhood of Paradise, and the lands of the Gog-Magogs, the Hyperboreans, and the Dog-Headed Folk.

[2] Thus 'Hereford' probably departs from the common original, as well as from Henry of Mainz, in making Jerusalem the centre of the world, and in adopting an absolutely circular instead of an oval form.

[3] And closer than K. Miller seems to recognise.

and the Psalter. We only possess the Eastern part of the *Orbis antiquus* in the Jerome examples, but here the likeness is marked; while the treatment, in the Mainz design, of the Twelve Tribes and their settlements corresponds with the well-supported tradition that the celebrated and sainted editor of the Vulgate, who passed so many years in Syria, himself composed a separate treatise and map[1] upon the subject.

The details in the Mainz design which are foreign to the Jerome tradition may be divided into three classes, respectively based upon Æthicus of Istria,[2] upon Solinus, and upon the contemporary knowledge of the Central Mediaeval period. Among these last we may notice the references to the Turks, the Danes, and the Saxons; the mention of the 'Lake of Nile'; and the names of Rouen, Pisa, Iceland, Lombardy, Frisia, and the Gulf of Venice.[3]

The so-called 'Jerome' maps, though belonging (as they stand) to the middle of the twelfth century,[4] and written in the script of that time, were possibly drawn under the direction, if not by the hand, of that Father, and were apparently intended to illustrate three of his lesser treatises.[5] Of these two map-sketches, one represents Palestine and Lower Egypt only; the other deals with the Levant in a wider sense, reaching out to the Far East; both are perhaps fragments of a lost *mappemonde*.[6] They were evidently designed in colour, but this

---

[1] Now lost.

[2] On Æthicus of Istria, cf. *Dawn of Modern Geography*, i. pp. 355-361; on Solinus, *ibid*, pp. 246-273.

[3] *Mare Veneticum*, unique in mediaeval maps. Among these names the first three are in Æthicus; the fourth in Solinus; the last six belong to Henry's own time more especially.

[4] c. 1150 A.D.

[5] *De Hebraicis Quaestionibus; De Nominibus Locorum;* and *De interpretationibus Nominum Veteris et Novi Testamenti*, composed in A.D. 388.

[6] But perhaps the real scope of the larger map is rather the *Bible World*.

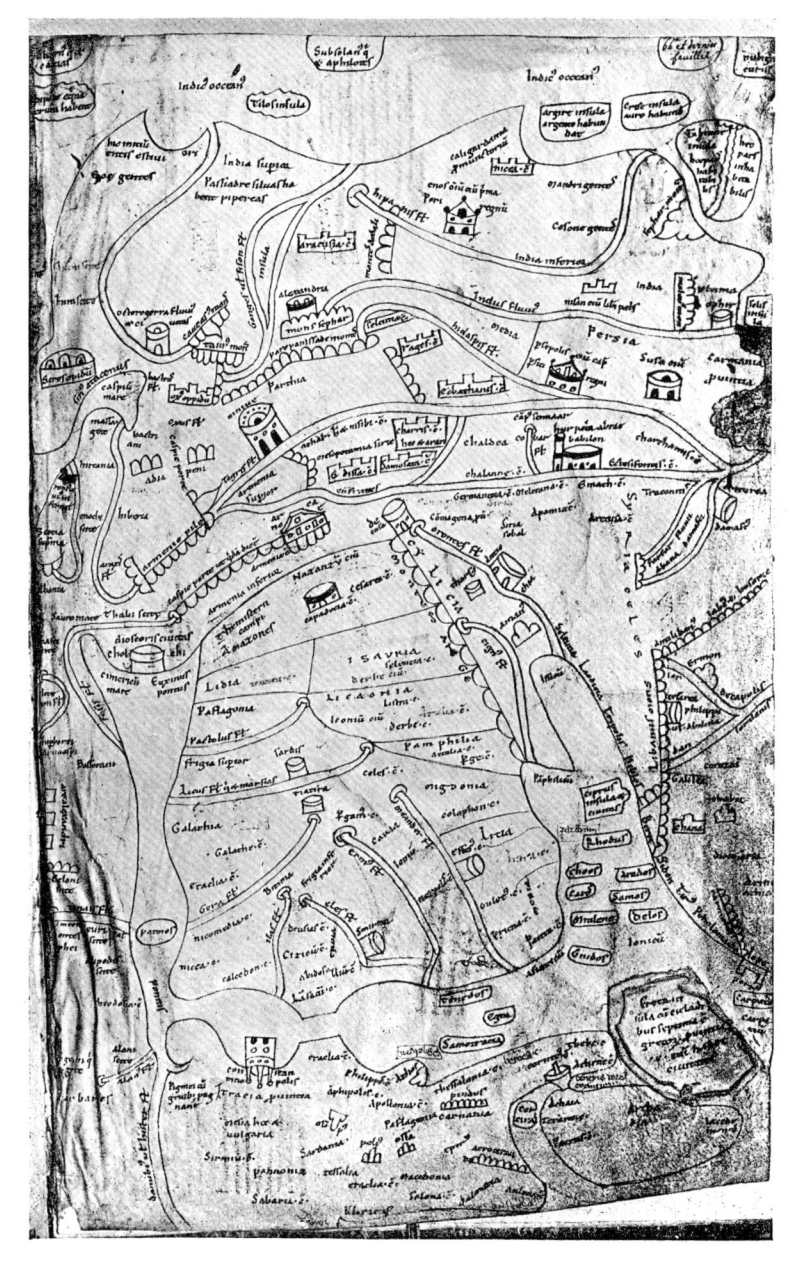

THE LARGER 'JEROME' MAP OF THE TWELFTH CENTURY (MIDDLE).

[*To face p.* 566.

has only been carried out in part; the shape of each is almost exactly square, the larger measuring about 12 inches, the smaller about 8, in length and breadth.[1] The Palestine map is probably illustrative of a translation of the *Onomasticon*[2] of Eusebius, made by Jerome; and the original work of the Greek historian seems also to have been accompanied[3] by a picture, or plan, of Jerusalem and the Temple, as well as by a genealogical scheme which showed the divisions and settlements of the Twelve Tribes. This scheme was apparently transferred (either by Eusebius himself, or by an early copyist of his writings) to a sketch-map, which served as a quarry for many later draughtsmen; among these we may perhaps include Orosius in the fifth century and (through the medium of Orosius) the designer of the 'Cottoniana' at the end of the first millennium.

The Venerable Bede, in his work *On the Place-Names in the Acts of the Apostles*, shows an intellectual relationship with the 'Jerome' maps so close as to support the belief that he knew and used these works. Their content, both 'sacred' and 'profane,' agrees closely with the fourth-century date required, and with the other geographical disquisitions of the great Roman Doctor; they are emphatically pre-mediaeval; indeed, they contain very few entries, and those easy of interpolation,[4] which refer to a time distinctly later than St. Jerome. The Earthly Paradise is not marked, nor the world-centre at Jeru-

---

[1] More exactly (a) 35.5 by 35.8 centimetres; and (β) 23.6 by 22.4 centimetres.

[2] Cf. also Eusebius, *On the Names of Races*, and the same writer's *Description of Judaea*, both illustrated in these maps to a less degree.

[3] According to Jerome.

[4] Chief of these is the identification of Bulgaria and Moesia ('Moesia haec et Bulgaria'), which refers to the Bulgar emigration (from the Middle Volga, etc.) about 679; cf. pp. 478-9 of this vol.

salem; and the richness of ancient geographical detail, untouched by Dark-Age legend, points clearly to the time of the Old Empire.

The draughtsman's original purpose was evidently the illustration of the Biblical localities described in the *Onomasticon*, and similar works; just as Beatus started with a purpose of depicting the Apostolic missions in various regions; or as the 'Macrobius' and 'Sallust' maps originated in the desire of explaining certain passages in those authors. But Jerome's wider design is carried into regions beyond the scope of Eusebius' descriptive catalogue and even in the sectional Palestine-map there is a certain difference of treatment. For whereas the Bishop of Caesarea had left out some of the larger Syrian towns, and enumerated many insignificant ones, St. Jerome's sketch supplies us with nearly all the capitals and omits some of the hamlets.

The extra-scriptural content of the 'Jerome' maps belongs to the oldest type of post-Ptolemaic *mappe-monde*.[1] Plinian material is largely used, sometimes in a manner that reminds one of Solinus, while some reference is apparently made to a Roman road map, similar to the Peutinger Table, which marked the imperial provinces, with two of the leading towns in each.[2]

The 'Psalter' Map of about 1250 is the last example we need take here of this family of cartographical designs; for 'Ebstorf' and 'Hereford' lie beyond the limits of the period now under review (900-1260), and 'Henry of Mainz' and the 'Jerome' maps have been already noticed. The

---

[1] The larger scheme has 278 legends; the smaller, 195; 66 being common to both; and among these only two or three appear to be later than the fourth century.

[2] On the 'Jerome' maps see Appendix, pp. 605-8.

THE 'PSALTER' MAP OF THE THIRTEENTH CENTURY (MIDDLE).

[*To face p.* 568.

closest relative of the 'Psalter' is the map of Ebstorf; but the former is more antique in character, and probably bears a closer resemblance to the common original (in everything except size and bulk of material) than any other of this group. Especially this would seem to be true of the contour and general delineation; the argument from the antique character of the legends and place-names, though it supports this conclusion, is weakened by the obviously weak scholarship of the draughtsman.

In the present design we have an extremely small circular map, less than 4 inches[1] in diameter, crowded with written matter, supplying no less than one hundred and forty-five inscriptions, but from a scientific point of view terribly debased. The Gog-Magog region, and the zone of monstrous races that runs along the southern coast of Africa, are important marks of the geographical mythology of the Middle Ages, like the trees of the Sun and Moon, which come into the map from the Alexander romance. Along with the 'Hereford' and the larger 'Jerome' map, the 'Psalter' is of peculiar value, as filling up a more complete mediaeval picture of the Far East.

We may compare this work, as an illustration of a Manuscript of the Book of Psalms, with the scheme of Beatus, at least in this, that both originate in the ornamentation or illustration of a certain portion of Scripture;—of the Hebrew poetry in one case, of the Apocalypse in the other. Illustrated Psalters[2] are very ancient; the British Museum possesses one of about A.D. 700; and the Albi map[3] of the eighth century, which occurs in a volume of *Glosses on the Gospels,* furnishes another parallel to this Biblical cartography.

---

[1] 8½ centimetres exactly.

[2] But in no other illustrated Psalter has a map been found; cf. also an illustrated 'Genesis' of the fifth century.

[3] Cf. *Dawn of Modern Geography,* i. pp. 385-386.

Another, more obscure, but deeply interesting, family of mediaeval maps falls within the limits of this period: it is composed of the inter-related[1] designs of Lambert of St. Omer, the 'Macrobius' and 'Sallust' map-illustrations, and the 'Climate' and T-O sketches.

Lambert, Canon of St. Omer, was the compiler of an Encyclopaedia, called *Liber Floridus*, composed of extracts from one hundred and ninety-two different works. In this he has left us a chronicle which reaches to the year 1119,[2] and which contains various maps, including a *mappe-monde*, originally[3] of a date at least earlier than 1125, and surviving in three forms.[4] In spite of a clearly-expressed intention of supplying a complete world-map, the oldest copy, that of Ghent, only gives us Europe, two Macrobian zone sketches, and a T-O design; the two later redactions,[5] though containing a less detailed Europe, both possess the complete *mappe-monde*, together with a special and interesting addition. Nowhere else in mediaeval cartography do we find greater prominence assigned to the Unknown Southern Continent, the 'Australian' land of the 'Fabled Antipods.' On the Paris manuscript a long inscription defines this 'region of the South,' as 'temperate in climate, but unknown to the sons of Adam, having nothing which belongs[6] to our race.' The Equatorial[7] Sea, which here divided the [great land masses or continents of the] world, was not visible, proceeds

---

[1] Consciously, according to Miller; but even if unconsciously, none the less truly inter-related.

[2] And alludes to a few matters subsequent to 1119.

[3] Like the text.

[4] In the MSS. of Ghent, Wolfenbüttel and Paris. The Ghent copy seems to have been written by Lambert himself, certainly not later than 1125. Its Europe contains some remarkable peculiarities.

[5] *I.e.* Wolfenbüttel and Paris of about 1150. These are simply different copies from the same original, which was doubtless of Lambert's own draughtsmanship.

[6] 'Or is related.'

[7] Lit., 'Mediterranean.'

the legend, to human eye; for the full strength of the sun [1] always heated it, and permitted no passage to, or from, this southern zone. In the latter, however, was a race of Antipods (as some philosophers believed), wholly different from man, through the difference of regions and climates.[2] 'For when we are scorched with heat, they are chilled with cold; and the northern stars, which we are permitted to discern, are entirely hidden from them.[3] ... Days and nights they have of one length; but the haste of the sun in the ending of the winter solstice causes them to suffer winter twice over.'[4]

The ideas here expressed are supplemented by the suggestion of two more unknown continents or earth-islands, one in the Northern and the other in the Southern hemisphere, lying in the expanse of an all-encircling and dividing Ocean. Four land-masses, therefore, are apparently assumed; of these, the first two were the ancient *Oikoumené*,[5] and the 'Australian' region just described; the others were on the reverse side of the globe (corresponding in some respects with the North and South America of later discoveries), and were divided by a tropical arm of Ocean, in the same way as the two 'islands' of the Eastern Hemisphere. This, at least, was the full theory of the ancient geographers, such as Krates of Mallos, to whom Lambert's scheme must be traced. The present map, however, only indicates the third and fourth continents by little circles placed in the margin of the Roman World, or Habitable Earth, and respectively

---

[1] 'Going just overhead by the milky way.'

[2] And 'of times and seasons.'

[3] The following curious words are added:—'Nulla alia astra sunt, quae illorum obtutibus denegentur. Et quae simul cum illis oriuntur, simul eveniunt in occasum.'

[4] Lastly, to the south of this temperate Australia, Lambert places a zone of extreme cold, uninhabitable by living creatures.

[5] Africa being supposed to end north of the Equator, and the Equatorial Belt itself to be covered by an impassably hot tropical Ocean.

entitled 'Paradise,' to the North-East, and 'Our Antipodes,'[1] to the South-West. The last term is clearly to be understood of the Continental mass exactly opposite to Europe on the other side of the globe, inhabited by living (but not apparently human) beings, and having day and night in an 'opposite relation' to our own; while the Paradise island is probably to be interpreted, in the same way, as precisely antipodean to the Australian Continent already described. The expression of this theory in Lambert's map was perhaps derived in the first place from Macrobius or Martianus Capella, but ultimately it depended on the speculations of a much higher antiquity, like the apparent indication of the Ecliptic[2] in this design, and the suggestion of a T-O form in the general contour of 'Our World,'[3] which is also to be found here.

But if Lambert's 'universal' conceptions are so narrowly dependent upon classical antecedents, it may be expected that the detailed material of the map will also display a markedly antique character; and indeed the relationship between the mediaeval geographers and those of the later Imperial time is seldom found in more complete expression. Most of the hundred and eighty inscriptions are entirely ancient, and must be referred to a lost design of the old Roman world: the chief additions to this pre-mediaeval

[1] As to these (in the Western Ocean) the legend here declares 'noctem diversam diesque contrarios perferunt, et [occasus astrorum]'; the last two words being Santarem's emendation for the 'estatorum' of Paris and the 'estatem' of Wolfenbüttel. We may contrast Lambert's classical and more scientific use of 'antipodes' with the merely conventional (i.e. mediaeval) employment of the term in the Beatus Maps. The idea of an under-sea course of rivers from a trans-Oceanic Paradise to the *Oikoumenê* is common to Lambert and Cosmas Indico-pleustês.

[2] In the form of a crooked line running over the Equator and marked by three star-pictures. The obliquity of the sun's path is clearly suggested.

[3] I.e. The *Oikoumenê*, the Northern or 'Roman' land-mass of the Eastern Hemisphere. See Appendix, pp. 621-4.

VII.]   LAMBERT AND THE 'MACROBIAN' SCHEMES   573

material are made from the geography of Lambert's own period. It does not appear that the writer derived his ideas or place-names from the classical authors[1] named in the *Liber Floridus* (with so little appearance of real knowledge); on the contrary, both ground-plan and detail probably come from a lost design of no small antiquity, in many respects similar to that from which Macrobius drew.

The geographical passages of the last-named author[2] were illustrated from an early period by certain zone—or climate—sketches, depicting the chief belts or parallels of the world from North to South, and adding special reference to various chapters of the *Commentary on the Dream of Scipio*. More particularly, these sketches illustrated the fifth chapter of the second book of this Commentary—where the question of the climates was discussed,—the close of the first book—devoted to the attraction of the earth and the existence of Antipodes,—and the seventh chapter of the second book—which explained the celestial zones and ocean currents.

Ambrosius Aurelius Macrobius, who filled high offices of state under the Emperor Honorius, was probably a Greek by birth and a Pagan by religion; his writings were very popular in the mediaeval world, and he shares with Sallust the honour of special map illustration, designed for him, if not by him, and connected in his case (if not in that of Sallust) with other groups of cartographical designs.

Among the Macrobian sketches,[3] some give us nothing but the five, or seven, zones: others picture the two earth-islands of the Eastern hemisphere, which we have noticed

---

[1] *E.g.* Sallust, Lucretius, and Ptolemy. Among later writers Isidore, Capella, and Orosius seem to have the nearest relationship to Lambert's thought, and are oftenest illustrated by citations in the *Liber Floridus*.

[2] Ambrosius Aurelius Theodosius Macrobius, fl. c. A.D. 410.

[3] See Appendix on Maps, pp. 625-6.

in Lambert of St. Omer. Here the encircling ocean covers most of the earth's surface, and the land-masses are reduced, in Cicero's words, to the position of 'specks' upon the water, while the chief currents from equator to pole are indicated, apparently as the chief cause of the tides.

It is doubtful how soon the Macrobius plans were altered by mediaeval copyists to the uncertain orientation which we find in the manuscripts. But Macrobius himself, in his own mind, certainly put the north at the top, for in one place [1] he says that the *Upper* temperate zone was inhabited by men of our race. In one of these climate-maps,[2] a distinction is drawn between the domesticated folk of the same temperate zone and the wild men of the woods, who inhabited arctic and torrid lands.

The *Liber Floridus*[3] of Lambert contains, besides the *mappemonde*, two Macrobian zone-maps; and indications of the same character, written or sketched, may be found in many other mediaeval authors. Thus the Venerable Bede, in his *De temporum ratione*,[4] discusses the five climates, and perhaps supplies the accompanying map, in which the equinoctial belt and the four great segments of the earth's circle are described in strict agreement with Macrobius.[5] Again, the *Imago Mundi* of Honorius of Autun reproduces Macrobian ideas, both in text and illustration;[6] while another twelfth-century work, the *Philosophy of Nature*[7] of the Parisian teacher, William of Conches,[8] contains three Macrobian maps,

---

[1] *Dream of Scipio;* or, in full, *Commentarius ex Cicerone in Somnium Scipionis*, ii. 5.

[2] Given in the Venice edition of 1489.

[3] Cf. fols. 24, 225 in Ghent MS. Also in the Hague MS.

[4] Chapter xxxiv.

[5] *E.g. Comment. Somn. Scip.*, ii. 6; each segment measures 63,000 stadia.

[6] Cf. the Macrobian zone-map in the Paris MS. of the *Imago*.

[7] *Magna philosophia naturae*.

[8] Near Evreux. He taught in Paris c. A.D. 1150. Of the three maps named, the first has the North, the second has the East, and the third the West, at the top. Sacrobosco's map hereafter referred to, p. 575, has the South uppermost.

THE MACROBIAN 'COTTON' ZONE-MAP OF THE TENTH CENTURY.

[*To face p.* 574.

—one sketching the five zones and the zodiac, another showing the two earth islands of the Eastern hemisphere, the third indicating the 'Habitable World,' in a T-O form. Yet again, the Abbess Herrade of Landsberg, in her *Garden of Delights* (about 1180 A.D.), gives us a slight Macrobian zone-sketch with the ecliptic; and another of the same kind is to be found in the *De Sphaera Mundi* of John Halifax of Holywood, in Yorkshire, the famous *Sacrobosco*, who flourished and wrote in Paris about 1220.

Lastly, we may notice in certain copies of Hyginus, one [1] of which is perhaps of the sixth century, a zone-map which depicts the four land-masses in full Kratesian fashion.

The remaining climate-maps [2] are not always easy to distinguish, except by the absence of definite Macrobian reference, and the addition of non-Macrobian matter, from the zone-schemes just noticed. But the sketch of the Spanish Jew, Petrus Alfonsus of Huesca (of about the year 1100), is obviously designed to illustrate the Hindu and Arabic conception of the world-centre called 'Arym,' [3] and thus has a special interest. 'Arym' was sometimes considered as a mathematical centre-point for the Habitable World, for the Eastern hemisphere, or for the whole earth-circle; sometimes as 'the throne of Iblis,' a home of accursed spirits; and sometimes as a mysterious and lonely mountain in the midst of the Indian Ocean. In the eleventh-century writings of Gerard of Cremona, if not earlier,[4] it passes into Latin thought; it is very prominent in Roger Bacon : and here in Petrus Alfonsus, and other 'schematists,' it is adopted as a geographical axiom of equal importance with the climates and celestial directions.

[1] Now at Wolfenbüttel, once at Bobbio.
[2] See Appendix on Maps, pp. 626-7.
[3] Otherwise Aren, 'Arim,' or 'Arin.'
[4] *E.g.* in the tenth-century Plato of Tivoli,

Undoubtedly these climate-maps had their origin in Greek science; Marinus of Tyre, according to Masudi, composed some of especial excellence, still existing in the tenth century; and among these classical examples two varieties may be distinguished. In the former, the climate-scheme was only part of a map-framework, and was combined with an immense amount of other matter, as in the extant work of Ptolemy, and probably in the lost designs of Marinus. In the latter, this scheme was abstracted from all else, and sketched in roughly outlined maps for the use of beginners. It is the latter form that passes into mediaeval use.

Lastly, the St. Omer map is suggestive, not only of the 'Macrobius' and other climate-sketches, but also of that curious variety of mediaeval cartography, known as the T-O maps.[1] These are very numerous, but at the same time very similar in character; at least eighty manuscripts, reaching from the eighth to the fifteenth century, contain designs of this type; and the conception of one and all is fully expressed in the lines of Dati :— [2]

'Un T dentro a un O mostra il disegno
Como in *tre parte* fu diviso il mondo.'

In some of the earliest examples, however, the T and O formations are not combined, and the 'dividing letter' is associated with square and oblong as well as with round, *enceintes*.

As early as the fifth century before Christ, some of the Ionic philosophers hit upon this as a convenient way of indicating the chief divisions of the *Oikoumené*, or Habitable World, and in spite of Aristotle's contempt, it survived as a popular favourite, like the climate-sketches and Kratesian schemes

---

[1] See Appendix on Maps, pp. 627-31.   [2] *Sphera*, iii. 11.

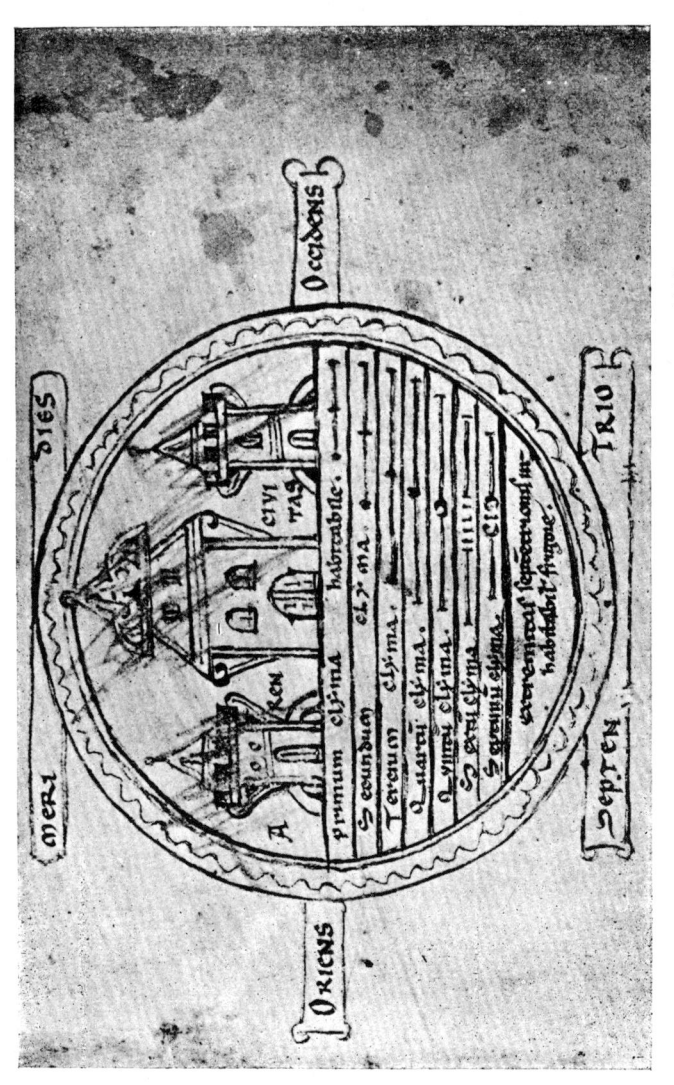

THE CLIMATE MAP OF PETRUS ALPHONSUS C. A.D. 1110.

[*To face p.* 576.

already referred to. There were many differences of detail, though Greece was usually placed at the midst of the circle,[1] and Delphi, or Delos, in the midst of Greece; some of these T or T-O designs made Europe—some Asia—the largest of the Continents, but no one gave the predominance to Africa, then commonly believed to end on this side of the Equator. These simple plans, grouping the chief land-masses of the known world in an easily recognisable shape, were generally associated with the allied conception of a centre for the circuit of the earth, for the infinitely extended horizon. They did not of necessity deny the theory of a globular earth; but they were concerned, and only concerned, with its aspect as a surface, flat or slightly curved, as apparent to the ordinary observer.

In some of the T-O family, traces may also be observed of the three-cornered representations which, like the square, or four-sided, sketch-maps, were in favour in ancient schools.[2] For the one, the threefold division of continents, of seas, of the Christian Godhead; for the other, the four quarters of the heavens, the four chief winds, even the four Gospels, were quoted; and to each variety there are numerous allusions. A clear description of a T scheme is given by St. Augustine, who probably used a work of this kind, though at the same time he can hardly have been ignorant of a very different style of chart—the road-map, in ribbon form, of which we possess a copy in the Peutinger Table.

The earliest among the T-O examples of the Dark-Age and Crusading Periods are to be found in the works of St. Isidore;[3] and in a measure these may be considered as typical of all others. Here, besides the three continents, we

---

[1] Where the O form was combined with the T.
[2] Cf. Orosius, II. ii.
[3] *Etymologies*, xiv. 2, 3; of these two are here specially referred to as archetypal; eleven others in St. Isidore's works are quite unimportant.

have the names of the three sons of Noah, one patriarch being attached to each continent; the East is at the top; and the 'Great' or Mediterranean Sea occupies the whole of the T-formed intersection of the land.¹ Other specimens develop the simple titles of *Asia*, *Shem*, and the like, by explanatory inscriptions which declare (for instance) that Asia is named after a Queen Asia,² and is inhabited by 27 peoples; that Africa is derived from Afer, a descendant of Abraham, and has 30 races with 360 towns; and that Europe, so called from the Europa of mythology, is overspread by the 15 tribes of the sons of Japhet, possessing 120 cities.

Besides the (sixth century) Isidorian examples, one other T-O map—and only one—deserves special mention. This is the Byzantine sketch of 1110, now at Oxford, which contains some features of high antiquity, and is one of the earliest plans where Jerusalem appears in the centre of the earth. Both Greek and Latin titles are here given for the quarters of the heavens; the Twelve Tribes of Israel and the locations of the Twelve Apostles are indicated; and from these indications, combined with the barbarous confusions of Levantine place-names, we may with some assurance infer that we have here a copy, poorly executed by an ignorant scribe in Western Europe, of a Byzantine work brought home by some of the earliest Crusaders about the year 1100.³

The threefold division⁴ of the world is expressed in some⁵

---

[1] Hence these are sometimes called Noachic maps.

[2] 'Of the posterity of Shem.'

[3] St. John's Coll. Libr., Cod. membr. fol., xvii., fol. 6. For other T-O maps, *e.g.* the Strassburg of c. 870, or the St. Omer of 1010 see Appendix, pp. 628-30.

[4] *Trifaria Orbis Divisio*.

[5] Especially those of 'Görlitz' and 'Rome.' In these the T has lost its rigidity; the idea of a central point

of the 'Sallust' maps[1] far better than in the T-O plans of the usual type. Here we have a partition of the *Oikoumené* into fairly equal continents, but with less rigid symmetry, and in a way more reconcilable with scientific views, Asia having a slight preponderance, and definite allusion being added to the 17th, 18th, and 19th chapters of the *Jugurtha*.[2]

It was probably at an early date, long anterior to our oldest surviving Sallust *manuscript*, if not in the lifetime of the author himself, that the original Sallust *map* was inserted; but this plan was either admitted into very few copies, or was replaced in most by an ordinary T-O map, lacking all definite reference to Sallust materials. The oldest example, the 'Leipzig,' of about 980, probably shows us certain features of this primitive type, a pre-Christian scheme, without Jerusalem, and with an overshadowing Rome. For this primitive type the authorship of a priest in North Italy, somewhere between 600 and 700, has been suggested; but it may fairly be referred to a time before the destruction, if not before the conversion, of the Roman Empire in the West.[3]

Next in order we may glance at some examples[4] of *Chorography*, of the detailed representation of limited areas; and among these, the most elaborate belong to the school we may call *Sionist*, as being concerned, mainly or exclusively, with the topography of Jerusalem. The best-known of these Sionist plans perhaps belongs to the Crusading period, and goes by the name of the *Situs*

---

is not expressed; and the general conception refers rather to a three-*cornered* world than to an exact tripartite division of the same.

[1] See Appendix on Maps, pp. 631-2.
[2] Especially ch. 17. None of the copies are important enough to be described, except in the Appendix on Maps, pp. 631-2.
[3] All existing Sallust maps conform more or less closely to the T-O type.
[4] See Appendix on Maps, pp. 633-8.

Hierusalem; but a far more important specimen has lately come to light from a much earlier time.

The Mosaic Map of Madaba,[1] which is here in question, though a discovery of the last two decades, belongs to the later time of Justinian (about 550), and may claim to be the oldest existing specimen of Christian cartography. For although the original designs of Cosmas Indicopleustes may have been executed twenty or thirty years earlier, the earliest manuscript we possess of the *Christian Topography* does not carry us back further than the ninth century. Compared with the latter, the Madaba Mosaic is of much greater merit, as far as we can judge from its broken and damaged state; it is not unworthy of the Restored Empire and its glories in arms, in legislation, or in architecture; and it increases our regret for the loss of so much Byzantine work in this as in other fields.

The Madaba map covers Palestine and parts of Egypt and Arabia; one of its special objects is clearly to delineate the Holy City; and it probably reproduces with fair exactitude the condition of Jerusalem half a century before the Persian sack in 614.[2] The extension of the design is somewhat similar to the smaller 'Jerome' map, where Neapolis or Nablûs lies in the middle, and the Mediterranean forms the Western limit. In its pictorial character, and especially in its town-vignettes, the Madaba plan recalls both the Peutinger Table, and some of the Beatus examples.[3]

Here then we have one of the oldest pictures yet discovered of Jerusalem (outside the Egyptian and Assyrian

---

[1] It should have been noticed in a former volume, but that it has only been described by European scholars since the publication of *The Dawn of Modern Geography*, vol. i., in 1897.

[2] Viz. c. A.D. 545-565.

[3] Notably the 'Paris' of 1150 ('Paris iii.'), whose illustrations show little resemblance to the sketches of a flat design, but much to those of a mosaic plan.

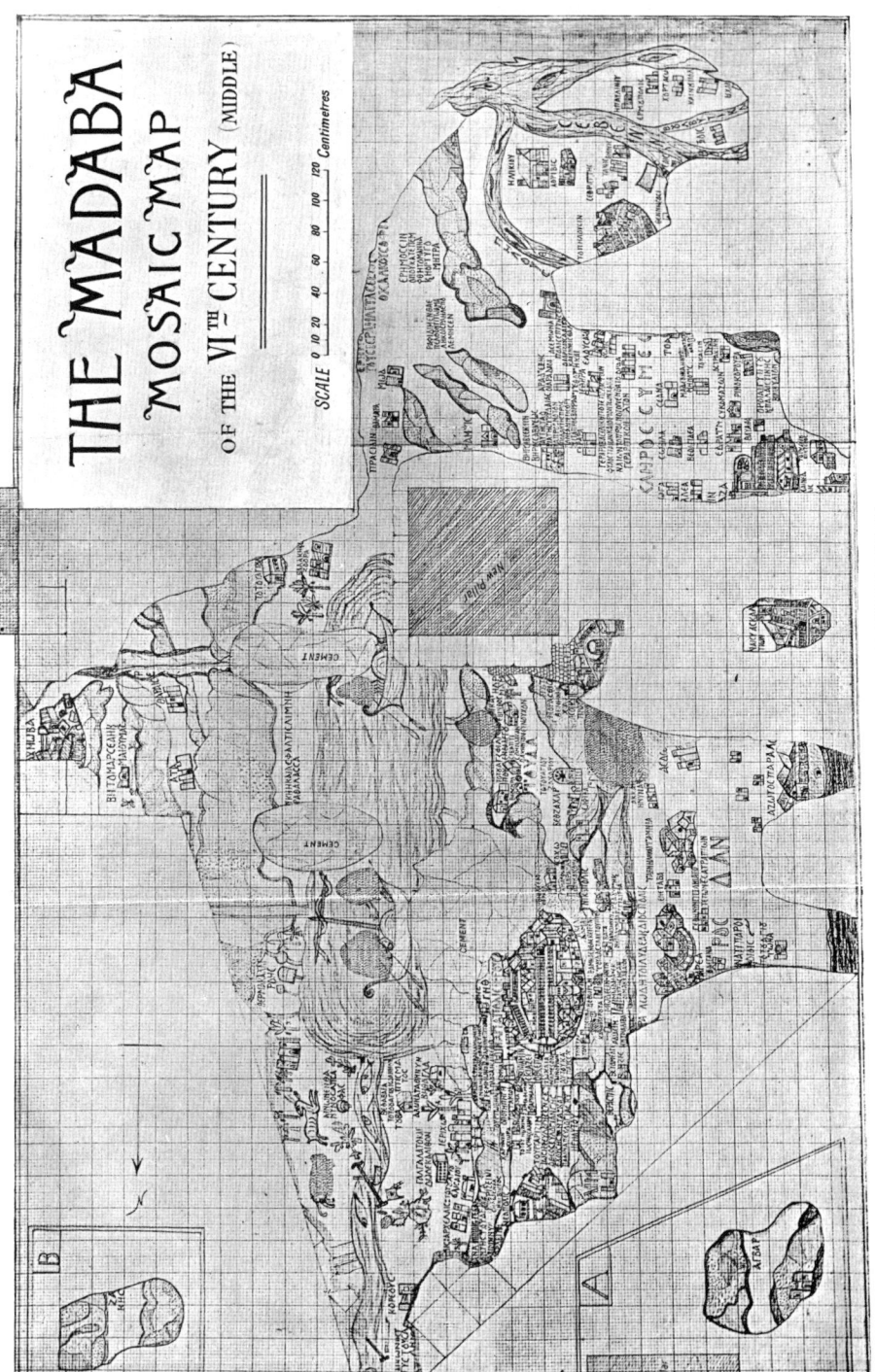

THE MADABA MOSAIC MAP OF THE SIXTH CENTURY (MIDDLE).

monuments), one of the first specimens of Christian map-making, and apparently one of the earliest illustrations of the division of Palestine among the Twelve Tribes of Israel.[1] It is of course strictly ecclesiastical in type, for it was designed for the pavement of a church. Most of its one hundred and thirty place-names correspond with the notices of Eusebius;[2] but one entry appears to refer to a monastic foundation[3] of the early sixth century on the east side of Jordan; and in some places the text is unique.[4]

From the fragments discovered it is clear that the original once occupied a space of about fifty by twenty feet; the length is from North to South; but the east is at the top, as in most primitive and mediaeval Christian designs. All the north part of the mosaic has been destroyed, except two morcels and what remains, in all about half of the complete scheme, is mainly concerned with the country between Nablûs and the Nile. The orientation has been greatly disturbed by the assumption of the Levant coast (from Alexandria to Acre) as a base, supposed to furnish a line running almost due north and south.

In general, this plan may be regarded as a decorative, freely conceived, and well-informed, but not closely scientific, illustration of Bible history. Names and objects are not kept in any strict proportion; the perspective is conventional; but the rich store of inscriptions furnishes not a few details of interest. In the larger towns, such as Jerusalem, Pelusium, and Gaza, an attempt is made to

---

[1] This is conjectural, but very probable, from the fragments as yet recovered.

[2] Especially in the *Onomasticon*, i.e. the early-middle of the fourth century.

[3] The Convent of St. Saphas or Sapsaphas, a saint contemporary with Elias, Patriarch of Jerusalem, A.D. '494-518.' This place is named as Sapsas by Johannes Moschos in the seventh century.

[4] Sometimes older and newer forms are given together.

represent the principal streets by elaborate colonnades, and even to portray some of the chief buildings, sometimes in round, sometimes in angular forms. Cities of the second class are indicated by sketches of walls flanked by round towers. Each of the Tribes of Israel seems to have originally appeared—the names being marked by great red letters, accompanied in some cases by a text of Old Testament prophecy. But in what remains only six tribes can still be found, and several of these (*e.g.* Symeon) survive in a very fragmentary state.

In the plan of Jerusalem the two great streets, marked by covered galleries, which run across the city from north to south, probably represent the markets or bazaars of Justinian's time, sacked by the Persians of Chosroes in 614; while the Church near the Western Gate, with its round or apsidal end facing the Mediterranean, and its staircase communicating with the colonnade street in the centre of the town, represents the Holy Sepulchre of Constantine and Helena, the first of the successive Christian sanctuaries on the site of the Passion.

In Lower Egypt and the Desert of Sinai are a number of places and legends; while three arms of the Nile are enumerated—the Pelusiac to the north (east), an unnamed and partially destroyed channel to the south (west), and the Sabennitic in the middle, with three ramifications—Saitic, Bucolic, and Bolbitic. These names are all inscribed along the course of the stream.

The city of Madaba, Medaba, or Medeba,[1] beyond the Dead Sea, south of Heshbon, and south-east of Nebo or Pisgah, was an important station of the Old Empire on its Arabian frontier; it lay upon the Roman road which connected Damascus with Petra and the Red Sea; and it was close

---

[1] Medaba usually in Roman writers; Medeba in the Bible.

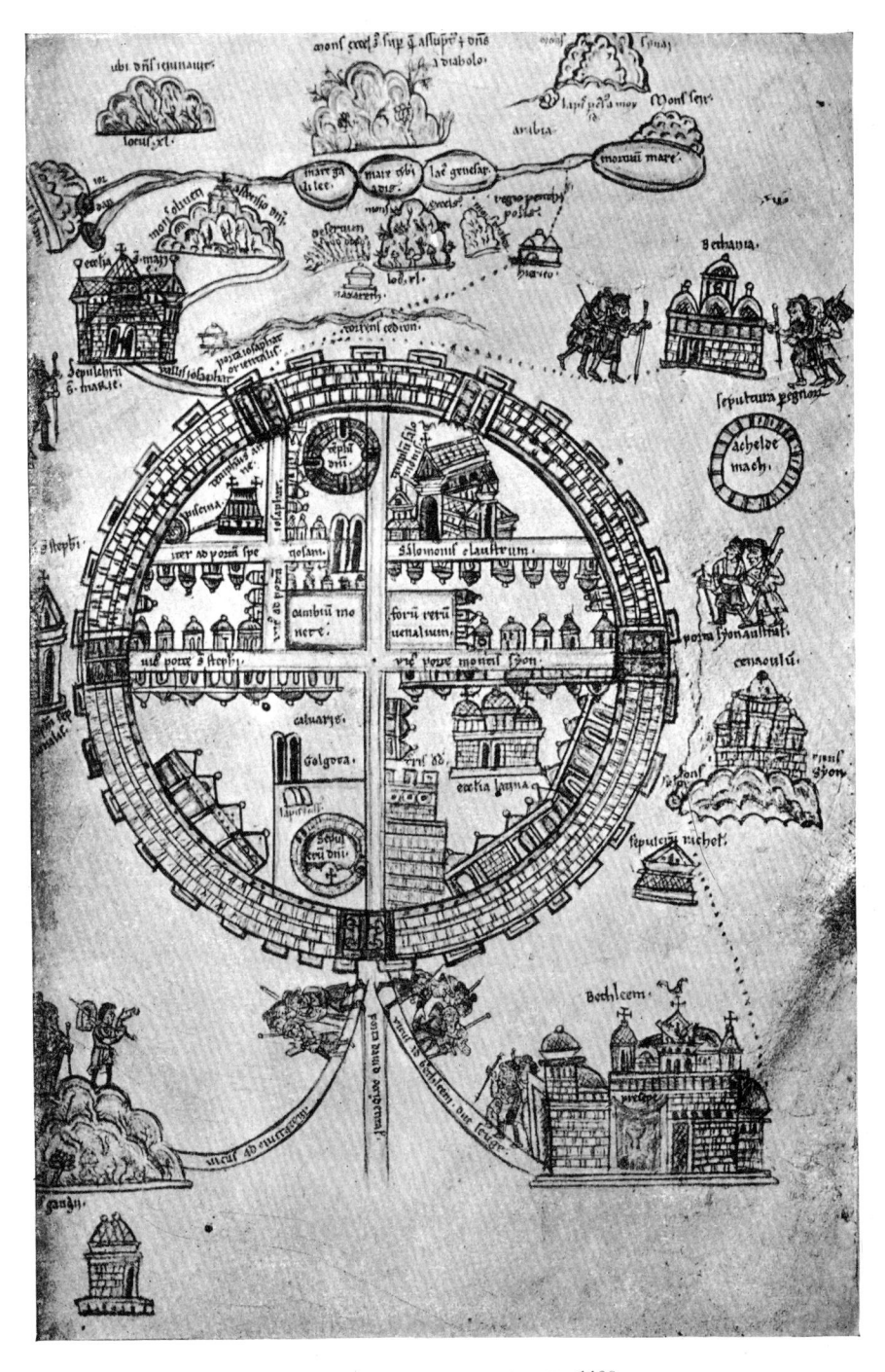

THE SITUS HIERUSALEM, C. A.D. 1100.

[*To face p.* 582.

to another road from Jerusalem and Jericho, which crossed the Jordan and united with the Damascus-Petra route at Heshbon. Once it belonged to the Tribe of Reuben; from the fourth to the sixth century—from the time of Constantine to the Moslem invasion—it was the seat of a Christian bishop; and thus it was not without remains of some pretension when, after twelve centuries of neglect, the Christians of Kerak migrated thither, and in clearing away the rubbish,[1] revealed the unsuspected treasure of this mosaic-map.

The famous Plan of Jerusalem, or *Situs Hierusalem*,[2] of the early twelfth century,[3] has been often coupled with the tract *Qualiter sita est civitas Hierusalem*, commonly supposed to be a paraphrase of the material provided by the *Situs*. It is probable, however, that Tobler and Molinier are right in referring this pamphlet[4] to the last quarter of the tenth century, the era of the short-lived and partial Byzantine reconquest of Syria under John Tzimiskes. One copy of the *Situs* appears to have been sketched by the anonymous compiler of the *Gesta Francorum*[5]: but the material of the sketch, in relation to the Holy City itself, is far more ample than what is given us in the *Gesta*. Professedly, but not literally, the latter follows the narrative of Fulcher of Chartres, who described the First Crusade as an eye-witness; but at times the *Gesta* agrees with the *Situs*, and differs markedly from Fulcher. The tract on the *City of Jerusalem*, however, is quite independent of the *Gesta*, and assuming its aforesaid tenth-century origin,

[1] 1880. The map was found in December 1896.
[2] See Appendix on Maps, pp. 636-8.
[3] Some put back the date to 1099.
[4] The *City of Jerusalem*. For the text of the *Qualiter sita est civitas Hierusalem*, cf. Société de L'Orient Latin, série geographique, i. 2, 345; edition by Tobler and Molinier.
[5] Probably finished before 1109.

584 GEOGRAPHICAL THEORY AND DESCRIPTION [CH.

under Byzantine inspiration, it would seem that the draughtsman of the Crusading *Situs* used the written description of Tzimiskes' time; while the author of the *Gesta*, ignorant of the *City of Jerusalem*, is indirectly connected with the same through the medium of the sketch-map[1] he uses and transcribes.

From the insertions of the two leading copies[2] (at St. Omer and Brussels), it appears that the *Situs* was but a section of a more extensive original, representing parts of Galilee,[3] the Upper Jordan, and the Way of the Israelites from Egypt through the desert. The later forms of this plan, such as we have in the Copenhagen transcript, have lost all these traces of a wider outlook; even the surroundings of Jerusalem are here wanting; and various additions appear, of a date clearly subsequent to the First Crusade. Thus the Temple and Sepulchre of the Lord no longer show the round form which is noticeable in the early copies of the *Situs*, referring us to the buildings of the pre-Crusading period.

Lastly, the group of maps which own the common authorship of Matthew Paris[4] fitly closes the cartography of the Central Mediaeval time.

In the twelfth and earlier thirteenth century, the monastery of St. Albans possessed what may be called an historical school, or institute, which was then[5] the chief centre of English chronicle, and with different environment might have become the nucleus of a great university. Among the writers of this school, the greatest was Matthew Paris (1195-1259), whose three chief works contain various maps

---

[1] The *Situs*.
[2] Of these copies one belongs to a twelfth-century manuscript of the *Gesta*, the other is bound up with the Crusading records of Fulcher.
[3] *E.g.* Tabor Magdalum, etc.
[4] See Appendix on Maps, pp. 638-42.
[5] And in all for about two hundred years.

MATTHEW PARIS' ENGLAND, OF THIRTEENTH CENTURY (MIDDLE).
(*Most perfect Cotton Library Copy*).

[*To face p.* 584.

and plans unsurpassed in mediaeval geography, before the rise of the Portolani. Thus, in the *Historia Major*, or *Cronica Majora*, we have the so-called *Itinerary to the Holy Land*, or *Stationes a Londinio ad Hierosolymam*, as well as a *mappe-monde*, a map of Palestine, and the first of Matthew's four maps of England. Again, in the *Historia Minor*, or *Historia Anglorum*, there is another form of the *Palestine Itinerary*, the second and third maps of England, and the *Situs Britanniae*. Lastly, in the *History of St. Albans*, a portion of the supposed Pilgrim-road, as far as South Italy, is given in another shape, together with the *Schema Britanniae*.

Matthew Paris, then, appears as the author of six geographical designs; a world-map, in two slightly different forms; a map of England, in four variants; a purely conventional sketch of the Heptarchy, in the form of a *Rose des Vents*;[1] a plan, or *schema*, of the Roman roads of the same country; a 'routier' to Apulia from the English Court; and a map of Palestine, which tradition has wrongly joined with the former, to make a Pilgrim Itinerary from London to Jerusalem.

Among all designs of purely mediaeval origin, Matthew's plans of England, as he knew it, show the best evidence of critical study, the most systematic attempt at an exact delineation of a particular country. They are, in fact, remarkable instances of what Ptolemy called *Chorography* or regional geography; and they would be even more important, but for another consideration. At the very time that the English chronicler was draughting these maps, compass-charts (based on coast-surveys, of a minuteness hitherto unknown) were beginning in the south of Europe. In comparison with these Portolani, Matthew's work is but secondary; for at his best he only represents approximate and traditional know-

[1] The *Situs Britanniae*. This has the East at the top.

ledge; the only bases of true geography, the fixing of terrestial positions by celestial observations, and the independent and detailed examination of limited areas by practical travellers,[1] were first given us on an adequate scale by the new school of Mediterranean pilots and map-makers. The monk of St. Albans was, after all, a student and a bookman, rather than an independent investigator of nature.

His world-map,[2] unlike his 'England,' is of small value, though it is curiously different from all other mediaeval designs; perhaps its most interesting feature is an inscription (placed in the neighbourhood of Mount Taurus) which alludes to the three great wall-maps existing in or near London at this time (c. 1250). One of these is ascribed to a certain Robert of Melkeley; another is called the *mappemonde* of Waltham in Essex; the third is termed the property of the Lord King at his Court in Westminster. This last (whether or no it hung in the Exchequer, as some have thought) had been 'figured by the direction' of Matthew Paris himself; and perhaps the same authorship may be assumed for the Waltham map. Some features of these lost examples may have survived in the fourteenth-century work of Ranulf Higden; in size they must have resembled the 'Hereford' and 'Ebstorf' designs; but they probably showed a better draughtsmanship, a wider and sounder knowledge, and a less fabulous spirit than the latter. Yet compared to Matthew's England, his surviving *mappemonde* is a disappointment; and if we were to assume that his wall-maps at Westminster and elsewhere presented merely

---

[1] This, repeating infinitely the actual experience of wayfarers, and expressing the result in extremely minute detail, was but faintly appreciated by the Classical World and was wholly unknown in the earlier Middle Ages.

[2] This measures 34.8 by 23.6 centimetres, and contains seventy-nine legends.

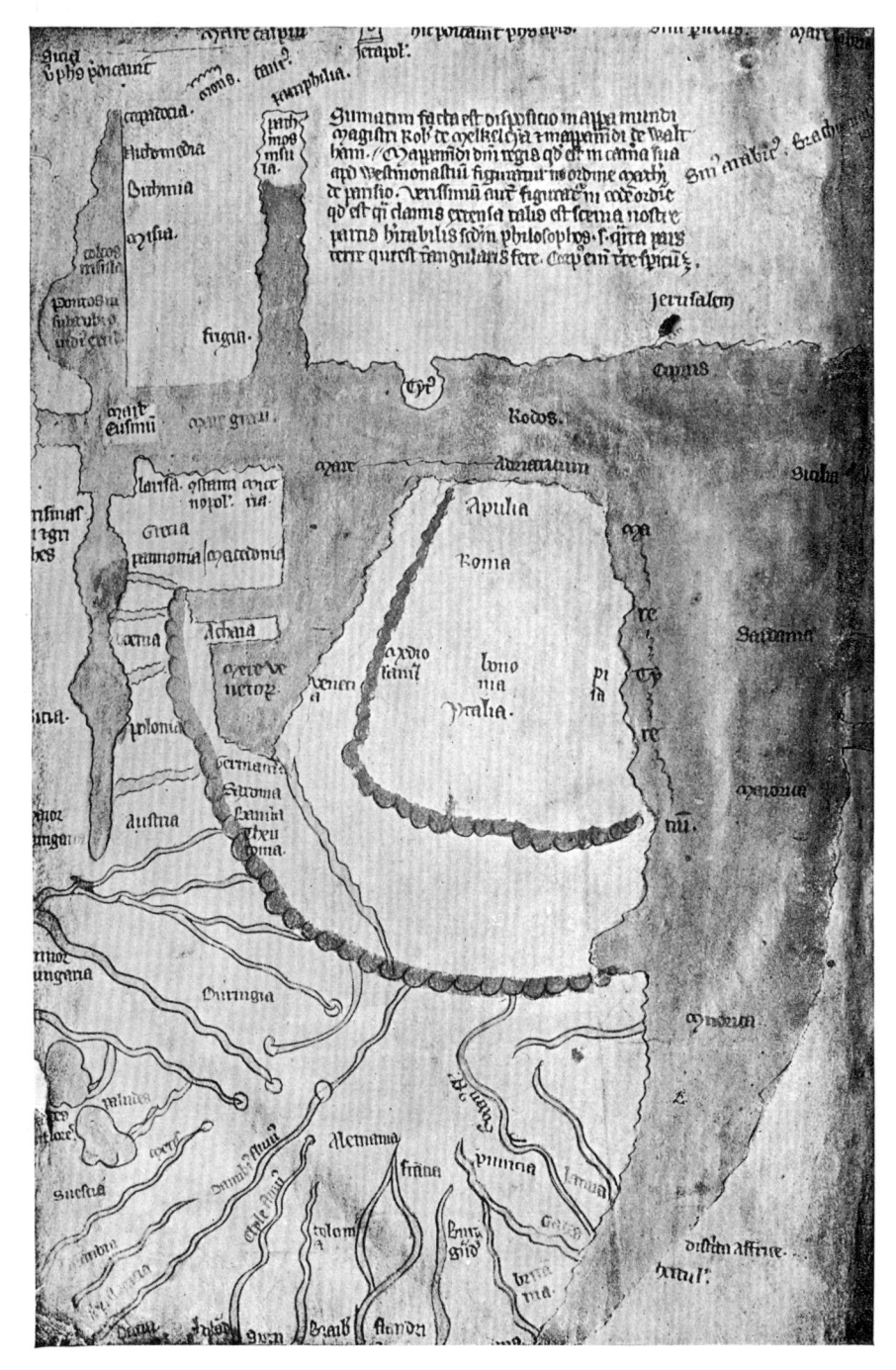

M. PARIS' WORLD-MAP OF THE THIRTEENTH CENTURY (MIDDLE).

[*To face p.* 586.

the same features on a larger scale, there would be less reason to regret the loss of these *Orbes picti*.

The chief thing worthy of remark in this world-map is its limitation. For it is not really a *mappe-monde*, but rather a sketch of Europe and the adjacent coasts; only the extreme northern edge of Africa is portrayed; as to the parts of Asia here given, the author has so little intention of working them out in detail, that he covers most of the spaces with the inscription just noticed, about the three wall-maps. Even the Europe of this example is not finished; its northern coast is absolutely straight, and apparently follows the requirements of the sheet or page without attempting to represent the true shore-line. The western littoral is scarcely better; England, which Matthew knew so well, is entirely omitted; and it would be difficult to rate the compiler's geography at a high level, if we only possessed this design, and could not also refer to the four maps he has left us of his native country.

These last (and especially the two examples in the Cotton Library) are the finest achievements of mediaeval student-geography.[1] In Wales, Devon, and Cornwall, the Humber estuary, the East Anglian peninsula, and the line of the Severn, the execution is so good as to suggest modern accuracy: but, on the other hand, highland, or *ultra-marine*, Scotland, is treated as an island wholly separated by the Firths of Forth and Clyde from the southern region, to which it is united only by a bridge.

Here also, for the first time in Northern Europe, we have a map with the North at the top; and in this we may see a victory of revived scientific feeling over the ecclesiastical

---

[1] They appear to be from the annalist's own hand in the manuscripts we possess. In the best copy the map measures 33.8 by 22.3 centimetres.

preference for the East, and of North-European feeling over the Arabic and other influence which had made the South the primary quarter of the heavens. But the Ptolemaic arrangement, here reproduced by Matthew, was also better adapted for a sketch of the long and narrow island of Britain, tapering towards the North,[1] and hence perhaps its reappearance in this map.

The *Situs* and *Schema Britanniae* are works of extremely slight interest; but it may be remarked that the latter, which deals with the four chief Roman roads[2] of England, and makes them intersect at Dunstable, has a peculiar orientation, with the West at the top.

Matthew's so-called *Itinerary from London to the Holy Land*, with which has usually been reckoned his map of Palestine, is not really a connected whole; it is the result of combining two different works, a pictorial representation of the route between London and Apulia, and a sketch of the Holy Land. In the former, the chief stations or halting-places are indicated in sections of the route, beginning at the North, the chief rivers and mountains of the journey also appearing; in the latter, we have a map of Palestine, with the East at the top, in the ordinary ecclesiastical manner. The connection of the two is simply in the fact that both are by Matthew Paris; that they are of almost the same date; and that each is written in Old French intermixed with Latin.

The *Itinerary to Apulia* is not, therefore, part of a pilgrim guide to the Holy Land; but rather appears to be a

---

[1] The East at the top would have made Matthew's England run on to two pages of the MS., broken in the middle by the fold. Again the North (from the Pole Star and Great Bear) had obviously greater scientific claims. The Anonymous Geographer of Ravenna, in the seventh century, claims to have composed a special map of Britain, as well as his *mappemonde*, and this also is said to have had the North at the top.

[2] Ermine St., Watling St., Icknield St., and the Fosse Way.

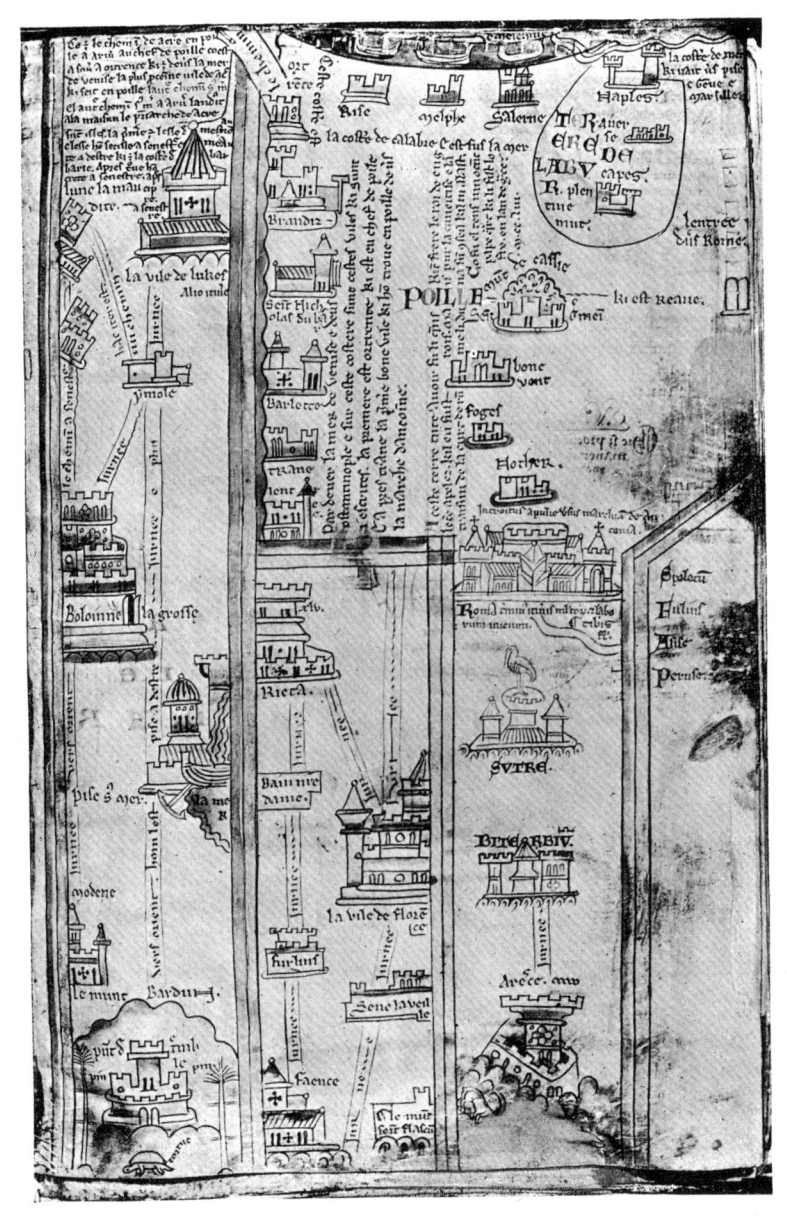

MATTHEW PARIS' ITINERARY FROM LONDON TO APULIA (SOUTH ITALIAN
SECTION) OF THE THIRTEENTH CENTURY (MIDDLE).

MATT. PARIS' ENGLAND, PALESTINE, ETC. 589

political sketch, with the following history. On St. Martin's Day, 1252, one Master Albert, a Papal notary, appeared at the English Court, and offered to Earl Richard of Cornwall the Kingdom of Apulia, on behalf of Pope Innocent IV., titular overlord of that realm. The Earl himself looked on the gift as a 'dominion in the moon'; but his brother, King Henry III., and the whole English Court party, were eager to accept the offer; and this Itinerary was probably composed[1] during the abortive negotiations on the matter, with the view of informing and fomenting English ambitions as directed on South Italy.[2] This is confirmed by the fact that the Itinerary proper reaches only to Rome; at this point it assumes a new character, and portrays the Norman lands of *Pouille*, or Apulia, in considerable detail, enumerating all the greater towns. No list of stations is given from South Italy to the Holy Land, and it is pretty clear that none was intended.

Matthew's sketch of Palestine,[3] so often connected with the Itinerary just mentioned, is in some of its more general aspects parallel to the smaller 'Jerome' map;[4] but in details there is a great difference. For 'Jerome' gives us ancient names throughout, while the English chronicler inserts many indications of thirteenth-century nomenclature and history; such as the fortress-enclosures of the Templars, the Teutonic Knights, the Pisan and Genoese merchants, and other Western corporations, in Acre; or the dwelling of the Old Man of the Mountain, or Chief of the

[1] From descriptions, not always with proper orientation.
[2] This is all recorded in the long inscription or legend over 'Pouille'; cf. *Historia Minor*, iii. 126. Another inscription at Trapes or Trapani in Sicily tells how Earl Richard called here on his return from a Crusade.

[3] This is accompanied by some vague indications of roads, which, taken together with the camel picture, points to a commercial object in this design.
[4] Both have the North at the top, Palestine in the middle, and Egypt on the right.

Assassins, 'far towards the North.' The text of this map is, in fact, closely related to various Itineraries of the period of Latin domination in Syria, such as *Les pèlerinages pour aller en Jérusalem, Les chemins et les pèlerinages de la Terre Sainte*, or *La dévice des chemins de Babiloine* (viz. Babylon of Egypt). The first of these is of 1231, the second of 1265, the third of 1289-1291, but it is probable that earlier redactions of the two last already existed in Matthew's time, and were used by him.[1]

---

[1] Thus *La dévice* mentions the arm of the river which the 'French King's people passed over', and herein appears to refer to St. Louis' Crusade of 1251. Cf. *Société de L'Orient, série geographique*, iii. vi. 87; x. 177; xiii. 237. Perhaps in his Palestine, whose form is traditional, while its content is largely novel (as in other works of the same author), Matthew Paris has worked up an older *mappe-monde*; with his Acre we may compare Marino Sanuto's plan of 1320—a later development on a similar basis. From 1229 to 1291 Acre was the capital and the only important relic of the Kingdom of Jerusalem.

On Hasan Ibn Saba, founder of the Assassins, cf. pp. 239-40 of this volume; the destruction of this sect by the Mongols (in 1258) is one of the last facts noticed in the *Historia Major*.

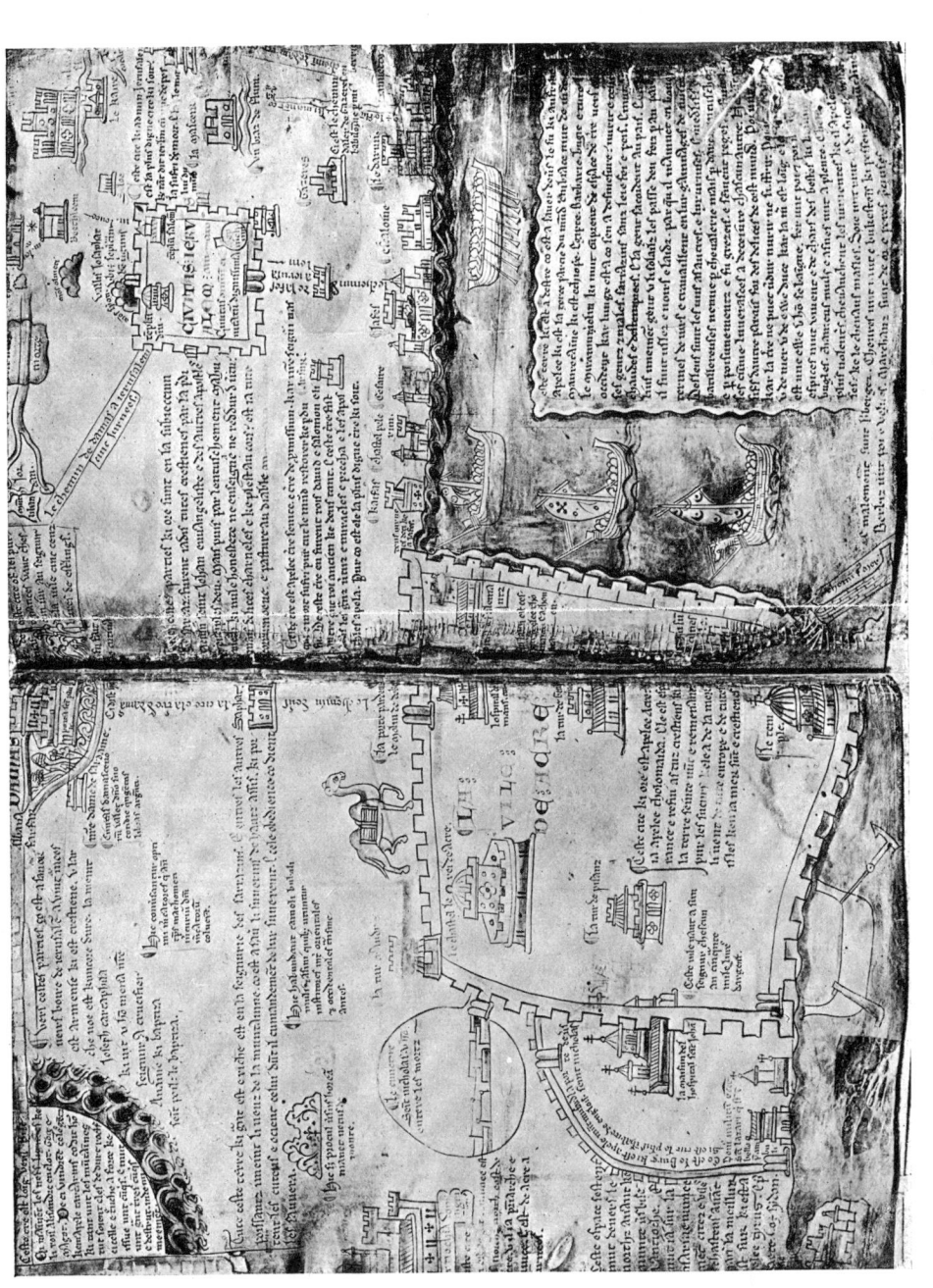

M. PARIS' MAP OF PALESTINE OF THIRTEENTH CENTURY (MIDDLE).

[*To face p.* 590.

(591)

APPENDIX ON MAPS

I.—IN the case of the Beatus maps, there is evidence for the existence (besides the Original of c. A.D. 776) of two primitive but vanished copies from which all existing examples are derived. The Commentary and its World-Picture were perhaps composed at Valcavado. Of the two primitive copies, one appears to have been made in the same house before the year 800, the other at Osma or Uxama in Old Castile, otherwise famous as an early home of St. Dominic.

In all probability, moreover, the primitive 'Osma,' of about 800, passed through another intermediate stage, which has also perished, before reaching its oldest surviving form, in 'St. Sever.' As descendants of the primitive Valcavado we must likewise admit some intermediates, especially (A) *one of about 900*, the source of 'Ashburnham,' 'Valladolid,' 'Madrid,' and 'London'; and (B) *two of the tenth century*, sources of 'Gerona,' 'Turin,' and 'Paris iii.' (of c. 1150), as well as of a lost copy of the twelfth century ('Las Huelgas').

Of the ten existing specimens three have been known for some time; the remaining seven have only been noticed lately. St. Sever, Turin, and London were all described before 1850; Turin, indeed, as early as 1749; the other two were dealt with by Santarem in 1849. For the first adequate account of the Ashburnham, Valladolid, Madrid, Gerona, Osma, and later Paris copies we have to thank Professor Konrad Miller, whose work has already made a revolution in the study of mediaeval maps. See p. 549, *n.*, of this vol.

As to the more important details of the various copies other than St. Sever ;—

(A) The *Paris* of 1250 ('Paris ii.'). In this Jerusalem is quite separate from Palestine; and on the eastern rim of its world occur Armenia, Arabia, Greece, and Pandonia (Pannonia).

Paradise, originally at the top because that was the *Eastern* quarter, has been left in the same position when a *Southern* interpretation has been given to it, although the Christian Eden is never conceived as being in the South (cf. Genesis ii. 8). This map forms a link between St. Sever and Osma, but is not derived from either. In its external form and colouring it recalls St. Sever; but, for instance, the red striped bands on a yellow ground which in the Aquitanian copy designate North and South, have lost their meaning in the Paris example, remaining only as ornaments; while the Red Sea has retained its form and colour, though its direction has been changed. The town-pictures, especially in Crete and Cyprus, also resemble St. Sever, and these and other features support the belief that the original work of Beatus is best represented in the French copy of the eleventh century. From the size and prominence of Astorga in Spain, encircled with a brown and yellow band, we may suppose this was the place of draughtsmanship. [Original in Paris, Bibl. Nat., MSS. Lat., nouv. acq., 1366; the page measures 350 by 230 millimetres.]

(B) The *Osma* of 1203, first published by K. Miller, may be called a more distant relative of St. Sever, which it almost equals in value and with which it has common ground in its general delineation of lands, rivers, and seas. It represents Paradise simply by the *springs* of the Four Sacred Rivers. Osma or Uxama is 56 kilometres to the south-west of Soria, in the upper valley of the Douro. It was an ancient stronghold of Sertorius, and St. Dominic was once a canon of the Church. The manuscript measures 38 by 30 centimetres.

(C) The *Valcavado* or *Ashburnham* map, of about 970, was also first published by K. Miller. It is, perhaps, one of two copies made about this time from two earlier and slightly different transcripts of the Beatus original. The present example was drawn and written in Valcavado by Obeco, under Abbot Sempronius, between the 8th of June, and the 8th of September in one year; the lost contemporary was made at Tabara by Emeterius, between the 1st of May and the 27th of July in a single summer. In the Valcavado of Obeco, the later Ashburnham, a very inferior intermediate has been used, and the new reproduction has made things worse. Most rivers have been left out, and the only two that appear, the Nile and the Danube, are represented like inland seas. Of the town-pictures of the original only Jerusalem remains. The canoes and fishes swimming in the encircling ocean are carefully preserved, like the mountains, though the latter are conventionalised. The chief peculiarity of this copy is its

APPENDIX ON MAPS 593

absolutely right-angled form; other copies approach, but none equal, it in this respect. The West-Gothic character of the Ashburnham script is also found in the next three copies, 'Valladolid,' 'Madrid,' and 'London'—(Brit. Mus. of 1109)—all of which may be treated as in part derivatives from the transcript of Obeco. This example is, however, too faded and damaged for us to estimate its value fully. [Ashburnham MSS., XV; the page measures 38 by 28 centimetres.]

(D) The *Valladolid* copy of 1035 is in various ways less complete than Ashburnham. For instance, it has not the two trees (?) given by the latter in Asia, the similar tree in Ethiopia, the names of the three Continents, or Paradise. On the other hand, like the two next copies, it gives us the Pyrenees, which are wanting in Ashburnham; and it returns to the original *oval* form, in preference to the right-angled. Ashburnham, Madrid, and London portray woods upon the mountains; Valladolid only ornaments upon the hill-ranges. But, like Ashburnham, Valladolid marks the Province-boundaries of Africa by ornamental trees. [Original in Valladolid University Library, MS. 229, parchment; the page measures 335 by 225 millimetres.]

(E) The *Madrid* [Nat. Libr.; measures 47 by 31 centimetres] is simply a copy of the Valladolid, with a few different readings; and the same may be said of—

(F) The *London* or so-called 'Spanish-Arabic' of 1109 [Brit. Mus., Add. MSS. 11695; measures 37.5 by 24.4 centimetres], which is, however, the best preserved of all Beatus maps, and shows some signs of copying both from Ashburnham and Valladolid.

The remaining copies all belong to the Tabara, or Emeterius, subdivision. There are apparently *two* lost intermediates of this stem; one already noticed, finished by Emeterius on the 27th July, 970, was the same as a work begun by Magius in 968. A twelfth-century copy of this existed till 1869 at the Monastery of Las Huelgas. The other Emeterius copy seems to have been made in 975 or 978, one Senior writing the script, and Emeterius painting in the design, both working under a certain Abbot Dominicus. The work of Emeterius, in these copies, was probably better than that of Obeco in his, just as the Gerona and twelfth century Paris are superior to the Ashburnham and Valladolid.

(G) The *Gerona* copy, of about 1100, has preserved the original oblong oval, as well as the boats and fishes in the ocean; [it measures 36 by 52 centimetres].

(H) The famous *Turin* of the twelfth century [Turin Libr. MS. I., ii. 1; measures 39 by 27.5 centimetres], once supposed to be a mediaeval design of the first importance, is thought by Miller to be a derivative of Gerona. It leaves out the boats and fishes, and has some gross mis-readings. In the *text* of this manuscript (viz. The *Apocalypse Commentary*, folios 45, 46), though not in the *map*, stress is laid on the twelve Apostles.

(I) The *Paris* of the twelfth century, 'Paris iii.', a direct copy from Emeterius, shares with Turin the peculiarity of a circular shape. Though much mutilated, its town-pictures are striking, and far excel anything in this line from the other Beatus Maps. But to the copyist the pictorial part was evidently more important than the geography, and the text is very imperfect and disfigured. The Mediterranean cuts the earth in two halves; the Mediterranean Islands are put in the ocean. The longer inscriptions of the original are all omitted; but three new names are given us, '*Anglia*,' '*Irlanda*,' and '*Sevilia*,' and two names, 'Maiorga' and 'Toletum,' otherwise missing from all the Valcavado copies. [In Paris, Bib. Nat., MSS. Lat., nouv. acq., 2290; measures 450 by 310 millimetres.]

The value of these designs lies chiefly in their high antiquity. Four of them certainly belong to the pre-Crusading Period, namely, St. Sever, Ashburnham, Valladolid, and Madrid; and both from their age and size, these plans are well worthy of notice. We have scarcely anything from Latin Christian cartography of so early a date; and the few specimens which carry us back to a still older time are much slighter sketches, such as the Albi map of the eighth century. From the close similarity between all members of the Beatus family, we may be pretty certain of the character, not merely of the primitive copies or intermediates, but of the original itself, as drawn by the 'obscure hill-man and cave-dweller' in about 776.

The map of St. Sever is, except for the Valcavado-Ashburnham of 970, the oldest of our surviving examples; the Valcavado-stem copies for the most part show us a slighter and more inaccurate type of sketch than those of the Osma family, lacking the all-important Apostolic pictures and various other primitive features.

Of the ten Beatus copies, seven are oval, somewhat inclining to the oblong; the oldest one (Ashburnham) is right-angled; two of the latest (Turin and 'Paris iii.') are circular. What was the original form? This question can only be answered by remembering the conditions under which it was drawn.

All the copies before us are drawn on two pages; each page gives half the map, or half the world; and perhaps the oblong so often to be noticed is due to the copyist lengthening two halves of a circle to fill up his space and give his work more room. The height of the map is, of course, the height of the manuscript in all instances, thus supporting the theory that the elliptical form was accidental. The comparatively short, upright axis, from top to bottom of the single page, represents the Longitude, or East to West prolongation, of the earth; while the breadth, the comparatively long horizontal axis reaching across the two pages, represents the Latitude, or North to South extension, of the world. But neither in classical antiquity, nor in the Middle Ages, do we meet with any geographer who believes the latitudinal extension of the *Oikoumené* to be greater than the longitudinal. If so, the very terms of 'latitude' and 'longitude' themselves would have been disputed; but, on the contrary, they were always accepted. Hence it will not do to use the Beatus maps as a proof that the ancient *Orbes picti*, and especially the world-map of Agrippa, were oblong or elliptical.

It is probable that on the original Beatus both the Four Sacred Rivers and the Ancestors of Mankind were depicted, as on the Hereford and Ebstorf maps of the thirteenth century. The rivers of Paradise come from Genesis ii. 11-14. The first three are usually identified with the Indus or Ganges, the Nile, and the Tigris. We have already described at some length in *Dawn of Modern Geography*, i., pp. 332-334, 385-386, 391, the patristic and mediaeval views on the inaccessibility of the earthly Paradise [cf. also Müller, *Geographi graeci minores*, ii., 513-514; the Ravennese Geographer, Isidore, and Cosmas; and the *Expositio totius mundi* in the version of Junior Philosophus, c. A.D. 350]. In the Beatus maps there is no clear evidence of a T-O design; the horizontal line dividing Europe and Asia is pushed up towards the top, and thus deflected from the actual middle; only the Osma map is here an exception. Whether this deflection is intentional or not is doubtful. The T pattern is prominent in mediaeval cartography for ages, from the time of St. Augustine and Orosius; it is certainly not, however, the usual type of later classical map.

Beatus seems to have followed Isidore in limiting Africa to this side of the Equator, like many of the classical geographers, Cicero, Pliny, Mela, etc. [cf. Cicero, *Tusc.*, i. 28; Pliny, vi. 22; Mela, i. 1; Solinus, liii., 1; Isidore, *Orig.*, xiv. 5, 17]. All the copies, except the Paris of 1250, mark the 'Antipod' region, and even 'Paris ii.' gives us a relic of the Australian Continent by indicating in a corner the Skiapod or shadow-footed monster whom the Osma Map of 1203 shows us in the Southern Land; this last was doubtless the original position.

The map of St. Sever evidently confuses Taprobane with the real Antipodean Land of some ancient geographers, and here are uncritically mingled ancient traditions of Ceylon, Sumatra, and other islands off the South of Asia [cf. the language of Pliny and Solinus on Taprobane, *often and for long time supposed to be the other Hemisphere or Land of the Antichthones.* P. Mela also speaks of the Antipodean Antichthones; cf. the map of Lambert of St. Omer, in the Paris and Wolfenbüttel copies].

The scientific champions of the Antipodean theory were usually Greeks, and always men who believed in the globular conception of the earth, like the Pythagoreans, Eratosthenes, Posidonius, M. Capella, Krates, etc. The doctors of the Church, such as Augustine and Isidore, who construct their system of geography after an exclusively Latin model, either avoid any acknowledgment of this globular conception (and its possible corollary of 'feet to feet' Antichthones), or, as in the case of Lactantius, absolutely reject such ideas as impious and absurd. Of Antipodean peoples in the sense of inhabitants—monstrous, however, rather than human—of a Southern temperate zone, there was greater tolerance; and it is this conception which is embodied in the shadow-footed race of Osma and 'Paris ii.'

As to the circumambient fish and boats: the fish occur in every example except the Turin map; the boats are found on St. Sever, Ashburnham, Valladolid, Gerona, and the Paris of the twelfth century ('Paris iii.').

In the original Beatus, as in most mediaeval maps, the Red Sea appears to have been coloured according to the name; but in the Paris of 1250, Valladolid, and Ashburnham, this tint is confined in a more modern sense to the Arabian and Persian Gulfs; both these gulfs, on the Ashburnham and Valladolid examples, are depicted, rather like mountains than seas, in red colour; in the Madrid, only the Arabian Gulf is still red; in the London the mountainous appearance is still more striking; in the Gerona, Turin, and twelfth-century Paris, both gulfs are tinted with the ordinary hue of the sea. On every Beatus map K. Miller recognises a trace of the original legends in this part.

The Black Sea, the Sea of Azov, and the Caspian occur on St. Sever and on Osma, but are wanting in all maps of the Valcavado group. The mutilated Mediterannean is no better than on other works of the earlier Middle Ages.

Only on our two best examples, St. Sever and Osma, do we meet with mountains, namely the Pyrenees, Alps, Riphaean Hills, Caucasus or Taurus, Sinai, and Atlas. The maps of the Valcavado Group mark more clearly the Pillars of

APPENDIX ON MAPS 597

Hercules—with the *two peaks opposite to one another* [*Alpes* in Madrid and London, *Calpes* in Ashburnham and Valladolid; *i.e.* apparently Calpe and Abyla], which play so great a part in classical geography, and here are both placed in Africa. Osma marks four African Alps, two on the south coast, two towards the west [*Alpes duo* as one mountain, and *Atlas* as the other].

The rivers of the world are best portrayed on our best copies (St. Sever and Osma); on our other examples, and especially on the Paris of the thirteenth century, the representation of streams may sometimes be used for restoring the probable contents of the original Beatus, which apparently contained no Spanish rivers, but marked the Rhine, the Rhone, the Danube, the Euphrates, the Tigris, the Jordan, the Nile with its delta, and certain affluents of the Caspian.

As to towns, the Paris of 1150 shows unquestionable developments on the original Beatus type, giving us vignettes of Constantinople, Chalcedon, Ascalon, Toledo, Rome, Thessalonica, Babylon, Antioch, Toulouse, and Tangier or 'Tingi.' All the great islands are illustrated with a battlemented town, like several continental countries and regions, to which no city-name is attached;—*e.g.* England, Scotland, Ireland, Sardinia, Sicily, Judaea, Pamphylia, Cappadocia, and Phrygia ('Frigida'). The original Beatus, probably marked four cities with pictures (viz. Rome, Antioch, Constantinople, and Jerusalem), to say nothing of the lighthouse-towers at Alexandria and Brigantia.

In St. Sever, as elsewhere, we must distinguish the additions of the copyists from the original plan; and to this original plan we may fairly assign everything that is common to the two chief families of the Beatus Group, to Osma and to Valcavado. Where all the chief copies agree, we may suppose that we are dealing with material from this original. Happily, the coincidences between all the ten derivatives are so numerous that we can from these alone form a pretty detailed picture of the fundamental draught.

As to interpolations;—In St. Sever, the *Ecclesia S. Severi* in the south-west of France is of course a prominent addition, being indeed the largest single picture in the map. Several other peculiar place-names occur in the same part of the world; but these additions have been to the detriment of other regions, *e.g.* the North-West, or Galician, corner of Spain, and South Italy. The place-names between the Maeotis or Sea of Azov and the Hellespont or Dardanelles are pushed rather to one side, towards the right of the map; and there are several dislocations in or near the Bosphorus, Propontis, Hellespont, and Ægaean. Similar displacements occur in parts, and especially in certain rivers, of Western Asia Minor, a region where St. Sever is inferior to Osma. Some critics have thought that the longer legends on the St. Sever example may be later additions, but this does not seem to be the case. The inscriptions on the Nile, Ethiopia, the Southern or Australian Continent, and Taprobane, are probably in accord with the original; and as to the rest of these elaborate legends, a great portion may be assigned to the same source.

On Osma, as on Ashburnham, Valladolid, Madrid, London, and 'Paris ii.', there is very little that can be regarded as interpolation, though the displacements and dislocations due to inferior draughtsmanship are more prominent than on the Map of St. Sever. On Osma *Troja* is perhaps an addition, like *Alpes Galliarum* in the four first examples of the Valcavado family. On the London map of 1109 (here in contrast with St. Sever), *Scocia* is given as a separate isle from *Brittannia* and to the *South* of the latter. St. Sever also marks *Britter. Ins.* as well as *Brittannia*, while Osma puts *Scocia inferior* on the mainland, E.-N.-E. of *Germania superior*. But this is rather a dislocation than an interpolation.

On the last three of our examples, viz. on Gerona, Turin, and Paris iii., *Seville*, the *Sea of St. George*, and *St. James of Compostella* are certainly additions of a time later than Beatus; the position of St. James in Galicia is also to be found in the Osma copy of 1203. Paris ii. also appears to interpolate *Astorga, Sol et Luna, Mare Magnum* (in the four corners), *Mare Rubrum* (turned into the Arabian Gulf), *Fines Romanorum . . . Francorum*. Here the Franks appear on the left bank of the Rhine, not on the right or east, as in St. Sever.

As to Beatus' chief sources;—With St. Jerome and Orosius there is no direct correspondence, though the indirect relation is unquestionable. The rediscovery of Isidore's lost World-Map would perhaps show us the immediate inspiration of the work of 776. Beatus apparently copies four long legends of one class and eight of another, from Isidore verbatim. The four in question are upon *Armenia, Scythia Major, Albania,* and *Hyrcania;* while the eight are on *India, Mesopotamia, Babylonia, Arabia, Taprobane, Nabatea,* the *Skiapodes,* and the *Quarta Pars trans Oceanum*. Beatus' legend about the Nile is not from Isidore, in so many words, but from Orosius; it is, however, of a nature common to all the later classical geographers, and is possibly connected with an ancient map. No definite reference occurs in Beatus either to Solinus or to Augustine, although three passages, which are copied verbatim from Isidore, are to some extent paralleled in the *De Civitate Dei*, viz. on the Skiapodes, the Antipodes, and the Springs among the Garamantes.

South France and Middle Italy offer certain other matters of remark, *e.g.* the reference to the name and the sixth-century migration of the Basques from the Pyrenees into the Lowlands, as far as the Garonne; the names of Septimania and Provence; the Lombard Duchies of Spoleto and Beneventum; the general correspondence of Beatus' Italy with the known history of the seventh and eighth centuries; and the contrast here shown with the Anonymous Geographer of Ravenna. Beatus' *Vasconia* (between the Pyrenees and the Garonne) agrees with the Roman Province of *Novempopulana* or with the Anonymous Geographer of Ravenna's *Spano-Guasconia*, but not with the latter's *Guasconia*, which is first marked out by Gregory of Tours (c. A.D. 581). The migration of the Basques here noticed was about 587. The name of *Septimania* first appears (c. A.D. 473) in Sidonius Apollinaris, for the West Gothic Lands;—viz.; Narbonne, Carcassonne, Toulouse, and the country to the Rhone, otherwise the coast region of *Narbonnensis Prima*. Beatus' *Provincia* is the south part of the *Viennensis*, as in Gregory of Tours.

A few minor points may be noticed separately, especially as to the sources of various details.

The *points of the compass* are only to be found in the St. Sever and the Paris of 1250; the *Wind-rose*, so frequent in the larger mediaeval maps of later time, only occurs in St. Sever. The classical world used two kinds of Wind-rose; one of these was eight-fold, the other twelve-fold. Both of course were based upon the four cardinal compass points. In the twelve-fold division, which was the favourite one among Greek scientists from the time of Aristotle, a classification of *seven* intermediate winds played a part. It is this twelve-fold division which is also to be found in Isidore, on the map of St. Sever, and in most of the circular designs of the later Middle Age. The eight-fold arrangement was derived from Eratosthenes, and was accepted by Pliny, by Orosius, and by Isidore of Seville, as a complement to the twelve-fold partition.

An important feature on several of the Beatus maps (though not on St. Sever) is the *Faro*, Pharos, or Lighthouse, on the coast of Spain, which in several cases is accompanied by a vignette or picture of the same. This is certainly the classical beacon of Brigantia in Galicia, described by Orosius as looking over towards Britain [cf. Orosius, *History*, ch. 33, *Brigantia*

Gallaeciae civitas altissimam Farum et inter pauca memorandi operis ad speculam Britanniae erigit].

The 'Oscorus' River in Asia is certainly meant for the Oxus, and this reading gives us perhaps an independent form of the classical name. The two sources of the Jordan are in the ordinary mediaeval manner, but especially recall the language of Isidore and Eusebius [cf. Isidore, *Etymol.*, xiii. 21, 20; Eusebius, Onomasticon, 169, 1-7]. The fountain, cold by day and hot by night, in the Land of the Garamantes in North Africa, and the vast salt marshes in the same region, are perhaps from Pliny; the 'tawny race who eat no bread' are otherwise unknown, though the idea may be borrowed from the 'Wild-' or 'Wolfish-eaters' of Solinus [cf. Pliny, *Hist. Nat.*, v. 36; Solinus, 144; Augustine, *De Civ. Dei*, xxi. 5; Isidore, *Etymol.*, xiii. 13, 10. The salt-marshes are referred to in Dicuil, 8, 7, 1, who says that he had read about them in a 'cosmography']. The Lighthouse at Alexandria may be derived from various classical writers or maps, but among existing sources, the Peutinger Table is perhaps the nearest parallel, nearer even than those in Ammianus Marcellinus and the second map of St. Jerome [cf. Tab. Peut., ix. 3]. Of course Beatus must be understood to use or consult, not the Tab. Peut. in the stage we have it (of a thirteenth-century copy), but in an earlier form, or in a parallel example of the Roman Imperial cartography. In the Tab. Peut. we are dealing with the only important survival of the cartography in question, and it must therefore stand for the whole class of Roman itinerary-maps from the time of Julius Caesar. The inscription at the mouths of the Nile (describing the course of the Great River) and that which deals with the nomade Gaetulians are again parallel to the language of the Peutinger Table and to Orosius; the two Alps (or 'Calpes') of West Africa, already noticed, are perhaps likewise from Orosius; while the picture and description of the Australian or Antipodean Land recalls the language of Isidore and Augustine [cf. Tab. Peut., vii. 2, viii. 5, ix. 3; Orosius, *Hist.*, chs. 12, 47; Isidore, *Etymol.*, xiv. 5, 17; Augustine, *De Civ. Dei*, xvi. 9-18].

As to the old Roman Province-map from which Beatus seems to have drawn so largely, a similar original was doubtless before Isidore, Orosius, and Julius Honorius, who have given us short extracts therefrom. The same map or maps served the Anonymous Geographer of Ravenna in good stead; while the Peutinger Table is, of course, a later form of the very same classical cartography applied to a special purpose, the Government of the Roman Empire.

Some have thought that Beatus used a map-design by Orosius, now lost; the existence of such a work is inferred from various points in the Cotton or Anglo-Saxon map, in the map of Henry of Mainz, and in that of Hereford. But a comparison shows certain differences which support the belief that Beatus could not have used a map which was simply an illustration of the Orosian text. Thus the *Asiatic Scythia*, and the lands east of the Caspian Sea, as Beatus gives them, differ greatly from Orosius and the Orosians (*e.g.* Henry of Mainz, and the so-called Richard of Haldingham), and agree much more nearly with Isidore, the Ravennese Geographer, and the Peutinger Table.

Some of the names which we find in Beatus' Syria and Asia Minor (*e.g.* Laodicea, Philadelphia, Thyatira), occur here for the last time in mediaeval cartography, and point strongly to an archaeological use of such classical place-names as may be found in the Table. The so-called Jerome maps, we may notice, show the same tendencies and survivals. In the East two Indian towns (Elimaida, Antiochia-Tarmata) are put in the exact positions given by the Table, and are named by no writer of later date, except the Ravennese, who is himself, of course, strongly Peutingerian. There is an odd contrast between Britain and Gaul, as treated by Beatus. The first-named, corresponding (for the most part) to the lost western section of the Table, is thoroughly antique in character; and its five towns, all marked by small pictures, agree with the chief British cities in 'Castorius,' as quoted by the Ravennese—'Castorius' being probably an older form of the Peutinger Table, as known in the seventh century. On the other hand, Beatus' Gallic towns have very different names from Peutinger's; the latter are the old city-names of the Empire; while Beatus, for the most part, gives only the more modern cantonal or tribal names which by the second half of the fourth century became attached to the towns of Gaul. These modern names are often given by the Tab. Peut. along with the ancient.

Two other correspondences of less extent and moment have been inferred—with Ptolemy and with Julius Honorius. The former depends upon one race-name otherwise unknown (the Macusienses) and various names of seas which Isidore, Orosius, and Mela also copied from the Geographer of Alexandria. The latter rests on a single curious misreading in the name of a Syrian river [*Orestes* for Orontes]. But we must not suppose that Beatus made use of a large number of ancient sources.

The text of some of the longer legends gives a fairly good idea of the mental outlook of Beatus and his copyists. Thus, *e.g.* *Albania*, so called from the whiteness of its people, and the colour of their hair, extended from the East, close by the Caspian Sea, and the shore of the Northern Ocean (into which the Caspian was believed to flow) to the Maeotid Lakes (Sea of Azov), through desert regions where the dogs were so strong and fierce that they could take lions and kill them. *Hyrcania*, so called from the Hyrcanian Wood (a confusion with the 'Hercinia Sylva' of Germany) which lay 'under Scythia,' was full of tigers, panthers, and pards. Many races lived here and in Scythia, among them cannibals and blood-drinkers. *Scythia*, stretching from the extreme East and the Seric Ocean to the Caspian Sea (at the setting of the sun) and southward to the ridge of Caucasus, abounded in gold and gems, in the best emeralds, and in the most pure crystal; but all these treasures were guarded by Gryphons, and no man could approach thereunto. *Armenia*, between the Taurus and the Caucasus, between Cappadocia and the Caspian, was divided into two parts, the Greater and the Less, and contained the source of the Tigris. *Arabia*, the land of incense and perfumes, of myrrh and cinnamon, of the phoenix and the sardonyx, was also called 'Saba,' from the son of Chus. *The Dead Sea*, so named because it produced nothing living, and received nothing from the race of living things, was in length 780 stadia [or furlongs] and in breadth 150. [This attempt at measurement is a very unusual feature in mediaeval maps, and shows a curious, if unfortunate, precision, which at any rate, marks a spirit of enquiry; the figures given are twice too great (cf. Isidore, xiii. 19, 3-4). Beatus also gives measurements, in Roman miles, for the Islands of Britain, Corsica, and Sardinia, as well as for Taprobane or 'Tapaprona.'] *India*, containing many peoples and tongues, men of dark colour, great elephants, and precious products, such as gems, ivory, aromatics, ebony, cinnamon, and pepper, was also famous for its parrots, its dragons, and its one-horned beasts [rhinoceros]. It was amazingly fertile, with crops twice a year; among its gems were diamonds, pearls, burning carbuncles, and beryls; it also possessed mountains stored with gold, and guarded by dragons and monstrous men. Among its islands were 'Chryse' and 'Argyre,' the isles of gold and silver, and Taprobane, which lay far to the South, was divided by a river, was only in part inhabited by men, had ten cities, and was full of jewels and elephants. *Ethiopia*, stretching to the borders of Egypt, abounded in races of diverse colour and monstrous form. It possessed multitudes of wild beasts and serpents, precious stones, cinnamon, and balsam. *The Nile* was said by some authors to rise far from Mount Atlas, and thereafter to be speedily lost in the sands. But soon it emerged from the desert, poured itself out into a vast lake, and thence flowed to the Eastern Ocean, through Ethiopia. Here again, bending to the left, it descended upon Egypt. *The Austral Continent* was across the ocean in the Far South, but was unknown to the men of our world from the heat of the sun. There lived the Antipods of fable.

Of the legends in the Beatus maps most, as we have said, are to be found in Isidore, but some must have been ultimately derived from far more ancient map-sources. Thus the notice of *Parthia* plainly refers to a time before the Persian revival of A.D. 226; while the dimensions of the Dead Sea and Lake of Gennesaret, in *stadia*, prove a considerable antiquity, and perhaps throw us back to a source of the time of Pliny. Again, the legends as to the Hellespont and Bosphorus correspond in substance with the descriptions of Pliny, Mela, and Solinus; and the measurements of the greater Islands (Britain, etc.), in Roman miles, seem to be a reminiscence of an Imperial itinerary.

As to the text which the Beatus maps are drawn to illustrate, *Commentaria in Apocalypsin* is the title of the work in its various manuscripts, without any mention of the author. Most catalogues call it anonymous or give impossible authorships (*e.g.* Rhabanus Maurus or Raban Maur, Apringius of Badajoz, St. Amandus, St. Victorinus of Pettau. Apringius lived at the beginning of the sixth century; he is one of Isidore's *viri illustres*, and wrote a *Commentary on the Apocalypse* which begins exactly like Beatus, *Biformem divinae legis historiam*, etc.).

But Morales (1577), Antonio, and Avezac saw that Beatus was the author from (1) the dedication to Etherius; (2) the tradition of the Valcavado manuscript; (3) the express assignment on folio 9 of Valladolid and Madrid;—*Esta obra es de Beato sobre el Apocalipsi* [cf. Avezac, *Annales des Voyages*, 1870, ii. 206-210; Gutierrez, p. 20; Antonio, i., 277, on Madrid MS., *Beato Etherii Uxamensis episcopi presbytero tribuit antiqua ejusdem inscriptio*].

The Beatus Commentary has only once been printed—at Madrid in 1770, ed. Florez. Of this one copy was used by Avezac and Delisle; another is in the British Museum; another is in Rome. [Bib. Casanatensis, D., v. 34 in C.C.; cf. vol. xxxiv., pp. 378-89, of *España sagrada;* also mentioned by Alcazar in 1604 as *Commentaria seu Catena Patrum in Apocalypsin*. It is fairly described as a Catena; for it names and quotes from, *e.g.*, Jerome, Augustine, Ambrose, Gregory, Isidore, Irenaeus, Fulgentius, and Apringius or 'Abrigius.']

The Commentary has three chief parts: (1) Introduction;

(2) Commentary Proper; (3) Appendix. (1) gives, with pictures, the history of St. John and some sacred genealogies; (3) gives extracts from Jerome, *On Daniel*, and from Isidore, *Etymol.*, vi. xiii, 14, 6; ix. 5, 6; while (2) gives the Dedication to Etherius, ascribes the Commentary mainly to the writings of Augustine and Jerome (perhaps based on Victorinus), and supplies twelve books of comment on the *Revelation*. It possesses numerous miniatures in West-Gothic and semi-Byzantine style, *e.g.* 102 on the 292 pages of St. Sever. Among these, in the Prologue to Lib. ii., is the MAP; which follows extracts from Isidore, *Etymol.*, books vii. and viii., especially vii. 9, 1-25, *De Apostolis;* cf. the MS. of St. Sever, folios 45, 46. Then the text proceeds : *These are the twelve Disciples of Christ, Preachers of the Faith, and Doctors of the Nations, Who, though they are all One, yet each of them received a special assignment for his preaching in the World, to wit; Peter, Rome; Andrew, Achaia* [etc., as before, p. 551]. *These are the twelve hours of the day, which are illuminated by Christ, the Sun; these are the twelve Gates of the Heavenly Jerusalem; . . . these are the first Apostolical Church; . . . these are the twelve Thrones . . . this is the Seed* [of the Word] *. . . this Church we believe and hold fast. . . . And they will reap with their sickles these grains of seed throughout the fields of this world . . .* [etc., as before ; see p. 550-1].

The *Divisio Apostolorum* given by Beatus is antique in character, differs from some later arrangements, and preserves the tradition as in Isidore, *De ortu et obitu Patrum ;* and in Freculf, *Cronica*, ii. 110 (Edn. of 1589); Augustine, *On Psalm* 86 ; Pope Gregory I., *Evangelical Homilies*, 17.

As to the *date :*—On Apoc., viii. Beatus speaks of the Sixth Age of the World, ending in 838 of the Spanish Era (= A.D. 800). He also gives the years from Christ to the *Present Era*, a computation which unfortunately varies in the MSS.

The original text of this *Present Era* seems to have been 814 of the Spanish reckoning [= A.D. 776] ; this is the reading of the two manuscripts, St. Sever and St. Millan ii. ['S. M. de la Cogolla']. The other manuscript readings, viz. 784, 785, 786, are probably wrong; for these were the years of Beatus' controversy with Elipandus, and in the *Commentary* there is no reference to the Adoptionist controversy. On the other hand, in Beatus' controversial writings there are many references to the *Commentary*. The Dedication to Etherius also appears to belong to the time while Etherius was still an Abbot and not a Bishop.

Probably Beatus compiled two chief copies of his Commentary ; one for himself and his monastery, one for Etherius. In these two are the originals of the Osma and Valcavado stems.

Of the Commentary there are fourteen existing manuscripts (ten with map). Sixteen other manuscripts are recorded, but now lost; two of these are perhaps still recoverable.

II.—The so-called 'Jerome' maps [see pp. 566-8 of this volume] are drawn and written in a uniform hand, with initial letters in pure Romanesque character. The designs cover the whole leaf without margin. A number of small, half-obliterated marks show that the scribe had tried his hand upon several names on the same sheet which afterwards bore the fair copy. We may infer that the more general map of the East, covering most of the *Bible Lands*, was but a part of a larger scheme, from the following, among other, points. In the middle of this design are faint traces of red text, and in several places there are relics still discernible of longer legends (*e.g.* one of eight lines under Constantinople); while immediately to the left of the Ganges is a picture of a house or temple, probably representing a city. Now, if this map were solely a mediaeval product, and originally the work of the scribe of the twelfth-century manuscript, which once belonged to the Monastery of St. Martin, at Tournai, it would be very difficult to explain these; and still more to explain another small but curious point. For the copyist of the map of Palestine has transcribed the figure LX in the neighbourhood of the Nile; this number is quite meaningless as it stands; but it seems likely that it is a page-mark of an older manuscript which was being mechanically copied with the common mediaeval absence of critical intelligence. This paging does not agree with that of the manuscript we possess, where the map occurs at folio 64, verso, and no leaf has been inserted. It would appear, therefore, that in the earlier copy, Map No. 2 stood on the right hand, or on folio 60, recto; Map No. 1 on the left, at folio 59, verso.

St. Jerome himself, in his preface to the *Liber de Locis*, declares how he translated in due order Eusebius' description of the Land of Judaea and of the City and Temple of Jerusalem, together with his Enumeration of the cities, mountains, villages, rivers, and other places mentioned in Scripture. The Church History of Eusebius comes down to A.D. 324, and his geographical works, which were 'published' later, were probably not composed before this year. The *Onomasticon* referred to by Jerome is dedicated to Paulinus, Patriarch of Antioch, who died in 324 or 325. Along with this went two other geographical writings, a *Book on the Names of Races*, and a *Description of Judaea*.

These writings of Eusebius perhaps contained several map-sketches omitted by later copyists. At the conclusion of the

Liber de Locis there follows in Migne's edition an explanation of the *Geographica Tabula Palaestinae; . . . in hac exstruenda Mappa supplere debui . . .* an explanation which is unintelligible as it stands, except as implying that other manuscripts contained maps at this place.

The connection between the 'Jerome' maps and Bede's writings appears especially from the English scholar's placing of various islands; from his description of Cilicia, and other parts of Asia Minor, such as Bithynia; from his interchanging of Neapolis in Macedonia with Neapolis in Caria; from his fixing of Lydda on the coast of the Great Sea or Mediterranean; and from his account of the threefold Ethiopia. We may compare in detail the agreements, *e.g.* of Cos, Cnidus, and Mitylene, the *Insulae contra Asiam*, and *Samothracia in Carpathico Sinu;* of Cilicia as a *Provincia Asiae quam Cygnus amnis intersecat, et Mons Amanus, cujus meminit Salomon, a Coelesyria separat;* of Bithynia, bordered by the stream Jera or Hiera; and of the Island of Chios lying over against Bithynia. The note upon the River Cygnus Isidore has also (xiv. 3, 45 of *Etymologies*); but the agreement with the map of 'Jerome,' including the form *Cygnus* for *Cydnus*, can hardly be accidental. The same is true of Bithynia. As to the *Neapolis* mistake, cf. also Pliny, *Hist. Nat.*, iv. 42; v. 107; and *Acts* xvi. 11. On Ethiopia, see Isidore, *Etymol.*, ix. 2, 128. The last-named country Bede and 'Jerome' agree in defining as divided into the lands of (1) *East India*, (2) *Tripolis and the Garamantes*, and (3) the *Hesperian West;* it starts, they both declare, from the River Indus.

As to the origin of the 'Jerome' Palestine map, this is fairly well established by its agreement with the text of the *Onomasticon*, and by its internal indications of date.

On the larger design Crete is coupled with the Cyclades as the *Seventh Province of the Greeks*, which was apparently an arrangement of the fourth century, and perhaps of Constantine the Great. Again, Emmaus, identified with Nicopolis, exactly agrees with a definition of another and unquestionably original passage of Jerome's. Lastly, the division of the Roman Provinces of *Lycaonia, Lycia, Dardania, Commagene, Isauria, Mesopotamia Syriae*, and *Syria Sobal*, as shown on these maps, agrees well with the known position of affairs at the end of the fourth century. Thus *Lycaonia* was made a Province shortly before A.D. 373; *Lycia*, between 313 and 325; *Dardania* between 385 and 400; *Commagene* in 297, under Diocletian; *Isauria* in the fourth century; *Syria Sobal* shortly before 400. The identification of these names in every case with Provinces is perhaps unwise. For, *e.g. Lycaonia* may be merely a reference to *Acts* xiv. 6, and *Mesopotamia Syriae* to the story of Jacob and Laban; while *Syria Sobal*, or *Syria Salutaris*, recalls *Judith* iii. 1. The Province of Isauria is noticed by Laterculus Veronensis, Ammianus Marcellinus, and the *Notitia Dignitatum*.

On the 'Jerome' maps the name-forms are not exclusively those of the *Onomasticon*; for besides the similarities with other works of Jerome there are credible resemblances with St. Paula and St. Silvia of Aquitaine; while these designs also show certain correspondences with Ptolemy, the Peutinger Table, Orosius, Mela, the *Alexandreis*, or fabulous *History of Alexander*, Isidore, Pliny, and Solinus. The 'Jerome' draughtsman was perhaps guided in his choice of extracts from the *Natural History* by a map-sketch of Plinian geography, similar to that which Solinus possibly used.

We may compare the present designs with Peutinger for, *e.g.* various towns of Greece and the *Bosforani*; with Ptolemy for *Oxus oppidum*; with Orosius for the *Chuni* (Huns) and the *Bactriani*; with Mela for the *Euri*; with the *Alexandreis* for the *Arae* and *Columnae Alexandri*, *Pori Regnum*, and *Alexandria Ultima*, with the oracle, etc.

On the larger 'Jerome' map there are traces of the old Province boundaries, *e.g.* of Isauria (whose *Metropolis* is given) as defined in the later fourth century A.D. We may notice also the following peculiarities in the present designs :—(1) *Gog* and *Magog* are of course identified with the Scythians, but not (as Ambrose thought) with the Goths. This view is controverted by Jerome (*Hebr. Quaest., in Genesim*, 318; 10, 2), and in the map by the inscription *Goti qui et Gite* (*Getae*). (2) *Barbaries* here specially refers (Map 1st) to the Slavs South of Danube; cf. *Colossians* iii. 11. (3) *Theodosia* here is, of course, Theodosia or Kaffa in the Crimea. (4) *Acheron* on the Caspian agrees with Æthicus, ch. 59 ; it is mentioned in the *Argonautica*, and was often placed in Asia Minor. (5) *Clitheron* is a river probably identical with the 'Gesclithron' ($\gamma\hat{\eta}s$ $\kappa\lambda\epsilon\hat{\iota}\theta\rho o\nu$) of Pliny (vii. 10) and Solinus (15, 20) who place it among the Arimaspians, connected with the old story, found in so many forms, of an opening into the under world. K. Miller suggests that its original was the petroleum springs at Baku. (6) *Albania* here forms the land on the northern limit of the *Oikoumenê*, as often in classical geography. (7) The River *Alanus*, close by the Maeotid Marsh, is not in any writer earlier than Jerome, but the people of the Alani are put in the same part by Ammianus Marcellinus and Dionysius Periegetes; cf. Peutinger Table, ix. 3.

Nothing prevents us from assuming the authorship of a draughtsman of St. Jerome's time for the two works which we have attempted to describe. No mediaeval writer could have compiled designs of so purely classical a nature, except by simple transcription of an older scheme; on the other hand, it is clearly an educated Christian who has here added to the material of the Roman road-maps a certain number of Scriptural names and allusions.

III.—As to the 'Cotton' or 'Anglo-Saxon' map, we may notice that *Jerusalem* is well away from the central part of this square design; that *Taprobane* occupies the place usually given to the *Terrestrial Paradise;* that of the *Nile's* three sections, the uppermost is called *Dora;* that here we find *Gog et Magog* to the West of the Caspian, the *Turchi* adjoining; and that the draughtsman places the *Bulgari* between the Danube and the Arctic Ocean. See pp. 559-563 of this vol.

The manuscript which contains the map [Cotton MSS., Tib. B. v.] is made up of various pieces, collected by Robert Cotton in 1598. The map is on folio 58, and is immediately followed by a copy of Priscian's Latin version of the *Periegesis* of Dionysius,—*De situ terrae Prisciani Grammatici, quem de priscorum dictis excerpsit Ormistarum,*—written in the same hand as appears in the map. The *Periegesis* is followed on folios 75-77, by an Anglo-Saxon manuscript, of far older date than any other in this volume; and, from a Donation Formula given on folio 88, it is fairly conjectured that the manuscript (at any rate of part of this volume) in the time of Henry II. belonged to Battle Abbey. But neither the *Periegesis* nor any other manuscript, as far as folio 74, is of later date than Archbishop Sigeric, whose Itinerary we must couple with the *Cottoniana*. Sigeric's predecessor died in 989, Sigeric himself in 994. On folio 20 of this volume he is referred to as 'Our Archbishop,' and this entry (which implies that he was still alive) is in the same handwriting as the *Periegesis*.

It is doubtful whether the handwriting apparent on the map is of the same date and authorship as the folios from No. 88 onwards, which are perhaps later than 1000.

The handwriting is small and difficult, with peculiar formations of various letters; *e.g. C.* written like *R., O.* like *A., R.* like *P.* and *A.* Cortembert's reproduction, made in 1830 for the Bibliothèque Nationale in Paris, is the best edition before K. Miller's, and gives the most accurate readings; Bevan and Phillot, *Mediaeval Geography,* 1874, p. xxxiv., come in second-best in text readings, but commit more than forty mistakes;

while Santarem, *Essai sur l'Histoire de la Cosmographie* . . ., ii. 47-76, is terribly disappointing. As in the case of some of the Beatus maps, certain lines appearing through the parchment from the text on the other side have given rise to the mistaken belief that the copyist used some kind of rude graduation for his design. Or if not this, these lines were perhaps, some have thought, intended to help the scribe in the more exact following of his original, even if they were not in any sense a survival of the graduation-net of Eratosthenes. But all this is imaginary [cf. Santarem, *Recherches*, and *Atlas*, Plate i.; *Penny Magazine*, vi. 278-280, 1837; Brit. Mus. MS., *Catalogue*, 11, 1844; Cortembert, *Trois des plus anc. monuments de Géographie* in *Bulletin* of Paris Geographical Society, vi. 14, 337, etc., 1877; Jomard, *Monuments de Géog.*, Plate xiii., 1862; Walleser, *Welt-tafel des Ravenaten*, i. 4, Mannheim, 1894].

A great part of the names that are not from Orosius are later additions. For the most part the Orosian names would appear to come directly, not from the Orosian text as we have it, but from the lost Orosian map. For one thing, the position of countries and places does not agree in all respects with the Orosian text. Again, the name-forms on our present map are in many ways different from the Orosian text, but often remind us of the early World-Maps closely related thereto.

The connection with Mela appears rather in the general idea of the *Oikoumené* and of various countries than in the legends or place-names in detail. Both Mela and the scribe of the *Cottoniana* conceive of the Habitable World as an oblong. They also show a likeness in the following points :—The general contour of Spain, Italy, the Gulf of Aquitaine, the north coast of Europe and Asia, the Caspian Sea with its bays and islands, the position of Britain,—just opposite a great gulf or indent of the Continent,—the Scythian Islands, the Burning Mountain in the extreme south-east of Africa, the Promontory of the Western Horn, with the twin Gulfs and the Seven Brother Mountains adjoining, the two Syrtes, the Peninsula of Tyre, and the position of the Mediterranean Islands. Both also appear greatly to exaggerate the breadth of Asia Minor. Elsewhere the *Cottoniana* agrees with Orosius in opposition to Mela, as in the north-east parts of the *Oikoumené*, in the course of the Nile, and in the position of the Pillars of Hercules.

Some of the Cotton names not to be found in Orosius occur in Jerome ;—such are Gog and Magog, the Ark of Noah, Mount Sinai, Nineveh, the Mountains of Armenia, Galilee, Jericho, Ephesus, and other 'Bible places.' But here, of course, it is not necessary to assume a direct use of the Father's writings.

610   APPENDIX ON MAPS

The Twelve Tribes of Israel, so clearly marked in our present design, do not occur in Jerome's Palestine map, as we have it; but they were possibly transcribed (in the first instance) from Eusebius, or directly from the Old Testament, by the Doctor of the Church of Bethlehem, either in an earlier form of the Palestine map than that which we possess, or in another special map, now lost.

With *Isidore* there are several, not very peculiar, coincidences; among these are the Griphi, Gryphons, or Griffins, Pentapolis in Palestine (which is neither Biblical nor Eusebian), the Seven Mountains (also in Mela), Tingis or Tangier, Zeugis, the City and Promontory of the Hesperides, Byzacena, and the Dog-headed Folk. Perhaps the scribe drew from the lost Isidorian map.

Some of the most modern names in the *Cottoniana* occur first among the writers of the next age, in *Adam of Bremen* (c. A.D. 1067). Such are the Turks, the Huns, the Slavs, the Scrito-Finns [*Scridefinnas*, located in *Island*], and the countries of Iceland (*Island*), Sleswig, and Norway. *Turci* are also in Æthicus of Istria, and re-appear in the Ebstorf map, like *Island*, which may also be found in Henry of Mainz. The *Bulgari*, perhaps, come to the Cotton map from the Ravennese Geographer, and are found again in the Hereford map. *Hunnorum gens* perhaps refer to the Magyars of our Hungary.

With *Julius Honorius* there is very slight connection. The name of the River Hypanis (which is also in Mela) ceases for many centuries with Honorius, and is only revived on this map. Also the cities of Salerno, Verona, and Tarsus offer a certain analogy. A similar likeness with the Anonymous Ravennese may be found.

The *Beatus* maps contain nearly all the Orosian names, and most of the non-Orosian, which are to be found on the *Cottoniana*. A similar resemblance may be traced between our present design and the later maps of Lambert of St. Omer, Ebstorf, the Psalter, and Henry of Mainz.

The *Biblical* connections, traceable especially in the centre of the Cotton map, include (directly or indirectly) the Twelve Tribes, Jerusalem, Bethlehem, Babylon, Tarsus, Caesarea Philippi, and the Ark of Noah (among the Vignettes). Most of the Biblical names of Hereford, Lambert, Henry of Mainz, the Psalter, and Ebstorf are also to be found on the Cotton map, and are perhaps in many cases borrowed directly by these later map-designs from the earlier 'Anglo-Saxon' work [cf. *Decusa civitas*, near the Euxine, in Cotton, Henry of Mainz, and Hereford]. For earlier parallels, cf. *Evilath* in the Ravennese Geographer; *Montes*

Aurei in Æthicus and St. Isidore; and the *Desertum ubi Filii Israel erraverunt* xl. *annis* in the Peutinger Table.

Lastly, there are several names and several features which show striking independence of any other known map-authority of the earlier Middle Age. Among these are the five names in Britain—*Camri* or Cambria, and *Marinus portus* in the north-west; Kent, London, and Winchester on the southern shore (Kent being in the position of Devon); and Arama or Armagh in Ireland;—the South Bretons (*Sud-Bryttas*) in Northern Gaul, the Golden Mountain of the Far East, and the *Boreani* and 'abundant lions' of the north-east of Asia.

The bulbous projection of land on the coast north of Jerusalem is perhaps meant for *Carmel*. Some idea (though exaggerated) of the *Syrtes* on the North African coast is evidently possessed by the draughtsman.

Of purely inland geography unconnected with the coast there is not much in the Europe of this map; cf. the *Huns, Dalmatia, Dardania, Histria,* and *Tracia*, all circling round *Pannonia*. What is now European Russia is contracted to a mere neck of land.

In Asia there is much more inland geography, chiefly connected with the Twelve Tribes and Biblical history. In *Africa* the lakes east and west of the *Lacus Salinarum* near the north coast are noteworthy; like Brigantia (of lighthouse fame) in the north-west of the Spanish peninsula. The Caspian, opening into the Northern Ocean, is of unusual size. *Mons Clinax(-max)* in the middle of the South African coast, is perhaps a misty reference to the 'Chariot of the Gods' as described by Hanno and the Greek and Latin geographers who copied him; while the two small unnamed isles west of Mount Atlas are probably intended for the *Insulae Fortunatae*.

England, Scotland, Ireland, Denmark [*Neronorweci* or *Neronorroen*], and *France* are better drawn on the Cotton than on any other early mediaeval map; there is also no small comparative merit in the land of the *Scrito-Finns* and *Island*, representing our Scandinavia, and in *Sicily*, whose three angles appear; the north coast of *Asia Minor* is likewise good. On the other hand, the West Mediterranean is poor and very contracted; the islands, especially Sardinia (?), are unsatisfactory. France is so squeezed between Spain and Italy that its south coast almost disappears, except for the Gulf of Lyons, which is fairly well indicated. In *Italy*, Ravenna on the south-west coast is curious; Verona appears in the position of Aquileia. In *Greece* the name *Macedonia* seems written over the *Morea;* Athens and Attica are widely separated.

The early date usually, and, as far as we can judge, rightly, assigned to the Cotton map, has been questioned on the ground of its correctness of form and multiplicity of detail. It is too good, some have said, for the tenth century. But this objection, as far as content goes, falls to the ground when we compare it with the map of St. Sever and with the written data still remaining as the basis of the lost map of Pomponius Mela.

III. (A).—With the Cotton Map may be associated, perhaps in a very intimate relation, the Itinerary of Archbishop Sigeric of Canterbury of about A.D. 992-994. [This occurs on folio 22 verso of the same Cotton manuscript.] The only part of this which is given in any detail is the return journey from Rome to England; and here we have a fairly adequate list of the pilgrims' stopping-places [*sub - mansiones*], which must refer to the night- or sleeping-stations. Hence, perhaps, the omission of several towns, through which Sigeric's route must have led, such as Langres. The day's march varies between seven and forty kilometres; the average gives us a little over eighteen, with four hours' travelling as the mean. In the whole reckoning of eighty days rainy ones are counted, during which very slow progress was made; so for fine weather we may perhaps raise the average to about thirty kilometres *per diem*.

Sigeric's ordinary mode of progress was probably either on horseback or on foot; it is not likely that he could have made a carriage-journey, at least over part of the route. From the omission of many of the larger towns in his 'routier,' we may infer that he was travelling incognito, and did not avail himself of his position as Archbishop of Canterbury to bespeak the hospitality of the great city prelates and religious houses in France or Italy.

All Sigeric's reckonings are made with a certain 'consciousness of his goal,' and a good general appreciation of the route he was following, of its distance as a whole, and of its separate sections in their parts. The road from Canterbury to Rome must have been fairly well known; thus Dunstan, Sigeric's second predecessor in the See of Canterbury, had been to the Apostolic See for his *pallium* in 960.

The eighty days' journey of Archbishop Sigeric is divided into eight main sections, viz. from Rome to Forcassi; from St. Valentine (by Siena and Lucca) to Camajore; from Luna by Parma to San Donnino; from Fiorenzola to Piacenza; from Piacenza to Ivrea and over the Great St. Bernard to Lausanne; from Lausanne to Besançon; from Besançon to Rheims; and from Rheims to the Sea. In the first section, to Forcassi, the pilgrim goes along the *Via Cassia;* on the second section, along the *Via Clodia*. In this last piece of the road our Itinerary seems to supply us with really valuable information, helping us to supplement or correct the gaps or misreadings of the Antonine Itinerary. The direct road from Lucca by way of Siena to La Storta, though much frequented in the Middle Ages, 'is not recorded in any ancient writing that we possess'; even the Peutinger Table gives us only a short piece from Materno to Foro Clodio [cf. the *Itinéraire Brugeois;* Lelewel, *Epilogue,* 296-298. Matt. Paris gives the same route in 1253;—from Siena by Aquapendente, Bolsena, and Cesano to Rome; and from Bologna by Siena to Rome]. Whereas later travellers from Rome usually journeyed on the *Via Cassia* as far as San Lorenzo, and first struck the *Via Clodia* at Aquapendente. Sigeric, during this part of his route, seems to have kept to the Clodian road all the way. This road appears to have led the wayfarer into the neighbourhood of Luna, where it joined the *Via Litoralis* near a certain Forum Clodii, which probably marked the north end of the *Via Clodia*. [At the south end of the *Via Clodia* there was another Forum Clodii.] The road between Luna and San Donnino, noticed by Sigeric, was in great use during the Middle Ages at various times, and we find references to it in the *Itinéraire Brugeois,* and in Matthew Paris [cf. Matthew's 'Munt Bardun' and 'Punt Tremble'; and the *Itin. Brug.,* xviii.;— Lucca—'Pont Tremoli'—Borgo San Donnino—Piacenza. Pont Tremoli or Punt Tremble is at Parma]. Between San Donnino and Piacenza the traveller goes along the *Via Emilia;* and between Piacenza and Lausanne his route corresponds with one that is marked upon the Peutinger Table. Beyond Lausanne Sigeric's way leads through Arbe, Pontarlier, and Nods, to Besançon, not along the present high road through Ornans, but somewhat north of this. Between Besançon and Rheims the Archbishop appears to keep closely to an old Roman military road through Cussey over the River Oignon, through Seveux over the Saône, and straight through Grenant to Langres and Ormancey, where he rested after a heavy day's march of thirty-two kilometres. At Bar-sur-Aube Sigeric appears to have met with the present country road to Brienne, where again he passes on to a Roman highway leading through Châlons to Rheims. The whole of this section is quite in agreement with one of the routes marked on the Peutinger Table, though the stopping-places are usually different. Lastly, the concluding portion of the Itinerary, between Rheims and the sea, offers certain difficulties in identification, but it would appear to lead through Corbèny and near Peronne, by Therouenne (the ancient 'Tervanna') to Guisnes and some point on the coast which is named Sumeran, and which must clearly be sought somewhere between Calais and Cape Gris Nez.

In all this the route probably runs, for the most part, along Roman roads. But the ancient Itineraries only assist us between Therouenne and Arras [Sigeric's 'Aderats'], giving no direct route from Arras to St. Quentin and Rheims.

Sigeric's journey shows us that in the tenth century men could still use large sections of the old Roman roads which in later time fell into almost complete disuse. It also gives us various finger-posts for the course of the ancient travel and trade-routes, even where ancient records leave us in some doubt. More generally, it helps to illustrate the detailed geographical knowledge of a very ignorant period along one of the great highways connecting the 'Mother and Mistress' of all Western churches with one of its most famous daughters.

IV.—For the map of Henry of Mainz;—cf. MS., lxvi. (earlier D., xii. 1) in the C.C.C. Library, Cambridge, which contains not only the *Imago Mundi* of Honorius of Autun, but also various geographical sections of Pliny, a text of Rubruquis, an *Itinerarium ad Paradisum terrestrem*, an *Epistola Presbyteri Johannis*, and many other different writings, twenty-four in all. Nasmith's Catalogue of 1787 dates the *Imago* manuscript in this volume to the thirteenth century; Bevan and Phillot think it is of about 1180; it once belonged to *Sta. Maria de Salleia*, otherwise the Monastery of Sawley in Craven, Yorks, not the House of Salem in Swabia, as Walleser suggests, *Welt-tafel des Ravennaten*, Mannheim, 1894, p. 3. On Henry's career;—cf. the Mainz *Acta* of 13th December, 1111, and see Böhmer, *Regesta Archiep. Mogunt.*, Innsbruck, 1877, i. 244. An Archdeacon Heinrich of Mainz is mentioned in 1104, and Böhmer thinks this is the same as the canon and chaplain, and also the same as the Archbishop and the draughtsman of our map. It is perhaps doubtful whether we have not here four different persons—(1) The Archbishop; (2) the Archdeacon; (3) the Canon; (4) the Map-maker. K. Miller would identify 1, 3, and 4. For some time the Archbishop was the tutor of Henry, son of the Emperor, Conrad III., who died in 1150. This prelate was buried at Eimbeck. He was of good family; the Counts of Wartberg were his cousins; and he himself is once referred to as Henry 'of Nassau.'

The *Imago Mundi* which this map illustrates has been sometimes wrongly ascribed to Anselm of Canterbury. The copy made by Henry is older than any we possess of Honorius' work; the earliest known is of 1123 [cf. Brit. Mus., Cotton MSS., Cleopatra, B. iv.]. It is pretty certain that Honorius spent the last twenty years of his life (1120 to 1140) in a German Monastery; some think in a Religious House on the Austrian March. But most of his scientific works, and the *Imago* among them, were composed while he was still a teacher at Autun, in the beginning of the twelfth, or even at the end of the eleventh, century.

The map which Henry drew to illustrate his transcript of the *Imago* of Honorius has remained in comparative obscurity till recent times; the text has been badly edited; and the connection of this work with others has been very dimly perceived. [Cf. Stanley, *Catalogue of the Library of Archbishop Parker*, 1722; Nasmith, *Catalogue* of the *C.C.C. Library*, as above; Gough, *British Topography*, 1780, i., 60; ibid, *Essay on the Rise and Progress of Geography in Great Britain*, p. 6; T. Wright, *St. Patrick's Purgatory*, 1894, p. 93, etc.; Santarem, *Essai sur . . . Cosmographie*, ii. 242-244; iii. 463-498, and *Atlas;* Bevan and Phillot, *Mediaeval Geography*, 1874, pp. xxxvi.-xxxix. Santarem's reconstruction of Mediaeval Map Legends is notoriously faulty, and there is no exception here to the general rule. But even Bevan and Phillot have more than thirty misreadings.] In the four corners (instead of winds, or wind-blowers) are four angels, whom Santarem regards as pointing to Gog-Magog Land and to Paradise, and blocking the way through the Straits of Gibraltar—perhaps too elaborate an explanation. However, the angel in the left-hand top corner is certainly pointing to 'Gog and Magog, an unclean race.' All these angels have golden halos, and are variously coloured in green and red; while the figure on the upper left hand carries something which has been variously interpreted as a cube or die, a box, or a church. His clothes are green, except for an upper cloak, which is red like the wings. Exactly the opposite arrangement of colour is adopted with the angel that fronts him on the right.

All the seas, save the Persian and Arabian Gulfs, are light green; the 'Red Seas,' the mountains, and certain of the more important names, are rubricated. Excluding the 'Jerome' maps, this is perhaps the richest in content and the best preserved among all twelfth-century examples of cartography.

As to the want of exact correspondence between Henry's map and the *Imago*, we may notice that the former's selection of European cities is not represented in the latter, and that the interchange of 'Thile' and 'Tilos' which we find on the *Imago* is not in the map.

Santarem has well pointed out that the Hereford scheme was a working up of Henry's design. As to this, we may compare the varied outline of the coast, on the north of Europe and Asia, and the position and outlines of the Baltic Sea, of the Scandinavian peninsula, of the Caspian, and of the lands of the Gog-Magogs, the Hyperboreans, and the Dog-headed Folk. The coast line near

Paradise may also be compared, and the islands adjoining this coast, such as 'Taraconta'; likewise the position and outline of the Persian Gulf, the Red Sea, and the island of Taprobane. The peninsular form of Italy is more developed on Henry of Mainz than in the Hereford, but the delineation is not unlike.

The Nile of Henry of Mainz resembles the Cotton or Anglo-Saxon map, as well as the Hereford, in adopting the theory of three sections, viz. (*a*) a short one springing from a Lake (*Nilidis Lacus*) near the Atlantic; (*b*) a long stretch from a larger Lake (*Lacus Maximus*) running parallel to the Southern Ocean, to a second point of submergence (*hic mergitur*); (*c*) the Nile of Egypt, springing from a *Fons Fialus* near the Red Sea, penetrating the *Montes Nibiae* [*i.e.* Nubiae], and thence flowing in a south-west direction to the Mediterranean. Both Hereford and Henry of Mainz also introduce a Lake and River of *Triton* flowing into the Middle Nile (in a south-west direction) from the *Altars of the Phileni*, which are wrongly placed, far from the Mediterranean.

In Central Africa Orosius is probably the source of Henry of Mainz' (and Hereford's) *Euzareae Montes*. To the east of these are the *Montes Ethiopiae*, Mount Atlas being near the Atlantic, and *Mons Hesperus* further south. Henry of Mainz also agrees with Hereford in the Mountains of Syria, East Asia, and Bactria, and in the Caspian Gates.

The rivers of Asia also agree closely;—*e.g.* the Hydaspes, Acesines, and Hypanis (drawn as independent of the Indus); the Ganges (on the other side of Paradise, towards the north, flowing due east); the Acheron and Oxus (flowing into the Caspian); the two unnamed rivers on the west side of the Caspian; the Pactolus (flowing into the Euxine); and the Cobar (Chebar (?), flowing into the upper Euphrates).

Among other coincidences are, (*a*) in Asia;—the wall shutting off the peninsula of the Gog-Magogs; and the description of the same people as unclean, of the Hyperboreans as untroubled by disease and discord, of the Gryphons, Griffons, or Griffins as most wicked, and of the Dog-headed Folk as adjoining the Arctic Ocean. Also the notices of Amazonia, the Golden Mountains (cf. the 'Anglo-Saxon' map), the Port of Cotonare, Mount Sephar on the Indian Ocean, and the Tower of Enos just outside Paradise. (*b*) In Africa;—the Burning Mountain and the Seven Mountains (here also cf. the 'Anglo-Saxon' map); the Troglodytes near the Middle Nile; the River Lethon near Cyrene; St. Augustine's Hippo; the Basilisk between Triton and the Nile; the horseshoe-formed Temple of Jupiter-Ammon; the Monasteries of St. Antony, near the end of the Middle Nile; and the Pepper Wood near the Red Sea; together with other oddities which are common in mediaeval cartography, *e.g.* the Pyramids as Barns, etc. (*c*) In Europe;—the Church of Santiago at Compostella, and near it a Pharos (of Brigantia ?); the Danus, tributary of the Ebro, unnamed in ancient geography; the boundary of the Danes and Saxons; and the heart-shaped town of Cardia near Constantinople.

As to *Islands*, Taraconta, Rapharrica, and Abalcia, on the north coast of Asia, are from Æthicus; Ganzmir (for Scanza or Scandinavia) is a remarkable misreading, also found in Hereford. Hister, Asia Minor, Galilea, Sinus Persicus, and some other names, wanting on Hereford, but supplied by Henry of Mainz, are probably from the common original.

Various peculiarities of nomenclature, *e.g. Mene, Island, Jabok,* etc., are also common to both works; but of course the Hereford map is far larger, and contains much more, especially in relation to classical material. The 229 legends of the one are overshadowed in the 1021 of the other. In the same way, among the other relatives of Henry's map, Ebstorf (a work on the scale of Hereford and a most close parallel to the latter) dwarfs its elder cousin of Mainz with 1224 legends; 'Jerome' supplies 407; while the little Psalter map, Henry's younger brother, in spite of all its crowding, can only give us 145. We may notice that, among other works of similar nature, the Cotton map gives us 146 legends; Lambert of St. Omer, 180; Matthew Paris' world-map, 81; the Beatus group, 477; while the vast scope of the Peutinger Table offers 3400 inscriptions.

It is plain from the great number of nameless rivers, mountains, and cities in the Mainz example, that the work may well have been taken from a larger original. There is another proof of the same in the eight half-circles which occur (apparently without reason) along the oval margin of Henry's ocean; from other works we may recognise these as representing the places of the eight intermediate winds.

In the draughtsmanship of Asia Minor, the Gulf of Issus, and the Black Sea, the most striking analogies may be found between Henry and 'Jerome'; and from a study of these particulars we may feel practically certain that some correspondence may be assumed. Henry's agreement with 'Jerome' is only, of course, partial, even in the Eastern World; but it is far closer than the likeness between 'Jerome' and the Hereford, Psalter, or Ebstorf maps; and we may believe it to be a true and conscious relationship.

V.—The whole design of the 'Psalter' map [Brit. Mus. Add. MSS., 6806] is, like the Hereford, a highly developed but scientifically debased example of semi-mythical Geography—an elaborate exposition of strictly mediaeval habits of thought, applied to Geography. It shows us 'World Knowledge' removed as far as possible from the comparative science of the Ancient Imperial World, and as yet quite untouched by the new light of the later Middle Age. In its debasement it may be to some extent compared (like Hereford and Ebstorf) with the map of Cosmas, 'the Indian traveller'; but its perversions are of course very different, both in kind and degree. They are in some ways more elaborate, and so more mischievous; on the other hand, they imply, taken as a whole, a more detailed knowledge of many parts, and a higher artistic sense.

At the top of the world-circle is the Saviour with uplifted hands; in His left He holds the globe of earth; the latter has the well-known T-O formation of the Continents sketched upon its surface. On both sides of the Saviour stand angels swinging censers; below are two dragons facing one another. On the reverse of the page the dragons are again sketched below the earth-circle, and crushed beneath the feet of the Saviour, whose form thus serves as a background and support, as in the Ebstorf example and in so many other pictures, to the circuit of the earth. The border that surrounds the map is almost identical in design with that of Hereford; but the 'Psalter' border is executed in pure Romanesque, the Hereford in Gothic. This helps us to date the former about fifty years earlier than the latter, *i.e.* c. 1250 A.D.

The ocean appears as a watery zone, of equal breadth in every part, encircling the world. The various winds, each represented by a head, as in the Hereford map and on the Beatus of 1250 ('Paris ii.') are designed in suitable places along the outer rim of ocean. [This sort of plan is prominent in later works, like the great *mappe-monde* of Ranulf Higden.] In the titles of these, the draughtsman of the Psalter map is unusually and severely classical, giving us the famous old names of *Aquilo* and *Septentrio* for the North, *Zephyrus* for the West, *Auster* or *Nothus* for the South, and *Eurus* or *Euro-Nothus* for the East and South-East. The term *Vulturnus*, usually applied by classical writers to the South-East wind, is assigned rather to the North-North-East by the Psalter draughtsman. The Mediterranean, Black Sea, Propontis, Caspian, and Red Sea are all represented; the waters of the Levant show unusual exaggeration; the Euxine is brought (as often elsewhere) very close to the Northern Ocean. The coast from the delta of the Nile round to Caesarea is monstrously distorted, almost resembling the shore of a lake. The Caspian appears as a narrow indent of the Northern Sea, divided in two by a long peninsula (in the extreme north-east of Asia), and encircled by the greatest mountain-wall in the world, pierced apparently at one point by the Gates of Alexander.

Paradise, in the Far East, is conceived in a somewhat exceptional manner. The sun pours out of its mouth the flood of waters which flows through the Garden of Eden, and supplies the *five* sacred rivers; for the draughtsman has entered both Ganges and Phison in this list. The heads of Adam and Eve appear within the enclosure, which seems to be marked off with lofty and symmetrical mountains. The Tree of the Temptation is roughly drawn between the two faces. [Bevan and Phillot, *Mediaeval Geography*, xlii., suggest the *Arbre Sec*, which they make identical with the Tree of the Knowledge of Good and Evil; and Yule, *Marco Polo*, ii. 397, refers us to legendary language about the Dry Tree which would perhaps support such an identification;—'in the midst of Paradise was a fountain, whence flowed four rivers, and over the fountain a great Tree bare of bark and leaves.'] The trees of the Sun and Moon are here separately indicated, close to Paradise on the south; while the Tigris flows direct from Paradise to the Indian Ocean, and the Euphrates (or rather one of two rivers so named) enters a mountain chain west of Paradise, named *Orcatoten*, and thence flows to the Persian Gulf. Of the Nile only the Egyptian portion is given. The *Arae Liberi et Colimae Herculi*[s] occur near the Indus, but the *Arae Alexandri* are near the border of Europe; Albania, in North-East Asia, recalls the 'Anglo-Saxon' or 'Cotton' map; *Cyropolis*, near the Caspian, is perhaps for *Cyreschata* on the Jaxartes, famous for Alexander's siege; *Sclaveni occidentales*, near the Black Sea, are suggestive of much more modern times, like the Island of *Norvegia*. The Arabian and Persian Gulfs appear to be melted into one by the draughtsman of the Psalter map, and in the same great indent he has put the ocean, off the coast of India, filled with large islands. The Ganges has an utterly false direction, flowing from the northern mountains, not into the sea, but to Paradise, like one of the *two* Euphrates rivers, here delineated. North-West Africa is marked off (like the north-east of Asia) by a belt, which was perhaps intended for mountains, as in the other case, but remains as a mere linear mark with the legend, *Sandy and Desert Land*.

Among the monsters of South Africa are Dog-headed Folk and people with heads in various stages of aggressiveness, having either descended between their shoulders or else absorbed the entire trunk of the body. Besides these there are Cannibals, a race with six fingers, Troglodytes, Serpent-eaters, Skiapodes, and a nation that obtained shadow from the hugeness not of their foot but of their lip ; tribes also without tongues, without ears, or without noses; others who, having only a little hole for mouths, were forced to suck their food through a reed ; 'Maritime Æthiops' with four eyes ; and beings who never walked, but crawled on hands and feet. [These races, fourteen in all, come mostly from Solinus ; many of them occur also on Ebstorf, on Hereford, or on both.] The Ark of Noah appears very clearly on a mountain of Armenia, and a large fish swims in the middle of the Sea of Galilee, perhaps as a reminiscence of New Testament history. The Barns of Joseph, close to Babylon of Egypt, show us that our artist has heard of the Pyramids. The most famous cities of the ancient world, and the most famous sites of the Bible Story, are nearly all represented ; while the immense and symmetrical Jerusalem, in the very middle of the world, forms a perfect centre to an exact circle. [One might almost compare the ideal world of the Psalter, Hereford, and Ebstorf maps, in its general shape, with the contour of real Moscow, Jerusalem answering to the Kremlin.]

The closest relation of the Psalter map is the Ebstorf, which is probably junior by at least half a century ; but the former is remarkable for a number of old names which do not occur on the maps of Ebstorf or Hereford. Its delineation of the monstrous races of the south show a more antique character, and so probably a closer relationship to the common eleventh-century (?) original. That original probably contained many names and legends, attached to various indications of cities and natural features, which have only partially survived in the derivatives. In its text the Psalter map seems to be a very imperfect copy of this original, both in amount and style, though it gives us an astonishingly large mass of matter for its size. In its delineation of the World-Picture our present example perhaps reproduces its model better than in its text; the scribe was presumably better as a draughtsman than as a scholar.

The Psalter and Ebstorf have a curiously similar treatment of the Caspian Rampart (otherwise 'Alexander's Wall,' the 'Hyrcanian Mountains,' or 'Barrier of the Jews'), shutting in the Gog-Magogs and other monsters of the North ; but the Gates of Alexander are more clearly marked on our present copy than anywhere else in this Family of Maps. The two bays which run off northward from the 'Erythraean' indent of the ocean are somewhat unusual in their position and conception ; one corresponds to the upper part of the Persian Gulf, the other to the sea at the mouth of the Indus, the Gulf and Runn of Cutch, or perhaps the Gulf of Cambay. On the Psalter,

'Jerome,' Hereford, and Ebstorf maps alike, Africa stretches round very close to the neighbourhood of India; and further similarities may be observed in the unnatural abridgment of the three peninsulas of Southern Europe, both Greek, Spanish, and Italian.

With the Hereford map the textual correspondence is almost as noticeable as with Ebstorf; and the differences in cartographical form are often mere arbitrary eccentricities of the designer. We may consider this little circular plan, so minute in scale, so immense in the quantity of its details, as a sort of bridge between the types of Ebstorf, Hereford, and Mainz. At the same time, like Ebstorf and Hereford, it stands much further away from the 'Jerome' maps than does the work of Henry; but, with the 'Jerome' map of the Orient, it helps us to fill up that great gap which has been left in the Far East of the Ebstorf example. [Perhaps the trees of the Sun and Moon, as shown in the Psalter, correspond to the Pillars of Alexander and of Hercules in the original design.] Outside its own family the Psalter map has some points of agreement both with Lambert of St. Omer and with Beatus. Of modern names it gives us several in Europe, one in Africa, none in Asia. The most interesting of these are Damietta, in a wholly wrong position; the 'Ruscitae' or Russians, perhaps derived from the *Ruzzia* of Adam of Bremen, the 'Olcus' or Volga, the 'Land of the Western Slavs,' 'Ala' or Halle in Germany, and three names in Britain, viz. Scotland, Wales ('Walni'), and Cornwall. [Bevan and Phillot were the first to describe and photograph this map. See *Mediaeval Geography*, p. xli. Cf. also Anton Springer, *Die Psalter-Imitationen*, Leipzig, 1880.]

VI.—Lambert of St. Omer was a son of one Onulf, canon of that church, who died 26th January, 1077; he composed his *Liber Floridus* at different dates, as well as from different authors (cf. his own words;— '*Ego Lambertus canonicus S. Audomeri librum istum de diversorum auctorum floribus contexui*'). In his chronicle he alludes to Pope Calixtus II. (1119-1124) as then reigning. Besides the world-map and other designs noticed in the text, pp. 570-3, he gives us lists of the winds and chief towns of the world, and four astrological or astronomical schemes. [The Ghent MS., apparently an autograph of Lambert's, is noticed in 1248 as belonging to the Cathedral Library of Ghent (St. Bavon). Its map content is as follows :—(1) On folio 19, a T-O sketch (*Sphaera triplicata gentium mundi*); here the names of peoples agree with Julius Honorius, in the Æthican Recension. (2) On folio 20, a *Sphaera Minotauri* and a *Domus Daedali*. (3) On folio

24, an *Ordo Ventorum.* (4) On the same folio, a *Sphaera Macrobii de V. Zonis.* (5) On folio 28, the promise of the unexecuted *Mappa vel Or[m]esda Mundi.* (6) On folio 49, a list of great towns, especially Rome, Nineveh, Babylon, and Ecbatana. (7) On folio 88, a *Terrae Globus* to illustrate Baeda, *De Astrologia.* (8) On folios 136-139, various genealogies, with a picture of Caesar Augustus on folio 138. This is related to cartographical ideas through the taxation, survey, and official mapping of the Roman world attributed to Augustus. (9) On folio 225, another Macrobian Zone map. (10) On folios 226-228, three astronomical sketches, of Macrobian type. (11) On folio 241, the map of Europe above referred to; cf. Santarem, *Essai sur . . . Cosmographie,* ii. 154, 155; Pertz, *Archiv,* vii. 540; *Serapeum,* 1842, 145-161, etc.; 1845 (vi.), 59, etc.; also the *Messager des Sciences historiques de Belgique,* 1844, 473-506.

Besides the world-map the Paris manuscript contains (with certain differences) several of the smaller designs which we find in the Ghent copy of the Lambertian encyclopaedia. Thus we have Augustus Caesar holding a T-O world in his left hand, an astronomical sketch, and an outline figure of the 'earth-globe.'

But on the Paris map all names of seas are wanting; the Mediterranean is indistinguishable from a river; and the continents and countries lack all clear differentiation. The script, moreover, is exceedingly difficult; and Lambert's material has been so much re-arranged that it is not easy, in some cases, to find any agreement with the indications of the Ghent copy.

The Wolfenbüttel manuscript is now known as 1 Gudiana lat., of the twelfth century; Warnkönig, *Geschichte von Flandern,* thinks this manuscript is older than that of Ghent, but this cannot be maintained; cf. Bethmann, *Serapeum,* vi. 64. The 'Paris' is in Bibl. Nat., Suppl. Lat., 10 bis, and belongs to the end of the thirteenth century; it contains some beautiful miniatures (cf. Santarem, *Essai . . . sur Cosmographie,* ii. 163-172, 198; and *Atlas*). Other references to the Lambert maps, especially at Ghent, may be found in Mone, *Anzeiger für die Künde der . . . Vorzeit,* Karlsruhe, 1836, p. 36, and Tafel i.; St. Genois in *Messager,* Ghent, 1844, p. 602, pl. iii.; Santarem, *Essai sur . . . Cosmographie,* ii. 44, 82, 95, 157-160, 185, 197, and *Atlas;* Lelewel, *Atlas,* pl. 8.]

On the Lambert maps the seas and rivers are usually green, the mountains red; but each of the three copies offers peculiarities of its own. *Ghent,* though it only supplies us with Europe, gives much more detail in this part of the world. *Wolfenbüttel* alone gives Philistia, Palestine, Bactria, and the Mountains of Taurus and Caucasus. *Paris* alone contains Gallia Comata, Troy, and the Australian inscription. A similar but shorter description of the Southern Ocean occurs in one of the small Zone maps of the *Liber Floridus*.

The conception worked out by Krates was shared to some extent by Agathemeros and Martianus Capella, as well as by Macrobius, Eumenius, Nonnus, Ampelius, and Kleomedes, and was widely diffused in the later Imperial time. From the fourfold Kratesian world, Berger, *Erdk. d. Griechen*, iii., 123, 129, plausibly derives the imperial ball ⊕ with its two circles.

Lambert's exact language here is not without interest. First we have the Terrestrial Paradise, with the unique addition of Enoch and Elias, and from the same flow, as is customary, the four sacred rivers of the Eastern World. Like Cosmas, Lambert evidently intends these streams to have a subterranean course between Paradise and 'Our World.'

Henry of Mainz offers some points of comparison (for instance, in the form of Spain and the position of Italy); the Psalter map, the 'Cottoniana,' and Guido also supply occasional analogies; but Lambert has no direct relation with any of these, is based on a far older original, and can only be connected for any useful result with the Macrobius-, Zone-, and Climate-maps.

The modern names are scanty; Norway, Flanders, Bavaria, the *Arm* or Strait of St. George, and a few others, make up the list. With Isidore there are some points of agreement which mean nothing; for they are such as may be found on all old maps, and afford no sufficient proof that Lambert used the Doctor of the Church of Seville. The Orosius connections are perhaps a little more definite, but they mainly depend on a specialised but not identical use of two rare names—'Samara' and 'Octogorra' or 'Ottorogorra.' [On Lambert Ottorogorra is an island instead of a river.] With Julius Honorius, on the other hand, we may confidently assume a certain, though slight, relationship; for Lambert himself gives us in the *Liber Floridus* a T-O map in which the names of peoples and races to a large extent agree with the Æthican Recension of Honorius; even some of the oddities are reproduced, such as the placing of the Germanic folks in Asia. On the Lambert *mappe-monde*, however, these names are differently written. Many of them are also found in other old writers, so that we must not press unduly the Honorian connection with Lambert's World-Picture.

With the Anonymous Ravennese there are some surprising points of contact, as, for instance, in the mention and placing of 'Tuscia,' Neustria, 'Scanza,' Venetia, Sclavinia, Emilia, Istria, Burgundia, the Huns, the Lazi, and Thyle (*West* of Spain). With Martianus Capella there are certain possible resemblances, which, however, mostly apply just as well to Pomponius Mela and Solinus;—*e.g.* Magnesia as a *Province;* Pontica and 'Margiane' or Merv as *Regions;* the Islands of the Sun and the Gorgades. With Mela we may compare the 'Oceanus Scythicus'; with Solinus 'Abalcia.' 'Margiane' is also a Ptolemy coincidence. With Beatus and the Peutinger Table, we have the common peculiarity of the 'Vandals.' With Æthicus of Istria certain doubtful coincidences may be found. But the Gryphons or Griffins occur also in Isidore, and even earlier; while the language of Lambert about the Gog-Magogs, and the thirty-two nations imprisoned by Alexander, is not identical with Æthicus, though it recalls the language of the Istrian romancist, especially in the matter of the *number* of these races of evil men here mentioned. For only Æthicus, among all writers from whom Lambert could have borrowed, speaks of *gentes xxxii*.

Lambert's Natural History is often curious, derived as it is from various ancient works, somewhat imperfectly understood, and often combined with apparent originality. Instances of the latter are the fauns of India, the parrots and elephants of Arabia, and the apes of Parthia. Lambert's Hyrcanian tigers are perhaps from Ammianus Marcellinus: his Arabian lions from Strabo or Agatharcides; his Indian pygmies and Arabian cinnamon from Isidore; his trees of the Sun and Moon from the Alexander Romance of the Pseudo-Kallisthenes; while his Griffins of the North might be derived from many authors.

We must not, however, suppose that Lambert's *mappe-monde* is a compilation from a large number of writers. It is not impossible that Lambert's map, with the exception of a few more recent names, was taken bodily from an ancient world-sketch of the fourth or fifth century. But even if it is the outcome (in its general outlines) of a lost original from the latter days of the old Empire, yet it has been greatly modified by its twelfth-century redactor, and in part at least it truly belongs to the central mediaeval time. As to this, we may notice especially some of the islands in Lambert, such as *Tritonia*, apparently a fancy name from the River Triton in Æthiopia; *Betania* or Britain, placed over against the Pillars of Hercules or Straits of Gibraltar; the Balearics, defined as 'over against Spain,' but located in the ocean; and the *Orcades* or British fringing-islets, thirty-three in number, lying over against Britain and *Gothia*. For Lambert's *one* 'Fortunate Island' off Mt. Atlas, and his Island of the Blessed (*Beata*) off Mauritania, we may refer to the Anonymous Ravennese [ed. Pinder and Parthey, 325, 5; 444, 3]; for his similar blunder in the double Hibernia the Beatus maps furnish parallels [cf. Santarem, *Essai sur . . . Cosmographie*, ii. 158-159].

VII.—Macrobius was also the author of *Saturnalia*, which deal with historical, critical, antiquarian, and mythological matters; and of a Grammatical Treatise which has perished, except in an abridgment. Though he says that Latin was a foreign tongue to him, he probably wrote these works in that language [cf. *Dawn of Modern Geography*, i. pp. 343, 344].

The chief examples of Macrobian maps are (1) two of the tenth century, mentioned by Santarem, *Essai sur* . . . *Cosmographie*, ii. 41, but without proper reference. (2) One of the tenth century, at Paris, Bibl. Nat., 7585 (Santarem, *ibid*, ii. 47). (3) One of the tenth or eleventh century (Santarem, *ibid*, iii. 460). (4) One of the ninth century, at Munich. (5) One of the tenth or eleventh century, at Munich. (6) One of the twelfth century, at Munich. (7) One of the eleventh or twelfth century, at Bamberg. (8) One of the thirteenth century, at Berlin. (9) One of the thirteenth century, at Berne. (10) Two of the eleventh to thirteenth centuries, at Leyden (Vossian MSS.). Nos. 4 to 11 are noticed in Jan's edition of Macrobius, 1848; which Eyssenhardt, 1868, has made great use of. Many other Macrobius manuscript-maps doubtless exist, but have not yet been noticed.

A zone map of Macrobian type is found in the tenth-century manuscript of Bede's *De temporum ratione*, bound up with the *Periegesis* of Priscian and the Cotton or 'Anglo-Saxon' map; in every important respect it may be assumed to reproduce Bede's original. It occurs in Cotton MS. Tib., B. V., folio 29; is uncoloured, except for a green border; is circular in form; measures ten inches across; represents the *globus terraqueus;* and seems intended to show the movement of the oceans as well as the position of the zones or climates. It has a slight sketch of the Mediterranean, with two city-pictures, probably of Rome and Jerusalem, and with the Pillars of Hercules. The continents are indicated by inscriptions ;—*Africa, Asia, Æquitania* (sic); but no drawing of their outlines is given. In the middle of the sketch is an inscription, which gives an estimate of the four great divisions of the world's circuit, with special reference to the ebb and flow of the ocean, and a description of the equinoctial zone, 'almost entirely washed above and below by the sea, which flows round about through the midst of the land, dividing the whole earth into four islands, as it were; which islands are inhabited . . ; and as through the midst of the land, so also round about the circuit of the same, runs the sea, which from heat or cold is untraversable ':—' *Æquinoctialis Zona hic incipiens pene tota alluitur superius et inferius mari, quod dum per medium terrae circumlabitur, in IIII quasi insulas totus orbis dividitur, quae inhabitantur ; est enim solstitialis superior et inferior habitabilis ; similiter superior et inferior hiemalis ; sicque fit ut per medium et in circuitu orbis mare currat quod calore vel frigore est intransmeabile : estque deprehensus totius orbis ambitus in stadiis ducentis quinquaginta duobus millibus.*' Besides this there are only the five climate labels : *Arcticus, Æstivus, Æquinoctialis, Hiemalis, Antarcticus*, and the four-fold inscription round the rim, '*Hinc refluit Oceanus ad septentrionem* [or, *austrum*] *per LXIII stadiorum*,' twice repeated in each form.

This map goes with chapter 34 of the *De temporum ratione ;* which again is based on St. Isidore, *Etymol.*, iii., 43 ; xiii. 6 ; *De Nat. Rer.*, x. It shows some relationship to the Dijon map of 1064 (MS. 269), which is, however, more astronomical [cf. K. Miller, *Kleineren Weltkarten*, 124].

[On Macrobian maps in the *Imago Mundi* of Honorius of Autun, cf. Santarem, *Essai sur . . . Cosmographie*, ii. 237, 239, etc. ; Lelewel, *Géog. Moyen Age*, i. 53 ; and *Atlas*, plate viii. The *Magna de Naturis Philosophia* of William of Conches is sometimes wrongly ascribed to the Abbot William of Hirsau. There is a thirteenth-century manuscript of this in the Stuttgart Public Library, *Cod. med. and phys.*, No. 15. The maps are on folios 13, 25, 29 ; cf. Santarem, *Essai*, iii. 499-505.] The second of William's designs shows round the earth-circle, as in the Baeda example, the dimensions of the four segments ; the four great tidal movements are also inscribed. Above the Earth-Island, towards the left, is the Indian Sea ; below is the Mediterranean ; to the left of the same are Europe and *Calpes ;* to the right, Africa and *Atlas*.

In the *Garden of Delights* of Abbess Herrade, the Macrobian map is on folio 10 A. In the manuscript of Hyginus, referred to in text (p. 575), once at Bobbio, later at Wolfenbüttel, Figure No. 160 is specially notable. Macrobian maps are also in the Naples MSS., V. A. 12, V. A. 10 (12 bis) of eleventh and twelfth centuries ; and in the Leipzig MS. xliii. (City Library) of thirteenth century, at the conclusion of Walter's *Alexandreis ;* cf. Wuttke, *Serapeum*, xiv. 249. According to Philippi, *Zur Rekonst. d. Weltkarte d. Agrippa*, i. 2, Cecco d'Ascoli (1257-1327 A.D.) wrote a commentary to the *De Sphaera Mundi* of Sacrobosco, and reproduced the zone map.

VIII.—The *Climate map* of Petrus Alphonsus is in MS. 1218, Suppl. Lat., of the Bibliothèque Nationale, Paris. Arim, Arin, or Aren occurs on Persian maps of the twelfth century as the umbilical point of the earth, just as it is described by earlier writers of the ninth and tenth centuries ; Roger Bacon identifies this spot with Syene.

Alphonsus clearly shows in this particular the fact of Arab influence; a confirmation of the same is seen in the orientation of his little map, which gives the south at the top. Below the Arim picture, which is here portrayed as a *city* of superhuman size and grandeur, he designates seven habitable climates, and at the foot of all a northern zone uninhabitable from the cold.

These Climate maps were probably common in the Moslem Geography which arose under Al Mamûn, and attained so great a development in the ninth and tenth centuries. According to Masudi they were coloured; one of special merit was issued by order of the Caliph Al Mamûn; but another standard example was to be found in Ptolemy. This branch of cartography among the Arabs and Franks was based on Greek originals; many of these originals, such as those examined by Masudi, are now lost; but we still possess enough in the way of written descriptions to trace the development of the Climate-scheme, and to see how greatly the primitive conception was modified in the course of time [cf. Santarem, *Essai sur . . . Cosmographie*, i. 92; iii. 332]. At first the term *Climate* implied merely the κλίμα, or supposed slope of the earth from a higher North to a lower South, or *vice versâ;* secondly, from the time of the great astronomer Hipparchus (c. B.C. 160), the different belts or zones of the curved or spherical earth-surface (as determined by the different lengths of the longest day at Syene, Alexandria, Byzantium, and so forth) are understood. It was to illustrate this leading feature of our world, especially for the aid of mathematical and astronomical geography, that the Climate maps arose; in the wider sense of the word, all ancient designs of truly scientific character, such as Ptolemy's, were climatic; but of course these works paid attention to many other things besides climatic distinction. In the narrower sense, with which we are here concerned, this last is the sole object.

IX.—Along with the elaborate and comparatively scientific geography of Eratosthenes or Ptolemy, the ancient world also had a popular system, represented in some of its phases by Krates of Mallos, by Macrobius, and by the zone- or climate-sketches we have already had to consider; another side of the same is portrayed in the T or T-O designs. Thus Hipparchus (in the second century B.C.) recommends students of Geography to glance at the ancient map sketches as well as the scientific plan of Eratosthenes. Again, Agathemerus says the Ancients constantly drew the Habitable World in circular form; putting Hellas Proper in the midst of the *Oikoumené*, and Delphi in the midst of Hellas. Among Roman geographers a more favourite middle point was found in the Ægaean Sea, and especially in the Isle of Delos; but the complete agreement of the T maps of the earlier and later time is expressed by Sallust in a famous passage of his *Jugurtha* [ch. 17, '*in divisione Orbis Terrae plerique partem tertiam Africam posuere; pauci tantummodo Asiam et Europam esse, sed Africam in Europa;* cf. S. Berger, *Geschichte d. Erdk. d. Griechen*, 1887, i. 10, 85; ii. 64; also Riese, *Geographi Latini Minores*, p. 90. Other descriptions confirm Sallust's language, and show us how differently the area of the different continents was estimated by various writers. Thus Pliny, dividing the whole land area into sixty parts, makes Europe occupy twenty-eight, Asia nineteen, and Africa thirteen of these. Again, Orosius, at the beginning of his cosmographical chapter, refers to the triple estimate (of three continents) as well as to a double estimate, based on the supposed preponderance of Europe. [*Our ancestors made a three-fold division of the Orb of the whole earth surrounded by the limbus of ocean, and named those three parts Asia, Europe, and Africa; but some considered it better to treat only of two parts, —Asia on one side, and Africa with Europe on the other.*] The latter reckoning usually made Asia equal to Europe and Africa together. In the cosmography of Æthicus (*i.e.* of Julius Honorius) the three-fold division of Orosius is contrasted with the four-fold arrangement of another tradition.

To the examples of T-O designs mentioned in the text pp. 577-8, we may add;—another in St. Isidore's *De Natura Rerum* (found in manuscripts even of the eighth century); the Roda-Madrid sketch of the ninth century; the Strassburg of about the same date; the St. Omer of the tenth or eleventh century; and one of the eleventh century in Hermannus Contractus' *Chronicle of the Six Ages of the World*. Besides these there are eleven lesser sketches from various manuscripts of St. Isidore's *Origins* or *Etymologies*, all before the end of the thirteenth century; a description of a T Design (illustrating the *Divisio Terrae*) in Bede, *De Natura Rerum*, ch. 51, which was perhaps accompanied by a sketch with astronomical additions, as reproduced in Migne. We have also from Raban Maur [in his *De Universo*, xii. 2; copying Isidore, *Etymol.*, xiv. 2] a repetition, not schematic but descriptive, of the chief Isidorian plan of this kind, c. A.D. 844; and various T Maps in Honorius of Autun, in Guido, in Lambert of St. Omer, in a Mons manuscript of Lucan's *Pharsalia*, in Walter of Metz, and in Nikephoros Blemmidas. We must briefly examine some of these.

APPENDIX ON MAPS 629

(1) The lesser Isidore T maps are in the following manuscripts:—(a) Paris, Bib. Nat., 7683 Lat. of tenth century; (b) same Libr., Bib. Nat., Lat. S. G., 538, of tenth century (this manuscript contains *two* T maps, one round, one square); (c) Bib. Nat., Lat. Nav., 87, of twelfth century; (d) Bib. Nat., Lat. 7592, of eleventh to twelfth centuries; (e) Bib. Nat., Lat. 7590, of thirteenth century (this manuscript also contains *two* T maps, one round, one square); (f) Bib. Nat., reference not yet known, mentioned by Santarem, *Essai*, ii. 284; (g) Bib. Nat., Nav. 6, of thirteenth century. Besides these nine Paris manuscripts (all falling within the period A.D. 900 to 1260), there are two in (h), a manuscript of the Library of Metz, of thirteenth century, cf. Santarem, *Essai sur . . . Cosmographie*, ii. 235-236, 283-284, 287; ii. 505-506; Otto's edition of Isidore, p. 434, dealing with *Etymol.*, xiv. 2, 3; and Riese, *Geographi Latini Minores*, 160, on the Isidorian T sketches in the *De Natura Rerum*. (2) Santarem, *Essai sur . . . Cosmographie*, ii. 102, refers to a non-Isidorian plan of this kind, but with marked peculiarities, in the Paris Bib. Nat., Lat. 5371, of eleventh century. (3) Raban Maur merely repeats the Isidorian data relative to the T-shaped *Divisio Terrae;* and the same is true of (4) the Roda-Madrid map, except that the latter supplies a ninth-century sketch instead of a merely descriptive summary, and adds, upon the three Continents, that Shem received the Temperate Land, Japheth the Cold, Ham the Hot [cf. Santarem, *Essai sur . . . Cosmographie*, ii. 32]. (5) The ninth-century Strassburg T plan [in the Strassburg MS. numbered c. iv. 15] is rather more elaborate, from a geographical point of view. For it attempts a rough estimate of the nomenclature of the principal countries. Thus, for instance, Greece, Italy, Frisia, Saxony, Germany, Dacia, Gothia, and Alamannia are found in Europe; the Amazons and the Moabites, Paradise and Egypt, India and Galilee, in Asia; Carthage and Mauritania, in Africa. Jerusalem is marked by a Greek cross, but is not at the centre of the circle. (6) The St. Omer plan comes from the last leaf of a manuscript collection of homilies written about 990-1010. Besides the names of Europe and Africa, we find here England and Hibernia, Thule, and Scandinavia [*Scandza*]. On the Strassburg and St. Omer plans, cf. Mone, in *Anzeiger für Kunde d. teutsch. Vorzeit*, 1836, p. 113, Taf. ii.; Santarem, *Essai*, ii. 81; Lelewel, *Géog. Moyen Age*, i. 49, and *Atlas*, plate vii. The map is in the St. Omer MS. No. 97. (7) Hermannus Contractus gives us only the simplest kind of sketch, with 'Noachic' description; but (8) the Oxford of 1110 [in St. John's Coll. Libr., Cod. Membr. fol., xvii., folio 6; cf. Bevan and Phillot, *Med. Geog.*, xxxvi.] is very notable,

e.g. for its Greek words. The diameter is about 6½ inches. The Terrestrial Paradise does not appear. Though Africa is included under Europe, the dividing arm of sea (Mediterranean) is given as the stem of the T. The number of 32 races given in Armenia by this map may be compared with the 34 Albanian races of Orosius, i. 2, 1, and the 32 Albanian and Armenian peoples of the Anonymous Ravennese, ii. 12. Perhaps the map was primarily intended to portray the 72 races of 'Greater Asia,' the 27 races of Shem, the 15 of Japheth, the 30 of Ham, the 33 of Armenia (given as if in South Asia), Jerusalem or Sion, the Cities of Refuge, the Tribes of Israel—8 only given—and other favourite mediaeval traditions. Also, there is a commencement of a *Divisio Apostolorum;*—Paul in Athens, Peter in Caesarea, Andrew in Achaia, John in Ephesus. The 72 races of Greater Asia may be compared with Gervase of Tilbury, *Otia Imperialia*, ii. 1, an alteration from the 70 of the Mosaic table. But the most interesting point about this map is its evidence of Hellenic or Byzantine authorship. *Anatole, Mesembrios, Disis* (*i.e.* δύσις), and *Arcton* appear, along with Latin equivalents, at the Cardinal Points. Many of the most important names are out of place; thus Constantinople is in Asia Minor, Athens north-east of Jerusalem, while Judaea, Palestine, and Carthage stand in adjoining plots of what is labelled Europe, but which is drawn as if belonging to Africa. With singular perversity Carthage is again repeated in the extreme north-west, roughly in the position of the Pyrenees. (9) Honorius of Autun's *Imago Mundi* is illustrated by a few T Maps of no special interest, like the (10) Guido manuscripts in Brussels and Florence, (11) some of the codices of Lambert of St. Omer, and (12) the Mons text of Lucan's *Pharsalia* [cf. Santarem, *Essai sur* . . . *Cosmographie*, ii. 176, 229, 237; *Recherches*, plate iii.; Lelewel, *Géog. Moyen Age*, ii. 7; and *Atlas*, plate viii.] These sketches immediately refer to the Archetypal passage of Isidore, *Etymol.*, xiv. 2, 3. (13) Walter of Metz, who about 1245 wrote a poem entitled *Liber Mappae Mundi*, or *Images du Monde* in 5900 French verses, several times employs the T-O scheme, but in the rudest manner; and the same may be said of (14), Nikephoros Blemmidas, c. A.D. 1250. The latter's T plan (called by Santarem a Sallust map) is accompanied by short Greek indications designating the three Continents and the Nile, and places the West at the top. [For Nik. Blem., see Paris Bib. Nat., 1414, folio 3; Santarem, *Essai sur* . . . *Cosmographie*, ii. 245, 248-254, 275; Spohn, *Nik. Blem. Opusc.*, 1818; for Walter of Metz, see Bib. Nat., 7991, 79, 29. The crucial passage for T-O cartography is Isidore, *Etymol.*, xiv. 2; *Orbem dimidium duae tenent, Europa et Africa, alium vero dimidium sola Asia; sed ideo duae partes factae sunt, quia inter utramque ab Oceano Mare Magnum ingreditur, quod eas intersecat. Quapropter si in duas partes, Orientis et Occidentis, orbem dividas, Asia erit in una, in altera vero Europa et Africa.*]

X.—As to the Sallust maps, this curious little family of mediaeval sketches was first properly noticed in its various connections by Wuttke ; although Spohn had already drawn attention to some of them. Lelewel, Philippi, and Konrad Miller have somewhat developed the study of this group, but it can hardly be doubted that it is capable of further expansion. As yet we know of eight larger, and five smaller, Sallust maps; of the former category, five only fall within our present view. The first of these is the tenth-century Leipzig ; two others, one of the eleventh, the second of the thirteenth century, are now in the same place ; the remaining two, both of the twelfth century, are at present in the Libraries of Florence and Görlitz.

The oldest example, of about A.D. 980, in the University Library at Leipzig, occurs in a fragment of a manuscript of the *Catilina* which has been used for bookbinding. It is extremely faded and difficult to read ; but the traces of a rather elaborate city-picture of Rome, some smaller sketches, and various names, can still be discerned. This map conforms to the T-O type, places the East at the top, and employs the Tanais and Nile in the division of continents. It was first noticed by Wuttke, *Zur Geschichte der Erdkunde des Mittelalters*, p. 6.

The Leipzig of the eleventh century [City Library, MS. No. xl., once the property of St. John in Magdeburg] from a manuscript containing not only Sallust, but Lucan and Capella, is the handsomest of all the Sallust maps, and gives pictures of Rome, Troy, Babylon, Carthage, Cyrene, and Jerusalem. Among mountains the Alps, the Lebanon, the Riphaean Hills, the Atlas, and the Pyrenees are given ; among rivers the Nile, Tanais, Danube, and Rhine ; for the general design a rather free rendering of the T-O form is adopted [cf. Santarem, *Essai sur . . . Cosm.*, ii., 93 ; Lelewel, *Atlas*, plate ix. ; Wuttke, *Zur Geschichte der Erdkunde ;* and *Serapeum* for 1853, xiv., pp. 261-270].

The thirteenth-century Leipzig [City Library, No. lxviii.] is of no particular interest, but appears most closely related to the Görlitz example of the twelfth century, which is one of the best and least conventional specimens of the Sallust type [in the Library of the *Oberlausitzische Gesellschaft* at Görlitz], with Jerusalem nearer to a central position than in the earlier members of this family.

632 APPENDIX ON MAPS

The Florence of the twelfth century [Bib. Laur., Plut. 64, Cod. 18] is only remarkable for its clear and excellent script, and its emphatic indication of the Four Cardinal Points. Of the lesser Sallust maps nearly all, as we have said before, are simple T-O sketches; one is of the tenth century, and bears underneath the inscription, 'Julius, the Emperor, divided the whole world in its several parts' [*particulatim*].

Two chief branches of this type may perhaps be traced back earlier than the tenth century. From these all place-names come, except for a few additions of later date, such as are found on the eleventh-century Leipzig, and the twelfth-century Florence.

XI.—In 1119 a certain Guido, of whom nothing more is known, compiled a work of extracts, mainly geographical, in six books. The first of these contained a description of Italy, with extracts chiefly from the Anonymous Geographer of Ravenna, but also from the Antonine Itinerary and the *Notitia Urbis*. The second book gave only extracts from Isidore of Seville; while the third, dealing with the general geography of the earth, combined Isidorian excerpts [*Etymol.*, xiv. 3-5] with passages from the Ravennese. The last three books gave a chronicle reaching down to A.D. 1108, with lists of the Lombard, Frankish, and German kings, and special notices of the deeds of the Pisans and Genoese against the Saracens. [See Brussels, Bib. Burg., 3897-3918; Florence, Codex Riccardianus, 881, mostly written A.D. 1277-1280. From the archetypal passage in Isidore, *Etymol.*, xiv. 2, 3, comes folio 44 of the Brussels manuscript, and folio 20 of the Florence.]

Two of the existing manuscripts of this work (those at Brussels and Florence) contain map-pictures, one being a T-O map of Isidorian type, and the other two works of a much higher value. These last are only found in the Brussels manuscript, and are devoted respectively to Italy and the World. The world-map belongs to the passage where Guido roughly defines the boundaries of the three Continents; it is coloured, and measures 13 centimetres across. The manuscript once belonged to Cardinal Nicholas 'of Cusa,' who obtained it in Italy, and presented it to a hospital which he founded in his native town of Cues on the Moselle. Later on it passed through the hands of the Bollandists into the 'Library of the Dukes of Burgundy.' The *mappe-monde* is perhaps a copy of a larger work, of which a piece is better represented in the sectional plan of Italy, which occurs in the same manuscript. For the most part, the text agrees with Orosius, the Ravennese Geographer, and Isidore; but the names of Barcelona, Lyons, Samaria, and others (which do not occur in

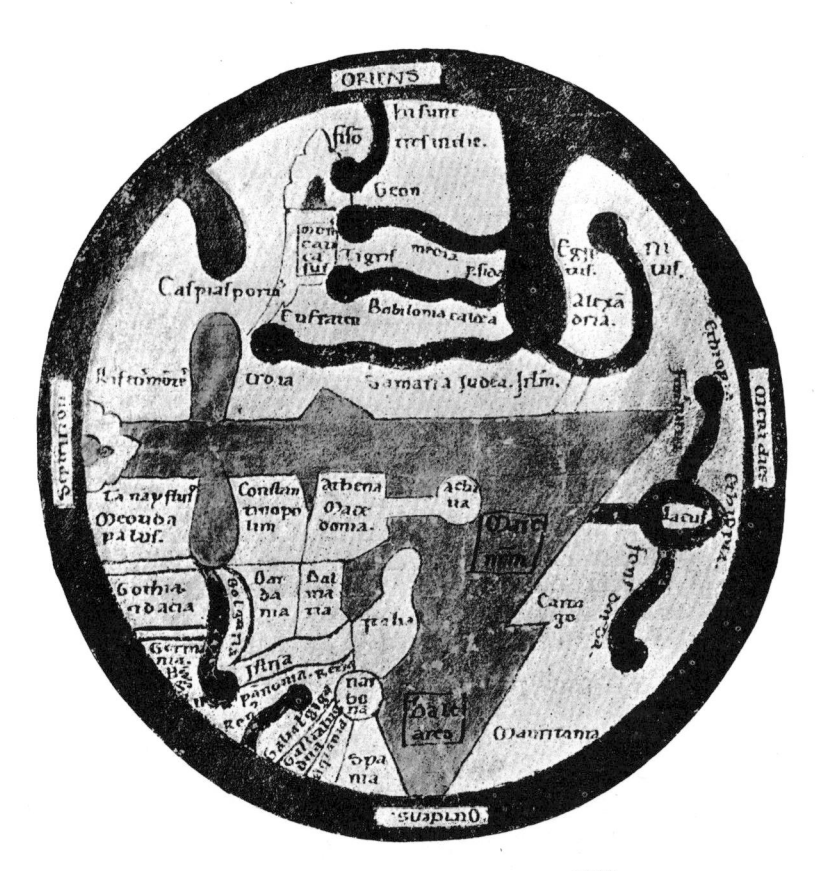

THE WORLD-MAP OF GUIDO, C. A.D. 1119.

Guido's chief sources) show that this is no mere illustration of these authors. The form of the Mediterranean is very peculiar, and the size of the rivers and inland waters of the Continents is more exaggerated than in almost any other mediaeval map. The strange triangular formation, in which the Mediterranean from one side, and the Black Sea and Ægaean from another, run southward almost to the Sea of Æthiopia, naturally affects the shape of Africa, whose northern coast has an unusual inclination to the south-east. [For the world-map, see p. 547 in Pinder and Parthey's edition of the Ravennese Geographer and Guido, and folio 51 verso, of the Brussels manuscript; for the Italy, p. 452 in Pinder and Parthey, and folio 1 verso, of the Brussels manuscript.]

The Italy map, giving a large city-picture of Rome, is accompanied by square indications (as if taken from a statistical plan) of the neighbouring Provinces, from Barcelona to Macedonia.

Besides these, the Brussels manuscript has a picture of an ancient man of learning, with a scroll in his left hand. Some identify this with Castorius, some with the Ravennese Geographer, some with M. Agrippa. At the end of Guido's Maritime Itinerary (Book iii.) there is a figure of Antonine, also with the Cap of the Learned on his head, and sketches of Rome, Ravenna, the Pharos of Alexandria, Augustus, and Vespasian with a globe. [These additional sketches are on folios 6, 31, 32, 51, of the Brussels manuscript, cf. Bethmann in Pertz, *Archiv.*, vii. 537-540 (1839); Reiffenberg in *Bulletin de l'Acad. Belg.*, 1843, x. 1, 468; 2, 73; 1844, xi. 1, 15, 99; Bock in *Annuaire Bib. Roy. Belg.*, 1851, pp. 41-204; Santarem, *Essai sur . . . Cosmographie*, ii. 212, 238, and *Atlas;* Lelewel, *Géog. Moyen Age*, i. p. 89, and *Atlas*, plate viii., figs. 29-30; Philippi, *Zur Rekonst. d. Weltk. d. Agrippa*, 1880, Taf., i. 5.]

XII.—On the Madaba map;—cf. the *Révue Biblique* for April, 1897; Germer-Durand, *La Carte Mosaique de Madaba*, Paris (Maison de la bonne Presse), 1897; K. Miller, *Rekonstruierte Karten, Anhang*, pp. 148-154, 1898; A. Schulten, *Die Mosaikkarte von Madaba*, 1900. To Father Kleophas, librarian of the Greek Patriarch at Jerusalem, the discovery and preservation of this map are chiefly due; and some of the happiest suggestions for its further reconstruction have also come from him. The town of Madaba, mentioned by Josephus, Ptolemy, and Stephanus of Byzantium, lies close to, a little south of, Heshbon, but south-east of Nebo or Pisgah, and on the Roman Road which came south from Damascus, passed through Rabbath Ammon or Philadelphia, Heshbon, Madaba, Rabbath Moab or Areopolis, and Kerak, and ran on to Petra and the Red Sea.

At Heshbon it crossed the Roman Road from Jericho, which ran south-east to the outposts on the edge of the desert.

The mountains of Palestine in this Mosaic are marked by variegated lines. The Dead Sea here is remarkable for its exaggerated size, its blue wave-lines, and the two great ships that float upon its waters. In the Jordan and the Nile swim fishes; a piece of the Arabian Gulf is indicated in the extreme south. In the desert fringing Palestine and Egypt the oases are marked by palms; in the desert itself lions pursue gazelles. Among other features, the Serbonian Bog between Palestine and Egypt is marked.

Out of the twelve tribes of Israel, only six can still be found —Dan, Symeon, Juda, Ephraim, Benjamin, and Zabulon—and several of these survive in a very fragmentary state. In connection with *Dan*, the apostrophe of Deborah from Judges v. 17, is quoted (ἵνα τί παροικεῖ πλοίοις;) as to *Benjamin*, the blessing of Moses from Deut. xxxiii. 12 is given (σκιάζει ὁ Θ[εό]ς ἐπ᾽ αὐτῷ, καὶ ἀνὰ μέσον τῶν ὁρίων αὐτοῦ κατέπαυσεν), probably in allusion to the holy city, lying as it did within the territory, but on the frontier, of Benjamin. *Ephraim* is associated with the blessing of Jacob upon Joseph, from Gen. xlix. 25, and with that of Moses upon the tribe of Ephraim from Deut. xxxiii. 13 (Ἰωσήφ . . . εὐλόγησέν σε ὁ Θεὸς εὐλογίαν γῆς ἐχούσης πάντα καὶ πάλιν ἀπὸ εὐλογίας Κυ[ρίου] ἡ γῆ αὐτοῦ). As for *Zabulon*, there now only remain five letters from a clause of the prophecy of Jacob in Gen. xlix. 13; but the whole sentence was fortunately copied before this piece of the mosaic perished: Ζαβουλὼν παράλιος κατοικήσει, καὶ παρατενεῖ ἕως Σιδῶνος. The remaining details of the map may be dealt with under the tribes above mentioned, under the trans-Jordan region, and under Lower Egypt and Sinaitic Arabia.

In the territory of *Dan*, Modeim or Moditha, 'whence came the Maccabees'; Anob or Betoannaba, the Nob of Saul, according to St. Jerome; Lod, Lydda, or Diospolis; Geth or Gath, 'one of the five satrapies of former days'; Ashdod in the upland; and Azotus on the sea, are the leading places indicated.

In *Benjamin*, the plan of Jerusalem naturally occupies the chief place. Here the most prominent feature is the great street flanked by colonnades which runs across the city from the north gate to the south; before the north gate there seems to be a paved square, and on it a column, recalling the modern title of 'Gate of the Pillar.' A second colonnade-street runs from this north gate to the south-east, crosses a lesser artery which enters from the eastern gate, and then proceeds due south, just within the east wall, and parallel to the main street.

These two roads are marked all along by the covered galleries or colonnades already noticed, and probably correspond to the δύο δημόσια, the two markets or bazaars, mentioned in an Arab account of the Persian storm of 614. Three gates are indicated, to north, east, and west; the first two being apparently of much greater size and importance than the third, which has rather the appearance of a postern. To the south, at the end of the great centre colonnade, appears what is probably the Church of Holy Sion or the *Coenaculum*; at the north-east extremity of the city is another church, perhaps that of the Nativity of the Virgin (now 'St. Anne'); while beyond the walls, to the east, is the Church of Gethsemane and the Virgin's Tomb. The territory of Benjamin appears undamaged upon this map; and the localities of Jericho, surrounded by palm-trees, in the midst of which appears the 'sanctuary of Elisha'; of Ephrata, 'where the Lord journeyed'; of Ælamon, 'where the moon stood still at the bidding of the son of Navê'; of Bethabara, with the Convent of St John Baptist; of Rama and Gabaon; and of two road-stations respectively called 'the fourth' and 'the ninth' (τὸ τέταρτον . . . τὸ ἔννα[τον]),—appear together with a few less important spots. Here, as elsewhere, the likenesses with St. Jerome's topographical indications are very marked; those with Eusebius, Jerome's master in Palestine geography, being somewhat less close than in other places.

In *Ephraim*, part of the space (*e.g.* around Nablûs) has perished; on what remains we have Ænon, 'near to Salem'; Silo, 'where stood the ark'; the tomb of Joseph; the well of Jacob; 'Garizim'; and Sichem, 'otherwise Sikima and Salem' (an exact repetition of a peculiar notice in Eusebius' *Onomasticon*). Like the latter, again, the map places the mountains of blessing and cursing near Jericho, though it has already marked them in the right position near Nablûs. The Aramaic term of 'tûr,' or mountain (Τουρ), is annexed to both Γαριζίμ and Γωβήλ, Gerizim and Ebal. The towns of Theraspis and Betomelgesis, apparently assigned to Ephraim by the present design, are unique; no trace of them has elsewhere been found.

In *Zabulon* nothing remains beyond what has been already noticed; but in *Symeon* we have Gerara, 'once a royal city of the Philistines and southern limit of the Canaanites.' Ἔνθα τὸ Γεραριτικὸν σάλτον, *here is the wood* [?] *of Gerara*, adds the map, which also gives, in the territory of this tribe, Arad, 'whence came the Aradians,' and Asemona, 'a town of the desert on the frontier of Egypt.' Arad is the Adar of the Vulgate, and Jethor, or Jethera, to the west of it, is the Jether of the *Onomasticon*, which notes it as wholly inhabited by Christians.

In the west of the land of Symeon, the mosaic records several peculiar names (Edrain, Sobila, Seana, and Bethagidea; cf. the Edrai-Bethaglaim of Eusebius).

In *Juda* much of the original detail has been lost, but to the south of Jerusalem is Akeldama, which is placed considerably to the north of the common or traditionary site, supported by St. Jerome; the unusual position is no doubt due to Eusebius, whose language here agrees with the map. Bethlehem Ephrata, Rama, Bethsura (of the Maccabees), the Church of St. Philip near the fountain where he baptised the Ethiopian, the terebinth-oak of Mamre near Hebron, Ascalon, Gaza, and Beersheba, or Berossaba, are the chief Judaean names; and 'so far,' says the map of the last-mentioned, 'came southward the limits of Judaea, even from Dan, near Paneas, which is the border in the north.' Gaza and Ascalon, though fragmentary, are evidently planned as large towns, and a good deal of these city-pictures still remain. Beyond the Jordan appear Ænon, 'now Sapsaphas'; the Baths of Kallirrhoe, marked as three in number; Kerak of Moab [Κρ]άχμωβα, on a height, as in nature; Betomarsea, 'otherwise Maiumas,' found only in this mosaic; Aïa, perhaps the Aïê of the *Onomasticon*, the ancient Rabbath Moab, the Greek Areopolis; and Balak, 'or Zoora,' to the south of the Dead Sea (Lot's Zoar). The principal inscription of this part refers to the same Dead Sea, also called the Salt or Asphaltic Lake, whose enormous size and detailed execution suggest that it was well known to the designer. Finally, in Lower Egypt and the desert of Sinai are a number of places and legends. Among these are 'the Desert where the brazen serpent healed the Israelites'; 'the Desert of Sin, where the manna was sent down and the quail [*sic*]'; and Raphidim, 'where Israel fought Amalek.'

According to a tradition at Madaba itself, the map, when first uncovered, had a picture of the holy family flying into Egypt, not along the coast as in later stories, but well inland across the desert. Rhinocorura and several towns lie on the edge of a curiously exaggerated stream marked as the frontier of Egypt and Palestine; in the Delta, Pelusium, though in fragments, appears as a city of great size and splendour, second only to Jerusalem; Tanis, Sais, Xois, Athribis, Hermoupolis, and five other Egyptian towns are also figured.

XIII.—As to the *Situs Hierusalem;*—Fulcher of Chartres (born 1048), accompanied the First Crusade, and described it, in the course of a narrative ending in 1126; but long before this extracts from his record must have found their way into Western

Europe. Thus, the text of the *Gesta Francorum* (1095-1106), is for the most part in close agreement with Fulcher, and often scarcely more than an excerpt from him. It also agrees closely with the *Plan of Jerusalem*, which we have now to consider; but in the topography of the Holy City neither the *Gesta* nor the *Situs* are in accord with Fulcher. It is clear that the chronicler of Chartres is not acquainted with this map, which has been termed 'certainly older' than the Anonymous *Gesta*, but really seems to have been draughted at the very same time that the *Gesta* was being written (1099-1109). The latter can hardly be brought later than 1109; for the chronicler finishes his work before Tripoli had fallen into the power of the Christians (1109). So far as the two differ, the *Situs* is plainly the original, and the *Gesta* the derivative, as is shown by the far richer content of the map, and by some of its legends, neither referred to nor explained in the *Deeds of the Franks who took Jerusalem*. [The Anonymous Author of the latter prefaces his description of Jerusalem as follows :—
'*Situs ipse Civitatis Sanctae, qui nunc est, murorumque ambitus, licet a prisca et illustri veteris compositione . . . valde discrepet, quaedam tamen temporis illius* [Xti] *adhuc monumenta continet, quibus . . . prae cunctis clarior civitatibus . . . esse debet . . .* cf. the edition in *Recueil des Historiens des Croisades orientaux*, tom. iii., Paris, 1866, p. 487 etc.].

Eight copies of the *former* are known, of which only three are of much importance, namely at St. Omer, at Brussels, and at Copenhagen. The first of these, of the twelfth century, is from a manuscript containing the Anonymous *Gesta*. The map is on folio 15; cf. *Zeitschrift d. deutsch. Pal.-Vereins*, xv., Taf., 1-5; p. 510, etc., 1892. The second is bound up with the writings of Fulcher *On the Deeds of the Pilgrims*, and of Robert the Monk, *On the Occupation of Jerusalem*. This example, like the former, shows us at the top of the sheet the whole course of the Jordan, with a number of other place-names taken from, or at least agreeing with, the ordinary description of Jerusalem, but sometimes wrongly placed. These, perhaps, result from compression, and refer us to a wider original. At the end of the 31st chapter of the *Gesta*, the Anonymous Author says, '*Nazareth autem ubi conceptus est* [Xtus,] *et Galilea et Mare Tyberiadis, et Mons Thabor . . . longe a Hierusalem remota sunt.*' It is just these places (excepting 'Galilee') which are here interpolated without any particular connection.

The Copenhagen copy, of the thirteenth century, is, like its predecessors, a work of art, but has a much scantier content. It comes from an Icelandic manuscript in the Danish Royal Library, No. 736, 4 b.; cf. Lelewel, *Géog. Moyen Age*, ii. 4, and

638   APPENDIX ON MAPS

n. 5. Among the five later copies of the *Situs* the chief are Count Riant's and the Montpellier example [Montpellier Library, Cod. H., 142, folio 67] of the fourteenth century. The former is perhaps transcribed from a Florentine manuscript, older than any now existing. The city of *Masphat*, here both pictured and named, appears without name on our other transcripts, which are all executed with less care than this. Some of its legends outside the city appear to be additions of the copyist, and remind us of the twelfth-century pilgrim, John of Würzburg (c. A.D. 1165).

The Montpellier copy is in a manuscript which contains the anonymous *Gesta*, but the map itself is fitted into the text of Petrus Tudebodus, *Historia de Hierosolimitano itinere*, xiv. 1. Here the older or round form of the city-plan is made square, and the North is at the top, both signs of later authorship.

The other copies are of little interest; the twelfth-century Stuttgart (which alone contains *Magdalum*) is probably a copy of the Brussels manuscript. It is also similar to the St. Omer copy, and may conceivably have been transcribed from a common original. Very different from all the types of the *Situs* hitherto noticed are two other copies;—one at Cambrai, of about 1180, whose shape is square; and one in London, of about 1200, which is round.

XIV.—Of *Matthew Paris'* map-work the chief manuscripts are;—(1) Cambridge, C.C.C., xxvi. [=earlier cix.] of thirteenth century; the *Map of Palestine* is on folio 3 b. (coloured); the *mappe-monde* is at the end of this manuscript. (2) In the Brit. Mus. ;—Cotton MSS., Nero, D. v. (with the *world-map* on folio 1), perhaps a copy of the Corpus MS. (3) *Ibid*, Cotton, Claudius D. vi., with the *Pilg. Itin.* on folios 2-5, the *Situs Brit.* on folio 6, the most finished *England* on folio 8. (4) *Ibid*, Reg. 14, C. vii. (2), with *England* on folio 5, verso. All these are manuscripts of the *Historia Minor*. (5) *Ibid*, Cotton, Nero, D. i., is the chief manuscript of the *Historia S. Albani*; on folios 183, 184 is part of the *Pilg. Itin.*; on folio 185 a chart of the winds; on folio 187 the *Schema Brit.* (6) For another form of the *Map of Palestine* and the first *England* cf. Cambridge, C.C.C., MS. xvi. [=earlier cv.]. (7) Brit. Mus., Cotton, Julius D. vii., folio 49, verso, has an interesting and advanced form of Matthew's *England*.

The colouring of the *mappe-monde* is mostly red for place-names, except those in the Mediterranean, such as Tyre, which lie to the right of the Adriatic; these are black. Mountains are portrayed in ochre, rivers in blue, for the most part; the Mediterranean Sea in green.

The inscription about the three world-maps is as follows: *Summatim facta est dispositio Mappa-Mundi Magistri Rober[ti] de Melkeleia et Mappa-Mundi de Waltham. Mappa-Mundi D[omi]ni Regis quidem [sic] est in camera sua apud Westmonasterium figuratur in ordine* [v. l., *ordinali*] *Mathei de Parisio.* . . . *Figuratur in eodem ordine* . . . *quasi* [*clamis*] *chlamys extensa. Talis est scema nostre*[*ae*] *partis Habitabilis, secundum philosophos, scilicet quarta pars terre*[*ae*], *que*[*ae*] *est triangularis fere: corpus terre*[*ae*] *SP*[*H*]*ERICUM est.* The chief peculiarity in the natural features of the map is the broad arm of sea running west from the Euxine. The *Palus Maeotis* is represented by two lakes near the North Ocean, into which they throw a river. Many unnamed rivers occur in Europe; the only ones named are Rhone, Danube, and 'Elple' (Scheldt). For contemporary names, cf. *Hungaria, Polonia, Austria, Saxonia, Bavaria, Theutonia, Braibe* (for Brabant); *Dacia* (for Denmark); and the towns of Cologne, Pisa, Bologna, and *Janua* (Genoa).

The text has some resemblance to the Hereford and Ebstorf maps (especially the latter), and to Lambert of St. Omer, Henry of Mainz, the Psalter, and the 'Cottoniana.' Most of the newer names may be found on Ebstorf, as, for instance, Holland, Burgundy, Flanders, Austria, Poland, Venice, Bavaria, 'Metis,' 'Hierapolis,' 'Teutonia'; but, after all, the great mass of name-forms in this *mappe-monde* are old.

The form of the design is, on the contrary, novel and peculiar, it has some relation to Henry of Mainz and Lambert of St. Omer;—the former of whom is not unlike Matthew in his islands, his Italy, and his Balkan Peninsula with its curious western projection; while the latter gives a similar course to the Danube flowing into the *North* Sea; but the present scheme must not be regarded as a derivative of either of these, but rather as itself a stem-form not directly borrowed from any other plan that has come down to us.

There are, as we have seen, no fewer than four copies of Matthew's *England* in the manuscripts:—(1) Cambridge, C.C.C., xvi. [=earlier CV.]. (2) London, Brit. Mus., Reg., 14, C. vii., folio 5, verso. (3) Brit. Mus., Cotton MSS., Claudius, D. vi., folio 8; on these three, cf. Gough, *Brit. Top.*, i., 61-64, 66-71, and plates ii., iii., and iv. (4) Cotton, Julius, D. vii., folio 49; this has been much damaged by verdigris. The first and second copies (Cambridge and King's Library) both contain a similar note, *Hunc Librum dedit*, etc.; both manuscripts probably belonged to the Monastery of St. Albans; and after Matthew's death, this note was inserted in both by the same scribe. The Cambridge copy is certainly not from any of the

London examples. It is much richer in content than one of its rivals (the King's Library map), and in various ways shows something of original work. We may therefore suppose that it also comes from Matthew's hand. It exists on a mutilated leaf, of which the lower third is wanting. Its measurements are 23 by 23.5 centimetres; its execution is careful, and its detail is ample; but in all respects it is inferior to the Cotton redactions. The King's Library example is a rougher draught, without any ornamental border, and with little more than half the names of the Cotton. It is all in black except the sea, which is green; the execution is rather careless; the measurements are 35 by 23 centimetres.

The third or Cotton form of Matthew's England is the best and most complete. The execution is admirable, the colouring detailed and systematic. The sea on west and east is green, like the inland gulfs and salt waters; the rivers are blue or red; the province-divisions are marked (in some cases) in red and blue; the mountains and Roman walls (of Hadrian and the Antonines) are yellow; the legends and inscriptions are by turns red and black. To the north the sea has been left uncoloured. At the edges this map is somewhat bent inwards, and on the left border something has been cut away.

As to the so-called *Itinerary to the Holy Land*, the London MS., Reg. 14, C. vii., folios 2-5, gives the complete Itinerary to Apulia. Here the distance between the stations is usually called *jurnée;* cf. Gough, *Brit. Topog.*, p. 85, plate vii. (for the English part); Lelewel, *Atlas*, Tab. 24; *Catalogue Brit. Mus. MSS.*, p. 14; Jomard, *Monuments de la Géog.*, Tab. v. 1, 2, 3 [but Jomard wrongly calls this a pilgrimage-map of 1318]. The Cotton MS., Nero, D. i., folio 183-184, though without the map of Palestine, gives the same matter as the seven pages of the *Reg. MS.* The Cambridge MS., C.C.C. xxvi., supplies the title of *Stationes a Londinio ad Hierusalem*, and on folio 3 b., a sketch in colours of the Holy Land [cf. Gough, *Brit. Topog.*, p. 85, plate vii., 2]. The Cambridge MS., C.C.C. xvi., has on folio 5 a rich 'station' map of Palestine in colours. The London Cotton MS., Tiberius, E. vi. (of thirteenth century) would be valuable from its age, if not ruined by fire; its map of Palestine in colour is on folio 2. The Brit. Mus. Lansdowne MS., 253, is only a copy of the sixteenth century in the handwriting of Camden, the antiquary; the whole Itinerary is on folios 228-231.

For Matthew's map of Palestine cf. the edition of the text by Michelant and Raynaud, *Soc. de l'Orient Latin, Série Géographique*, iii. 123-139 (1882). Two redactions are distinguished by these editors, who regard Matthew's authorship as certain.

XV.—Among the remaining and least important maps of this period there is a little sketch which perhaps should be associated with the T-O designs. This is the work of *Asaph the Jew*, and is of the eleventh century; it is the last of twenty figures of cosmographical or astronomical significance. The three Continents are roughly sketched, separated by the *Mare Maius* or Mediterranean, the Tanais or Don, and the Nile; they are of nearly equal size; but, as in Pliny, Europe is slightly the largest, Africa the smallest. Here, therefore, we have another example of the *Triquetra* or *Trifaria Divisio*, of which we have already heard so much. The Arabic influence in this work is clearly shown by the position of the south as the primary quarter, at the top of the map. The original is in Paris, Bib. Nat., MSS., Lat., No. 4764; cf. Santarem, *Essai sur . . . Cosmographie*, i. 54, 319-321.

There are certain representations of the world on the coins of the Middle Ages, which also call for a brief remark; though they are perhaps less frequent during this later period than at an earlier time, when classical conceptions exercised a more direct influence. Under the Old Empire the globular form of the earth was clearly realised and often portrayed. Surmounted by a cross this globe first appears on the coins of Valentinian III., in the fifth century; and in the sixth and seventh we have the same expressed on Merovingian money. Thus Theodebert I. [d. 539], Theodebert II., and Childerich [d. 674] offer good examples; and on the seals of various mediaeval emperors (among them Charles IV. and Wenzel) and of certain kings, such as William the Conqueror of England (and Normandy), the same is expressed.

The lost maps of this period are doubtless numerous, though perhaps not in so great proportion as in the former period [A.D. 300-900]. Among these we may notice the following, some of a rather traditional character.

The famous Gerbert, afterwards Pope Sylvester II., tells us in his letters (under the date of about 983) that he had caused a map to be painted [cf. Gerbert's *Letters*, No. 148. We may add, though it belongs to the previous period, that the Abbey of Reichenau possessed a still more ancient *Mappa-Mundi*, in two rolls ('Rotuli'); so Reginbert's Catalogue of the Abbey Library declared about A.D. 842; cf. Neugart, *Ep. Const.*, i. 1, 540]. The Catalogue of the Munich Library contains vague references to several maps now unknown, two at Tegernsee, and one at Werinher among others [cf. Munich Library, Codex Teg., 541; the *Munich Catalogue* of A.D. 1500-1504, marked as Cod. 476 of the same library, folio 159 (mentioning a map marked B 47); K. Miller, *Weltkarte des Castorius*, p. 9; *Serapeum*, 1841, pp. 248-268. Albert the Great, *De Natura Locorum*, iii. 1, declares *nos etiam brevem orbis mappam huic operi adiungemus;* but this has not yet been found]. Also the Monastery of Weihenstephan, according to a book-list of the twelfth century, possessed a *mappe-monde*, which may have passed into some great library.

ADDITIONAL NOTE TO PAGE 391.

As to the names of Carpini and Rubruquis :—the former in Latin is usually *Johannes de Plano Carpini;* emended forms, *Carpinis* or *de Carpine*, are suggested by the editors of *Analecta Franciscana* (iii. 266). *Planum Carpinis* or *Planum Carpi* is the Latin form of Pian di Carpina, the modern [Pian la] Magione, about 14 miles from Perugia; cf. F. Liverani, *Giovanni di Piano di Carpine*, Siena, 1878. Rubruquis' name, in five MSS. used by Michel and Wright in *Receuil* text of the Paris Geographical Society, 1839, is spelt Rubruc or Rubruk. He was probably a native of Rubrouck, in old French Flanders. As to Ascelin's journey, Simon of St. Quentin's account is in Vincent of Beauvais, *Spec. Hist.*, xxxi. Cahun, *Introd. à l'Histoire de l'Asie*, p. 350, suggests that in destruction of Kiev and its trade Mongols acted partly under influence of Venetians, jealous of competition with their Black Sea and Crimean commerce.

SHORT INDEX OF NAMES

Abderrahman III., Caliph of Cordova, 493, 495, 496
Abraham, 244
Abraham ben Chija, 224
Abraham ben Meir ben Ezra, 224
Abraham Halevi ben David [Daud], 224
Abulfeda, 7, 451
Achardus of Arroasia, 207
Adalbert, Archbishop of Bremen, 70, 72, 515–521, 532
Adamatus of Salerno, 517
Adam of Bremen, 9, 71, 111, 465, 466, 482, 514–548, 561
Adam of Paris, 514
Adelard of Bath, 4, 9, 183
Adelheid, Queen of Hungary, 483
Adelward, Bishop, 522, 545; Dean, 522
Adernesch, 504, 509
Adhemar of Chafanais, 127
Adolf of Cologne, 196
Adosinda, Queen, 550
Ælfred the Great, 114, 119, 523, 543, 546, 560
Æthelred II. of England, 85, 86, 87
Æthelstan [and Sighelm], 119
Æthicus of Istria, 545, 566
Alaëddin Kaïkobad, Sultan, 447
Alberic, the friar, 277
Alberic Trois-Fontaines, 275, 342
Albert, Papal Notary, 589
Albertus Magnus, Albert of Ballstadt, 9, 466, 547
Albiruni, 14
Alcuin, 550
Aldobrandini, House of the, 447
Alebrand, Bishop, 548
Alexander III., Pope, 231

Alexander the Great of Macedon, 261
Alexander, the friar, 277
Alexander Nevski of Russia, 312, 313
Alexius Comnenus, 147, 180, 402, 427, 499
Alfonso VI., 136, 177
Al Heravi of Herat, 13
Al-Mamûn, Caliph, 120, 183
Almanzor, 135
'Aloha,' 316
Amalfi, trade of, 396–400
Amalrich, Amalric, or Amaury, King of Jerusalem, 197, 198; 'of Ascalon,' 428, 429
Anastasius I., Pope, 120
Anastasius of Cherson, 503
Andrew, pilgrim, 126
Andrew, Prince of Tchernigov, 283
Andrew of Longumeau or Lonciumel, 277, 278, 317–320, 339, 364
Anonymous Pilgrims or 'Innominati,' 138, 152, 167, 184, 203-207, 274
Ansgar, St., 516, 520
Antoli (Jacob) of Provence, 274
Antoninus Martyr, 167
Arabuccha, Mongol Prince, 357
Arculf, pilgrim, 262, 556
Are Frode, 72, 82, 83
Are Marson, 72, 73
Arghun, Mongol General, 352, 372
Aristo, the Greek, 517
Aristotle, 576
Armenians, in Byzantine service, 468, 512, 513
Arnold, Bishop, 70
Arnold, the Chronicler, 540
Arnulf of Carinthia, 479

643

644 SHORT INDEX OF NAMES

Arsenius, Patriarch, 125
Ascelin or Anselm, friar-traveller, 277, 318
Aschod [III.], 'King' of Armenia, 123, 504, 506, 511
Askold and Dir, 487
Assaf, 224
Augustine, St., 557, 577
Avaldamon, 69
Avang, 77

Bachu (Baiju) Noian, Mongol General, 318, 372, 384, 389
Bacon, Roger, 9, 10, 275, 276, 466, 575
Baeda, the Venerable Bede, 208, 522, 524, 526, 540, 574
Baitu, Mongol General, 277
Baldwin I., 140, 153, 156, 161, 170, 171, 180, 420
Baldwin II., 279
Baldwin III., 428, 433
Baldwin IV., 429
Baldwin of Hainault, 279, 321, 337, 359
Bardi, the Icelander, 37
Bartholemew of Cremona, friar-traveller, 321, 324, 340, 367
Basil I., Emperor, 468, 472, 499, 506, 511, 513
Basil II., Emperor, 122, 492, 498, 506, 508, 512, 513
Basil, the priest, 385
Basil, the prisoner of the Mongol, 360
Batu, Mongol, Prince and General, 283, 300, 301, 314, 316, 334, 338–41, 345, 346, 365, 369, 384, 385, 390, 391
Batuta, Ibn; *see* Ibn Batuta
Beatus, 11, 549–559, 591–605
Belardus of Esculo, pilgrim, 207
Benedict IV., Pope, 121
Benedict, the Pole, friar-traveller, 277–280, 302, 303, 390, 391
Ben-Hadad, 246
Benjamin of Tudela, 6, 169, 185, 186, 218-264, 303, 435, 460, 461, 488
Berengar, King of Italy, 496
Berké, 451
Bernard, friar-traveller, 372
Bertram of Toulouse, 424
Biarne Asbrandson, 72
Biarne Grimolfson, 63, 68, 70
Biarne Heriulfson, 49, 50, 51
Biarne Thordarson, 72

Bibars, Sultan, 240
Biorn, Champion of the Broad-vikings, 37, 73, 74
Biorn, Marshal of St. Olaf, 100
Bjorgolf of Halogaland, 32, 33
Bohemund (various), 153, 238, 379, 420, 422, 429, 440
Boleslav, 293
Bolli Bollison, 106
Boniface of Molendino, 374, 448
Botund, 482
Bovo, 517
Brancaleone, Francesco, 418
Brian Born, 94, 115
Brynjolf Bjorgolfson, 32
Bulan, King of the Khazars, 222, 223
Bulgai, 352, 355, 362, 363
Bulgarians, 478–481
Buri, 305, 345
Burislav, King of 'Gardar,' 26
Burislav, King of the Wends, 42, 44, 46
Burkhard [or Gerhard] of Strassburg, 207
Burkhard, envoy of Frederic Barbarossa, 460
Byzantine Agents in Armenia [Sinutes, Constantine Krinites, Constantios, Theodoros, etc.], 505

Caliphs, 225, 246, 250
Capella, Martianus, 522, 530, 542, 546
Carpini, John de Plano, friar-traveller, 6, 7, 10, 266, 275, 276, 278, 279–317, 339, 374, 375–381, 390, 391, 421, 446, 450
Cassiodorus, 523
Castile, trade of, 437, 438
Catalans, Catalonia, trade of, 435–437
Catan, 299
Catherine II. of Russia, 455
Chang-chun, 14
Charles of Anjou, Count of Provence, 449
Charles the Great, Charlemagne, 120, 534
Chasdai [Chisdai] ben Isaac, 220, 222, 463, 493, 496
Chingay, protonotary of Mongol Court, 311, 314
Chosroes I. of Persia, the Just, 222
Chosroes II. of Persia, 149

SHORT INDEX OF NAMES 645

Clavijo, 449
Cnut the Great, 85, 87, 96-102, 539
Coelestine I., Pope, 120
Conrad, Emperor, 539
Conrad, Duke of Lenczy, 293
Conrad of Montferrat, 429
Conrad of Wittelsbach, 382
Constantine the Great, 148, 582
Constantine Monomach, 106
Constantine Pogonatus, 477, 478
Constantine V., 491
Constantine VII., Porphyrogennetos, 9, 116, 122, 126, 421, 452, 453, 465, 467-514
Constantine, the Patrician, 515
Conti, Nicolo, 418
Cosmas, 523, 524, 580
Cotton Map, 559-563, 608-612
Crusades, 1, 2
Cyril, Apostle of Slavs, 503

Dagobert [Daibert], 157, 427, 428
Daniel of Kiev, 4, 5, 148, 155-174
Daniel, Bishop of Suriev (? same as preceding), 157
Daniel [Danil], Prince of Galicia, 294, 295, 317
Dati, 576
David-el-roy, 255
David, Prince of Georgia, 292
David, Nestorian priest, 352
David Sviatoslavitch, 173
David, pseudo-envoy, 278, 318, 364, 367
Dermot [Dermatius], 207
Dicuil, 524, 525
Dietrich [Theoderich ?], 190, 195
Diogenes, Governor of Cherson, 499
Dionysius Periegetes, 560
Dizabul, the Turk, 283, 345
Donizo, 427
Dosa, Rabbi, 220
Dubarlaus, Jacob and Michael (Attendants of Yaroslav of Russia), 314

Ealdred, Archbishop of York, 4, 128, 129
Edrisi, 13, 248, 257, 260, 442, 443, 444, 451, 452, 459, 461, 493, 494, 504
Edward the Confessor, 103
Egil Skallagrimson, 37
Einar Thambarskelvir, 47
Einar Wry-mouth, 94, 95

Eindrid Oddson, 72
Einhard, 522, 531, 533, 534, 536, 537
Eldad, the Danite, 220
Eldegai, 300
Elias, Patriarch of Jerusalem, various, 118, 119, 121, 581
Elipandus of Toledo, 549, 550
Emeterius of Valcavado, 554
Emma, wife of Æthelred II., 87
Emmerad of Anzy, 125
Eric, Bishop, 70
Eric Blood-axe (two), 3, 32, 33, 93
Eric, brother of Hakon the Good, 34, 35
Eric Emundson, 33
Eric Hakonson, 46, 47
Eric Haraldson, 3, 31
Eric of Normandy, Earl, 87
Eric the Red, 3, 28, 29, 46, 48, 49, 50, 51
Erling Sigvatson, 72
Ernoul, 208, 211, 212
Etherius, Bishop of Osma, 549
Eupeterius of Cherson, 503
Euphrosyna, 215
Eusebius, 567, 568, 581
Eustathius, 216
Eustochium, 151
Eutychius, Exarch of Ravenna, 402
Eyvind Skald-spiller, 36
Eyvind Urochs-horn, 93-95
Ezra, the Scribe, 248

Falcandus, 236
Faravid, King of the Kvens, 33
Felix of Urgel, 549
Feodor, 283
Ferdinand III. of Castile, 435, 437
Fetellus or Fretellus, 184, 186-189
Fidelis, Abbot, 549
Finn Arnison, 98
Flavio Gioja (Gisia), 398
Frederic I., Emperor (Barbarossa), 460, 564
Frederic II., Emperor, 276, 402, 429, 461
Frederic of Verdun, 124
Freydis, daughter of Red Eric, 50, 59-61, 63, 67, 68
Frisian Explorers, 548
Fulcher of Arras, 129, 130
Fulcher of Chartres, 156, 161, 583
Fulcher, Patriarch of Jerusalem, 197
Fulk the Black Count of Anjou, 4, 122, 125

SHORT INDEX OF NAMES

Ganuz Wolf, 531, 532
Gauzlin of Fleury, 126
Genoa, trade of, 418-426
Gerard of Cremona, 575
Gerard of Nazareth, 208
Gerbert of Sens, 320
Gerbert (Pope Sylvester II.), 4, 124
Gervase of Tilbury, 216
Geysa, King of Hungary, 483
Ghenghiz Khan, 6, 287, 306, 307, 319, 350, 445
Gilli, 'the Russian,' 37
Gisbert and Reinard, 119
Gisela, Queen, 483
Gissur Hallson, 208
Gleb of Minsk, 173
Godfrey of Bouillon, 428
Godric, the Pirate, 148
Gorm, 'Duke in Sweden,' 26
Gorm the Old, 26
Gorodislav Mikhailovitch, 173
Gosset, 321, 324, 340, 369
Gregory I., Pope, 120
Gregory IV., Pope, 120
Gregory VII., Pope (Hildebrand), 131, etc., 520
Gregory IX., Pope, 275, 276
Gregory, the Daronite, 507
Gregory of Tours, 523
Gualdo, 'the Frenchman,' 517
Guaymar of Salerno, 399
Gudleif Gudlaugson, 72, 73
Gudleik of Novgorod, 96, 97
Gudrid, 56-59, 61
Guest Thorhallson, 106
Guido 'of Italy,' 517
Guido 'of Ravenna,' 11, 632
Guido of Tripoli, 429
Guiscard of Cremona, 277
Guiscard, Robert, 399
Gunnar of Lithend, 33, 37
Gunnbiorn, 3, 21, 28
Gunstein, 88-92
Günther of Bamberg, 129-131
Guthorm [Guthrum] of Dublin, 110
Guy of Brabant, 126
Guyot de Provins, 10
Guzman, 513

Haji, Mohammed, 444
Haki and Hakia, 64
Hakim, Caliph ,126 ,127, 149, 241
Hakon the Good, King, 34, 35, 36
Hakon the Wise, Earl, 42, 43, 44
Haldor Snorrison, 110

Halfdan the Black, King, 26
Halfdan, 'White' and 'Black,' brothers, 34
Hallvard the Easterling, 33
Hanno of Cologne, 517, 519
Harald Fairhair, 26, 27, 30, 31
Harald Gormson, 44
Harald Greycloak, 34
Harald Hardrada, 3, 100, 101, 103-111, 175, 525, 528, 529, 531, 532, 548
Harek of Thiotta, 96, 100, 102
Harun-al-Rashid, 120
Hasan Ibn Saba, 239, 590
Hayton, King, 362, 368, 369, 374, 382-391
Helgasons (brothers), 70
Helge and Finnboge, 59-61
Helinand of Laon, 123, 129
Henry III., Emperor, 519
Henry IV., Emperor, 488, 517
Henry V., Emperor, 564
Henry I. of England, 176, 564
Henry II. of England, 564
Henry III. of England, 276, 589
Henry of Burgundy and Portugal, 177
Henry of Mainz, 11, 553, 560, 563-566, 614-617
Heraclius, Emperor, 150, 491
Herberstein, Sigismund von, 453
Herbert 'le Sommelier,' 320
Heriulf of Drepstok, 49
Hermann of Bamberg, 4
Herrade of Landsberg, 575
Hialte Skeggieson, 77
Hictarius, pilgrim, 126
Hierotheos, 483
Hiouen-Thsang, 345, 442
Homo Dei Turgemanus, 321, 324, 368
Honainus, 514
Honorius of Autun, 564, 574
Hoskuld, the Viking, 37
Hugh of Provence, 496
Hugh of Tuscany, 124
Hugo of St. Victor, 208
Hulagu, 362, 387
Hungarians, Magyars, or 'Turks,' 474, 476, 481-484
Hyginus, 575

Ibelin, House of, 422
Ibn Batuta, 7, 257, 258, 451, 494
Ibn Butlan, 441
Ibn Dasta, 458

SHORT INDEX OF NAMES 647

Ibn Foslan, 158, 223, 458, 463, 485, 494, 504
Ibn Haukal, 504, 510, 511
Ibn Khordadbeh, 488, 493, 504
Ibn Mohalhal, 13
Ibn Said, 435, 451
Igor, Prince of Russia, 157, 485, 486, 491
Ilchikadai, 278, 317, 318
Ingolf and Leif, 3, 28
Innocent I., Pope, 120
Innocent III., Pope, 216, 460
Innocent IV., Pope, 275, 276, 589
Iomala, 90, 91, 98
Isaac, the Jew, 219
Isiaslav Ivanovitch, pilgrim, 173
Isiaslav, Prince, 156, 488
Isidore of Seville, 551, 555, 557, 560, 577, 578
Isleif, Bishop, 72, 517
Ismael, 239
Ivan II. of Moscow, 38
Ivan III. of Moscow, 39, 41
Ivan IV. of Moscow, 41

Jacob ben Sheara, 219
James I. of Aragon, 434, 435, 436, 460
James of Edessa, 149
James de Vitry, 170, 208, 212–214, 459, 460
James, the Abbot, 385
Jehan Goderiche, 320
Jerome, St. [and Jerome maps], 151, 564, 565, 566–568, 569, 589, 605, 608
Johannes Moschos, 581
Johannes Phokas, 184, 199–203
John VIII., Pope, 117, 118, 119
John X., Pope, 496
John of Beaumont, 339
John of Beneventum, 123
John of Carcassonne, 320
John Gurgen [Kurkuas], Byzantine General, 505, 513
John of Parma, St., 123
John Pilatus, 496
John of Salisbury, 440
John Scotus Erigena, 517
John Tzimiskes, Emperor, 122, 123, 484, 486, 488, 491, 506, 507, 508, 511, 513, 583
John of Würzburg, pilgrim, 150, 152, 166, 184, 190–195
John VI., Patriarch, 505, 506
John, Bishop and Martyr, 539

Jonah of Tudela, 228
Jordanes [Jornandes], 455
Joseph, Rabbi, 250
Joseph, King of the Khazars, 492, 493
Josippon ben Gorion, 224
Julian [alias Hugo, alias William], Embriaco, 240
Juliette of Tuscany, 124
Julius I., Pope, 120
Julius Honorius, 558
Justin, 511
Justinian I., 450, 454, 500, 501, 511
Justinian II., 491, 499, 500

Kadak, 314
Kadan, 305
Kalf Arnison, 101, 102
Kallinikos of Heliopolis, 478
Kalokyr, 502
Karli of Halogaland, 88–92
Karl-o'-mere, 98
Kashkitch (the two), 173
Kazwini, 495
Kharmath, 239
Khazars, 472, 474, 491–494
Khoja Noian, 389
Kiartan, 45
Kirakos Gandaketsi, 385, 389
Klerk, 36
Koiak, 337
Kolskegg of Iceland, 37, 107
Kosmas, 313
Krates of Mallos, 571, 576
Krum, Tsar of Bulgarians, 480
Kublai Khan, 7
Kurancha, 297, 316
Kurd, the Armenian, 389
Kuyuk Khan, 277, 278, 292, 306–315, 383, 390, 391

Lambert of St. Omer, 12, 551, 570–573, 574, 621–624
Leger, St., Dean of Auxerre, 126
Leif and Ingolf, 3, 28
Leif Ericson, 3, 45, 46, 49, 51–54, 56, 57, 64, 77, 93
Leo II. of Little Armenia, King, 382, 456, 457
Leo III. the Isaurian, Emperor, 491, 510
Leo IV., 'the Khazar,' Emperor, 475, 480, 491, 492, 499
Leo VI., Emperor, 479, 499, 510
Leo I., Pope, 'the Great,' 120

648   SHORT INDEX OF NAMES

Leo III., Pope, 120
Leo IV., Pope, 119, 397
Leo of Monte Cassino, 123
Leo, the physician, 507
Leon, Governor of Daron, 123
Lietbert of Cambrai, 123, 129
Liudprand [Liutprand, 'Luitprand'], 481, 484, 487, 495, 496, 498, 502, 512, 514
Liutfrid, 484
Lodin, merchant, 45
Lorenzo, 277
Lothair, Emperor, 182
Louis Lewis, Ludwig, the 'Pious,' 'Debonnair,' etc., Emperor, 120, 480, 487
Louis IX., St. Louis of France, 278, 317-321, 590

Macrobius [and Macrobian Maps], 526, 570, 573-576, 625-626
Madaba Map, 580-583, 633-636
Magnus Barefoot, King of Norway, 175, 176
Magnus the Good (Olafson), King of Norway, 101-103, 107, 109
Maimonides, Moses, 224, 247, 264
Maius of Zara, 498
Makko of Constance, 125
Malacene, Bishop, 121
Malcolm, King of Scots, 94
Malefrid, Count, 123
Mangu Khan, Mongol Emperor, 278, 321, 340, 345, 346, 354-369, 384-386
Manuel Comnenus, Emperor, 193, 198, 199, 201, 202, 421, 422, 430, 450, 453, 454, 499
Manuel of Venice, 317
Margaret, Queen of France, 320, 321
Margath [or Margad], 110
Mari, Abba, 229
Marinus of Tyre, 576
Marseilles, trade of, 432-434
Martin I., Pope, 119, 120
Masudi, Arabic Geographer, 87, 99, 223, 458, 463, 488, 489, 493, 494, 510, 511, 544, 576
Matthew of Edessa, 490, 506, 513
Matthew Paris, 2, 6, 12, 145, 216, 217, 275, 461, 551, 559, 560, 584-590, 638-641
Mauci, 299, 302, 316
Maximus, Archbishop, 148
Mekhitar of Skerra, 385

Melic, 292
Melkorka, 37
Meshech, 271
Michael III., the Drunkard, Emperor, 460, 513
Michael Palaeologus, Emperor, 426
Michael of Cherson, 502
Michael of Genoa, 317, 446
Michael of Tchernigov, 283
Michael Sviatopolk, 156, 173
Michael Sviatoslavitch, 173
Micheas, 296
Moawiyah, Caliph, 145
Moktafi, Caliph, 226
Montpellier, trade of, 434, 435
Mostanshed or Mostanieh, 226
Mozaffer, Caliph, 129
Muntaner, Gregory de, Abbot of St. Sever, 552
Myrkiartan, Irish King, 37

Nabal, the Carmelite, 242
Nadodd, Norse Explorer, 3, 27
Narbonne, trade of, 434, 435
Nasir-i-Khusrau [Arabic traveller], 13, 126
Nathan, 220, 264
Nathan ben Jechiel, 224
Neckam, Alexander, 10
'Nestor,' Russian Chronicles of, 25, 156, 485, 487, 488, 492, 495, 502
Nestorians, 2, 347, 348, 352, 353, 362, 364
Nicolas, Abbot of Thing-Eyrar, 430
Nicolas, companion of Rubruquis, 324
Nicolas, Patriarch of Constantinople, 505
Nicolas 'Pisanus,' 317, 446
Nicolas of Santo Siro, 374, 447
Niger, Radulphus, 216
Nikephoros I., Emperor, 479
Nikephoros II., Phokas, Emperor, 484, 495, 497, 502, 508, 513
Niketas, 122
Northmen, 2, 3, 17-111
Norva, 23
Nur-ed-din, 245, 248

Obeco, 552
Odin, 24, 25
Ogul Gaimish, Mongol Empress, 278, 319, 367
Ohthere, Norse explorer, 3, 31, 32, 34

SHORT INDEX OF NAMES 649

Okkodai Khan, Mongol Emperor, 284, 289, 305, 306, 307, 308
Olaf Haraldson, St. Olaf, King of Norway, 3, 39, 84-88, 91-101, 104, 109
Olaf Kuaran, King of Dublin, 42
Olaf Tryggveson, King of Norway, 3, 36, 37, 41-48, 64, 87, 94
Olaf, King of Sweden, 47 [96]
Olaus Magnus, 543
Oleg, Prince of Russia, 157, 163, 482, 487
Oleg Sviatoslavitch, 173
Olga, Princess of Russia, 482, 483, 485, 487
Omar ('Amor'), Caliph, 166
Ordericus Vitalis, 547
Orestes, Patriarch of Jerusalem, 124
Orosius, 529, 535, 543, 562, 563, 577
Orso Ipato, Doge of Venice, 402
Othmar, 208
Otto I., the Great, Emperor, 483, 484, 495, 539
Otto III., Emperor, 518
Otto IV., Emperor, 418

Palnatoki, Viking Leader, 35
Pancrati Sviatoslavitch, 173
Paquette of Metz, 357
Paul, the Deacon, 542, 543
Paul of Zara, 498
Paula, St., 151
Pegolotti, Balducci, merchant and traveller, 454
Pepin, King of Franks, 120
Petachia, Rabbi Moses, 229, 247, 264, etc.
Petchinegs, 473, 476, 489-491, 492, 493, 494
Petronas, 494
Petrus Alfonsus or Alphonsus of Huesca, map of, 575, 626, 627
Philip de Toucy, 321
Pisa, trade of, 427-432
Plato of Tivoli, 575
Polo, Marco, 7, 275, 279, 451, 456, 459
Polo, Marocca, 451
Polo, Matteo, 7, 451
Polo, Nicolo, 7, 451
Pomponius Mela, 560
Poppo, Archbishop of Treves, 122, 129
Poppo of Stavelot, 124
Prester John, 319

Priscian, 560, 562
Procopius, 543
Psalter Map, 568, 569, 617-621
Pseudo-Baeda, 203-205
Ptolemy, Claudius, astronomer and geographer, 523, 530, 576
Pytheas of Marseilles, navigator, 547

Raban Maur, Archbishop of Mainz, 515
Rabban Çauma, 352
Rafn, the Limerick trader, 72, 73
Raganarius, 120
Ralph Glaber, 125, 126, 127
Ranulf Higden, 580
Rashid-ed-din, 305
Raven Floke, 28
Ravennese Geographer, 523, 524, 543, 588
Raymond, Count of Toulouse, 140, 229
Raymond II., Count of Tripoli, 240
Raymond III., Count of Tripoli, 397 [445]
Raymond Berenger III., 424, 435
Raymond Berenger IV., 424
Richard, Earl of Cornwall, 589
Richard, Abbot of Verdun, 122
Richard, Duke of the Normans, various, 122, 127, 128
Rimbert, St., 529
Robert, the Clerk, 320
Robert, Duke of Normandy, 4, 128
Robert Guiscard, 141, 176
Robert of Melkeley, 586
Roerek (Rurik) Vassal King 93
Roger I. of Sicily, 176
Roger II. of Sicily, 179, 180, 399, 400
Roger, Duke of Apulia, 233
Roger of Figeac, 126
Roger of Wendover, 216
Rolf, Hrolf or Hrodolf, the Ganger, 30, 31, 87, 176
Roman, 316
Romanus I., 511
Romanus IV., 508
Romanus Lecapenus, 492, 504, 506, 513
Rubruquis or Rubruck, William de, Friar-traveller, 7, 10, 266, 278, 281, 298, 299, 303, 320-382, 421, 449, 450, 451, 452
Rudolf, Bishop of Bethlehem, 433
Rupen, the Armenian, 382, 456
Rurik, 21, 25, 157, 176

SHORT INDEX OF NAMES

Saadia Gaon, 220
Saabh, Hassan Ibn, 239
Sabbas of Servia, 215
Sacrobosco (John Halifax of Holywood), 575
Saemund, Bishop, 72
Saewulf, pilgrim, 4, 5, 139-155, 160, 162, 163, 165, 166, 169
Sahensa, Prince of Ani, 372
Saladin, Sultan, 423, 439, 461
Salamon, Byzantine envoy, 495
Sallam the Interpreter, Arabic traveller, 223, 504
Sallust maps, 578, 579, 631, 632
Sanjar, Sultan of Merv, 226, 255
Samuel ben Simson, 274
Sanuto, Marino, 590
Sapsas or Sapsaphas, 581
Sartach, Mongol Prince, 278, 319, 320, 323, 331, 334-337, 369, 385, 388
Saul, Joseph, and Jacob ben Eliezer, 220
Saxo Grammaticus, historian, 526, 540, 543, 545, 546
Scatai, 331, 332
Seljuks, 226
Sempad, Constable of Armenia, 383, 384
Sergius, monk, 358
Sergius IV., Pope, 122
Shah Rokh, 444
Sherira ben Khanina, 224
Siegfried, Archbishop of Mainz, pilgrim, 4, 129, 130
Sigeric, Archbishop of Canterbury, traveller, 12, 560-562, 612-614
Sighvat the Skald, 96, 98, 101, 102
Sigrid of Sweden, Queen, 46
Sigurd of Norway, King, Pilgrim, and Crusader, 5, 157, 176-183
Sigurd of Orkney, 94
Silo of Oviedo, King, 550
'Silvia of Aquitaine,' St., 165
Simeon, Nestorian doctor, 362
Simeon, Tsar of Bulgarians, 479, 480, 492
Simon of St. Quentin, Friar-traveller, 277, 318
Skopti Ogmundson, Norse traveller, 23, 175
Skraelings, 55, 57-59, 66-69, 78, 81-83
Snorri(e) Sturleson, ch. ii. passim, especially p. 104
Snorri(e) Thorbrandson, 63
Snorri(e) Thorfinnson, 58, 61, 70

Solinus, 523, 524, 529, 541, 542, 546, 568
Stefnir Thorgilson, 123
Stein of Iceland, 98
Stephanus Garcia Placidus, 552
Stephen, Byzantine envoy, 495
Stephen of Hungary, St. and King, 127, 128, 483
Stephen, the Bishop, 385
Stephen of Bohemia, Friar-traveller, 302, 390
Strabo, 163, 546
Sueiro Mendes, Count, 177
Svein Godwineson, 4, 128
Svein Estrithson [Astridson], 108-110, 516, 520, 521, 532, 541, 544, 547
Svein Haraldson, 'Forkbeard,' 44, 46, 85
Sviatoslav, Russian Prince, 157, 450, 452, 486, 488, 491, 503
Symeon [Simeon], the Armenian, 125
Symeon, St., recluse of Treves, 121, 122, 129

Tancred of Antioch, 428
Temer, 'knight,' attendant of Yaroslav of Russia, 314
Thangbrand [Dangbrandt], priest and missionary, 45, 46
Themistos, Governor of Cherson, 499
Theoderich [Dietrich], pilgrim, 184, 195-199
Theodora, Empress, 122
Theodore, 262
Theodosius, Patriarch, 118
Theodosius, David, and Sabas, envoys, 118
Theodoulos [Raymond], 357, 358
Theophanes, historian, 475, etc.
Theophano, Empress, 518
Theophilus, Emperor, 487, 494, 501
Theophylact, 483
Theotonius, pilgrim, 207
Thietmar, 215
Thomas the Usurper, 510, 512
Thorarin of Iceland, 93
Thorbrand Snorrison, 67
Thorbiorn Angle, 106
Thorbiorn Vifilson, 62, 63
Thogarma, Thogarmim, 237, 269
Thorfinn Karl, 73
Thorfinn Karlsefne, explorer, 3, 48, 56-59, 61, 62, 63-70, 75, 76, 77

SHORT INDEX OF NAMES

Thorfinn Skull-cleaver of Orkney, 93–95
Thorhall Gamlison, 63
Thorhall, the Huntsman, 63, 64, 65, 69
Thorir Hund, 3, 88–92, 96, 98, 100–102, 175
Thorir, the Easterling, 56
Thorir Klakka, 43
Thorkell, brother of Gisli, 37
Thorkell Foulmouth, 37
Thorkell Gelleson, 73, 82
Thorleik, the Fair, 109, 110
Thorolf, heir of Bjorgolf and Bard, 33
Thoros, 238
Thorstein Dromund, 106
Thorstein Ericson, 56, 62, 63
Thorstein Stirson, 106
Thorvald Ericson, 54–56, 62, 63, 68, 69
Thorvald (Thorvard), of Gardar, son-in-law of Red Eric, 50, 59–61, 63
Thorvald Kodranson, 123
Thyri, wife of Olaf Tryggveson, 46
Titus, Emperor, legend of, 232
Toki, 'dux Winlandensis,' 546
'Tortushi' (Abubekr Mohammed), 495
Toucy Narjot de, 321 (*see* Philip de Toucy, 279)
Trasmundus, 517
Turakina, Mongol Empress, 308, 311, 312, 313, 314, 315
Turolf, 517
Tyrker 'the German,' 51, 53, 78

Uvaegi, 69

Vaetildi, 69
Valdidada, 69
Varlaam of Kiev, 155
Varthema, L. de, traveller, 443
Vassili Vladimirovitch, 173
'Vassilko,' Prince of Russia, 294, 317
'Vastacius,' 355
Venice, trade of, 400–418
Venerius ('Reverius'), 446
Victor II., Pope, 122, 129
Vidgaut of 'Samland,' merchant and traveller, 537
Vigdis, 71
Vilecinus, 516

Vincent of Beauvais, 10, 275, 277
Vioni [Viglioni], P., 445
Virgil, Bishop, 524
'Vissavald' of Novgorod, 46
Vivaldo, Benedetto, 462
Vivaldo, Sorleone, 462
Vladimir of Kiev and Novgorod, the Great, 36, 41, 47, 157, 492, 503
Vladimir Monomach, Prince of Russia, 453

Walid, Caliph, 262
Widukind, 539
Wilbrand of Oldenburg, 215, 216, 461
William of Angoulême, 4, 128
William of Apulia, 398
William III. of Aquitaine, 128
William of Conches, 574, 575
William, the Englishman, Bishop, 522
William of Malmesbury, 139
William, the Monk, 523
William of Normandy, various, 87, 127
William of Paris, 350, 355–357, 360, 364, 367, 368
William of Tyre, 126, 157, 166, 170, 180, 458, 460
William, companion of Andrew of Longumeau, 320
Willibald, pilgrim, 145, 151, 526
Wolf, son of Uspak, 110
Wulfstan, Bishop of Worcester, St., 139
Wulfstan, navigator, 3, 34, 546

Yakut, Arabic Geographer, 7, 13
Yaroslav, Prince of Russia, 283, 285, 291, 304, 309, 312, 313
Yaroslav, the Lawgiver of Kiev and Novgorod, Prince of Russia, 39, 96, 99–101, 106–108, 156, 157, 158
Yehuda ben Elia Hadasi, 224
Yuri [George], Danilovitch of Novgorod, 39
Yussuf, Mercantile House of, 457
Yusûf Ibn Tashfin, 136

Zemarchus, 283, 345
Zenghi, Atabeg, of Mosul, 197, 226
Ziani, P., Doge of Venice, 417, 460
Zurbaneles, 509